# Sustainable Development Goals Series

The **Sustainable Development Goals Series** is Springer Nature's inaugural cross-imprint book series that addresses and supports the United Nations' seventeen Sustainable Development Goals. The series fosters comprehensive research focused on these global targets and endeavours to address some of society's greatest grand challenges. The SDGs are inherently multidisciplinary, and they bring people working across different fields together and working towards a common goal. In this spirit, the Sustainable Development Goals series is the first at Springer Nature to publish books under both the Springer and Palgrave Macmillan imprints, bringing the strengths of our imprints together.

The **Sustainable Development Goals Series** is organized into eighteen subseries: one subseries based around each of the seventeen respective Sustainable Development Goals, and an eighteenth subseries, "Connecting the Goals", which serves as a home for volumes addressing multiple goals or studying the SDGs as a whole. Each subseries is guided by an expert Subseries Advisor with years or decades of experience studying and addressing core components of their respective Goal.

The SDG Series has a remit as broad as the SDGs themselves, and contributions are welcome from scientists, academics, policymakers, and researchers working in fields related to any of the seventeen goals. If you are interested in contributing a monograph or curated volume to the series, please contact the Publishers: Zachary Romano [Springer; zachary.romano@springer.com] and Rachael Ballard [Palgrave Macmillan; rachael.ballard@palgrave.com].

Jian Zuo • Liyin Shen • Ruidong Chang
Editors

# Circular Economy for Buildings and Infrastructure

## Principles, Practices and Future Directions

Springer

*Editors*
Jian Zuo
University of Adelaide
Adelaide, SA, Australia

Liyin Shen
Hangzhou City University
Hangzhou, China

Ruidong Chang
University of Adelaide
Adelaide, SA, Australia

ISSN 2523-3084 ISSN 2523-3092 (electronic)
Sustainable Development Goals Series
ISBN 978-3-031-56240-2 ISBN 978-3-031-56241-9 (eBook)
https://doi.org/10.1007/978-3-031-56241-9



This Springer imprint is published by the registered company Springer Nature Switzerland AG
The registered company address is: Gewerbestrasse 11, 6330 Cham, Switzerland

# Foreword

The book *Circular Economy for Buildings and Infrastructure: Principles, Practices and Future Directions*, edited by Professor Jian Zuo, Professor Liyin Shen, and Dr. Ruidong Chang, came in at an opportune moment in the implementation of the SDGs. Now at the midpoint for the 2030 Agenda for Sustainable Development, the world needs to reassess the work done on the ground, and plan accordingly to accelerate the efforts in achieving the goals. Buildings, infrastructure design and use, and its waste management remain top priority areas for circularity and sustainability in high-level discussions and professional roundtables. Housing and mobility represent not only half of the basic needs, but its provision locks in human resource consumption for decades, if not a century or more.

This book has holistically presented the technical issues of building and infrastructure resource management together with the social dimensions and economic impacts, which are all equally important in the pursuit of the global goals. It further sheds light to the temporal aspect of sustainability at a macro level. On a socio-political basis, a number of factors affect the continuity of grassroot initiatives and programs especially when one considers the stakeholders' "term of office" and also the "political will" to support the SDGs. At the minimum, scoping assessments and technological tools need to be both in place to operationalize circularity in infrastructure development. Not only the scientific community but also the policy makers will find this publication very helpful and informative in providing such operationalization tool.

Aside from the traditional scientific discussions on LCA and critical reviews, this publication emphasized essential elements of inclusiveness concept, high temperature exposure, cadaster, developing economies viewpoints, affordability, resilience, smart system, revit-plugin system, and stakeholders' behavior. They are very crucial to the successful implementation of SDG programs, especially goals 3, 6, 7, 9, 11, 12, and 13. In parallel to many relevant urban metabolism works published by the International Resource Panel, this book exposed these factors to much more details.

I concurred with the editors' future remarks that there is a significant demand for quality data related to a circular economy in the built environment, yet not only to be used for assessment, but also for forward thinking in resource conservation and resource management. Composite and many alloys have been a challenge to recycling and remanufacturing reuse, and these materials are getting more in complicating forms in the renewables technology. The future needs to identify better circularity strategies in the design, use, and reuse of these building and infrastructure materials. I also concurred with the editors' viewpoint on the aggressive integration of digital and artificial intelligence into our daily life, including optimal use of resources (determined by AI). Optimization will likely be handled by these new technology, while human's role will have to be more strategic and innovative in the future. I highly believe that this book is going to bridge us into the new industrial revolution era.

De La Salle University
Manila, Philippines

Anthony S. F. Chiu

# Contents

# Contributors

**Michael Atafo Adabre** Building and Real Estate Department, The Hong Kong Polytechnic University, Hung Hom, Hong Kong

**Laura Almeida** Centre for Smart Modern Construction (c4SMC), School of Engineering, Design and Built Environment, Western Sydney University Penrith, Penrith, NSW, Australia

**Mitchell Bacon** ResourceCo Pty Limited, Rose Park, SA, Australia

**Bree Bennett** School of Architecture and Civil Engineering, The University of Adelaide, Adelaide, SA, Australia

**Xin Bian** Shenzhen Research Institute (SRI), City University of Hong Kong, Shenzhen, China

Collaborative Innovation Center on Forecast and Evaluation of Meteorological Disasters/Research Center of Urban Sustainable Development/School of Geographical Sciences, Nanjing University of Information Science & Technology (NUIST), Nanjing, China

**Albert P. C. Chan** Building and Real Estate Department, The Hong Kong Polytechnic University, Hung Hom, Hong Kong

**Ruidong Chang** School of Architecture and Civil Engineering, The University of Adelaide, Adelaide, SA, Australia

**Jiayu Chen** School of Civil Engineering, Tsinghua University, Beijing, China

**Cathy Chesson** McConnell Dowell Diona Joint Venture, Adelaide, SA, USA

**Nicholas Chileshe** UniSA STEM, Sustainable Infrastructure and Resource Management (SIRM), University of South Australia, Adelaide, SA, Australia

Faculty of Engineering and the Built Environment, University of Johannesburg, Johannesburg, South Africa

**Yangxue Ding** College of Civil & Transportation Engineering, Underground Polis Academy, Shenzhen University, Shenzhen, China

**Liang Dong** Department of Public and International Affairs (PIA), City University of Hong Kong, Hong Kong, SAR, China

School of Energy and Environment (SEE), City University of Hong Kong, Hong Kong, SAR, China

Sustainable Development Lab, Centre for Public Affairs and Law (CPAL), City University of Hong Kong, Hong Kong, SAR, China

Shenzhen Research Institute (SRI), City University of Hong Kong, Shenzhen, China

**Huabo Duan** School of Environmental Science and Engineering, Huazhong University of Science and Technology, Wuhan, China

**Srinivasa Raghavendra Bhuvan Gummidi** SDU Life Cycle Engineering, Department of Green Technology, University of Southern Denmark, Odense, Denmark

**Mingyuan Guo** Centre for Smart Modern Construction (c4SMC), School of Engineering, Design and Built Environment, Western Sydney University Penrith, Penrith, NSW, Australia

**Tao Hu** School of Management Engineering, Shandong Jianzhu University, Jinan, Shandong, China

**Ruchini Jayasinghe** UniSA Online, University of South Australia, Adelaide, Australia

**Yongsheng Jiang** School of Management Engineering, Shandong Jianzhu University, Jinan, Shandong, China

**Kamyar Kabirifar** School of Built Environment, University of New South Wales, Sydney, Australia

**Neema Kavishe** School of Built Environment, Engineering and Computing, Leeds Beckett University, Leeds, UK

**Anna Kordahi** McConnell Dowell Diona Joint Venture, Adelaide, SA, USA

**Maud Lanau** Department of Architecture and Civil Engineering, Chalmers University of Technology, Gothenburg, Sweden

**Khoa N. Le** School of Engineering, Design and Built Environment, Western Sydney University, Penrith, NSW, Australia

**Liancheng Li** School of Architecture and Civil Engineering, The University of Adelaide, Adelaide, SA, Australia

**Wengui Li** Centre for Infrastructure Engineering and Safety, School of Civil and Environmental Engineering, The University of New South Wales, Sydney, NSW, Australia

**Chunlu Liu** School of Architecture and Built Environment, Deakin University, Geelong, Australia

**Gang Liu** SDU Life Cycle Engineering, Department of Green Technology, University of Southern Denmark, Odense, Denmark

College of Urban and Environmental Sciences, Peking University, Beijing, China

**Hui Liu** Collaborative Innovation Center on Forecast and Evaluation of Meteorological Disasters/Research Center of Urban Sustainable Development/School of Geographical Sciences, Nanjing University of Information Science & Technology (NUIST), Nanjing, China

**Zhiyu Luo** Centre for Infrastructure Engineering and Safety, School of Civil and Environmental Engineering, The University of New South Wales, Sydney, NSW, Australia

Department of Civil and Environmental Engineering, National University of Singapore, Singapore, Singapore

**Mingxue Ma** School of Engineering, Design and Built Environment, Western Sydney University, Penrith, NSW, Australia

**Ruichang Mao** School of Civil Engineering, Tsinghua University, Beijing, China

DTU Sustain, Department of Environmental & Resource Engineering, Technical University of Denmark, Lyngby, Denmark

**Zhongyi Men** School of Management Engineering, Shandong Jianzhu University, Jinan, Shandong, China

**Mohammad Mojtahedi** School of Built Environment, University of New South Wales, Sydney, Australia

**Ethan Nikookar** School of Business, University of Wollongong, Wollongong, NSW, Australia

**Daniel Oteng** School of Project Management, The University of Sydney, Sydney, NSW, Australia

**Srinath Perera** Centre for Smart Modern Construction (c4SMC), School of Engineering, Design and Built Environment, Western Sydney University Penrith, Penrith, NSW, Australia

**Sepani Senaratne** Centre for Smart Modern Construction (c4SMC), School of Engineering, Design and Built Environment, Western Sydney University Penrith, Penrith, NSW, Australia

**Yuchen She** School of Architecture and Built Environment, Deakin University, Geelong, Australia

**Liyin Shen** Hangzhou City University, Hangzhou, Zhejiang Province, China

Chongqing University, Chongqing, China

**Vahid Shobeiri** School of Architecture and Civil Engineering, The University of Adelaide, Adelaide, SA, Australia

**Tianheng Shu** School of Architecture, Tsinghua University, Beijing, China

**Andrew Silvestri** ResourceCo Pty Limited, Rose Park, SA, Australia

**Xiangnan Song** School of Management, Guangzhou University, Guangzhou, China

**Larissa Statsenko** UniSA STEM, University of South Australia, Adelaide, Australia

**Vivian W. Y. Tam** School of Engineering, Design and Built Environment, Western Sydney University, Penrith, NSW, Australia

**Olubukola Tokede** School of Architecture and Built Environment, Deakin University, Geelong, Australia

**Nilupa Udawatta** School of Architecture and Built Environment, Deakin University, Geelong, Australia

**Phillip Visintin** School of Architecture and Civil Engineering, The University of Adelaide, Adelaide, SA, Australia

**Cynthia Changxin Wang** School of Built Environment, University of New South Wales, Sydney, Australia

**Wei Wang** Southeast University, Nanjing, Jiangsu Province, China

**Huanyu Wu** School of Civil and Transportation Engineering, Shenzhen University, Shenzhen, China

**Ibrahim Yahaya Wuni** Department of Construction Engineering & Management, King Fahd University of Petroleum & Minerals, Dhahran, Saudi Arabia

**Minghui Xie** State Key Laboratory of Environmental Criteria and Risk Assessment, Chinese Research Academy of Environmental Sciences, Beijing, China

**Tianyu Xie** School of Architecture and Civil Engineering, The University of Adelaide, Adelaide, SA, Australia

**Zihao Xu** School of Management Engineering, Shandong Jianzhu University, Jinan, Shandong, China

**Wei Emma Zhang** School of Computer and Mathematical Sciences, The University of Adelaide, Adelaide, Australia

**Yanquan Zhang** School of Architecture and Civil Engineering, The University of Adelaide, Adelaide, SA, Australia

**Yunjia Zhang** School of Management Engineering, Shandong Jianzhu University, Jinan, Shandong, China

**Dong Zhao** School of Planning, Design and Construction, Michigan State University, East Lansing, MI, USA

Department of Civil and Environmental Engineering, Michigan State University, East Lansing, MI, USA

**Hanbing Zhao** Centre for Infrastructure Engineering and Safety, School of Civil and Environmental Engineering, The University of New South Wales, Sydney, NSW, Australia

**Jian Zuo** School of Architecture and Civil Engineering, The University of Adelaide, Adelaide, SA, Australia

# 1 Introduction

Jian Zuo, Liyin Shen, Ruidong Chang, and Yanquan Zhang

## 1.1 Introduction

Communities and cities are characterized by a large number of buildings and infrastructure. These buildings and infrastructure (e.g. roads, bridges, power stations, wastewater treatment facilities, etc.) provide necessary functions to the residents of the community. The circularity of buildings and infrastructure, and their components will largely affect the sustainability performance of these communities and cities. In the context of this book, circular built environment covers all efforts that aim to improve the sustainability performance of developments, including material efficiency and energy efficiency. In particular, circularity is taken into consideration throughout its life cycle, e.g. planning, design, use and reuse. Meanwhile, the connection to places and spaces are taken into account. Circularity meant that resources are put back into system as many time as possible so that the demand on virgin materials is minimized.

As one of the largest consumers of natural resources, buildings and infrastructure have a major role to play in a circular economy. The role of building and infrastructure in a circular economy is two-folded. On the one hand, the development of buildings and infrastructure demands a considerable amount of resources. Traditionally virgin materials were used that are not renewable. Indeed, the rapid depletion of raw materials is one of most critical issues in pursuing sustainable development. The traditional liner approach, i.e. make-use-disposal led to significant amount of construction and demolition waste. On the other hand, buildings and infrastructure have significant environmental impacts throughout their lifecycle. For instance, the manufacturing of construction materials involves a considerable amount of energy and water while embodied emissions have drawn a growing level of attention. Similarly, the construction process involves machinery and plants that consume energy (e.g. petrol and diesel) and emit greenhouse gases. The long service life of buildings and infrastructure means that the operational emissions will be very high, not to mention the emissions related to end-of-life (EoL) associated with demolition.

J. Zuo (✉) · R. Chang · Y. Zhang
School of Architecture and Civil Engineering,
The University of Adelaide, Adelaide, SA, Australia
e-mail: Jian.zuo@adelaide.edu.au

L. Shen
Hangzhou City University, Hangzhou, Zhejiang Province, China

## 1.2 IPCC Reports

The impacts of building and infrastructure are well recognized. For example, buildings and infrastructure featured in the Intergovernmental



J. Zuo et al. (eds.), *Circular Economy for Buildings and Infrastructure*, Sustainable Development Goals Series, https://doi.org/10.1007/978-3-031-56241-9_1

Panel on Climate Change, buildings and infrastructure reports. In the Sixth Assessment Report released by Intergovernmental Panel on Climate Change, buildings and infrastructure (e.g. transport, energy, etc.) are responsible for nearly 80% of global Greenhouse Gas (GHG) emissions therefore play a crucial role in achieving the 1.5 °C target by 2030 (Portner et al., 2022). In this report, sustainable material flow (e.g. replacing virgin materials by high quality recycled products) has been emphasized, e.g. "renewed attention to end-use demand, material efficiency and more and better-quality recycling measures" (p. 1165).

As a response to IPCC reports, a number of peak bodies, e.g. World Green Building Council and the Green Building Council of Australia have released reports on strategies to achieve the 1.5 °C target by 2030 target. These strategies share similarities, i.e. reducing the carbon emissions throughout the supply chain of the construction industry. This is to acknowledge the fragmented nature of supply chains in the construction industry, therefore a holistic approach will be required. These measures have been incorporated into various sustainability assessment tools such as Leadership in Energy and Environmental Design (LEED), Green Building Council of Australia (GBCA) Green Star and Infrastructure Sustainability Council's Infrastructure Sustainability (IS) rating.

## 1.3 UN SDGs

Meanwhile, the crucial role of building and infrastructure are well reflected in SDGs developed by UN member states. Within the SDGs there are a total of 17 goals and 169 targets. Considering its broader linkage with other sectors, sustainable building and infrastructure contribute to several UN Sustainable Development Goals (SDGs) such as:

- Goal 3: Good Health and Well-being
- Goal 6: Clean Water and Sanitation
- Goal 7: Affordable and Clean Energy
- Goal 9: Industry, Innovation and Infrastructure
- Goal 11: Sustainable Cities and Communities
- Goal 12: Responsible Consumption and Production
- Goal 13: Climate Action

Indeed, all UN SDG goals are underpinned by the resilience and sustainability performance of infrastructure (The Economist, 2019). In particular, sustainable infrastructure requires a system approach, i.e. as an interlinked portfolio of assets that will deliver not only environmental benefits but also cost savings. Providing various functions to human beings such as shelter and education, sustainable buildings are also closely related to UN SDGs (Omrany et al., 2022).

## 1.4 Circular Economy

The rapid depletion of natural resources has attracted wide attention globally. The last decades have witnessed a strong push from various stakeholders (e.g. public, government and industry) towards a more circular economy rather than a linear economy. The traditional approach of make-use-disposal has attracted significant criticism while a closed-loop approach has gained an increasing level of recognition.

Several definitions have been provided for the circular economy. For instance, according to Ellen MacArthur Foundation, the circular economy "tackles climate change and other global challenges like biodiversity loss, waste, and pollution, by decoupling economic activity from the consumption of finite resources." European Parliament (2023) defined circular economy as "a model of production and consumption, which involves sharing, leasing, reusing, repairing, refurbishing and recycling existing materials and products as long as possible. In this way, the life cycle of products is extended." The Economist (2023) highlighted circular economy as "a production cycle is fully self-sufficient and aims to keep materials and products in use for as long as possible, by reducing waste and reusing, repairing and refurbishing materials and products". According to United Nations Development Programme, circular economy "aims to minimize waste and promote a sustainable use of natural resources, through smarter product design, longer use, recycling and more, as well as regenerate nature" (UNDP, 2023).

These definitions share some commonalities. These include: minimizing the use of virgin materials, improving resource efficiency, and enhancing the circularity of materials and products (via reuse, repair and repurposing, etc.). As these definitions cover the wider supply chain, it is necessary to involve a large number of stakeholders in the design, manufacturing, construction, operation, recycling and remanufacturing businesses, hence a holistic approach is required. It also calls for collaboration between the public and private sectors to collaborate to tackle related challenges. Meanwhile, the general public plays a role in changing their attitude and behavior toward consumption and disposal.

Despite significant efforts on circular economy and its application in various sectors, there are some worrying statistics on the global scale. For instance, the new Circularity Gap Report showed that the global circularity has dropped to 7.2% in 2023, down from 9.1% in 2018 (Fraser et al., 2023). This meant that a large proportion of materials are not able to return to the eco-system for use/reuse such as in building and infrastructure. There is urgent need to showcase how the circularity of materials can be improved, including construction materials in buildings and infrastructure.

According to Holland Circular Hotspot (2022), high quality recycling and reuse plays a crucial role in circular building. To achieve this, it is necessary to establish and maintain a digital system that can track the use and flow of materials. In addition, an indicator system should be developed to measure circularity of each type of materials.

There are evolving practices on applying principles of circular economy in built environment. Traditionally efforts are placed on waste recycling and reuse within the same project site. The concept of Zero Waste Construction Sites was introduced that are encouraged to reuse and recycle (e.g. using a mobile crusher to make recycled aggregate from waste concrete) waste materials following on-site sorting. Later efforts were introduced to reuse or recycle construction materials between different project sites. The concept of material banks was introduced to allow sites and companies to exchange unwanted materials, therefore, reducing waste and diverting from landfill. There are also applications for using waste materials across different sectors. For example, the shampoo bottle from municipal waste streams could be used in the manufacturing of benches therefore minimizing construction and demolition waste streams (Li et al., 2021). Meanwhile, soft plastics and waste toners have been used in asphalt mix in infrastructure projects that have helped to deal with such problematic waste stream that were normally sent to landfill (Li et al., 2022). The entire built environment ecosystem for circular economy is shown in Fig. 1.1.

## 1.5 Related Policies and Initiatives

In this section, Australia is used as a focal point to illustrate policies and initiatives related to circular economy. According to KPMG, the built environment is one of three key sectors in Australia that will enjoy the benefits of a circular economy where the market of these three sectors is expected to contribute to AUD$23 billion in GDP by 2025. The new federal budget has set $83.1 million in the next 5 years to support the transition to circular economy such as emission reduction, cleaner energy, resource efficiency, durability of products, and right-to-repair policy (Parliament of Australia, 2022). The recycling of problematic waste streams such as plastics was given higher priority.

Australia is in transition to a circular economy via collaboration between public and private sectors (DCCEEW, 2023). The focus is placed on designing out environmental impacts (e.g. waste and pollution), facilitating the market and keeping resources in the loop. Waste plays a crucial role in circular economy-related initiatives

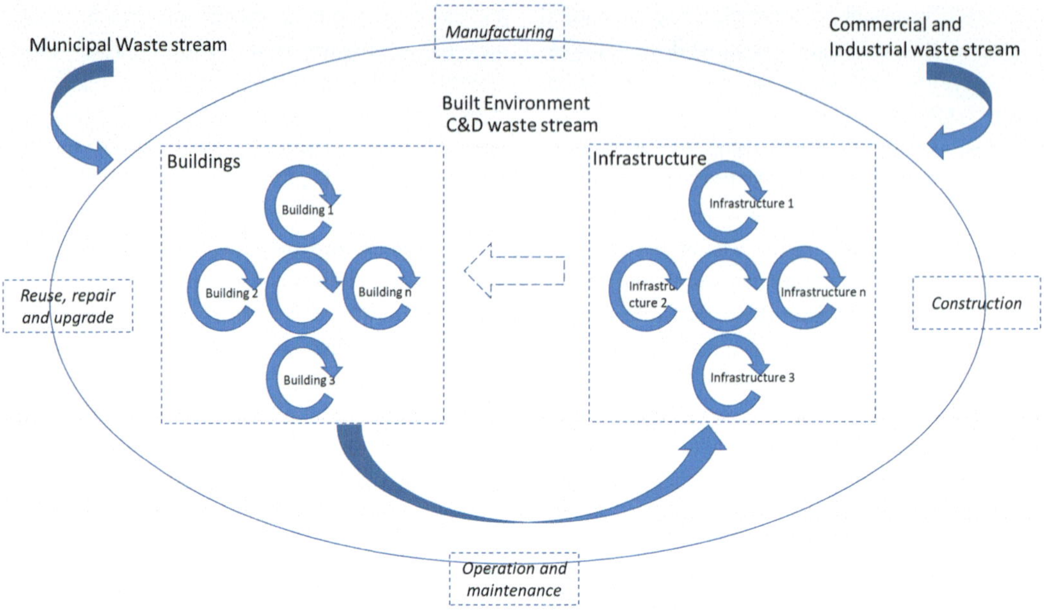


**Fig. 1.1** Built environment ecosystem for circular economy

in Australia that witness efforts to harmonise the waste management policies and practices nationwide through the national waste policy.

Waste prevention should be given higher priority as it is more effective than dealing with waste once it is generated. According to a recent study commissioned by the Department of Climate Change, Energy, the Environment and Water, waste prevention could be better achieved through the use of second products, change of ownership, better durability of products (including repairability and upgradability) and giveaway (e.g. donation) (Downes, 2022).

According to the National Waste Policy (Commonwealth of Australia, 2018), the following five principles are adopted to allow a better transition to a circular economy:

- Avoiding waste through design efficiency and allowing reuse and repair;
- Improving resource recovery through efficient collection, recycling and better quality of waste for remanufacturing;
- Facilitating the market of recycled materials and products;
- Addressing health, the environment and economic issues via better material flow; and
- Improving the data infrastructure to allow better decision making.

The National Waste Policy 2018 clearly highlighted the value of the waste recycling industry to achieve a circular economy. These include: A$ 12.6 billion p.a. worth of waste management services, A$ 6.9 billion of added-value to the economy per annum, 9.2 employment opportunities per 10,000 tonnes of waste being recycled.

The National Waste Policy Action Plan 2019 further set out a series of targets for waste management. These include: a waste ban of four materials (glass, tyre, paper and plastic), a waste reduction of 10% by 2030, an 80% recovery rate of all waste streams and the increased uptake of recycled products. The quality of data and its transparency were highlighted as crucial aspects to facilitate the action plan.

States and territories have released corresponding policies and initiatives for the circular economy. For instance, NSW Government defined key principles of circular economy (NSW Government, 2019):

- "Sustainable management of all resources
- Valuing resource productivity
- Design out waste and pollution
- Maintain the value of products and materials
- Innovate new solutions for resource efficiency
- Create new circular economy jobs
- Foster behaviour change through education and engagement"

In this policy statement, it was clearly stated that the roles of the NSW Government include: incorporating circular economy principles into the State Waste Strategy, implementing CE principles into the government operations and procurement, and advocating CE at local, national and international levels. In particular, the waste industry plays a critical role and specific recycling targets have been defined for various waste streams.

Office of Energy and Climate Change, NSW Treasury (2023) released Circular design guidelines for the built environment with four objectives, i.e. reducing embodied carbon, minimising the generation of waste, improving materials efficiency and increasing the circularity of materials. According to this guideline, a number of issues should be taken into consideration of circular design strategies such as: longevity, flexibility and adaptability, material circularity, material efficiency, operational waste management, re-use of materials or assets, specifying recycled products, design for disassembly, low-emission materials, green infrastructure, material database.

The Victorian Government defined four goals of the Circular Economy (Victorian State Government, 2020):

- Design to last, repair and recycle
- Use products to create more value
- Recycle more resources
- Reduce harm from waste and pollution

In addition, Recycling Industry Strategic Plan was released that aims to enhance the quality of recycled materials, facilitate the market of recycled products, and improve the efficiency of recycling businesses. It was made very clear that infrastructure and the market play crucial roles in facilitating the circular economy in Victoria. Targets include: 80% diversion rate from landfill by 2030, reduction of waste generation by 15% by 2030. This is further reinforced in Circular Economy (Waste Reduction and Recycling) Act 2021 (Victoria).

The importance of circular economy is greatly reflected in the recently released South Australia's Waste Strategy 2020–2025 (Fig. 1.2). The target set in this strategy is to achieve a 95% diversion rate for construction and demolition waste, the highest in all states and territories in Australia. Great emphasis is placed on reducing the demand for raw materials and increasing confidence in recycled materials. These include encouraging technological innovations via new products and systems that use recycled content. Reuse is encouraged, including repair and resale.

Similarly, strong efforts have been made to develop the circular economy market by engaging clients from both the public and private sectors. Sustainable procurement guidelines are developed to guide the state and local governments to use recycled products. One of the initiatives is The Buying it Back LGA Circular Procurement Pilot Project funded by the State Government agency Green Industries SA. Nine councils in South Australia (City of Charles Sturt, City of Norwood, Payneham and St Peters, City of Onkaparinga, City of Port Adelaide Enfield, City of Prospect, Mount Barker District Council, City of Burnside, Adelaide Hills Council and The Rural City of Murray Bridge) participated in this pilot project to address the dilemma: on the one hand each council struggles to deal with generated waste, on the other hand, each council demands a considerable amount of resources in their operations (e.g. printing, construction of local streets, etc.). In this pilot project, participating councils commit to buying back products containing waste generated from their administrated suburbs. For instance, recycled asphalt, plastic and toner have been used in the surfacing of local streets.

Meanwhile, Circular Economy Market Development Grants are awarded to councils, industry

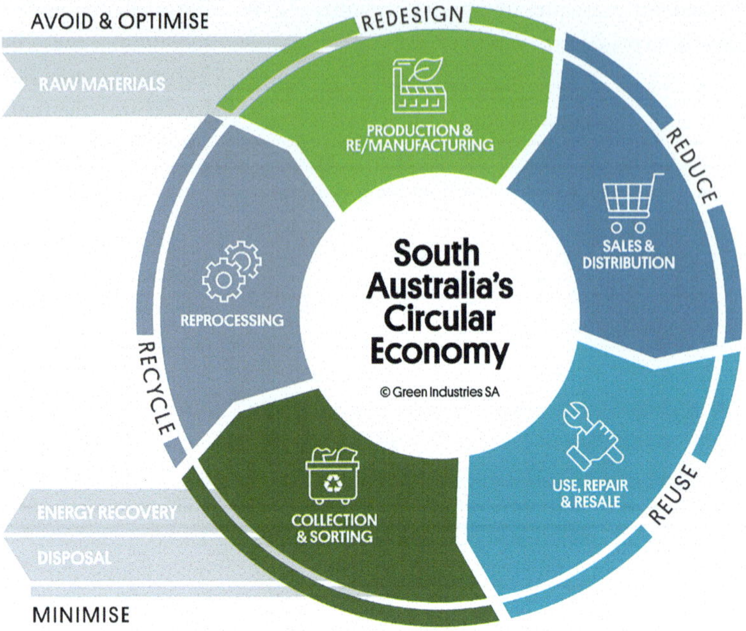


**Fig. 1.2** The Circular Economy Model of South Australia (GISA, 2020)

associations, not-for-profit organisations, research institutes, and businesses that design, apply and scale up innovative circular economy business models and practices or produce, manufacture, sell, or promote South Australian recycled materials and recycled-content products. One example is XFRAME™. This prefabricated lightweight wall framing system is flexible and adaptable to accommodate future changes. The components assemble into structural panels without the need for nails, glues, or screws—making this modular system easily reparable and reusable. Being designed for deconstruction reduces time, labour and costs at the end of life.

In collaboration with the Green Building Council of Australia, Green Industries SA recently released a discussion paper for the circular built environment. As reported in this discussion paper, concrete, steel, brick and aluminium combined for the vast majority of embodied emissions therefore were identified as a priority for the transition to a circular built environment through reuse, recycling and remanufacturing (GISA, 2022). Meanwhile, a large number of barriers were identified: disjointed governance across federal, state and local governments, different priorities and investment opportunities across various state agencies, lack of awareness of cleaner materials, lack of specifications for recycled products, the higher upfront cost of using recycled products, and lack of consistent metrics for effectiveness of circular built environment-related measures. This discussion paper continues to demonstrate national leadership in resource management and circular economy.

This discussion paper has led to an Action Plan for Circular economy in South Australia's built environment, which was developed by Green Building Council of Australia. Acknowledging South Australia's leadership in waste management, renewable energy and climate change mitigation, this Action Plan outlines opportunities related to circular economy in South Australia's built environment. The model covers the whole life cycle of any typical development process, i.e. manufacture, plan and design, construct, operate, renovate, reuse and recycle with fundamental principles of extracting less and minimizing disposal (GBCA, 2023). One of key actions is to further promote the collaboration amongst related stakeholders such as State Government, local governments, construction industry, universities and the resource recovery sector. In addition, government's leadership and support play a paramount role through measures such as legislation and regulation to identify innovation models, mandatory circular economy requirements in procurement processes and design guidelines, evaluating embodied emission via standardized metrics, establishing database for sustainable materials, and providing low-interest Green Loan scheme for energy retrofitting of residential buildings, etc.

Apart from technological factors, behavioural factors are equally important in the circular built environment. Ghafoor et al. (2023) examined the readiness of the Australian Architectural, Engineering and Construction (AEC) industry for circular economy through interviews and a questionnaire survey. Their study found that there is lack of adequate knowledge and guidelines for the adoption of circular economy in the AEC industry. Awareness was identified as one of most critical factors that can facilitate the implementation of circular economy in the industry.

## 1.6 Summary of Chapters

This book draws contributions from five countries and regions across 20 chapters. These chapters cover a wide arrange of topics, from both theoretical and practical perspectives. Some of the chapters focused on a micro scale while others on macro-level issues. A summary of the chapters is provided below:

**Chapter 2: Urban waste landscape and inclusive community management: a case study on Hong Kong.**

This study examines the issue of inclusive waste management, with a focus on the development of decision-aid tools. This study also builds an activity and location-based MFA approach, which can be used for constructing location-based inventories for waste generation. A case study is performed in Hong Kong.

**Chapter 3: Inclusive construction management: insights and recommendations by matching analysis on circular economy and sustainable development goals (SDGs).**

This study analyses how circular economy policies in construction would contribute to or affect the 17 sustainable development goals (SDGs). Policies based on accounting for carbon content in buildings in the Greater Bay Area (GBA) are discussed.

**Chapter 4: Carrier-load perspective evaluation on urban resources environment carrying capacity: theory and application.**

This study proposes a theoretical method, named URECC-LC, to evaluate urban resources environment carrying capacity from a load-and-carrier perspective. The assessment can help provide wise strategic planning and environmental management of natural resources and also help raise awareness of sustainable development.

**Chapter 5: Quasi-static to dynamic mechanical properties of recycled aggregate concrete-filled steel tube columns exposed to high temperature.**

The study investigates the degradation of mechanical properties of recycled aggregate concrete-filled steel tube (RACFST) columns from quasi-static to dynamic after exposure to high temperatures. The results indicate that there are no significant differences in failure modes between RACFSTs with varying recycled aggregate ratios, while the steel ratio is found more important in improving impact properties.

**Chapter 6: Developing an urban resource cadaster for circular economy.**

This study details the concept, methodology, and general implications of urban resource cadasters.

Through three case studies, the research finds that a spatially refined urban resource cadaster can assist with better urban planning and waste management.

**Chapter 7: Modeling Sustainable Housing for Sustainable Development in Cities and Communities: The Perspective of a Developing Economy.**

This study simulates the impacts of various factors on the achievement of sustainable development goals (SDGs) for sustainable urban and community housing, and thus reveals groups of significant barriers and contributors that may critically affect the attainment of the SDGs in housing.

**Chapter 8: Low-carbon sustainable development of 5G base stations in China.**

From the perspective of energy consumption, this study discusses the construction of 5G base stations in China and the application of life cycle assessment (LCA) in wireless networks. Energy saving and carbon reduction measurements are proposed accordingly.

**Chapter 9: Affordable Housing and Sustainable Communities in sub-Saharan Africa: Implementation Drivers, Challenges and Practices.**

This study focuses on addressing the first target and indicator of sustainable development goals (SDGs) 11, and provides insights to understand drivers, challenges, and practices of affordable housing through theoretical lenses such as Giddens Structuration Theory. Case studies are performed to propose some best practices for actualizing affordable housing.

**Chapter 10: Towards City 5.0: Building Resilient Communities through the Interplay of Circular Economy, Resilient and Sustainable Supply Chains, and Reverse Logistics.**

This study provides an overview of concepts such as reverse logistics (RL), circular economy (CE) sustainable and resilient supply chain (SRSC), evaluates the integration of these concepts, and explores how the application contributes to sustainable development goals (SDGs). The proposed conceptual framework may help achieve the targets established in SDG 11.

**Chapter 11: The impact of transportation and emissions allocation on the global warming potential of concretes containing alternative binders and fillers.**

This chapter demonstrates the need for a detailed life cycle assessment in conjunction with concrete mix design, and that such a life cycle assessment must fully consider the effects of emission allocations as well as transport modes and distances.

**Chapter 12: Costs and Pricing of Green Buildings.**

This study comprehensively reviews the effects of the incremental cost, price premium, and willingness-to-pay on green building development. The review shows that the largest obstacle to green building development is the low willingness-to-pay from the public, and incremental cost also provides negative contributions.

**Chapter 13: C&D waste recycling and management: Chinese practices.**

This study identifies six challenges in the development of construction and demolition (C&D) waste management by visiting recycling plants and interviewing industry practitioners, where the challenges include unstable sources, absence of subsidies and regulations, unregulated landfill activities, low-accuracy waste estimation and ineffective waste tracing system.

**Chapter 14: Architects' Behaviour Towards Implementing Construction and Demolition Waste Minimisation Strategies: The Case of Australia.**

This study explores the architects' real behaviours in implementing waste minimisation strategies. The results indicate that architects' attitude greatly impacts on behavioural intention, which can influentially explain architects' real behaviour, while subjective norms, perceived behavioural control, personal norm, and institutional pressures were revealed ineffective on architects' behavioural intention.

**Chapter 15: Building Occupancy and Smart Metering.**

The chapter discusses energy-saving and occupancy concerns by exploring occupancy resolution types and their applications in energy-saving projects, and by introducing the construction of Wi-Fi signal networks to sense building occupancy. It also investigates the connection between Wi-Fi-enabled occupancy sensing and building energy simulation and system control.

**Chapter 16: Building Eco-feedback System and Occupants' Energy Using Behavior.**

This study provides a conceptual approach to developing an eco-feedback system, along with the corresponding key factors and references. Two case studies are performed to validate the approach.

**Chapter 17: Towards the circular economy: improving construction and demolition waste management through stakeholder engagement.**

Through a traditional literature review, the chapter highlights the importance of stakeholder engagement in building and strengthening a circular economy culture in the construction industry.

**Chapter 18: Circular Economy Outcomes in Civil Infrastructure Projects—A Case Study at the Regency to Pym Street Protect.**

By presenting the Regency to Pym Street (R2P) project, this study shows some practices adopted to achieve emission reduction and circular economy objectives, along with some beneficial outcomes. Key lessons and future directions are drawn based on a detailed discussion of the R2P project.

**Chapter 19: Life cycle assessment of construction and demolition waste: A case study of recycled aggregate products in Australia.**

The study compares recycled aggregate products from recycling plants and virgin materials production and conducts several scenarios to evaluate their efficiency. The results show that recycled aggregate from recycling plants reduces carbon footprint by 57% compared to quarry production. Besides, alternative transportation sources can significantly affect the carbon footprint of recycled products and processes.

**Chapter 20: Towards circularity in roads infrastructure: A critical review.**

Through a critical review, this chapter explores the main advantages and disadvantages of implementing circularity in road infrastructure projects, and how circularity may reduce the depletion of natural resources.

**Chapter 21: Developing a new Revit Plug-in enabling building components-based construction and renovation waste quantification: An Australia study.**

This study proposes a prototype Revit-plugin enabling automated waste quantification at the early design stage. Based on available Revit models and rational assumptions, the plugin can estimate the waste generated at both the construction stage and operational stage due to material replacement in renovations.

## 1.7 Future Remarks

It is well recognized that the built environment plays a crucial role in facilitating circular economy due to its significant demand for resources on one hand, and the significant impacts on society, environment and economy on the other hand. This book captured some developments in theories and practices to achieve a circular built environment in different contexts.

Looking into the future, there are several opportunities for future research. There is significant demand for quality data related to a circular economy in the built environment. These include the generation, collection, transportation and remanufacturing of various waste streams. Such quality data will be extremely useful for evidence-based making. Similarly, the attitude and behaviour of various stakeholders could be further investigated for a circular built environment. The role of actors and their dynamic interactions could be influential to push the circular economy agenda. Meanwhile, the last decades have witnessed some unexpected consequences since the introduction of sustainable development and circular economy. The sustainability measures don't necessarily lead to positive results. For instance, the increased landfill levy has been linked to growing illegal dumping activities that have drawn wide attention due to its severe impacts on land use, human health and other environmental issues. Future research is required to examine the fundamental issues associated with illegal dumping for its prevention. Smart technologies such as computer vision, IoT sensing and blockchain could be introduced to assist in the improvement of the circularity of construction materials. In addition, the circular economy needs a system approach, which is more likely to involve multiple companies and industries working together. More efforts could be invested in exploring the potential linkages and interactions across enterprises and sectors,

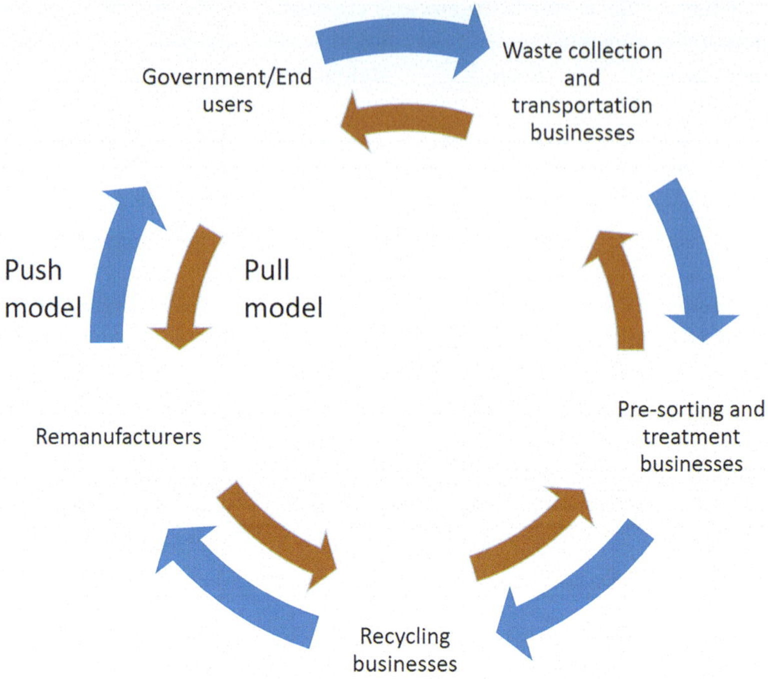


**Fig. 1.3** Push and Pull model for circular economy. (Source: Li et al. (2022))

and their collective impacts on circular built environment. There are different models such as push or pull models that depends on the role of various stakeholders such as government and clients (Fig. 1.3). Effectiveness of these models in facilitating circular built environment is worth further investigation. Future studies could be undertaken to examine how different forms of collaboration, including cooperation, competition and coopetition could facilitate the further uptake of recycled products and materials.

Circular economy will also need to make sense financially. It is not going to be sustainable to keep requesting subsidies from the government. The cost effectiveness of sustainable measures will be crucial for its wider uptake.

## References

Commonwealth of Australia. (2018). *National waste policy: Less waste, more resources.*

Department of Climate Change, Energy, the Environment and Water (DCCEEW). (2023). *Transitioning to a more circular economy*, https://www.dcceew.gov.au/environment/protection/circular-economy#:~:text=All%20of%20Australia's%20environment%20ministers,a%20circular%20economy%20by%202030. Accessed 15th March 2023.

Downes, J. (2022). *Framework for understanding, measuring & communicating waste prevention.* Prepared for the Australian Government Department of Climate Change, Energy, the Environment and Water. Behaviour Works Australia, Monash University.

European Parliament. (2023). *Circular economy: Definition, importance and benefits.*

Fraser, M., Haigh, L., & Soria, A. C. (2023). *The Circularity Gap Report 2023.*

GBCA. (2023, June). *Circular economy in South Australia's built environment – Action Plan, Green Building Council of Australia.*

Ghafoor, S., Hosseini, M. R., Kocaturk, T., Shooshtarian, S., Arnel, T., King, D. K., Bonsey, J., & Garofano, N. T. (2023). *Unlocking the power of the circular economy in the Australian AEC industry: A journey through attitudes, barriers and enablers.* Deakin University.

GISA. (2020). *Supporting the circular economy: South Australia's Waste Strategy 2020–2025.* Green Industries SA.

GISA. (2022, September). *Circular economy in South Australia's built environment.*

Holland Circular Hotspot. (2022). *Circular Buildings—constructing a sustainable future.*

Li, L., Zuo, J., Duan, X., Wang, S., Hu, K., & Chang, R. (2021). Impacts and mitigation measures of plastic waste: A critical review. *Environmental Impact Assessment Review, 90*, 106642.

Li, L., Zuo, J., Duan, X., Wang, S., & Chang, R. (2022). Converting waste plastics into construction applications: A business perspective. *Environmental Impact Assessment Review, 96*, 106814.

NSW Government. (2019, February). *NSW CIRCULAR economy policy statement too good to waste.*

Office of Energy and Climate Change, NSW Treasury. (2023, February). *Circular design guidelines for the built environment: Whole-of-life principles for transitioning buildings, precincts and infrastructure to a circular economy.*

Omrany, H., Chang, R., Soebarto, V., Zhang, Y., Ghaffarianhoseini, A., & Zuo, J. (2022). A bibliometric review of net zero energy building research 1995–2022. *Energy and Buildings*, 111996.

Parliament of Australia. (2022). Building Australia's circular waste economy, Budget Review 2022–23 Index.

Portner, H. O., Roberts, D. C., Adams, H., Adler, C., Aldunce, P., Ali, E., Begum, R. A., Betts, R., Kerr, R. B., Biesbroek, R., & Birkmann, J. (2022). *Climate Change 2022: Impacts, adaptation and vulnerability, contribution of working group II to the sixth assessment report of the Intergovernmental Panel on Climate Change.*

The Economist. (2019). *The critical role of infrastructure for the Sustainable Development Goals.*

The Economist. (2023). *Circular economy rhetoric needs a reality check.*

UNDP. (2023). *What is circular economy and why does it matter?* https://climatepromise.undp.org/news-and-stories/what-is-circular-economy-and-how-it-helps-fight-climate-change, accessed online 25th April 2023, United Nations Development Programme.

Victorian State Government. (2020, February). *Recycling Victoria: A new economy.* Department of Environment, Land, Water and Planning.

# 2 Urban Waste Landscape and Inclusive Community Management: A Case Study on Hong Kong

Liang Dong and Hui Liu

## 2.1 Introduction

As highlighted in the Sustainable Development Goals (SDGs), which are proposed by the United Nations (UN) (Castor et al., 2020; Dong et al., 2021; Kørnøv et al., 2020), sustainable and integrated municipal solid waste management is one critical issue for urban sustainability (Barros et al., 2021; Dong et al., 2022; Fidélis et al., 2021; Liu et al., 2022b; Sharma et al., 2021; Zhao et al., 2021). The concept of the circular economy (CE) offers an integrated solution to tackle the issue of urban waste management (Liang et al., 2023; Rios et al., 2021; Tong et al., 2023; Velenturf & Purnell, 2021). CE is defined as ***"a system innovation, in which, products and materials are recycled, repaired, and reused rather than thrown away, and waste from one process is utilized as s a valued input into another"*** (Bimpizas-Pinis et al., 2021; Clube & Tennant, 2020; Hu et al., 2018; Joensuu et al., 2020; Shah et al., 2020b). From the perspective of urban metabolism, CE provides both preventative and regenerative pathways to optimize material flows within cities (Dong et al., 2017; Fang et al., 2017; Hao et al., 2022; Kim et al., 2018). As a CE practice, in recent years, the concept of "Zero-waste Cities", which ***"aim to adopt comprehensive CE solutions to minimize urban waste generation"***, have been actively promoted in Japan, European Union, and Mainland China, and are driven by visionary local leaders and practitioners (Fidélis et al., 2021; Fujii et al., 2019; Liu et al., 2022a; Melles, 2021; Mies & Gold, 2021; Pizzi et al., 2021; Shah et al., 2020a; Zhao et al., 2021).

One challenge emerging as a concern for the transition towards CE is social impact (Clube & Tennant, 2020; Mies & Gold, 2021; Sharma et al., 2021; Wang et al., 2019; Xavier et al., 2019). From an environmental and economic system perspective, the transition from the traditional linear economy to a CE changes the material metabolism and, inevitably, will change the inter-sectoral relations within the production-consumption system, resulting in uncertain impacts on the social system (Baba et al., 2015;

L. Dong (✉)
Department of Public and International Affairs (PIA), City University of Hong Kong, Hong Kong, SAR, China

School of Energy and Environment (SEE), City University of Hong Kong, Hong Kong, SAR, China

Sustainable Development Lab, Centre for Public Affairs and Law (CPAL), City University of Hong Kong, Hong Kong, SAR, China

Shenzhen Research Institute (SRI), City University of Hong Kong, Shenzhen, China
e-mail: liadong@cityu.edu.hk

H. Liu
Collaborative Innovation Center on Forecast and Evaluation of Meteorological Disasters/Research Center of Urban Sustainable Development/School of Geographical Sciences, Nanjing University of Information Science & Technology (NUIST), Nanjing, China



J. Zuo et al. (eds.), *Circular Economy for Buildings and Infrastructure*, Sustainable Development Goals Series, https://doi.org/10.1007/978-3-031-56241-9_2

Bian et al., 2020; Gutberlet, 2015; Khoo, 2020; Tong & Tao, 2016). This leads to a series of "inclusive" problems, such as how our social system can adapt to and be affected by the transition towards becoming a circular city (Alessandria, 2016; De Vita & Oppido, 2016; Dou et al., 2018; Tong et al., 2018).

From a local perspective, the CE strategy is significant for high-density cities like Hong Kong. Hong Kong generates more than six million tons of waste every year; of this, only 30% (approximately) is being recycled (Hossain et al., 2021; Kuong et al., 2019; Steuer, 2022; Wuni & Shen, 2022). Despite being well recognized for its advanced economy and good urban management, Hong Kong has yet to achieve a milestone in CE practices. This, therefore, calls for Hong Kong's attention to promote CE practices with a view to help enhancing the city's brand of sustainability and competitiveness in the new round of global zero-waste cities' competition. For example, a typical phenomenon to illustrate the "social inclusion" issue, is the daily scene of "trolleys and the elderly" of in Hong Kong and many other Asian countries. Thousands of elderly individuals use ubiquitous "trolleys" to scavenge urban waste (Alessandria, 2016; Sharma et al., 2021; Steuer, 2022; Tong & Tao, 2016; Tong et al., 2023). This informal economy of "trolleys and elderly" is, no doubt, already an important part of Hong Kong's economy and reflects critical social challenges such as aging society, pension system failure, and new health concern in the post COVID period. The promotion of CE, in the short term, will inevitably cause q negative impact on these vulnerable populations. Therefore, there is a need for a new research agenda on an "inclusive" CE (Baba et al., 2015; Gutberlet, 2015; Haase et al., 2017; Morfeldt, 1993; Morone et al., 2016), which not only considers the urban metabolism optimization (material flows and waste flows) but also the social adaptability (Dong et al., 2018; Hu et al., 2018; Siu et al., 2017; Tong et al., 2018; Tseng et al., 2018).

This research agenda requires two fundamental studies: (1) how the mature material flows analysis (MFA) (urban metabolism) could better integrate with social analysis (actors, activities, behavior, and governance) and spatial analysis (activities with certain locations). (2) to evaluate the "social inclusion" issue in the transition towards a CE by assessing how the integrated environmental, economic, and social costs and benefits could be better evaluated under the SDGs framework. To the best of our knowledge, these two aspects have been under-studied. This study will hereby propose the following research developments for filling the abovementioned gaps: we conduct a study to conceptualize the urban system into three boundaries, namely: (1) general geography (GG), which encompasses the overall urban planning scope, such as land use, infrastructure planning, and facility location; (2) physical geography (PG), which represents the urban metabolism in terms of material and energy flows; and (3) human geography (HG), which encompasses the socioeconomic activities and agents that act as metabolism agents. Based on the urban boundaries, we will further propose a framework of activity and location-based MFA approach. We will build a location-based inventory for waste generation using statistical data on population, economic outputs, and waste generation in various categories. Geospatial data, such as population distribution, land use, housing shapes, and nightlight data, will be applied to construct the spatial inventory. This method and data spatialization will help in building the waste landscape. Inclusive waste management and policy implications are discussed and proposed based on the developed method and established waste landscape.

The organization of the study is as follows: after this introductory section, Sect. 2.2 reviews the current progress of waste landscape and agent, and location-based MFA as a decision aid method. Section 2.3 will present the framework of an activity and location-based MFA model. Section 2.4 will present our preliminary work in Hong Kong, focusing on building a waste landscape using the activity and location-based model. Finally, Sect. 2.5 will draw the conclusions and policy implications.

## 2.2 Literatures Review on Current Progress of Waste Landscape and Agent & Location-Based Material Flows Analysis

Like living organisms, urban systems operate through metabolic processes (Wolman, 1965). Urban metabolism is a very complex issue. Urban metabolism is a complex issue that has been analyzed by numerous scholars from the perspective of urban living organisms (Newcombe et al., 1978; Newman et al., 1996). With the development of industrial metabolism theory and social metabolism theory, MFA was proposed and gradually became the mainstream of research and application. MFA refers to the systematic analysis of material flow and storage in a specific system within a certain space-time range, mainly involving four dimensions: flow, time, active subjects, and space (Brunner & Rechberger, 2016). In the early research on MFA, scholars mainly focused on the two dimensions of flow and time and explained the inflow, stock, and output of materials in various systems. Hekkert et al. (2000) analyzed paper and wood flows in the Netherlands by combining the MFA method with supply and use tables to construct a paper and wood flow model for the year 1990 (Hekkert et al., 2000). Similarly, by taking Vienna and the Swiss lowlands as examples, Hendriks et al. (2000) reviewed previous studies through MFA and found that MFA is not only a research method but also an important tool for supporting environmental policy decision-making (Hendriks et al., 2000). Ščasný (2003) analyzed the overall material flow characteristics of the Czech Republic from 1990 to 2000 by combining MFA with various economic and social indicators (Ščasný et al., 2003). Huang and Hsu (2003) conducted a study starting from the major urban engineering projects in Taipei metropolis from 1981 to 1998. They combined MFA and emergent evaluation to evaluate the material flow (sand, cement, asphalt, and construction waste) of engineering projects and their impact on the ecological and economic system, thereby making valuable contributions (Huang & Hsu, 2003). In another study, Rotter et al. (2004) conducted a process assessment using MFA for RDF-production processes, i.e., material input, waste output, and product yield to develop a process that minimizes hazardous chemicals (Rotter et al., 2004). Barles (2007) took the example of Paris, including the city itself, its suburbs, and the entire region, to analyze the global material flow through MFA, and made a comparative analysis of the material flow in the three regions (Barles, 2007). Browne et al. (2011) analyzed the raw material input and waste flow in Irish urban areas on a long-term scale (1992–2002) through indicators, such as the Domestic Material Consumption (DMC) per capita and Direct Material Input (DMI) per capita, combined with MFA (Browne et al., 2011). Hodson et al. (2012) proposed to combine MFA and transitions analysis (TA) to analyze the relationship between urban material flow and the organization of cities, infrastructure systems, and resource flows. Their study focused on the concept of a "plastic city" (Hodson et al., 2012).

The focus on flow and time explains the internal operation mechanism of the city to a certain extent, deepening our comprehension of urban metabolism. However, it tends to overlook the various metabolic activities in the city, the connection between active subjects, and the flow of various substances in space. Therefore, spatializing the flow and analyzing the relationships between flow subjects are crucial for enhancing the MFA method and gaining a more comprehensive understanding of urban metabolism. In follow-up research works, many scholars have made many attempts to spatialize material flows and examine the connections between active subjects. Roy et al. (2015) proposed a potential decision support tool called Spatial allocation of material flow analysis (SAMFA), which explored the spatial characteristics of material flow in County Kildare, Ireland (Roy et al., 2015). ALBUJA (2019) combines ArcGIS methods with MFA to spatially visualize material flows for Brazilian commodity imports/exports(Albuja, 2019). Yeow & Cheah (2019) incorporated commodity flow and truck activity data into MFA research to explore the spatial flow and dispersion of goods and commodities (Yeow & Cheah, 2019). Chen and Liu (2021) calculated the generation and disposal

**Table 2.1** Summary of literatures

| Aspects of MFA | Studies (research scope) |
|---|---|
| Q | Hekkert et al. (2000) (Paper and wood)<br>Hendriks et al. (2000) (Comprehensive review)<br>Rotter et al. (2004) (RDF)<br>Barles (2007) (Economic, goods)<br>Hodson et al. (2012) (Material, Energy) |
| Q, T | Ščasný et al. (2003) (Economic, goods)<br>Huang and Hsu (2003) (Construction)<br>Browne et al. (2011) (Raw material, waste) |
| Q, A | Fan et al. (2021) (Phosphorus flows) |
| Q, S | Roy et al. (2015) (Buildings)<br>Albuja (2019) (Economic)<br>Yeow and Cheah (2019) (Goods)<br>Chen and Liu (2021) (C&DW)<br>Sileryte et al. (2022) (Waste) |
| Q, T, A, S | Geldermans et al. (2019) (C&DW)<br>Furlan et al. (2020) (Economic, waste) |

Note: *Q* quantity, *T* time, *A* activities (agents), *S* spatial (location-based)

of construction and demolition waster (C&DW) in a district through MFA and described the spatial flows of C&DW at the district-level with reference to the location of garbage generation and disposal points (Chen & Liu, 2021). Fan et al. (2021) examined the phosphorus load characteristics of Erhai Lake from 2000 to 2017 from the perspective of material flow, and analyzed the proportion of pollution emissions from different activities (Fan et al., 2021). Sileryte (2022) combined government agency data, business activity data, and road data to construct a waste stream originating in the city of Amsterdam in 2018 (Sileryte et al., 2022). However, these studies only added a variable to activity subjects and space when building upon flow and time, forming approaches such as AMFA and SMFA. However, a comprehensive analysis that fully integrates these is yet to be achieved.

In order to gain a more accurate understanding of the internal activities of urban systems, analyze the urban operation mechanism, and address the "black box" issue of traditional MFA. Moreover, a new method called the activity-based spatial MFA (AS-MFA) has been proposed and gradually cited. AS-MFA Link materials, spaces, activities, and activity subjects, and identifies the links between various systems by building a network system to find out the space, process, and key players of material flow (Geldermans et al., 2017). Geldermans et al. (2019) combined C&DW originating in the Amsterdam metropolitan area (AMA), and visualized it using the AS-MFA method (Geldermans et al., 2019). Furlan et al. (2020) utilized geographic big data on C&DW generated in the AMA from 2013 to 2018 and aggregated it by economic sector, depicting the economic activities, outward flows, and waste flows of the AMA using AS-MFA (Furlan et al., 2020). Table 2.1 has summarized the current status of literature in this field. At present, the application of MFA with consideration of activities (and the engagement of agents) is still in its infancy. Enriching the theory of AS-MFA and applying it is the key challenge to be addressed.

## 2.3 Framework of an Activity and Location-Based Material Flows Analysis Model

### 2.3.1 A Novel and Conceptual Urban Waste Landscape Model

Based on urban geography theories (Anderson, 2019; Chaudhary, 2021; De Meulder et al., 2022; Knight, 2018; Mason-Renton & Luginaah, 2018), a conceptual model called the novel urban

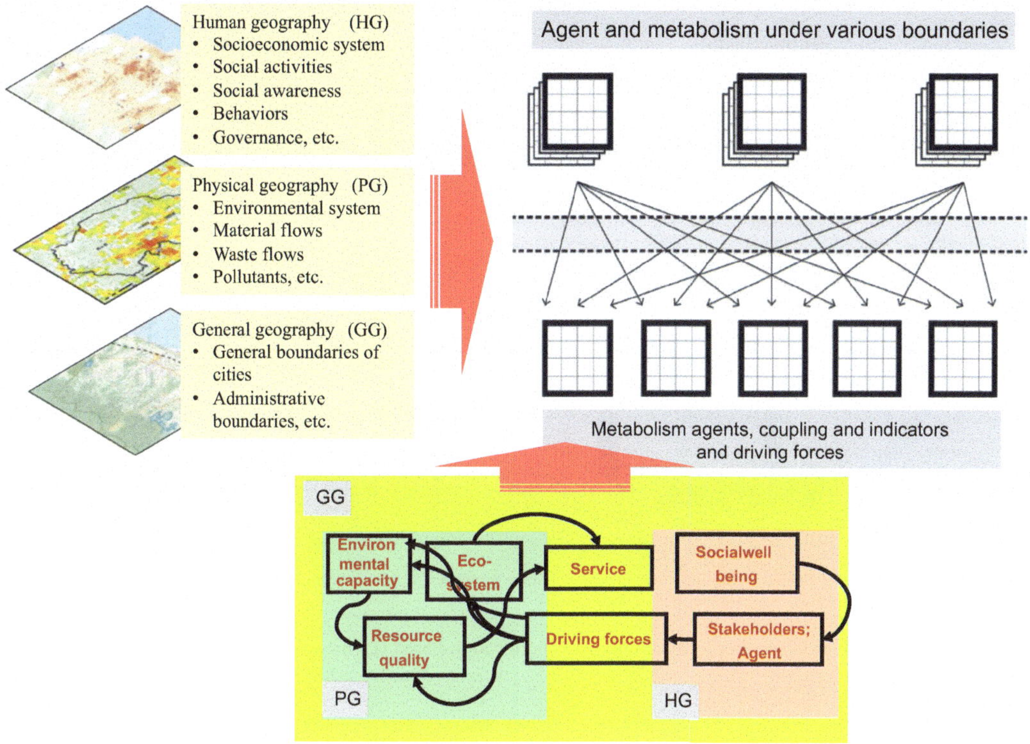


**Fig. 2.1** A novel urban waste landscape model as a conceptual model to support integrated material flows, social, and spatial analysis. The division of urban geography refers to. (Anderson, 2019; Chaudhary, 2021; Haining, 2009; Knight, 2018)

waste landscape model is proposed, as illustrated in Fig. 2.1. In order to characterize the different system boundaries and the corresponding actors of the city, we divide the city into three interconnected system boundaries, and allocate the various social activities and the urban metabolism into certain geographical aspects: (a) The general urban system boundary (General geography-GG) represents the general administrative boundaries of the city and encompasses basic urban planning and land use; (b) the environmental system (Physical geography-PG) represents the environmental system where urban metabolism occurs with characterized material flows and associated environmental impacts (resources consumption, wastes, and pollutions); (c) the human geography-HG represents the social and economic activities within certain spatial locations as the carrier of material flows. This conceptual model provides a systemic representation of urban waste-related activities, flows, social activities, and their spatial significance. It helps to identify the downstream and upstream activities throughout the stages of production, consumption and use discharge systems, which contribute to specific waste flows.

### 2.3.2 An Activity and Location-Based Material Flows Analysis (AL-MFA)

Based on the urban waste landscape conceptual model, this study proposes an activity and AL-MFA approach. This aims to identify the

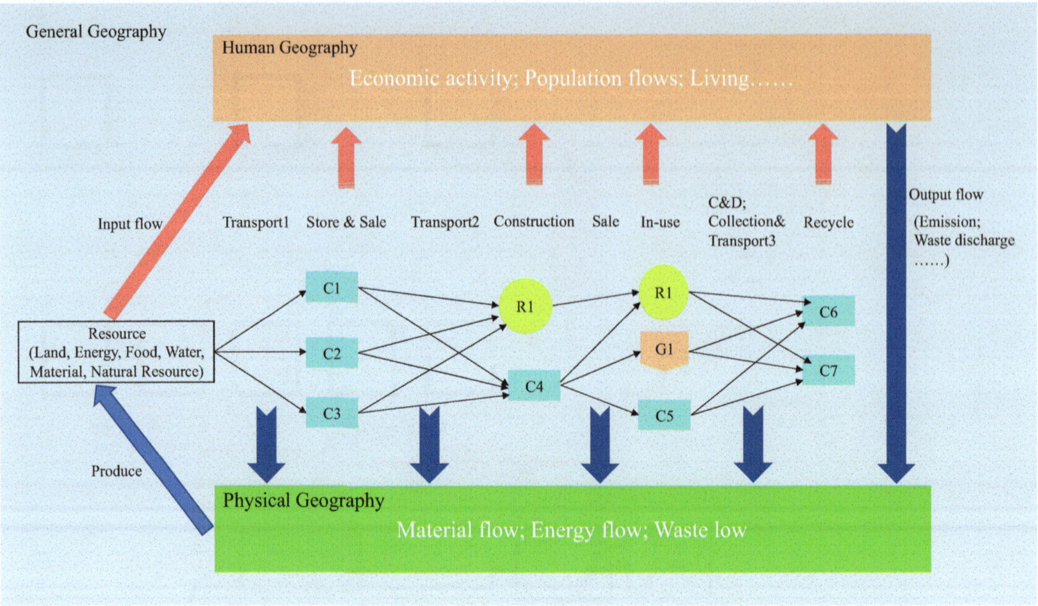


**Fig. 2.2** Schematic chart of the proposed activity (A) and location-based (L) material flows analysis (MFA) approach (AL-MFA)

material flows volume & quantity (Q) within a certain time span (T), considering the activities (and the agents related to the activity) that carry the material flows and generate waste (A), as well as their spatial locations (L).

Figure 2.2 presents the framework of AL-MFA to analyze the material flows not only in quantity and time, but also in activity and location. Each activity contains different subjects. The supply chain mainly includes the production, storage and sales of products, transportation, use, and end-of-life. *C* represents the actor or agent involved in the supply chain who carried the activities with material flows and emissions. *R* represents the residents who play a main player at the community level contributing to household waste emissions. *G* represents authorities and supervisors who don't directly cause material flows and emissions, but engage in the waste management system. With the AL-MFA approach, the study will analyze the locations and spatial characteristics of activities, agents, players, material flows, and emissions and will provide insights into the dynamics of waste generation and management, considering both the activity and location aspects.

### 2.3.3 Activities and Location-Based Inventory

Based on the urban waste landscape conceptual model, an activity and location-based inventory for waste flows and the associated social activities is constructed in Fig. 2.3. The database will enable the analysis on how the waste amount could be spatially explicit, the involved actors and social activities, as well as well as their interrelations in the given scope. The general data comes from statistics, maps, reports, and literature. Social science methods, such as on-site investigation, survey and questionnaires, and expert consultancy, will offer complementary data (example is illustrated in Fig. 2.4). In addition, observation and community experiments will offer the monitoring for the data and analytical results.

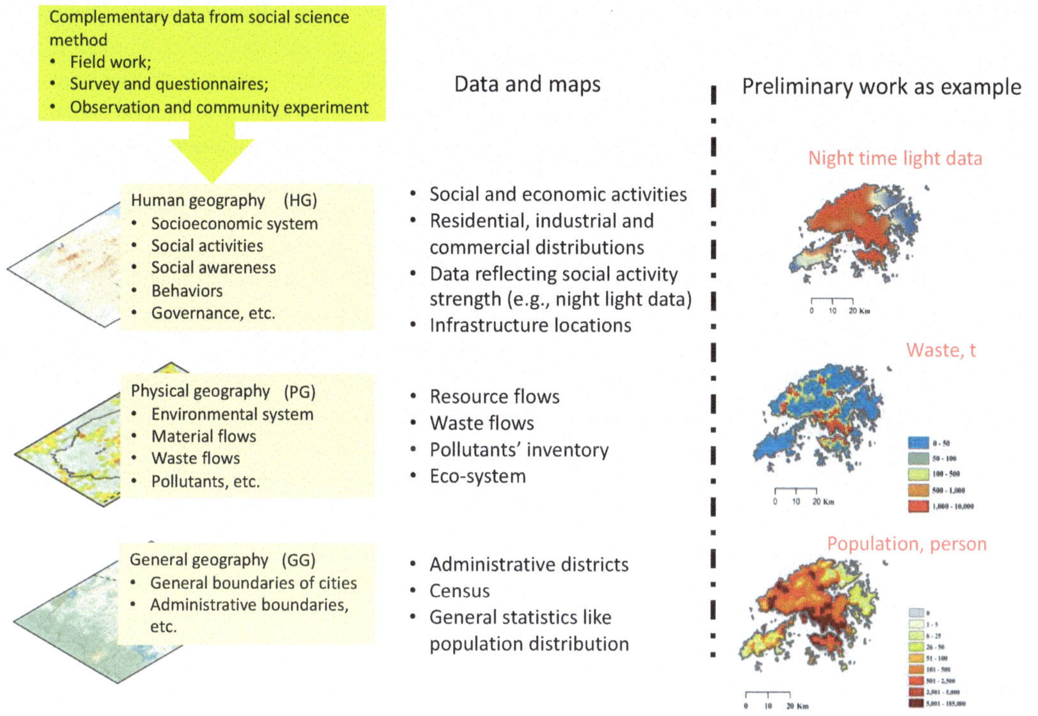


**Fig. 2.3** Construction of the location-based inventory to analyze urban waste landscape

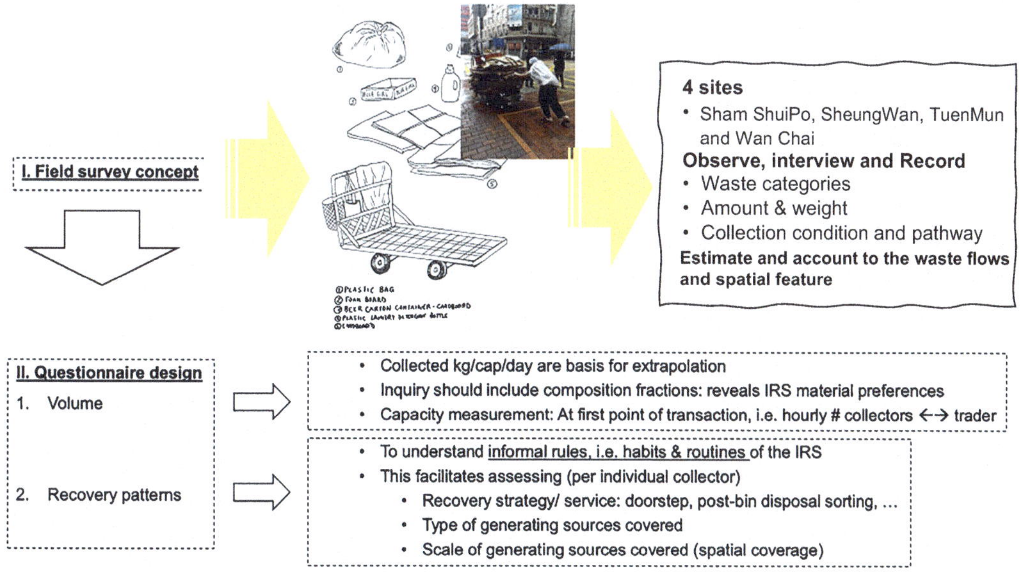


**Fig. 2.4** Preliminary work of social analysis on the inclusive waste management as an application of this developed location-based inventory

## 2.4 Case Study of Hong Kong

This study will use general data to build Hong Kong's waste landscape and employ social science methods to collect more detailed data on the informal waste sector in order to to understand extent of the economic, environmental, and societal challenges posed by the informal waste collection economy in Hong Kong.

### 2.4.1 Urban Metabolism in Hong Kong

Firstly, we analyzed the economic-wide material flows of Hong Kong based on typical EW-MFA framework and indicators, based on statistics in Hong Kong. The domestic material consumption (DMC, presented as: DMC = domestic resources extraction + imports − exports), and domestic material output (DPO, mainly consisting of emissions and pollutions) of Hong Kong was explored in various categories.

It was highlighted that fossil fuels and construction materials accounted for the majority of the material flows, and, therefore, become the major drivers of related emissions (Fig. 2.5).

### 2.4.2 Location-Based Inventory of Hong Kong

Based on the urban waste landscape conceptual network, as well as general material and waste flows, nigh-time light data (NTL), as well as general socioeconomic data, such as population and GDP, we have constructed the location-based inventory of Hong Kong, as illustrated in Fig. 2.6. The spatialization for location will be realized using NTL data, which represents a regressive relationship with human and economic activities. The activities related material and waste flows are calculated based on the per unit material or waste volume (for economic activities, per unit of GDP; for household waste, per capita), and are spatialized according to the location of these activities.

Based on the location-based inventory (Fig. 2.7, upper side) and in combination with social science methods, such as on-site investigation, survey & questionnaires, experts consulting, the inclusive waste management issues could be analyzed in Hong Kong. For example, the on-site data collection, observations, questionnaires and interviews were conducted in four areas of Hong Kong known for their concentration of trolleys and elderly activities. These four sites include Sham Shui Po, Sheung Wan, Tuen Mun, and Wan Chai in the Island district. They also exhibit high social diversity. This kind of research could enlighten the policy makers on the quantitative and spatial explicit figures on waste collectors, collection and recycling activities, as well as the material and waste flows associated with these activities and stakeholders in specific spatial locations.

## 2.5 Conclusions and Policy Implications

This study explored the inclusive waste management issue by developing a decision aid tool. We proposed an urban waste landscape with three boundaries: (1) general geography (GG) standing for the overall urban planning scope, such as land use, infrastructure planning, and facility location. (2) physical geography (PG) representing the urban metabolism, in terms of material and energy flows. (3) human geography (HG), representing the socioeconomic activities and agents, who play a role as the metabolism agents. Based on this conceptual model, we constructed an AL-MFA approach and used it for constructing a location-based inventory for waste generations. This inventory was based on the statistical data on population, economic outputs, and waste generation in various categories, as well as geospatial data, such as population distribution, land use, housing shapes, and NTL. The application of this approach was demonstrated in Hong Kong.

**Fig. 2.5** The domestic material consumption (DMC), and domestic material output (DPO, mainly consisting of emissions and pollutions) of Hong Kong from 1984 to 2018, in various categories. (**a**) DMC values. (**b**) DPO values

The AL-MFA model and the location-based inventory expects to contribute to the following policy implications:

1. They can help identify the key players and their locations in the life cycles of material and waste flows, allowing for a more in-depth discussion of social inclusion issues, such as the elderly and trolley issue in Hong Kong. The activity and location information for the waste flows provides crucial information for policy design and simulation.
2. The location for activities can help clarify the decision-making boundaries and players in it (the *G* variable in our AL-MFA model). This can contribute to the further development of a polycentric waste governance that addresses social inclusion concerns.

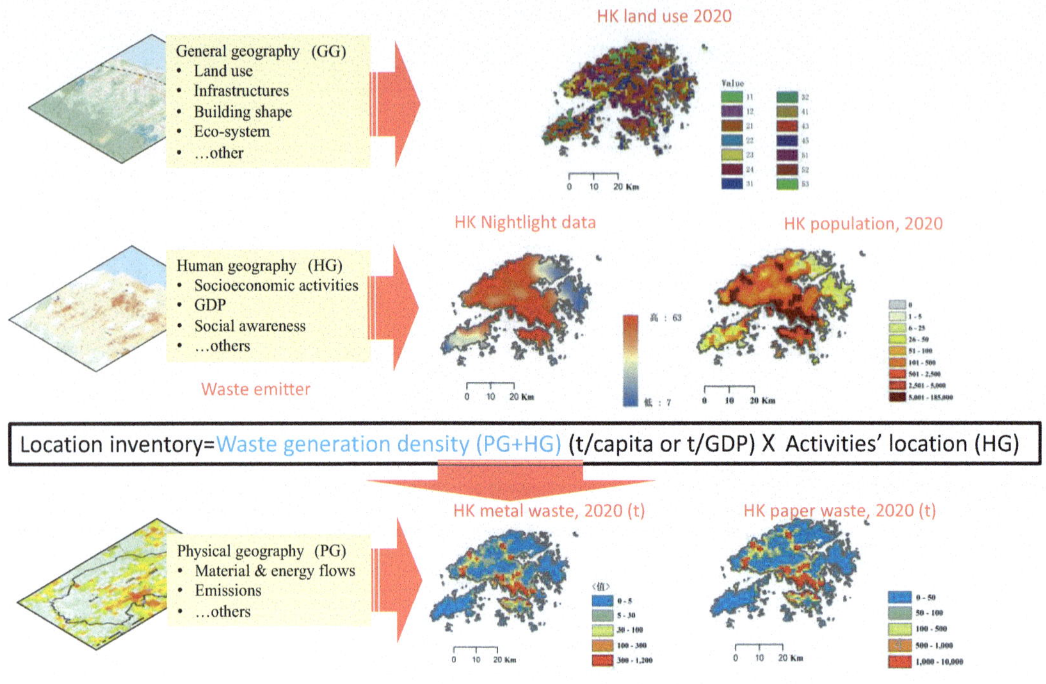


**Fig. 2.6** Workflow to spatialize urban waste (landscape) to support integrated material flows, and social and spatial analysis

3. The information on activities, location, and material and waste flows, together with economic data, is valuable for quantifying inclusion issues. This can facilitate the design of new recycling scenarios and business models that transfer the grey jobs (waste collectors under the informal economy) to green jobs (waste recycling activities in a well-organized manner).

**Declarations Funding** (information that explains whether and by whom the research was supported)

This research was supported by National Natural Science Foundation, China (NSFC), and the Dutch Research Council (NWO): 'Towards Inclusive Circular Economy: Transnational Network for Wise-waste Cities (IWWCs)' (NSFC-NWO, NSFC project number: 72061137071; NWO project number: 482.19.608), and the general research grant for young scientists from NSFC (NO. 41701636).

**Author Contributions** L.D. conceived the research idea, conceived the structure, conducted the work, and wrote the manuscript. L.H. conducted the literature review and constructed the preliminary framework of the method.

**Conflicts of Interest/Competing Interests** (include appropriate disclosures)

The authors declare no conflict of interest.

**Availability of Data and Material** (data transparency)

Not applicable.

**Code Availability** (software application or custom code)

Not applicable.

**Ethics Approval** (include appropriate approvals or waivers)

This paper is not submitted to more than one journal for simultaneous consideration.

This paper is original and has not been published elsewhere in any form or language. Re-use of material (one figure) is clearly marked with reference.

This paper doesn't involve splitting a single study into several parts to increase the quantity of submissions, nor has it been by submitted to various journals or to one journal over time.

This paper doesn't involve concurrent or secondary publications.

The results of this paper are presented clearly, honestly, and without fabrication, falsification or inappropriate data manipulation.

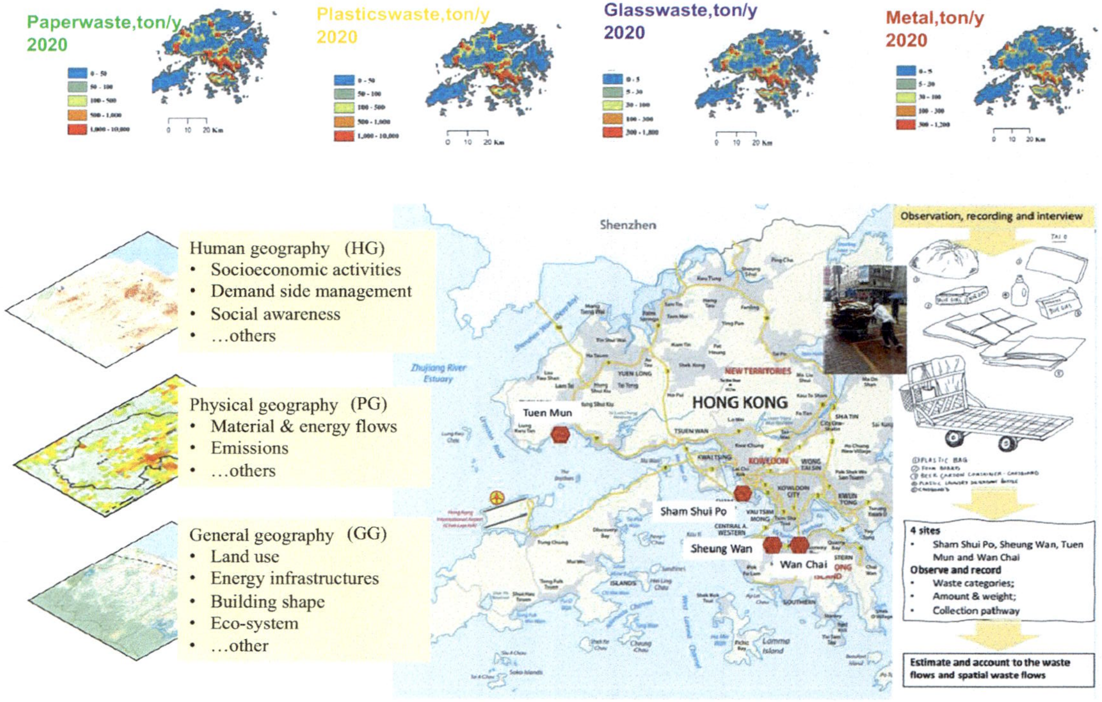


**Fig. 2.7** Image of the application of the location-based inventory in Hong Kong for supporting inclusive waste management

The authors of this paper adhere to the discipline-specific rules for acquiring, selecting, and processing data.

No data, text, or theories by others are presented as if they were the author's own, and proper acknowledgements are given to other works.

This research does not included human participants, their data, or biological material.

**Consent to Participate** (include appropriate statements)
Not applicable.

**Consent for Publication** (include appropriate statements)
Not applicable.

## References


Albuja, V. M. B. (2019). Integration of an urban metabolism framework with geographic information system software.

Alessandria, F. (2016). Inclusive city, strategies, experiences and guidelines. *Procedia – Social and Behavioral Sciences, 223*, 6–10.

Anderson, B. (2019). The city in geography: Renaturing the built environment.

Baba, A. N., Yusoff, N. B., & Elegba, E. O. F. (2015). Towards cities inclusiveness: The land use paradigm option for Nigeria. *Procedia – Social and Behavioral Sciences, 172*, 367–374.

Barles, S. (2007). *A material flow analysis of Paris and its region, international conference renewables in a changing climate-innovation in the built environment* (pp. 579–584). Ecole Polytechnique Fédérale De Lausanne Eplf.

Barros, M. V., Salvador, R., Do Prado, G. F., De Francisco, A. C., & Piekarski, C. M. (2021). Circular economy as a driver to sustainable businesses. *Cleaner Environmental Systems, 2*, 100006.

Bian, Y., Dong, L., Liu, Z., & Zhang, L. (2020). A sectoral eco-efficiency analysis on urban-industrial Symbiosis. *Sustainability (Switzerland), 12*.

Bimpizas-Pinis, M., Bozhinovska, E., Genovese, A., Lowe, B., Pansera, M., Alberich, J. P., & Ramezankhani, M. J. (2021). Is efficiency enough for circular economy? *Resources, Conservation and Recycling, 167*, 105399.

Browne, D., O'regan, B., & Moles, R. (2011). Material flow accounting in an Irish City-region 1992–2002. *Journal of Cleaner Production, 19*, 967–976.

Brunner, P. H., & Rechberger, H. (2016). *Handbook of material flow analysis: For environmental, resource, and waste engineers*. CRC Press.

Castor, J., Bacha, K., & Fuso Nerini, F. (2020). Sdgs in action: A novel framework for assessing energy projects against the sustainable development goals. *Energy Research & Social Science, 68*, 101556.

Chaudhary, T. (2021). 'Waste' to 'use'? Making of 'Narela' in periphery of Delhi Metropolis. *GeoJournal, 86*, 2105–2117.

Chen, C., & Liu, Y. (2021). Construction and demolition wastes in Beijing: Where they come from and where they go? *Waste Management Research, 39*, 555–560.

Clube, R. K. M., & Tennant, M. (2020). The circular economy and human needs satisfaction: Promising the radical, delivering the familiar. *Ecological Economics, 177*, 106772.

De Meulder, B., Marin, J., & Shannon, K. (2022). Evolving relations of landscape, infrastructure and urbanization toward circularity: Flanders and Vietnam. *Geojournal Library*, 107–121.

De Vita, G. E., & Oppido, S. (2016). Inclusive cities for intercultural communities. European experiences. *Procedia – Social and Behavioral Sciences, 223*, 134–140.

Dong, L., Liang, H., Zhang, L., Liu, Z., Gao, Z., & Hu, M. (2017). Highlighting regional eco-industrial development: Life cycle benefits of an urban industrial Symbiosis and implications in China. *Ecological Modelling, 361*, 164–176.

Dong, L., Wang, Y., Scipioni, A., Park, H. S., & Ren, J. (2018). Recent progress on innovative urban infrastructures system towards sustainable resource management. *Resources, Conservation and Recycling, 128*, 355–359.

Dong, L., Liu, Z., & Bian, Y. (2021). *Match circular economy and urban sustainability: Re-investigating circular economy under sustainable development goals (Sdgs)* (Vol. 1, pp. 243–256). Circular Economy and Sustainability.

Dong, L., Taka, G. N., Lee, D., Park, Y., & Park, H. S. (2022). Tracking industrial symbiosis performance with ecological network approach integrating economic and environmental benefits analysis. *Resources, Conservation and Recycling, 185*, 106454.

Dou, Y., Togawa, T., Dong, L., Fujii, M., Ohnishi, S., Tanikawa, H., & Fujita, T. (2018). Innovative planning and evaluation system for district heating using waste heat considering spatial configuration: A case in Fukushima, Japan. *Resources, Conservation and Recycling, 128*, 406–416.

Fan, Y., Ni, Z., Wang, S., Zhang, J., & Wu, S. J. (2021). Whole process phosphorus management strategy construction with phosphorus load characteristics, driver and efficiency from the material flow perspective. *Journal of Cleaner Production, 279*, 122896.

Fang, K., Dong, L., Ren, J., Zhang, Q., Han, L., & Fu, H. (2017). Carbon footprints of urban transition: Tracking circular economy promotions in Guiyang, China. *Ecological Modelling, 365*, 30–44.

Fidélis, T., Cardoso, A. S., Riazi, F., Miranda, A. C., Abrantes, J., Teles, F., & Roebeling, P. C. (2021).

Policy narratives of circular economy in the Eu – Assessing the embeddedness of water and land in National Action Plans. *Journal of Cleaner Production, 288*, 125685.

Fujii, M., Dou, Y., Sun, L., Ohnishi, S., Maki, S., Dong, H., Dong, L., & Chandran, R. (2019). Contribution to a low-carbon society from improving exergy of waste-to-energy system by upgrading utilization of waste. *Resources, Conservation and Recycling, 149*, 586–594.

Furlan, C., Wandl, A., Geldermans, B., & Sileryte, R. (2020). A refined waste flow mapping method. *Contesti Città, Territori, Progetti, 2020*(1), 74–89.

Geldermans, B., Bellstedt, C., Formato, E., Varju, V., Grünhut, Z., Cerreta, M., Amenta, L., Inglese, P., Van Der Leer, J., Wandl, A., 2017. *Repair: Resource management in Peri-urban areas: Going beyond urban metabolism: D3. 1 introduction to methodology for integrated spatial, material flow and social analyses.*

Geldermans, B., Tenpierik, M., & Luscuere, P. J. (2019). Circular and flexible indoor partitioning—A design conceptualization of innovative materials and value chains. *Buildings, 9*, 194.

Gutberlet, J. (2015). More inclusive and cleaner cities with waste management co-production: Insights from participatory epistemologies and methods. *Habitat International, 46*, 234–243.

Haase, D., Kabisch, S., Haase, A., Andersson, E., Banzhaf, E., Baró, F., Brenck, M., Fischer, L. K., Frantzeskaki, N., Kabisch, N., Krellenberg, K., Kremer, P., Kronenberg, J., Larondelle, N., Mathey, J., Pauleit, S., Ring, I., Rink, D., Schwarz, N., & Wolff, M. (2017). Greening cities—To be socially inclusive? About the alleged paradox of society and ecology in cities. *Habitat International, 64*, 41–48.

Haining, R. (2009). *The special nature of spatial data. The Sage handbook of spatial analysis* (pp. 5–23). Sage.

Hao, X., Liu, G., Zhang, X., & Dong, L. (2022). The coevolution mechanism of stakeholder strategies in the recycled resources industry innovation ecosystem: The view of evolutionary game theory. *Technological Forecasting and Social Change, 179*, 121627.

Hekkert, M. P., Joosten, L. A., & Worrell, E. (2000). Analysis of the paper and wood flow in The Netherlands. *Resources, Conservation Recycling, 30*, 29–48.

Hendriks, C., Obernosterer, R., Müller, D., Kytzia, S., Baccini, P., & Brunner, P. H. (2000). Material flow analysis: A tool to support environmental policy decision making. Case-studies on the city of vienna and the Swiss lowlands. *Local Environment, 5*, 311–328.

Hodson, M., Marvin, S., Robinson, B., & Swilling, M. (2012). Reshaping urban infrastructure: Material flow analysis and transitions analysis in an urban context. *Journal of Industrial Ecology, 16*, 789–800.

Hossain, M. U., Ng, S. T., Dong, Y., & Amor, B. (2021). Strategies for mitigating plastic wastes management problem: A lifecycle assessment study in Hong Kong. *Waste Management, 131*, 412–422.

Hu, M. M., Zhang, C. B., Dong, L., Xiang, P. C., & Zhang, Q. (2018). Economical pillar of sustainability assessment on resource circulation-development and application of life cycle costing approach. *Zhongguo Huanjing Kexue/China Environmental Science, 38*, 4788–4800.

Huang, S.-L., & Hsu, W.-L. (2003). Materials flow analysis and Emergy evaluation of Taipei's urban construction. *Landscape Urban Planning, 63*, 61–74.

Joensuu, T., Edelman, H., & Saari, A. (2020). Circular economy practices in the built environment. *Journal of Cleaner Production, 276*, 124215.

Khoo, S. L. (2020). Towards an inclusive creative city: How ready is the historic city of George Town, Penang? *City, Culture and Society, 23*, 100367.

Kim, H. W., Dong, L., Choi, A. E. S., Fujii, M., Fujita, T., & Park, H. S. (2018). Co-benefit potential of industrial and urban symbiosis using waste heat from industrial park in Ulsan, Korea. *Resources, Conservation and Recycling, 135*, 225–234.

Knight, J. (2018). Transforming the physical geography of a city: An example of Johannesburg, South Africa. In *Urban geomorphology: Landforms and processes in cities* (pp. 129–147).

Kørnøv, L., Lyhne, I., & Davila, J. G. (2020). Linking the un Sdgs and environmental assessment: Towards a conceptual framework. *Environmental Impact Assessment Review, 85*, 106463.

Kuong, I. H., Li, J., Zhang, J., & Zeng, X. (2019). Estimating the evolution of Urban Mining resources in Hong Kong, up to the year 2050. *Environmental Science and Technology, 53*, 1394–1403.

Liang, H., Bian, X., Dong, L., Shen, W., Chen, S. S., & Wang, Q. (2023). Mapping the evolution of building material stocks in three eastern coastal urban agglomerations of China. *Resources, Conservation and Recycling, 188*, 106651.

Liu, S., Dong, L., Han, L., Huan, J., & Qiao, B. (2022a). Efficiency versus system synergism: An advanced life cycle assessment for a novel decarbonized grid system innovation. *Energies, 15.*

Liu, Y., Ren, J., Dong, L., Jin, Y., & Man, Y. (2022b). Urban sludge to value-added products for promoting the development of circular economy: Supply network design and optimization. *Resources, Conservation and Recycling, 182.*

Mason-Renton, S. A., & Luginaah, I. (2018). Conceptualizing waste as a resource: Urban biosolids processing in the rural landscape. *Canadian Geographer, 62*, 266–281.

Melles, G. (2021). Figuring the transition from circular economy to circular society in Australia. *Sustainability (Switzerland), 13.*

Mies, A., & Gold, S. (2021). Mapping the social dimension of the circular economy. *Journal of Cleaner Production, 321*, 128960.

Morfeldt, C.-O. (1993). Landscaping with waste. *Engineering Geology, 34*, 135–143.

Morone, P., Lopolito, A., Anguilano, D., Sica, E., & Tartiu, V. E. (2016). Unpacking landscape pressures on socio-technical regimes: Insights on the urban waste management system. *Environmental Innovation and Societal Transitions, 20*, 62–74.

Newcombe, K., Kalma, J. D., & Aston, A. R. (1978). The metabolism of a City: The case of Hong Kong. *Ambio*, 3–15.

Newman, P., Birrell, B., Spessa, A., & Tait, D. (1996). Human settlements. *Transport, 3*, 6.

Pizzi, S., Corbo, L., & Caputo, A. (2021). Fintech and Smes sustainable business models: Reflections and considerations for a circular economy. *Journal of Cleaner Production, 281*, 125217.

Rios, F. C., Panic, S., Grau, D., Khanna, V., Zapitelli, J., & Bilec, M. (2021). Exploring circular economies in the built environment from a complex systems perspective: A systematic review and conceptual model at the city scale. *Sustainable Cities and Society, 80*, 103411.

Rotter, V. S., Kost, T., Winkler, J., & Bilitewski, B. (2004). Material flow analysis of Rdf-production processes. *Waste Management, 24*, 1005–1021.

Roy, M., Curry, R., & Ellis, G. (2015). Spatial allocation of material flow analysis in residential developments: A case study of Kildare County, Ireland. *Journal of Environmental Planning Management, 58*, 1749–1769.

Ščasný, M., Kovanda, J., & Hák, T. (2003). Material flow accounts, balances and derived indicators for The Czech Republic during the 1990s: Results and recommendations for methodological improvements. *Ecological Economics, 45*, 41–57.

Shah, I. H., Dong, L., & Park, H.-S. (2020a). Tracking urban sustainability transition: An eco-efficiency analysis on eco-industrial development in Ulsan, Korea. *Journal of Cleaner Production, 262*, 121286.

Shah, I. H., Dong, L., & Park, H. S. (2020b). Characterization of resource consumption and efficiency trends in Bangladesh, India and Pakistan: Economy-wide biotic and abiotic material flow accounting from 1978 to 2017. *Journal of Cleaner Production, 250*, 119554.

Sharma, H. B., Vanapalli, K. R., Samal, B., Cheela, V. R. S., Dubey, B. K., & Bhattacharya, J. (2021). Circular economy approach in solid waste management system to achieve un-Sdgs: Solutions for post-Covid recovery. *Science of the Total Environment, 800*, 149605.

Sileryte, R., Sabbe, A., Bouzas, V., Meister, K., Wandl, A., & Van Timmeren, A. J. (2022). European waste statistics data for a circular economy monitor: Opportunities and limitations from the Amsterdam metropolitan region. *Journal of Cleaner Production, 358*, 131767.

Siu, K. W. M., Wong, Y. L., & Lam, M. S. (2017). Inclusive play in urban cities: A pilot study of the inclusive playgrounds in Hong Kong. *Procedia Engineering, 198*, 169–175.

Steuer, B. (2022). The emergence of circular economy SMEs in Hong Kong: What is needed to invigorate the dynamic. In *Sustainable production, life cycle engineering and management* (pp. 63–72).

Tong, X., & Tao, D. (2016). The rise and fall of a "Waste City" in the construction of an "urban circular economic system": The changing landscape of waste in Beijing. *Resources, Conservation and Recycling, 107*, 10–17.

Tong, X., Wang, T., Chen, Y., & Wang, Y. (2018). Towards an inclusive circular economy: Quantifying the spatial flows of E-waste through the informal sector in China. *Resources, Conservation and Recycling, 135*, 163–171.

Tong, X., Yu, H., Han, L., Liu, T., Dong, L., Zisopoulos, F., Steuer, B., & De Jong, M. (2023). Exploring business models for carbon emission reduction via post-consumer recycling infrastructures in Beijing: An agent-based modelling approach. *Resources, Conservation and Recycling, 188*.

Tseng, M. L., Chiu, A. S. F., & Liang, D. (2018). Sustainable consumption and production in business decision-making models. *Resources, Conservation and Recycling, 128*, 118–121.

Velenturf, A. P. M., & Purnell, P. (2021). Principles for a sustainable circular economy. *Sustainable Production and Consumption, 27*, 1437–1457.

Wang, Y., Ren, H., Dong, L., Park, H.-S., Zhang, Y., & Xu, Y. (2019). Smart solutions shape for sustainable low-carbon future: A review on smart cities and industrial parks in China. *Technological Forecasting and Social Change, 144*, 103–117.

Wolman, A. (1965). The metabolism of cities. *Scientific American, 213*, 178–193.

Wuni, I. Y., & Shen, G. Q. (2022). Developing critical success factors for integrating circular economy into modular construction projects in Hong Kong. *Sustainable Production and Consumption, 29*, 574–587.

Xavier, L. H., Giese, E. C., Ribeiro-Duthie, A. C., & Lins, F. A. F. (2019). Sustainability and the circular economy: A theoretical approach focused on e-waste Urban Mining. *Resources Policy, 74*, 101467.

Yeow, L. W., & Cheah, L. (2019). Using spatially explicit commodity flow and truck activity data to map urban material flows. *Journal of Industrial Ecology, 23*, 1121–1132.

Zhao, R., Sun, L., Zou, X., Fujii, M., Dong, L., Dou, Y., Geng, Y., & Wang, F. (2021). Towards a zero waste city—An analysis from the perspective of energy recovery and landfill reduction in Beijing. *Energy, 223*, 120055.

# 3 Inclusive Construction Management: Insights and Recommendations by Matching Analysis on Circular Economy (CE) and Sustainable Development Goals (SDGs)

Liang Dong and Xin Bian

## 3.1 Introduction

Sustainable construction in the building sector is critical in the realization of the net-zero carbon society, taking into consideration the intensive energy consumption and significant emissions of greenhouse gases from the sector (Dong et al., 2017, 2018; Liang et al., 2014; Liu et al., 2022; Too et al., 2022). As a result, there have been numerous emerging efforts to promote the net-zero carbon building from multiple perspectives, particularly in the management and policy dimensions (Table 3.1) (Braulio-Gonzalo et al., 2022; Li et al., 2018; Pakdel et al., 2021; Qin et al., 2021; Tseng et al., 2018). Particularly, they highlighted the needs in developing new construction materials, investigating new space for carbon offset from materials and accounting for the embodied carbon emission in the supply chain (scope-3 emission). Hence, life cycle assessment and life cycle management on green construction could provide valuable decision aid support (Hao et al., 2022; Liang et al., 2023; Qiao et al., 2022).

From the perspective of the waste hierarchy (Calderón Márquez and Rutkowski, 2020; Fujii et al., 2019; Kim et al., 2018; Xavier et al., 2019; Zhang et al., 2018), circular economy (CE), which seeks a novel solution from a systemic perspective, offers solution to transit the current linear economy into a circular one via the promotion of the "circularity" of wastes into resources and emphasizing the "3R" principles, namely, reduce, reuse, and recycle (Al-Hamrani et al., 2021; Bian et al., 2020; Clube & Tennant, 2020; Fang et al., 2017; Joensuu et al., 2020). CE has promoted an economic system innovation that minimizes waste flows via the improvement of resource efficiency from the input flows side and via reusing and recycling from the output flows side (Baum et al., 2023; Dong et al., 2017; Dong et al., 2021; Shah et al., 2020; Wang et al., 2018). In addition, life cycle assessment and life cycle

---

L. Dong (✉)
Department of Public and International Affairs (PIA), City University of Hong Kong, Hong Kong, SAR, China

School of Energy and Environment (SEE), City University of Hong Kong, Hong Kong, SAR, China

Sustainable Development Lab, Centre for Public Affairs and Law (CPAL), City University of Hong Kong, Hong Kong, SAR, China

Shenzhen Research Institute (SRI), City University of Hong Kong, Shenzhen, China
e-mail: liadong@cityu.edu.hk

X. Bian
Shenzhen Research Institute (SRI), City University of Hong Kong, Shenzhen, China

Collaborative Innovation Center on Forecast and Evaluation of Meteorological Disasters/Research Center of Urban Sustainable Development/School of Geographical Sciences, Nanjing University of Information Science & Technology (NUIST), Nanjing, China



J. Zuo et al. (eds.), *Circular Economy for Buildings and Infrastructure*, Sustainable Development Goals Series, https://doi.org/10.1007/978-3-031-56241-9_3

**Table 3.1** Global promotion of net-zero carbon building

| Aspects of net-zero carbon buildings | Policies | Examples |
|---|---|---|
| Green building management & certification | Developing certificating system to guide the application of green materials with ow emissions and building energy efficiency approaches | LEED: United States<br>BREEAM: United Kingdom<br>CASBEE: Japan<br>Green Mark: Singapore<br>BEAMS Plus: Hong Kong SAR |
| "Advancing Net Zero" is World Green Building Council's global project | Calls for stakeholders and authorities in business, organizations, cities, and countries & regions to realize net zero operating emissions by 2030 and for all buildings to be net zero in operation by 2050. | European Union, United States, Hong Kong SAR, Singapore |
| Green construction materials | Establish recommended material inventory, which is low embodied emissions | British Standards Institution label for Construction products |
| Procure high-quality carbon offsets | Applied nature-based solutions to increase carbon sequestration effects and offset embodied carbon emissions. | The Green Roofs for Healthy Cities: North America<br>Nature Based solutions: European Union |
| Reduce Scope 3 emissions | Initiatives and information disclosure for construction companies to report their scope 3 emissions (in supply) and actions to reduce climate risk. | Climate related information disclosure guidance by Financial Stability Board and its Task Force on Climate-Related Financial Disclosures, since 2017 and updated in 2021. |

management can support the realization of waste hierarchy. The life cycle stages of construction—namely, material extraction, transport and production, use, and end of life of products— engage different activities; and the related players, therefore, will generate various impacts to CE and vice versa (Amoruso & Schuetze, 2022; Desai & Bheemrao, 2022; Dong et al., 2022; Zhang et al., 2019). Strategic action taken in the earlier stage in the life cycles, such as the design for products and materials, will significantly decrease the cumulative cost as it will contribute to the downstream stage as well. It, therefore, calls for more applications in sustainable construction based on the philosophy of life cycle management (Dong et al., 2018; Liang et al., 2014; Shafique et al., 2020; Sun et al., 2017).

On the other hand, the transition to CE will inevitably break down parts of the current economic system, presented in the way of the altered interlinkage of different sectors and the associated inputs and outputs of material, energy, and capital flows (Chen, 2021; Sharma et al., 2021; Velenturf & Purnell, 2021). This transition, hereby, will generate uncertainty and risk for some social groups (i.e., players in certain activities in different life cycle stages), hence, raising concerns on social inclusivity (Alessandria, 2016; Morfeldt, 1993; Tong & Tao, 2016; Tong et al., 2018). Therefore, it calls for more attention to inclusion issues in promoting CE, e.g., social equity to vulnerable groups in the transition. In this regard, a transition process under the United Nations (UN) sustainable development goals (SDGs) is important. The 17 SDGs emphasize economic efficiency and social equity beyond the environmental concerns (Castor et al., 2020; De Vita & Oppido, 2016; Huchzermeyer & Misselwitz, 2016; Liu et al., 2020).

From a micro level perspective, corporate social responsibility (CSR) and the emerging environmental, social, and governance (ESG) has provided regulation tools to motivate the proactive responses of enterprises to sustainability and inclusion (Pizzi et al., 2021; Roy, 2021; Taleb & Al Farooque, 2021). These enterprises are the key players in the different stages of the whole life cycles. Therefore, an investigation on the challenges and critical environmental impacts of SDGs (e.g., scope-3 emissions and embodied carbon emission) in each life cycle stage is important. To the best of our knowledge, this work has not yet been systematically explored.

This paper aims to perform an in-depth analysis on matching how CE policies in the construction sector will affect the SDGs in the supply chain (life cycles), respectively. The CE policies in the construction sector are explored based on the life cycles in the whole supply chain—from the mining of natural resources, transportation of materials, manufacturing of construction materials, on-site construction process, to final waste disposal and treatment, including construction-and-demolish waste (C&D waste) recycling. We analyze the policy or measures in each life cycle stage, investigating how they would contribute (positive impacts) or affect (trade-offs) the 17 SDGs, particularly embodied carbon being applied as the key indicator to provide critical policy insights.

The study is organized as follows: Sect. 3.2 reviews urban sustainability and SDGs and how CE and life cycle management could contribute to SDGs. Section 3.3 presents the method employed in conducting the matching analysis on each life cycle stage. Section 3.4 applies the method to Hong Kong and the Greater Bay Area (GBA) as an illustration of our proposed method. Finally, Sect. 3.5 draws the conclusions and policy implications.

## 3.2 Inclusive City, Circular Economy, and Sustainable Construction

With the rise of opportunities in employment and education, more convenience in healthcare and transportation, etc., city life has long been the pursuit for most of the population, which will account for 70% of the global population by 2050—more than double in just 100 years (Giles-Corti et al., 2020; Martos et al., 2016). Nonetheless, with rapid urbanization, several exposed risks have emerged far beyond the capability that a city can afford, such as overpopulation, limited resource and space, inequality, housing unaffordability, disasters, and waste emissions, which severely impact the sustainability and quality of life (Arslan et al., 2016; Azunre et al., 2019; Florida & Pedigo, 2019).

Facing a global challenge, sustainable development goals (SDGs) was put forward in 2015 by the UN, which almost all member states adopted. It consists of 17 goals and 169, sub-goals aiming to be the guide or the framework for global sustainable development towards 2030 (Li, 2020). Moreover, worldwide issues, such as climate change, mainly originate from cities. For instance, it is reported that around 80% of global greenhouse gas emissions come from urban locations (Martos et al., 2016). Therefore, cities should be the critical centers in improving the agenda of the SDGs. This proves that more concern should be paid on the sustainable urban problems. For instance, as one of goals, SDG11 (Sustainable Cities and Communities) is directly connected with the sustainable development of city.

Furthermore, there are also many other relevant SDGs, such as SDG3 (Good Health and Wellbeing), SDG4 (Quality Education), SDG6 (Clean Water and Sanitation), SDG7 (Affordable and Clean Energy), SDG12 (Sustainable Consumption and Production), and SDG13 (Climate Action) (Wiedmann & Allen, 2021). These targets contain the urban risks listed before. Therefore, the 17 SDGs should be taken into consideration when studying the issues on sustainability on the city. Table 3.2 summarizes the 17 SDGs and their main concerns.

Two concepts are highlighted in inclusive city research: Inclusive wealth (IW) was constructed by Dasgupta (2001) and Arrow et al. (2012) then gradually developed and adopted as the sustainable evaluation index (Inclusive Wealth Report 2012, 2014, 2018) (Barbier and Wealth Accounting: Natural Capital Project, 2012; Managi & Kumar, 2018; Programme, 2015). It contains three capitals: natural capital, human capital, and produced capital (Managi & Kumar, 2018). When inclusive wealth (V) doesn't decline with the time ($\partial V/\partial t \geq 0$), it is indicative of sustainable development (Manik, 2017).

Nonetheless, as reflected in the latest review of the inclusive wealth index and SDGs, it only contains SDG3, 4, 6, 8, 9, 13, 14, 15 (Dasgupta et al., 2021). In fact, considering the problems of

**Table 3.2** Details of SDGs[a]. This table summarizes the key objectives of each sustainable development goal and how they are related to the triple bottom line (TBL) of sustainability

| 17 SDGs | Content and target | Category of TBL |
|---|---|---|
| Goal 1 | No poverty | Economic |
| Goal 2 | Zero hunger | Economic |
| Goal 3 | Good health and well-being | Economic |
| Goal 4 | Quality education | Social |
| Goal 5 | Gender equality | Social |
| Goal 6 | Clean water and sanitation | Environmental |
| Goal 7 | Affordable and clean energy | Environmental |
| Goal 8 | Decent work and economic growth | Economic |
| Goal 9 | Industry, innovation and infrastructure | Economic |
| Goal 10 | Reduced inequalities | Social |
| Goal 11 | Sustainable cities and communities | Economic & Environmental |
| Goal 12 | Responsible consumption and production | Economic & Environmental |
| Goal 13 | Climate action | Environmental |
| Goal 14 | Life below water | Environmental |
| Goal 15 | Life on land | Environmental |
| Goal 16 | Peace, justice, and strong institutions | Social |
| Goal 17 | Partnerships | Economic & Social & environmental |

Source: The authors' previous study (Dong et al., 2021)
[a]Source: https://sdgs.un.org/2030agenda

rapid urbanization, Fuji and Managi (2016) has also applied inclusive wealth index in urban planning in 2016. Besides, Cook and Davíðsdóttir (2021) found that it contained SDG1, 2, 3, 8, 9, 11, 12, 13, 14, 15. Here comes the paradox. Therefore, it is reasonable to re-evaluate the interaction of inclusive wealth and SDGs especially in the urban dimension.

Inclusive prosperity (IP) is an emergent policy conception with advocates including the Center for American Progress (CAP) (Lowe et al., 2015), the Netherland-funded UNDP-DESA-UNEP Joint Program (Bann et al., 2015), Global Urban Development (GUP) international organization (Weiss and Schoenberger, 2015), World Bank (Anttiroiko & Jong, 2020), the Organization for Economic Co-operation and Development (OECD) (Anttiroiko & Jong, 2020), the Economics for Inclusive Prosperity (EfIP) group (Naidu et al., 2020), etc. As the material idea of an inclusive city, inclusive prosperity has more connections in urban dimension. For example, Senator Peter interpreted *inclusive prosperity* as "the democratization of welfare, or wellbeing, a state where every citizen was given the opportunity to realize their full potentialities" (Adeyemo, 2018) Florida et al. have taken several cities like Philadelphia (Florida, 2019), Denver (Florida & Pedigo, 2019), and Miami (Florida & Initiative, 2018) as examples to further analyze the urban crisis and provide critical pillars for inclusive prosperity such as affordable housing, good jobs, and inclusive public space, etc., based on the specific situations of different cities. Besides, D&L Partners (Partners, 2019) constructed the Inclusive Prosperity Cities Index framework (PICSA index) so as to measure the performance of cities and identify the best practices. The multi-dimensional definition of *inclusive prosperity* consists of three pillars (Prosperity, Social Inclusion and Spatial Inclusion). The United Nation's Sustainable Development Goals call for poverty elimination, equality achievement, and inclusive prosperity promotion (Mirvis, 2020). Hence, it's meaningful to have a comprehensive analysis between specific SDGs and inclusive prosperity especially in city dimension.

Under this new arena of urban sustainability, CSR and ESG are heavily promoted in the construction sector in the realization of an inclusive construction practices that contribute to the 17 SDGs.

## 3.3 Matching Circular economy to SDGs Based on the Life Cycles of Construction

In order to explore how the CE strategy will help to realize the SDGs, we conduct a matching analysis between CE and their potential impacts on SDGs in the social, economic, and environmental dimensions (as summarized in Table 3.3).

We further analyze the CE-related policies based on the life cycles of construction, as illustrated in Fig. 3.1. Generally, the life cycle

**Table 3.3** Details of CE and how they match to TBL

| Key elements | Content and target | Category of TBL |
|---|---|---|
| CE 1 | Prioritize renewable resources | Environmental & Economic |
| CE 2 | Extend lifetime of products | Environmental |
| CE 3 | Reduce | Environmental |
| CE 4 | Reuse | Environmental |
| CE 5 | Recycle | Environmental |
| CE 6 | Circularity | Environmental & Economic |
| CE 7 | Industrial symbiosis | Environmental & Economic |
| CE 8 | Circular business model | Economic |
| CE 9 | Team up to create joint value | Economic & Social |
| CE 10 | Design for the future | Economic & Social & environmental |
| CE 11 | Application of digital technology | Economic & Environmental |
| CE 12 | Knowledge creation | Economic & Social & environmental |
| CE 13 | Education and pro-environmental behavior | Environmental & Social |

Source: The authors' previous study (Dong et al., 2021)

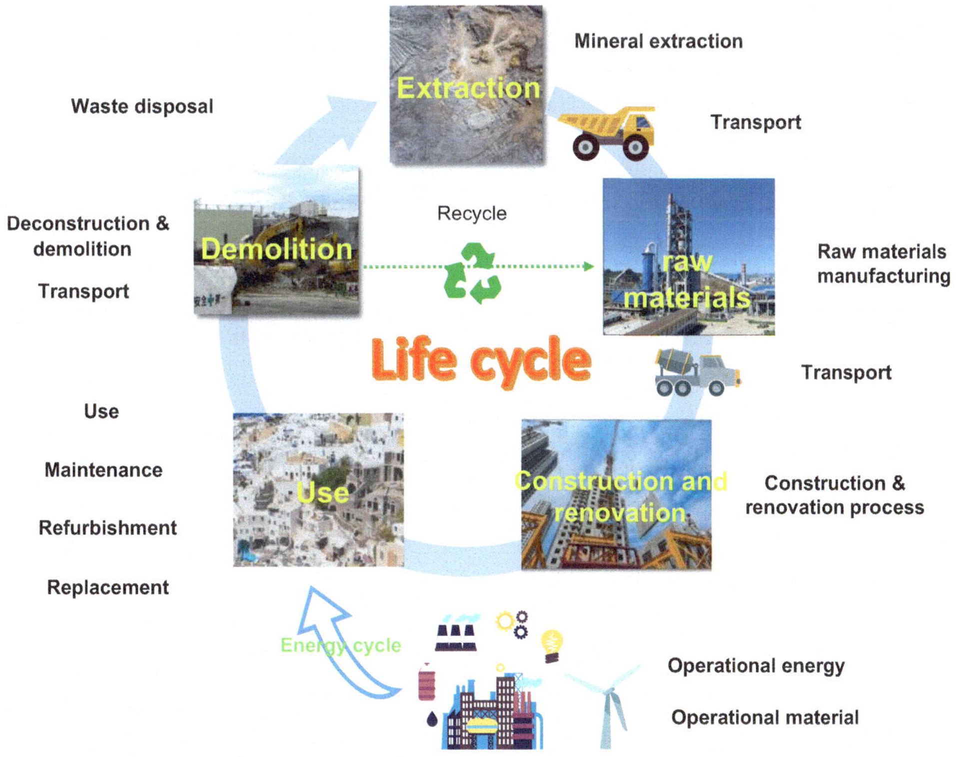


**Fig. 3.1** Life cycles of construction activity and building

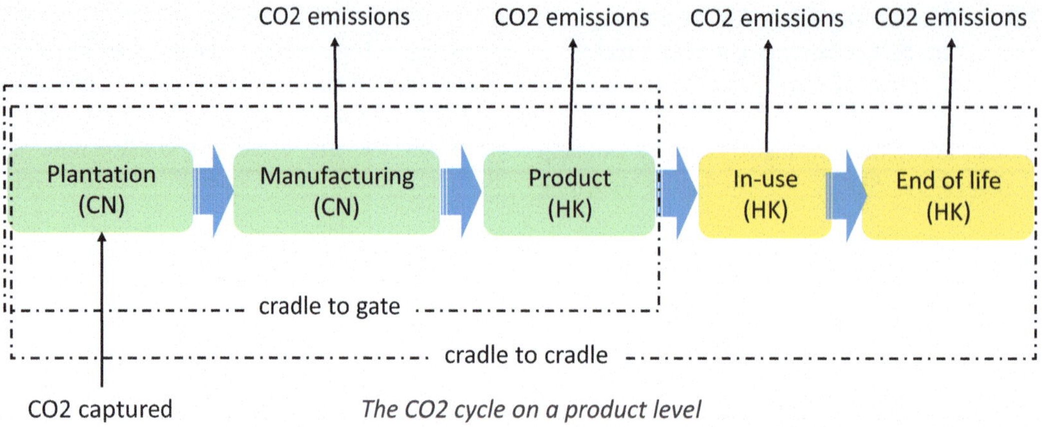


**Fig. 3.2** Concept of "cradle to gate" and "cradle to cradle" in the life cycles of construction activity and building

stage in the construction sector includes construction materials exploration; production and transportation; construction process; building use; operation and maintenance stage; end of life of the construction materials, such as demolition of building waste; waste reuse and recycle. Each stage involves the material and energy consumption and the related pollutants and $CO_2$ emissions. There are also various players (e.g., contractor, constructor, supervisor, regulator, etc.) engaged into the different activities in each life cycle stage, and therefore, CE strategies in different life cycle stages would have different impacts to SDGs and to players. The general CE strategies and their linkage to the TBL of SDGs is summarized in Table 3.3.

## 3.4 Case Study on Hong Kong and the Greater Bay Area (GBA)

The case study is applied in Hong Kong and the broad GBA. According to Hong Kong SAR Climate Action, on 2019, 66% carbon emissions was from electricity generation, and of which, 90% was consumed by the building sector, indicating a contribution of over 60% of carbon emission from the construction and building branch. It can be noted that this number didn't fully consider the carbon emission in the whole life cycles of building, e.g., embodied carbon in materials.

We, hereby, match the critical CE implications in each life cycle of the building based on a preliminary calculation on the embodied carbon. How these embodied carbon and CE implications influence each life cycle stage is further investigated in relation to their impacts on the TBL of SDGs.

The calculation of embodied carbon is based on the accounting to building materials stocks and the material-based emissions factors, which we have done in a previous study (Liang et al., 2023). Figure 3.2 presents a standard procedure in calculating the embodied carbon in life cycles of building materials, which was produced in mainland China and consumed in Hong Kong. In operation, the inventory is collected from both literatures and databases like Ecoinvent. Some calculation could be simplified if material-based carbon emission factor is available, which already presents the life cycle carbon emission. The overall results of embodied carbon in each life cycle is presented in Fig. 3.3.

## 3.5 Life Cycle–Based Policy Recommendations

Based on the embodied carbon, the critical CE implications and their contribution to SDGs were summarized in Table 3.4. The calculation of embodied carbon has highlighted the importance

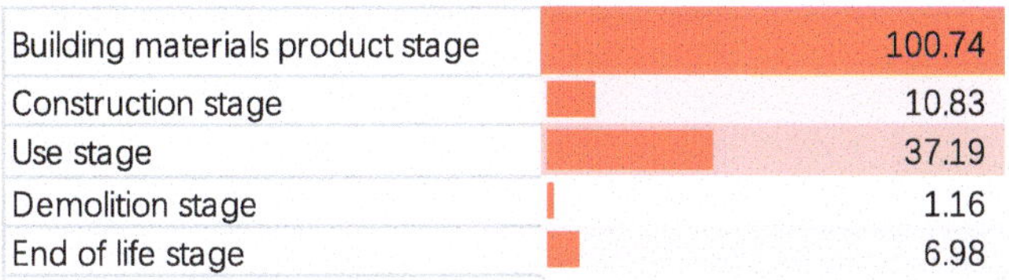


(a) Embodied carbon in each life cycle stage in GBA, Mt

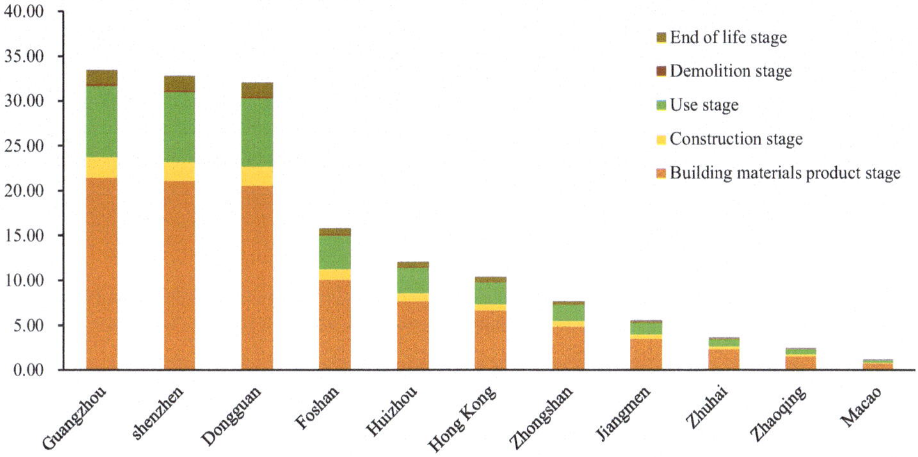


(b) Embodied carbon in cities of GBA

**Fig. 3.3** The embodied carbon emission share of different life cycle stages in the 2020s newly generated buildings in GBA. (**a**) Embodied carbon in each life cycle stage in GBA, Mt. (**b**) Embodied carbon in cities of GBA (Unit: million tons)

**Table 3.4** Policy analysis on each life cycle stage of green construction and how they match with the 17 SDGs

| Life cycle stages | Embodied carbon | Policy highlights under key CE elements | Impacts on SDGs |
|---|---|---|---|
| Raw materials manufacturing | 100.74 Mt. in GBA.<br>The material embodied carbon is significant in the total embodied carbon emission in building stocks. | CE 1: Prioritize renewable resources, e.g., promotion of biomass-based renewable construction material and green construction materials, which has lower or even negative carbon emissions.<br>CE 2: Extend lifetime of products: Extend the lifetime of buildings.<br>CE 4: Reuse & CE 5 recycle: Promoting demolition wastes recycle and recycle to reduce the consumption of raw materials.<br>CE 10: Design for the future creative building design to use more renewable materials.<br>CE 11: Application of digital technology: Integrating block chain and carbon footprint to monitor, report, and verify the carbon emission in the supply chain. | SDG-6<br>SDG-7<br>SDG-8<br>SDG-11<br>SDG-13<br>SDG-14<br>SDG-15 |

(continued)

**Table 3.4** (continued)

| Life cycle stages | Embodied carbon | Policy highlights under key CE elements | Impacts on SDGs |
|---|---|---|---|
| Transport | The embodied carbon emission mainly comes from the fossil fuel consumption in transportation. | CE 1: Prioritize renewable resources: Application of hydrogen and biodiesel.<br>CE 4: Reuse & CE 5 recycle: Waste to energy in reducing fossil fuel consumption in transportation. | SDG-7<br>SDG-13<br>SDG-14<br>SDG-15 |
| Construction & renovation process | 10.83 Mt. in GBA. The embodied emission mainly come from the material and energy consumption in construction and renovation activities. | CE 1: Prioritize renewable resources: Using more renewable materials and renewable energy.<br>CE 10: Design for the future: Develop and promote prefabricated buildings. | SDG-7<br>SDG-11<br>SDG-13<br>SDG-14<br>SDG-15 |
| Use | 37.19 Mt. in GBA.<br>Carbon emissions mainly come from the energy (electricity) consumed during building use. Particularly in GBA, energy consumption is still dominated by coal. | CE 1: Promotion of renewable energy and materials.<br>CE 4 reuse.<br>CE 5 recycle.<br>CE 6 circularity. | SDG-7<br>SDG-11<br>SDG-13<br>SDG-14 |
| Maintenance Refurbishment Replacement | Building renovation and interior decoration will consume corresponding building materials. The phenomenon of large-scale demolition and large-scale construction exists in many areas, and many buildings are forced to be demolished before reaching the life cycle, resulting in a waste of resources. | CE 1: Promotion of renewable energy and materials in green interior decoration<br>CE 3 reduce & CE 4 reuse: Ensure that short-lived components can be repaired or replaced without damaging long-lived components.<br>CE 2 and CE 10: Promoting the green transformation of existing buildings, encouraging the renovation of old residential areas in cities and towns, and the renovation of dilapidated houses in rural areas.<br>CE 2: Retrofitting existing buildings rather than demolishing and building new ones. | SDG-7<br>SDG-11<br>SDG-13<br>SDG-14<br>SDG-17 |
| Deconstruction & demolition | 1.16 Mt. embodied emissions.<br>The longer the life cycle of a building, the smaller the environmental impact per unit time. Compared with countries such as those in Europe and the United States, the average service life of buildings in China and GBA is generally shorter, only 20–30 years. | CE 2: Extend lifetime of products: Extend the building lifetime.<br>CE 8: Circular business model: Develop novel circular business model between constructor, contractor, and recycler.<br>CE 9: Team up to create joint value: Boost value chain in the circular business.<br>CE 11: Application of digital technology: Integrates carbon footprint approach and block chain to improve supply chain management. | SDG-7<br>SDG-11<br>SDG-13<br>SDG-14<br>SDG-17 |
| Waste disposal & recycle | 6.98 Mt. embodied emissions.<br>The recycling and reusing of construction waste can effectively reduce the consumption of original building resources and carbon emissions in the production stage of building materials. However, there is still sufficient room to enhance the recycling efficiency of building materials (e.g., currently, there is only 50% of steel that is full recycled in China). | Improve the recycling efficiency of building materials, strengthen the recycling of building materials, and promote the reduction of construction waste.<br>By changing the traditional landfill treatment method, the thermal energy and electric energy of biomass material combustion can be rationally utilized. | SDG-7<br>SDG-11<br>SDG-13<br>SDG-14 |

of green and low-embodied carbon materials, carbon audit (particularly scope-3 emission accounting) using the life cycle information, updating the current certification system (net-zero building calls for new attention on how to incorporate life cycle and embodied carbon emissions into current green building certification system) using the embodied carbon, how to reduce construction-related emissions from the perspective of carbon offset (e.g., Japan highlights their focus on carbon absorbed in concrete, and China highlights to carbon-sink measures), and how to better use new policy analytical tools like LCA and embodied carbon emission accounting with emerging technologies (like block chains) to strongly support these decision-making efforts. More details are highlighted in Table 3.4.

**Declarations** **Funding** (information that explains whether and by whom the research was supported)

This research was supported by the National Natural Science Foundation, China (NSFC) and the Dutch Research Council (NWO): "Towards Inclusive Circular Economy: Transnational Network for Wise-waste Cities (IWWCs)" (NSFC-NWO, NSFC project number: 72061137071; NWO project number: 482.19.608), and the general research grant for young scientist of NSFC (No. 41701636).

**Author Contributions** L.D. conceived of the idea of the research, made the structure, conducted the work; and wrote the manuscript. B.X. conducted the calculation on the embodied carbon in the building sector.

**Conflicts of Interest/Competing Interests** (include appropriate disclosures)

The authors declare no conflict of interest.

**Availability of Data and Material** (data transparency)

Not applicable.

**Code Availability** (software application or custom code)

Not applicable.

**Ethics Approval** (include appropriate approvals or waivers)

This paper is not submitted to more than one journal for simultaneous consideration.

This paper is original and not have been published elsewhere in any form or language. Re-use of material (one figure) is clearly marked with reference.

This paper doesn't involve the single study split-up into several parts to increase the quantity of submissions and submitted to various journals or to one journal over time.

This paper doesn't involve the concurrent or secondary publication.

The results of this paper are presented clearly, honestly, and without fabrication, falsification, or inappropriate data manipulation.

The authors of this paper adhere to discipline-specific rules for acquiring, selecting, and processing data.

No data, text, or theories by others are presented as if they were the author's own. Proper acknowledgments to other works are given.

This research does not include human participants, their data, or biological material.

**Consent to Participate** (include appropriate statements)

Not applicable.

**Consent for Publication** (include appropriate statements)

Not applicable.

## References

Adeyemo, P., 2018. *Citizen participation in governance and inclusive prosperity for Nigeria*. oer.ui.edu.ng

Alessandria, F. (2016). Inclusive city, strategies, experiences and guidelines. *Procedia – Social and Behavioral Sciences, 223*, 6–10.

Al-Hamrani, A., Kim, D., Kucukvar, M., & Onat, N. C. (2021). Circular economy application for a Green Stadium construction towards sustainable FIFA world cup Qatar 2022™. *Environmental Impact Assessment Review, 87*, 106543.

Amoruso, F. M., & Schuetze, T. (2022). Life cycle assessment and costing of carbon neutral hybrid-timber building renovation systems: Three applications in the Republic of Korea. *Building and Environment, 222*, 109395.

Anttiroiko, A., & Jong, M. D. (2020). *The inclusive city: The theory and practice of creating shared urban prosperity*. Springer.

Arrow, K. J., Dasgupta, P., Goulder, L. H., Mumford, K. J., & Oleson, K. (2012). Sustainability and the measurement of wealth. *Environment and Development Economics, 17*, 317–353.

Arslan, T. V., Durak, S., & Aytac, D. O. (2016). Attaining SDG11: Can sustainability assessment tools be used for improved transformation of neighbourhoods in

historic city centers? *Natural Resources Forum, 40*, 180–202.
Azunre, G. A., Amponsah, O., Peprah, C., Takyi, S. A., & Braimah, I. (2019). A review of the role of urban agriculture in the sustainable city discourse. *Cities, 93*, 104–119.
Bann, C., Anand, P.B., Iftikar, U., Scott, T., Bouma, G., Dutta, S., 2015. Towards green and inclusive prosperity: Building green economies that deliver on poverty reduction.
Barbier, E.B.J.E.S., Wealth accounting: Natural capital project. Stanford University. California, U.S., 2012. Inclusive Wealth Report.
Baum, S., Stadler, I., & Maas, A. (2023). The influence of regulation, taxes and charges on the energy supply of buildings. *Energy, 262*, 125318.
Bian, Y., Dong, L., Liu, Z., & Zhang, L. (2020). A sectoral eco-efficiency analysis on urban-industrial symbiosis. *Sustainability (Switzerland), 12*.
Braulio-Gonzalo, M., Jorge-Ortiz, A., & Bovea, M. D. (2022). How are indicators in green building rating systems addressing sustainability dimensions and life cycle frameworks in residential buildings? *Environmental Impact Assessment Review, 95*, 106793.
Calderón Márquez, A. J., & Rutkowski, E. W. (2020). Waste management drivers towards a circular economy in the global south—The Colombian case. *Waste Management, 110*, 53–65.
Castor, J., Bacha, K., & Fuso Nerini, F. (2020). SDGs in action: A novel framework for assessing energy projects against the sustainable development goals. *Energy Research & Social Science, 68*, 101556.
Chen, C.-W. (2021). Clarifying rebound effects of the circular economy in the context of sustainable cities. *Sustainable Cities and Society, 66*, 102622.
Clube, R. K. M., & Tennant, M. (2020). The circular economy and human needs satisfaction: Promising the radical, delivering the familiar. *Ecological Economics, 177*, 106772.
Cook, D., & Davíðsdóttir, B. (2021). An appraisal of interlinkages between macro-economic indicators of economic Well-being and the sustainable development goals. *Ecological Economics, 184*, 106996.
Dasgupta, P. (2001). *Human Well-being and the natural environment*. Oxford University Press.
Dasgupta, P., Managi, S., & Kumar, P. (2021). The inclusive wealth index and sustainable development goals. *Sustainability Science, 17*, 899–903.
De Vita, G. E., & Oppido, S. (2016). Inclusive cities for intercultural communities. European experiences. *Procedia – Social and Behavioral Sciences, 223*, 134–140.
Desai, A., & Bheemrao, N. (2022). Life cycle assessment of construction materials and its environmental impacts for sustainable development. *Materials Today: Proceedings, 65*, 3866–3873.
Dong, L., Dai, M., Liang, H., Zhang, N., Mancheri, N., Ren, J., Dou, Y., & Hu, M. (2017). Material flows and resource productivity in China, South Korea and Japan from 1970 to 2008: A transitional perspective. *Journal of Cleaner Production, 141*, 1164–1177.
Dong, L., Wang, Y., Scipioni, A., Park, H.-S., & Ren, J. (2018). Recent progress on innovative urban infrastructures system towards sustainable resource management. *Resources, Conservation and Recycling, 128*, 355–359.
Dong, L., Liu, Z., & Bian, Y. (2021). Match circular economy and urban sustainability: Re-investigating circular economy under sustainable development goals (SDGs). *Circular Economy and Sustainability., 1*, 243–256.
Dong, L., Taka, G. N., Lee, D., Park, Y., & Park, H. S. (2022). Tracking industrial symbiosis performance with ecological network approach integrating economic and environmental benefits analysis. *Resources, Conservation and Recycling, 185*.
Fang, K., Dong, L., Ren, J., Zhang, Q., Han, L., & Fu, H. (2017). Carbon footprints of urban transition: Tracking circular economy promotions in Guiyang, China. *Ecological Modelling, 365*, 30–44.
Florida, R.J.D.U., Lindy Institute for urban innovation, 2019. Philadelphia's Next Challenge: From Urban Revitalization to Inclusive Prosperity.
Florida, R., Initiative, M.U.F., 2018. Miami's new urban crisis.
Florida, R., Pedigo, S., 2019. Positioning denver for shared and inclusive prosperity.
Fujii, H., & Managi, S. (2016). An evaluation of inclusive capital stock for urban planning. *Ecosystem Health and Sustainability, 2*.
Fujii, M., Dou, Y., Sun, L., Ohnishi, S., Maki, S., Dong, H., Dong, L., & Chandran, R. (2019). Contribution to a low-carbon society from improving exergy of waste-to-energy system by upgrading utilization of waste. *Resources, Conservation and Recycling, 149*, 586–594.
Giles-Corti, B., Lowe, M., & Arundel, J. (2020). Achieving the SDGs: Evaluating indicators to be used to benchmark and monitor progress towards creating healthy and sustainable cities. *Health Policy, 124*, 581–590.
Hao, X., Liu, G., Zhang, X., & Dong, L. (2022). The coevolution mechanism of stakeholder strategies in the recycled resources industry innovation ecosystem: The view of evolutionary game theory. *Technological Forecasting and Social Change, 179*, 121627.
Huchzermeyer, M., & Misselwitz, P. (2016). Coproducing inclusive cities? Addressing knowledge gaps and conflicting rationalities between self-provisioned housing and state-led housing programmes. *Current Opinion in Environmental Sustainability, 20*, 73–79.
Joensuu, T., Edelman, H., & Saari, A. (2020). Circular economy practices in the built environment. *Journal of Cleaner Production, 276*, 124215.
Kim, H.-W., Dong, L., Jung, S., & Park, H.-S. (2018). The role of the eco-Industrial Park (EIP) at the national economy: An input-output analysis on Korea. *Sustainability, 10*.

Li, X. (2020). Transforming our world: The 2030 agenda for sustainable development: An appeal of global cooperation for building green civilization. In *Green civilization: Human consensus on global collaboration for sustainable development* (pp. 17–35). Springer Singapore.

Li, Y., He, Q., Luo, X., Zhang, Y., & Dong, L. (2018). Calculation of life-cycle greenhouse gas emissions of urban rail transit systems: A case study of Shanghai metro. *Resources, Conservation and Recycling, 128*, 451–457.

Liang, H., Tanikawa, H., Matsuno, Y., & Dong, L. (2014). Modeling in-use steel stock in China's buildings and civil engineering infrastructure using time-series of DMSP/OLS nighttime lights. *Remote Sensing, 6*, 4780–4800.

Liang, H., Bian, X., Dong, L., Shen, W., Chen, S. S., & Wang, Q. (2023). *Mapping the evolution of building material stocks in three eastern coastal urban agglomerations of China* (Vol. 188, p. 188). Resources.

Liu, Z., de Jong, M., Li, F., Brand, N., Hertogh, M., & Dong, L. (2020). Towards developing a new model for inclusive cities in China—The case of Xiong'an new area. *Sustainability, 12*.

Liu, S., Dong, L., Han, L., Huan, J., & Qiao, B. (2022). Efficiency versus system synergism: An advanced life cycle assessment for a novel decarbonized grid system innovation. *Energies, 15*(12), 4214.

Lowe, N., Tewari, M., & Lester, T. W. J. C. P. (2015). Planning for inclusive prosperity: Lessons from the North Carolina experience. *Carolina Planning, 40*, 6–13.

Managi, S., & Kumar, P. (2018). *Inclusive wealth report 2018*. Taylor & Francis.

Manik, Y., 2017. Evaluating scenarios in coupling growth and resource use for palm oil biodiesel policy in Indonesia.

Martos, A., Pacheco-Torres, R., Ordonez, J., & Jadraque-Gago, E. (2016). Towards successful environmental performance of sustainable cities: Intervening sectors. A review. *Renewable & Sustainable Energy Reviews, 57*, 479–495.

Mirvis, P. (2020). *From inequity to inclusive prosperity: The corporate role*. ncbi.nlm.nih.gov

Morfeldt, C.-O. (1993). Landscaping with waste. *Engineering Geology, 34*, 135–143.

Naidu, S., Rodrik, D., & Zucman, G. (2020). Economics after neoliberalism: Introducing the EfIP project. *AEA Papers and Proceedings, 110*, 366–371.

Pakdel, A., Ayatollahi, H., & Sattary, S. (2021). Embodied energy and CO2 emissions of life cycle assessment (LCA) in the traditional and contemporary Iranian construction systems. *Journal of Building Engineering, 39*, 102310.

Partners, D.L., 2019. Creating an inclusive prosperity cities index: Background and methodology.

Pizzi, S., Corbo, L., & Caputo, A. (2021). Fintech and SMEs sustainable business models: Reflections and considerations for a circular economy. *Journal of Cleaner Production, 281*, 125217.

Programme, U. N. U. I. H. D. (2015). *Inclusive wealth report 2014*. Cambridge University Press.

Qiao, L., Tang, Y., Li, Y., Liu, M., Yuan, X., Wang, Q., & Ma, Q. (2022). Life cycle assessment of three typical recycled products from construction and demolition waste. *Journal of Cleaner Production, 376*, 134139.

Qin, L., Kirikkaleli, D., Hou, Y., Miao, X., & Tufail, M. (2021). Carbon neutrality target for G7 economies: Examining the role of environmental policy, green innovation and composite risk index. *Journal of Environmental Management, 295*, 113119.

Roy, M. (2021). Chapter 7 - circular economy: A new sustainable management paradigm. In M. Roy (Ed.), *Sustainable development strategies* (pp. 189–214). Butterworth-Heinemann.

Shafique, M., Azam, A., Rafiq, M., Ateeq, M., & Luo, X. (2020). An overview of life cycle assessment of green roofs. *Journal of Cleaner Production, 250*, 119471.

Shah, I. H., Dong, L., & Park, H.-S. (2020). Tracking urban sustainability transition: An eco-efficiency analysis on eco-industrial development in Ulsan, Korea. *Journal of Cleaner Production, 262*, 121286.

Sharma, H. B., Vanapalli, K. R., Samal, B., Cheela, V. R. S., Dubey, B. K., & Bhattacharya, J. (2021). Circular economy approach in solid waste management system to achieve UN-SDGs: Solutions for post-COVID recovery. *Science of the Total Environment, 800*, 149605.

Sun, L., Li, H., Dong, L., Fang, K., Ren, J., Geng, Y., Fujii, M., Zhang, W., Zhang, N., & Liu, Z. (2017). Eco-benefits assessment on urban industrial symbiosis based on material flows analysis and emergy evaluation approach: A case of Liuzhou city, China. *Resources, Conservation and Recycling, 119*, 78–88.

Taleb, M. A., & Al Farooque, O. (2021). Towards a circular economy for sustainable development: An application of full cost accounting to municipal waste recyclables. *Journal of Cleaner Production, 280*, 124047.

Tong, X., & Tao, D. (2016). The rise and fall of a "waste city" in the construction of an "urban circular economic system": The changing landscape of waste in Beijing. *Resources, Conservation and Recycling, 107*, 10–17.

Tong, X., Wang, T., Chen, Y., & Wang, Y. (2018). Towards an inclusive circular economy: Quantifying the spatial flows of e-waste through the informal sector in China. *Resources, Conservation and Recycling, 135*, 163–171.

Too, J., Ejohwomu, O. A., Hui, F. K. P., Duffield, C., Bukoye, O. T., & Edwards, D. J. (2022). Framework for standardising carbon neutrality in building projects. *Journal of Cleaner Production, 373*, 133858.

Tseng, M.-L., Chiu, A. S. F., & Liang, D. (2018). Sustainable consumption and production in business decision-making models. *Resources, Conservation and Recycling, 128*, 118–121.

Velenturf, A. P. M., & Purnell, P. (2021). Principles for a sustainable circular economy. *Sustainable Production and Consumption, 27*, 1437–1457.

Wang, H., Dai, H., Dong, L., Xie, Y., Geng, Y., Yue, Q., Ma, F., Wang, J., & Du, T. (2018). Co-benefit of carbon mitigation on resource use in China. *Journal of Cleaner Production, 174*, 1096–1113.

Weiss, M. A., & Schoenberger, E. (2015). Peter Hall and the western urban and regional collective at the university of California, Berkeley. *Built Environment, 41*, 63–77.

Wiedmann, T., & Allen, C. (2021). City footprints and SDGs provide untapped potential for assessing city sustainability. *Nature Communications, 12*, 10.

Xavier, L. H., Giese, E. C., Ribeiro-Duthie, A. C., & Lins, F. A. F. (2019). Sustainability and the circular economy: A theoretical approach focused on e-waste urban mining. *Resources Policy, 74*, 101467.

Zhang, C., Hu, M., Dong, L., Xiang, P., Zhang, Q., Wu, J., Li, B., & Shi, S. (2018). Co-benefits of urban concrete recycling on the mitigation of greenhouse gas emissions and land use change: A case in Chongqing metropolis, China. *Journal of Cleaner Production, 201*, 481–498.

Zhang, C., Hu, M., Dong, L., Gebremariam, A., Miranda-Xicotencatl, B., Di Maio, F., & Tukker, A. (2019). Eco-efficiency assessment of technological innovations in high-grade concrete recycling. *Resources, Conservation and Recycling, 149*, 649–663.

# 4 Carrier-Load Perspective Evaluation on Urban Resources Environment Carrying Capacity: Theory and Application

Liyin Shen and Tianheng Shu

## 4.1 Introduction

Currently urban population shares over 55% of the world population, and this proportion is reported to be 68% owing to the fast urbanization process (United Nations, 2018). In line with the high densification of urban population and fast urbanization, urban environmental degradation and resources depletion appear to be more serious around the world, which poses a great threat to sustainable urban development (Shen et al., 2012; Shi et al., 2016; Świąder, 2018; Galli et al., 2020). It has been appreciated in other studies that the limited urban land resources have been struggling to cope with the boost of urban population, resulting in land shortage, threatening natural ecosystems, and changing agricultural based land use (Izakovičová et al., 2017). And this problem has become a huge challenge globally for urban managers and planners (Lane, 2010). According to Guan et al. (2018), there are more than 400 cities in China suffering from the threatening problem of water shortage, which has become one of the most tremendous challenges faced by human beings worldwide (Aznar-Sánchez et al., 2018). As Wei et al. (2015a) and Graymore et al. (2010) opined, urban ecological, social and economic systems would be gradually deteriorated even irreversibly destroyed if urban population and human activity intensity exceed the "limit of urban capacity". As a typical example, Jakarta, the capital city of Indonesia, has been suffering resources and environmental crisis due to the human activities induced consequences such as dramatically intensive population density, severe water pollution and heavy traffic congestion, besides the significant challenges brought by climate change (Delinom et al., 2009). It is considered that the city of Jakarta couldn't carry so much burden from human activities, thus nearly exceeding the "limit of urban capacity", and Indonesia even has decided to move the capital from Jakarta to East Kalimantan province (Delinom et al., 2009; Nugroho, 2019). Although human intervention, such as implementing circular economy, offers a solution to this problem aforementioned by reducing the demand for natural resources, lowering the environmental footprint of cities and improving the capacity of urban infrastructure (Panchal et al., 2021), this is far from enough due to the tremendous and long-standing human pressures exerting to urban resources and environment. In fact, urban ecological, social and

L. Shen (✉)
Hangzhou City University, Hangzhou, Zhejiang Province, China

Chongqing University, Chongqing, China
e-mail: shenliyin@hzcu.edu.cn

T. Shu
School of Architecture, Tsinghua University, Beijing, China
e-mail: shuth21@mails.tsinghua.edu.cn



J. Zuo et al. (eds.), *Circular Economy for Buildings and Infrastructure*, Sustainable Development Goals Series, https://doi.org/10.1007/978-3-031-56241-9_4

economic systems provide essential natural living conditions, health care, employment, education, and many other necessary materials and services for human beings, which underpin the sustainable urban development (Graymore et al., 2010; Kaur & Garg, 2019). It is therefore imperative to have a method that can help properly understand the interrelations between limited urban resources environment capacity and the pressures imposed by increasing population and human activities (Lane, 2010; Ren et al., 2016). In other words, it is crucial to examine whether urban population and human activities are in excess of the capacity that urban resources environment system can provide for the mission of sustainable urban development. Accordingly, as a measurement, urban resources environment carrying capacity (URECC) was introduced to help understand the conflict aforementioned during the fast urbanization development (Tian & Wang, 2013; Zhang et al., 2018).

URECC is a comprehensive measurement which embodies sustainable urban development attributes across social, environmental and economic aspects (Feng et al., 2018; Kaur & Garg, 2019; Liu & Borthwick, 2011; Ye et al., 2016). Previous studies have presented various definition and connotation on URECC (Feng et al., 2018; Irankhahi et al., 2017; Lin et al., 2011; Liu et al., 2008; Liu & Borthwick, 2011; Tian & Wang, 2013; Wei et al., 2015b; Ye et al., 2016; Zhang et al., 2018). For a couple of typical examples, Liu and Borthwick (2011) defined environmental carrying capacity as "the limit at which human activity will lead to undesirable changes to the environment, assuming there are certain limits the environment itself imposes on development". Ye et al. (2016) expressed URECC as "the social and economic pressure that can be placed on ecological surroundings on the condition that the ecosystem can maintain a stable structure and complete function". In essence, URECC is traditionally explained as the limit or maximum of urban population and human activities that a city can support under particular resources environment condition without causing urban irreversible deterioration and damage. However, considering that a city itself is an open and resilient system, this limit or maximum value is only a potential existence, and it is hard to find out the real value, in our opinion.

Although existing studies on the assessment of URECC have made great progress, there is no consensus on the theoretical definition and evaluation method. In this regard, this chapter attempts to exploit a new method for urban resources environment carrying capacity assessment from the load-and-carrier perspective, which is expected to provide a unified evaluating framework for the assessment of URECC and allow for comparison between cities. This method is named here as URECC with the load-and-carrier perspective (URECC-LC). With this new perspective, urban loads and urban carriers, which are originated from an engineering mechanical term "carrying capacity" (CC), are introduced to establish the measurement for URECC. This measurement highlights that not only the urban resources environment system itself but also what extent of loads imposed on the system should be taken into account when measuring the carrying capacity of the system. The method introduced in this chapter sheds light on the development of URECC-LC in light of solid theoretical analysis on urban resources environment.

## 4.2 Literature Review

The term "carrying capacity (CC)" is originated from engineering field, which comprises the two variables of load and carrier which support the load, and the carrying capacity refers to the interaction relationship between the load and carrier (Philpot, 2014; Feng et al., 2018). This concept has then been widely introduced to the subjects of biology, ecology, human ecology, human society science and many other research fields relating human activities and the resources environment (Carey, 1993; Chen, 2004; Peng et al., 2016). In the context of urban environment, URECC is a concept and instrument for studying urban resources environment management and planning (Świąder, 2018).

The concept of CC has been particularly popular in conducing animal ecology research

(Kessler, 1994). According to Carey (1993), CC refers to the maximum number of animal population within certain environment range and conditions, and such environment can sustain animal living. With the recognition of conflict between population growth and limited resources environment (e.g. food, clean water, environment, and land), the research of CC was extended to human population growth research, namely human carrying capacity (Cohen, 1995). For instance, Malthus' population theory is the extension of CC research (Seidl & Tisdell, 1999), and Verhulst (1838) introduced the logistic equation for assessing CC qualitatively. In 1980s, United Nations Educational, Scientific and Cultural Organization (UNESCO) first defined the concept of resources carrying capacity and used population scale as the measurement of resources carrying capacity (Tian & Wang, 2013; Ren et al., 2016). Centered on the adverse consequences of human population growth and urbanization, the concerns on CC were extensively extended to different particular fields, such as land resources carrying capacity (Cui et al., 2019a, b; Lane, 2010; Shi et al., 2013), water resources carrying capacity (Cheng et al., 2018; Naimi Ait-Aoudia & Berezowska-Azzag, 2016), tourism carrying capacity (Castellani et al., 2007), forest resources carrying capacity (Martire et al., 2015), atmospheric environment carrying capacity (Guo et al., 2018; Zhou & Zhou, 2017), ecological carrying capacity (Rees, 1996; Peng et al., 2019a, b; Budihardjo et al., 2013; Świąder et al., 2018a, b), urban carrying capacity (Oh et al., 2005; Wang et al., 2019), environment carrying capacity (Brown et al., 2000; Zeng et al., 1992; Widodo et al., 2015; Świąder et al., 2018a, b), and resources environment carrying capacity (Lane et al., 2014; Liu et al., 2008; Liu & Borthwick, 2011; Wang & Liu, 2019; Ye et al., 2016; Zhang et al., 2018).

In fact, the concept of URECC is the extension of CC in studying urban environment-society-economy activities. It is a comprehensive measurement for sustainable urban development, and has been referred more frequently in academic research recently (Tian & Wang, 2013). Previous studies associated with URECC are the results of research development from a single-factor research to a multiple-factors comprehensive research, which provides a more accurate and effective insight to characterize the status of a complex urban system (Feng et al., 2018; Zhang et al., 2018). The research works in recent years tend to pay more attention to the conflict and interactive relationship between resources environment and human activities from holistic perspective (Liu & Zhu, 2017; Feng et al., 2018).

As regards the assessment method or model for URECC, many scholars have developed diverse methods for taking into account different resources environment factors. Among these methods, Ecological Footprint (EF) based ecological carrying capacity method has been widely acknowledged. EF based method aims to evaluate how human demand against nature's capacity by comparing the difference between Ecological Footprint (EF) and Biocapacity (BC) (also indicating environmental carrying capacity, see Świąder et al., 2018a, b) (Wackernagel & Rees, 1996; Wang, 2010; Baabou et al., 2017). In the application of Ecological Footprint method, Peng et al. (2019a, b) explored the dynamic evolution of ecological carrying capacity in Jiangsu province of China; and Świąder et al. (2018a, b) estimated the EF and BC of food for the city of Wrocław, Poland, based on which an analysis of the city's environmental carrying capacity was carried out. Furthermore, Shi et al. (2013) investigated urban population carrying capacity by using a comprehensive evaluation indicator system with the assistance of GIS analysis. The demand and supply balance analysis (Naimi Ait-Aoudia & Berezowska-Azzag, 2016), entropy weight method (Cui et al., 2018) and system dynamic (SD) (Wang et al., 2014) have been used to investigate water resources carrying capacity. There are still other methods introduced to assess the carrying capacity from single factor perspective, such as DPSIR model for tourism carrying capacity (Castellani et al., 2007), a GIS-based decision support system for forest carrying capacity (Martire et al., 2015) and coupling simulated annealing-projection pursuit model for atmospheric environment carrying capacity (Guo et al., 2018).

In the development of single-factor carrying capacity models, some assessment methods for addressing comprehensive-factor carrying capacity have been introduced (Liu et al., 2019). For example, Irankhahi et al. (2017) assessed urban environment carrying capacity (UECC) using GISFM and TOPSIS by incorporating multiple factors of urban natural and man-made systems, and classified the UECC into different grades, including very low, low, medium, high and very high. Zhang et al. (2018) constructed a URECC indicator evaluation system by embodying the connotation of ecological civilization in China, in which AHP method was adopted to calculate the URECC values. The evaluation results of URECC were categorized into three levels, namely poor, good and excellent. This URECC indicator evaluation system encompasses seven specific dimensions, such as water, land, and energy. From ecological aspect and socioeconomic pressure perspective, Ye et al. (2016) assessed URECC in the northwest temperate continental climate ecotope in China using weighted comprehensive evaluation and statistical analysis, and the relationship between regional economic development and urban sustainable ecological capacity was examined as well. A square deviation decision-making method was designed by Cheng et al. (2015) for evaluating provincial URECC in China based on a four-dimension evaluation indicator system which includes water, land, environment and ecosystem aspects. Wang et al. (2017) established a regional URECC indicator system including the aspects of land resources, water resources, mineral resources and environment, and carried out a coupling coordination analysis to explore the "short plank" of URECC. There are still some other methods introduced by many researchers in studying comprehensive URECC, such as surplus ratio model by Liu and Borthwick (2011), carrying capacity dashboard method by Lane et al. (2014), component analysis and state-space method by Tang et al. (2015). To summarize, the existing documents on comprehensive URECC research mainly adopted a comprehensive indicator system to evaluate the carrying capacity state of urban resources environment.

It can be seen that previous studies have provided deep insights to cognize URECC either for its connotations or its measurements. Indeed, URECC is appreciated as an effective tool for examining the quality of urban resources environment management and for sustainable development based decision-making (Cheng et al., 2019; Świąder, 2018). However, the literature review shows that there is no unified definition for URECC. Some studies have criticized that the concept of carrying capacity is vague (Liu & Borthwick, 2011; Price, 1999). Resultantly, different scholars have introduced various evaluation systems or methods for measuring URECC according to their different understandings and research designs. However, these diversified interpretations on URECC would confuse urban managers or researchers. The major reason for this is that the existing studies lack a theoretical insight on URECC in a systematic way. Instead, most researchers attach more importance to the calculation or evaluation methods on URECC, which has been mainly referred to the maximum or limitation in the context of human population and resources environment capacity. With the traditional perspective on URECC, people are advised to use the limitation value to judge the status of URECC assumed on the historic data assessment. Nevertheless, urban system is an open system with strong resilience. That said, cities can absorb disturbance from human activities to maintain its function dynamically (Meerow et al. 2016). This indicates that URECC is not fixed but changeable in order to adapt for human economic evolution and social development as well as environmental changes (Arrow et al., 1995; Cheng et al., 2019).

In line with the discussion on these traditional studies, a new perspective for understanding the connotation of URECC is introduced in this study. In this new perspective, the cognition to URECC is conformity with the principles of sustainable development, which embodies environmental, economic and social sustainability. Conventional studies appear to overemphasize on URECC evaluation result but overlook the investigation on how the carrying capacity is built up, which will be adequately addressed in our study.

## 4.3 Understanding Urban Resources Environment System

In order to develop a new method for measuring the carrying capacity of urban resources environment system (URES), there is a need to understand clearly the components of URES and their interrelationships with a system view.

City is generally considered as an open, adaptive, nonlinear, resilient and complex system (urban system), characterized with a variety of uncertain internal and external factors such as urban economic, social, environmental and cultural factors (Chelleri, 2012; Finka & Kluvánková, 2015). Urban system maintains its functions by relying on the constant input and output of energy, material and information flow (Chelleri, 2012; Fan & Fang, 2019). For the maintenance of urban system, obviously human activities play a key role as human directly dominates urban development and evolution (Chelleri, 2012; Yu et al., 2018). City would be no longer a city without human activities. However, in recent decades, intensive human activities have been increasingly posing great threat to the sustainability of urban system, especially in terms of urban resources environment (Finka & Kluvánková, 2015; Liu et al., 2017; Zhao et al., 2013). The conflict and imbalance between human activities and urban resources environment has increasingly become a tremendous challenge. A systematic thinking is instrumental to explicate urban resources environment (Chapman et al., 2016). More clearly, urban resources environment can be considered as a comprehensive system. Thus, in this study, we highlight that urban resources environment system is comprehensively composed of three essential subsystems (Fig. 4.1), namely urban socioeconomic subsystem (USS), urban natural resources subsystem (URS) and urban natural environment subsystem (UES) (Chapman et al., 2016; Panagopoulos et al., 2016; Zhang et al., 2019). In this system, human activities become the driving force to the interaction between subsystems (Chelleri, 2012). On the other hand, human activities are subject to the function of urban resources environment system (Fig. 4.1).

The three subsystems are intimately linked as a whole through human activities (Fig. 4.1) such as energy consumption, urban construction, pollutant and waste discharges, and traffic flows

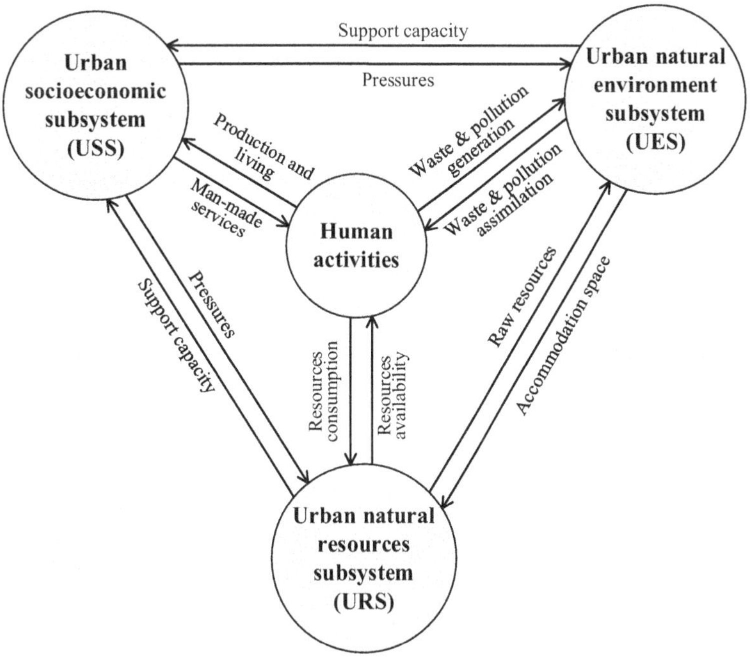


**Fig. 4.1** Interaction between human activities and urban resources environment system

(Graymore et al., 2010; Peng et al., 2016; Zhao et al., 2013). Urban natural resources subsystem (URS) which consists of numerous elements such as water resources, land resources, forest resources and mineral resources, provides available resources to support human survival. Urban natural environment subsystem (UES) which is composed of typical elements of natural water environment, land environment and atmosphere environment, assimilates sufficiently waste and pollution generated by regular human activities (Liu & Borthwick, 2011; Wei et al., 2015b). The two subsystems referred above actually constitute the ecosystem, and their provision of resources and environmental functions for human activities can be regarded as ecosystem services (Costanza et al., 2017). Ecosystem services across provisioning, regulating, cultural and supporting services (MEA, 2005), derived directly or indirectly from nature, contribute to human well-being, and well support human living and production activities in urbanized area (Mancini et al., 2018; Costanza et al., 2017). Nevertheless, urban socioeconomic subsystem (USS), the results of human production and living activities, can exert various pressures to URS and UES to different extents, that is, the pressures or negative effects to the provisioning ecosystem services (Shapiro & Báldi, 2014; Zhang et al., 2018). These pressures include over population, over consumption on resources, generation of excessive waste and pollution, etc. Meanwhile, UES and URS jointly provide support for the operation of USS, which in turn offers necessary man-made services such as urban infrastructure, transportation and public health services for human activities (Trubka et al., 2010; Wei et al., 2015b). Additionally, USS will assist in improving and promoting the capacity of URS and UES through technology advancement and policy improvement (Peng et al., 2016). A typical example is that the adoption of circular economy; for example, circular waste management systems, can help cities reduce the amount of waste sent to landfills. And waste reduction can help prevent environmental degradation and health issues (Joensuu et al., 2020), and this in turn indirectly enhances the capacity of URS and UES.

Ideally, the whole system URES itself will keep its function and configuration in a relatively dynamic equilibrium state (Chelleri, 2012). This complex system rarely stays within an equilibrium point or stable condition for a long time. The pressures or impacts posed by human activities will affect and break this equilibrium of the system when the corresponding carrying objects, URS and UES, are unable to offer matchable supporting capacity. As a consequence, the state of URES may arrive at a crossing threshold, namely tipping point, thereafter moving into an uncertain transitional process (Walker & Meyers, 2004). Obviously, the equilibrium state of URES depends on how and to what extent the outward manifestation of USS by human activities acts on URS and UES. In other words, the equilibrium state of URES is determined by the capacity that URS and UES resist against the pressures from human activities. This process can be further described by carrying capacity encompassing carrying subject (urban loads) and carrying object (urban carriers) (see Fig. 4.2). This argument is also supported in previous study by Tang and Ye

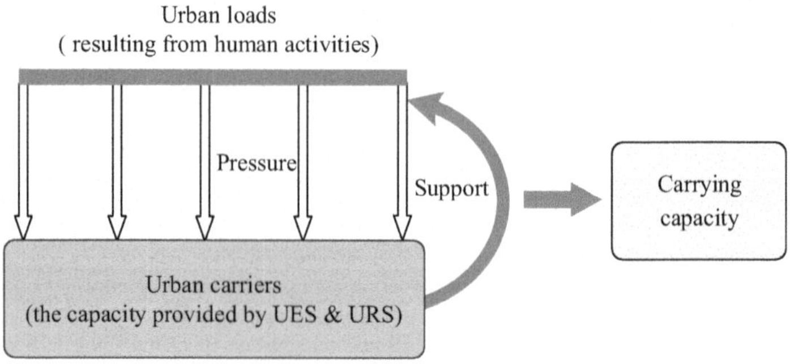


**Fig. 4.2** Relationship between urban loads and urban carriers

(1998), suggesting that urban resources environment carrying capacity is the external manifestation of the whole URES's functions, and it reflects the ability of URS and UES to resist the interference from human system and the capacity to re-adjust self-adaptive frame. With this understanding on URES and the interplay between these three subsystems of URES, an instrument to assess the state of urban resources environment carrying capacity will be developed in the following sections.

## 4.4 Development of URECC-LC Method

### 4.4.1 Identification of URECC Loads and Carriers

Known from the discussion in the previous section, whether urban resources environment system can maintain its function and configuration depends on urban loads and carriers, and the interplay between the two sets of elements. With reference to urban loads, they can be generally divided into two types, namely regular loads and accidental loads. Regular loads mainly refer to human living and production activities. Cui et al. (2019a, b) suggested that human activities are across agricultural, industrial, urban construction, transportation and daily life activities. With these loads, urban resources are continuously consumed and urban environment is imposed by tremendous pressures such as pollutants and waste discharges (Monte-Luna et al., 2010; Dijst et al., 2018). Accidental loads, on the other hand, refer to natural disasters such as earthquakes, floods and landslides, or some man-made emergent hazards such as leakage of toxic materials, ferocious fires and explosions (Bathrellos et al., 2017; Chen et al., 2019a, b). Urban accidental loads are usually unpredictable and occur less frequently. As a city has specific spatial, temporal, and attribute characteristics (Yu et al., 2018; McPhearson et al., 2016), urban loads can be characterized and determined properly by particular time, space and load attribute factors accordingly. Taking water as an example of urban resources, the consumption on water by both living and production activities have been increasing year by year due to fast urbanization (Lu et al., 2018), which reflects the increasing pressures on urban water supply system (temporal factor). Spatially, different cities consume different volumes of water as they are in different scales (spatial factor), thus leading to the variability of urban loads for water between cities (Lu et al., 2018). Furthermore, the factors of water consumption, such as industrial activities, domestic activities and agricultural activities (Song et al., 2010) should be considered when the loads of urban water resources are described (load attribute factor).

Following the above discussions, urban loads (UL) can be expressed as the following Equation:

$$\mathrm{UL} = f(l_1, l_2, \ldots, l_n, t, s) \tag{4.1}$$

where $s$ and $t$ are spatial and temporal factors respectively, and the two factors indicate that UL can be defined only at a particular space (city) and time point. $l_i$ represents the load factor $i$. $f$ describes the functional relationship about how these factors build up UL.

As to urban carriers (UC), they function to support urban loads resulting from human activities (see Fig. 4.2). For urban resources environment system, it possesses indispensable carriers (e.g. water, land, energy) to carry UL, ensuring that the system operates normally (Dijst et al., 2018). City is considered as a typical metabolism system and its corresponding human activities rely on the constant exchange of material and energy flows within urban resources environment system (González et al., 2013; Sevegnani et al., 2018). Actually, these flows in the metabolism system reflect the response of UC to UL. As stated above, urban loads include agricultural, industrial, urban construction, transportation and daily life activities (Cui et al., 2019a, b), and different loads need corresponding carriers to support. To be specific, agricultural activities need the biomass extraction; industrial activities need to obtain raw material from natural system; urban construction activities needs the construction materials extracted from natural system; transportation activities refer to vehicles usage and require energy consumption; daily life activities indicate the consumption on food and industrial products, and the needs for education, medical and other public services, and high-quality living environment (Costanza et al., 1997; Conke & Ferreira, 2015; Chapman et al., 2016; Cui et al., 2019a, b). Meanwhile, all these loads will produce various types of pollution and wastes to urban natural environment, such as water pollution, air pollution, and solid waste (Świąder, 2018). These pollution and wastes need to be handled by the provision of various socioeconomic infrastructures and natural environment.

In summary, various carriers are needed to support urban loads. Based on the understanding of UL, the urban carriers should be built up across land, water, energy, atmosphere, municipal infrastructure, transportation, public services and tourism (Liu & Borthwick, 2011; Belčáková, et al., 2017; Świąder, 2018). In fact, these carriers comprise social, economic and resources environmental aspect, which are the pillars for urban sustainability. The proper provision of these carriers can facilitate sustainable urban development (Liu & Borthwick, 2011). Similar to urban loads, urban carriers are dynamic in time, variable in space and also determined by particular carrier attribute factors. Taking land resource environment as a specific case, urban land resource is dynamic because land development occurs all the time and land use changes year by year (Basse et al., 2016; Furmankiewicz et al., 2019), thus the capacity of urban land is changing constantly (temporal factor). In terms of spatial factor, as cities have different economic, social, technological and environmental conditions, the utilization efficiency of urban land resources is variable, suggesting distinct land resources carrying capacity between cities even for an equal area of land (spatial factor) (Zhu et al., 2019; Shi et al., 2013). The factors of land attributes across urban construction, industrial development, agricultural production, rural living, green ecological and others will also affect the capacity of land carrier (carrier attribute factor) (Shi et al., 2019). Based on these discussions, urban carriers (UC) can be expressed as follows:

$$\mathrm{UC} = f(c_1, c_2, \ldots, c_n, t, s) \tag{4.2}$$

where $s$ and $t$ are spatial and temporal factors, respectively, which indicate a specific city and exact time point for portraying UC. And $c_i$ represents the carrier factor $i$, $g$ describes the functional relationship how these factors form into UC.

### 4.4.2 Measurement of URECC-LC Method

According to the discussions in Sect. 4.3, urban resources environment system (URES) will keep in a state of dynamic equilibrium with the coordination between human activities which present UL and resources environment which present UC. Therefore, urban resources environment carrying capacity reflects the degree that UC can carry UL, namely, the degree that resources environment responds to the pressure exerted by human activities. For measuring this degree, an index ρ is introduced to assess URECC, which can be defined as the following Equation:

$$\rho = \frac{UL}{UC} \tag{4.3}$$

where UL and UC have been described in Sect. 4.4.1. According to this Equation, URECC can be defined as the degree that urban carriers can support urban loads. This definition addresses both density impact and intensity impact imposed by human activities on urban resources environment, which is also suggested in the study by Barrett and Odum (2000) when studying carrying capacity. That is to say that URECC should consider not only the characteristics of UC, but also what extent the UL loads on UC. It is concluded that the measurement ρ is an index with temporal and spatial characteristics in light of the identification of UL and UC (because they have temporal and spatial factors) in Sect. 4.4.1. Moreover, ρ is a degree value (greater than 0), and neither a larger ρ indicate a better performance of URECC, nor a

**Fig. 4.3** Graphical intervals of ρ

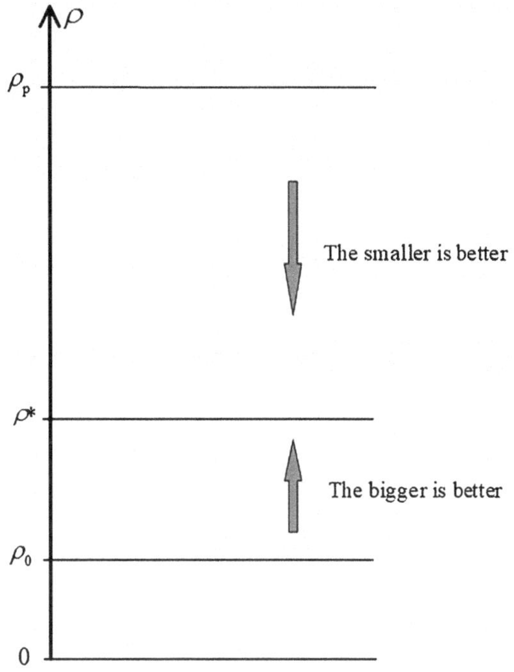


smaller ρ (Fig. 4.3). We argue that a suitable ρ, should meet the rules ensuring a relative balance between the capacity of urban resources environment system and the pressure induced by human activities, thus realizing sustainable utilization of urban resources. From another view, a much greater ρ commonly indicates that the possibility (risk) of URECC being overloaded is high, indicating a less sustainable state of urban resources environment in the long term. $\rho_p$ means the potential limit of URECC, over which the URES may be irreversibly damaged. Those typical values of ρ theoretically exist but are unable to accurately determined, and in practical application, we should choose alternative method to quantify them.

Further, the detailed interpretation for the connotation of ρ will be addressed in the following discussions from the perspective of urban system resilience.

### 4.4.3 Extended Justifications on URECC

This section will present insightful comments on URECC based on our proposed URECC-LC method and this will proceed with the rationale of urban system resilience. Urban system resilience refers to the ability that a city system and all of its socio-economic-ecological networks on spatial and temporal scales to maintain or quickly bounce back to desirable configuration and function when confronted with external disturbance (Meerow et al. 2016; Masnavi et al., 2018). When the disturbance is imposed, time is needed to return to equilibrium state for system (Cavallaro et al., 2014; Romero-Lankao et al., 2016), namely, resilience is referred to a timespan (Reed et al., 2009). For urban resources environment system, the disturbance is from UL, so the time process that UC responds to UL reflects urban resilience. At any point of time, a value of ρ can be derived by Eq. (4.3), which represents a particular state of carrying capacity. This value changes with time but will never arrive at a point where URES can't maintain its system function and configuration. This point is potential and would be the threshold of URECC if it does exist. In other word, the threshold is only a potential existence (Fig. 4.4) but can never be determined because URES is unlike a physical material and instead is an open, complex, resilient and nonlinear system. What we can capture is the

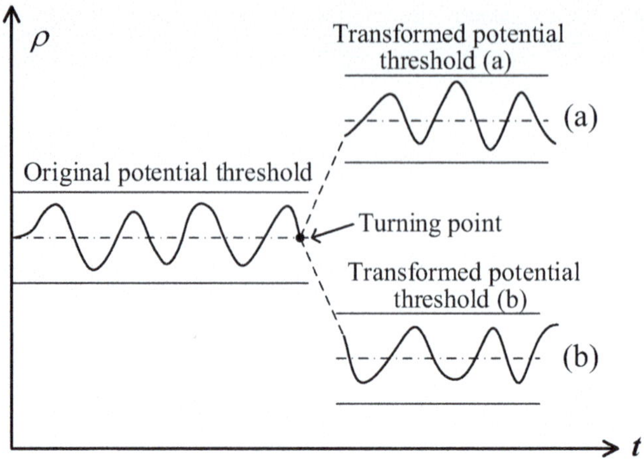


**Fig. 4.4** Change of ρ with time

state of ρ and its change trend. Observing the state of ρ helps keep tracking the actual situation of the intensity that human pressures imposed on urban resources environment, thereby providing useful information for improving environmental awareness and city's resources environmental management (Galli et al., 2020). In fact, although many large cities are facing tremendous loads from human activities, they still function without irreversible collapse (Chelleri, 2012). For example, Beijing has been suffering severe air pollution due to the dramatic human activities in recent years (Xu et al., 2019), and the PM2.5 concentrations in the air in Beijing was beyond 1.09 times the national standard in 2017 (Fu et al., 2019). However, Beijing still struggles to function normally with the assistance of its self-adaptive capacity and urban governance. That is what the urban resilience takes effects. This resilience effects can be delineated in a plot in light of the change pattern of ρ with time (Fig. 4.4). Originally, ρ changes under a potential threshold in a timespan, but with the transition of system under certain condition, this threshold may increase to (a) due to the advancement of technology, or decrease to (b) due to certain damage such as earthquakes and extreme climate change.

With the increase or decrease of the potential threshold for ρ, URES will arrive at a new equilibrium state, namely another resilient condition, which is a more sustainable state (Loorbach, 2010). In any case, the potential threshold represents a limit of urban resources environment carrying capacity, below which cities can achieve sustainable development (Hong & Ye, 1998; Chelleri, 2012). In order to further analyze URECC and apply this indicator properly in practice, there is a need for a value to measure the potential threshold. However, as mentioned above, it is impossible to find its actual value (noted as $\rho_p$ in Fig. 4.3). An alternative method to define this potential value is to introduce a virtual value for the threshold according to historic data and experience. This virtual value takes effects in the short term and should be adjusted dynamically in light of different urban development stages in the long term. As a matter of fact, the threshold of URECC given in previous studies is just a virtual value. For example, Zhang et al. (2018) set a threshold score of 60 for URECC to judge if Tianjin is overloaded during 2007–2015 according to historic data. Even the world-widely used Ecological Footprint method based carrying capacity calculation result (Rees, 1996, 2016; Świąder et al., 2018a, b; Peng et al., 2019a, b), in our argument, is also a virtual threshold of carrying capacity (by comparing EF and BC). In other words, the Ecological Footprint method based carrying capacity does not estimate the real carrying capacity and the ecological deficit (it shows the overshooting pressures on ecosystem rather than urban system) does not indicate the overload of urban resources environment carrying capacity due to the existence of urban system resilience (Graymore et al., 2010; Fatemi et al., 2018).

The introduction of ρ can help assess the state of URES and conduct a dynamic monitoring on the state of URES with time, which will provide early warning signals of resources environment carrying capacity (Li et al., 2018). In this way, actions can be taken to ensure sustainable URECC which relies on urban system resilience. According to Carpenter et al. (2009) and Graymore et al. (2010), urban system resilience is determined by system's properties across buffering, redundancy, adaptability, learning, social memory, diversity, spatial heterogeneity and human governance, and a stronger urban resilience means a higher carrying capacity because system can absorb more disturbances from human activities. The resilience could be changed (enhanced or weakened) if urban form and its ecosystem change (Ernstson et al., 2010).

Finally, there is an issue being noted. Based on the definition of URECC in this study, it is argued that the assessment of urban resources environment carrying capacity should focus on independent resources environment factors which are derived from urban carriers. URECC can be described in the following CC dimensions: land resources environment, water resources environment, energy resources, atmosphere environment, human resources, municipal infrastructure, transportation, public services and tourism resources. These nine dimensional CC actually meet with the three pillars of sustainable development, namely economic, social and environmental sustainability. It is important to assess each dimension's URECC individually instead of a comprehensive evaluation as each dimension is essential for supporting city's sustainable development. In the following contents, we will demonstrate the implication of our proposed method.

## 4.5 Application Demonstration of URECC-LC Method

This section will conduct a case study to demonstrate how to apply the introduced innovative method. For this, we select water resources as a typical dimension of URECC for conducting the demonstration. The case demonstration is schemed as following procedures: (1) define UL and UC for UWRCC, (2) collect data and design method to quantify UL and UC of UWRCC, and (3) calculate the results for UL, UC and ρ of UWRCC and perform results analysis.

Water resources is an indispensable and fundamental natural resources for human survival and urban sustainable development (Naimi Ait-Aoudia & Berezowska-Azzag, 2016; Song et al., 2010). The supply of water possesses the largest ecosystem services value among all provisioning ecosystem services (Costanza et al., 1997). Many large cities are facing severe challenge of water resources shortage due to the increasing growth of urban population and climate change, and sustainable water resources management has been attracting significant attention by urban managers around the world (Serrao-Neumann et al., 2019).

### 4.5.1 Identification of Carriers and Loads in Urban Water Resource System

According to the formula (4.3) introduced in Sect. 4.4, urban water resources environment carrying capacity, $\rho_w$, can be obtained from the following formula:

$$\rho_w = \frac{UL_w}{UC_w} \tag{4.4}$$

where $UL_W$ and $UC_W$ respectively denote the loads and carriers of urban water resources environment. Thus, to determine the value of $\rho_w$, we firstly need to confirm the measures of $UL_W$ and $UC_W$. For this, a water cycling system (WCS) can be referred. From a hydrological perspective, water cycling system refers to the cycling of natural water bodies, which includes water precipitation and evaporation, water-source conservation and infiltration, and runoff (Chen et al., 2019a, b). From a social perspective, the water cycling system focuses on the transformation activities and measures during the utilization process of water resources, which covers the whole

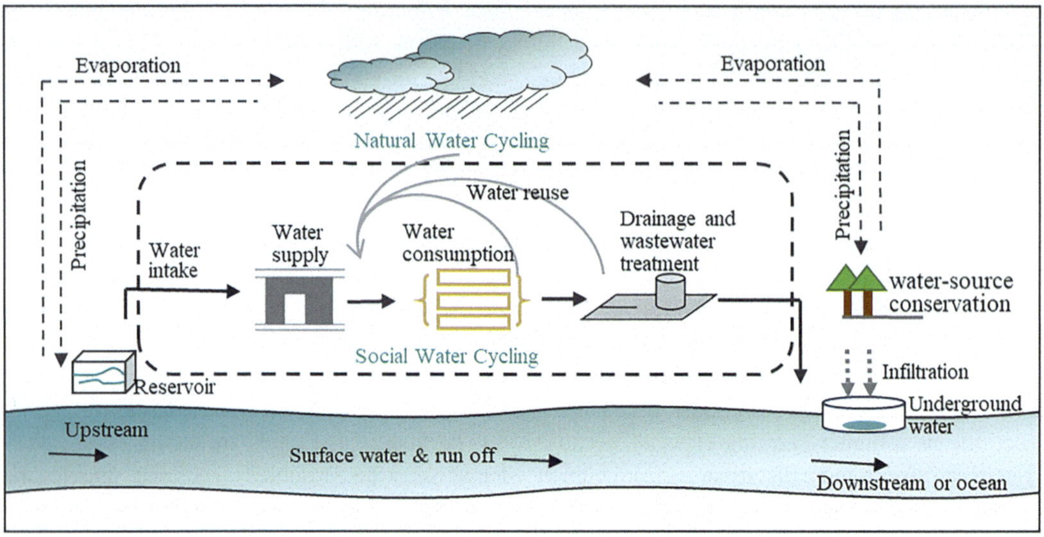


**Fig. 4.5** Regional water cycling system. (Marsalek, 2014; Chen et al., 2019a, b)

processes of water collection, water supply, water consumption, drainage and wastewater treatment (Marsalek, 2014). Accordingly, a regional water cycling system can be illustrated graphically as shown in Fig. 4.5.

In Fig. 4.5, RWCS consists of four subsystems: water sources, water supply, water consumption, and drainage. The four subsystems formulate a cycling chain in utilizing water resources and each subsystem assumes the role of a linkage which is indispensable and contains different elements. In this cycling system, the subsystem of water consumption indicates the water load which is supported jointly by the other three subsystems. In other words, the water carrier's capability in supporting the water load is determined jointly by the three subsystems of water source, water supply, and drainage. Water sources represent the natural water resource endowments in urban region; water supply is to collect and purify natural water (including freshwater and seawater) to make them meet certain quality standards and convey the quality water to various water users; Drainage system works to discharge rainwater and sewage to prevent urban from flooding and pollution, and to treat wastewater and prevent water environment from pollution.

Based on the above comprehension of RWCS, water carrier can be identified from three dimensions, namely, water sources, water supply and drainage, whilst the water loads will be identified from the perspective of water consumption which is imposed by human activities, including socioeconomic activities and environmental pressures. Previous studies have proposed various indicators for assessing RWRCC. The typical carrier indicators proposed in existing studies include the amount of surface water and underground water (Jia et al., 2018), annual precipitation (Meng et al., 2009), water supply coverage rate (Wei et al., 2016) and wastewater treatment rate (Sun et al., 2018). The load indicators include the resident population (Wang et al., 2018a, b), irrigated area of cultivated land (Cheng et al., 2018), and ecological protection water proportion (Ren et al., 2016).

By referring to these references, with together considering the previous analysis on the carriers and loads, the carrier indicators for assessing the capability of regional water resources are identified under three dimensions, namely, water sources, water supply, and drainage, as listed in Table 4.1. And the load indicators for measuring the pressure effects imposed by human activities are identified, as listed in Table 4.2.

**Table 4.1** The carrier indictors of UWRCC

| Dimension | Carrier indicator |
|---|---|
| Water sources | $C_1$-Amount of surface water resources ($10^8$ $m^3$) |
| | $C_2$-Amount of groundwater resources ($10^8$ $m^3$) |
| | $C_3$-Water production modulus ($10^4$ $m^3/km^2$) |
| | $C_4$-Annual precipitation (mm) |
| | $C_5$-Mean annual relative humidity (%) |
| Water supply | $C_6$-Integrated production capacity ($10^4$ $m^3$/day) |
| | $C_7$-Water coverage rate (%) |
| | $C_8$-Number of water plants (unit) |
| | $C_9$-Leakage rate of water supply network ($10^4$ $m^3$) |
| Drainage | $C_{10}$-Total treated wastewater ($10^4$ $m^3$) |
| | $C_{11}$-Number of wastewater treatment plant (unit) |
| | $C_{12}$-Wastewater treatment rate (%) |
| | $C_{13}$-Daily urban wastewater treatment capacity ($10^4$ $m^3$/day) |

Note: $C_9$ is a negative indicator

**Table 4.2** The load indictors of UWRCC

| Dimension | Load indictor |
|---|---|
| Social | $L_1$-Resident Population at Year-end ($10^4$ persons) |
| | $L_2$-Natural Growth Rate of Population (‰) |
| | $L_3$-Domestic water use ($10^8$ $m^3$) |
| | $L_4$-Water use of public service ($10^8$ $m^3$) |
| Economic | $L_5$-GDP ($10^4$ yuan) |
| | $L_6$-GDP growth rate(%) |
| | $L_7$-Industrial Water use($10^4$ $m^3$) |
| Environmental | $L_8$-urban green coverage($km^2$) |
| | $L_9$-Environmental water use ($10^8$ $m^3$) |
| | $L_{10}$-Amount of sewage discharge($10^8$ $m^3$) |

### 4.5.2 Models for Analyzing Urban Water Resources Carrying Capacity

As shown in Table 4.1, there are three carrier dimensions of regional water resources, namely, water sources, water supply, and drainage. It is therefore necessary to assess the three types of carrying capacity separately. By referring formula (4.4), the dimensional UWRCC can be obtained by following equations:

$$\rho_{(SO)} = \frac{UL_w}{UC_{(SO)}} \tag{4.5}$$

$$\rho_{(WS)} = \frac{UL_w}{UC_{(SU)}} \tag{4.6}$$

$$\rho_{(DR)} = \frac{UL_w}{UC_{(DR)}} \tag{4.7}$$

where $UL_w$ denotes the loads of urban water resources system, $UC_{(SO)}$, $UC_{(SU)}$ and $UC_{(DR)}$ represent the three-dimensional water carriers, respectively. The values of these variables can be determined by a set of indicators described in Tables 4.1 and 4.2. The calculation for these values will be through three processes: indicator normalization, weighting determination, and linear weighting. Further, the overall RWRCC performance can be derived through the integration of these three types of capacities, as expressed in Eq. (4.8):

$$\rho_w = \rho_{(SO)} + \rho_{(SU)} + \rho_{(DR)} \tag{4.8}$$

where $\rho_w$ are the overall index of RWRCC; $\rho_{(SO)}$, $\rho_{(SU)}$ and $\rho_{(DR)}$ represent respectively the RWRCC performance of water sources, water supply, and drainage. Further, by combing the weight value of each indicator, we can derive the value of $\rho_w$ for a certain city, as presented in Eq. (4.9).

$$\rho_w = \frac{\sum_{k=1}^{10} l'_{i,k} W_k^L}{\sum_{j=1}^{5} c'_{i,j} W_j^C} + \frac{\sum_{k=1}^{10} l'_{i,k} W_k^L}{\sum_{j=6}^{9} c'_{i,j} W_j^C} + \frac{\sum_{k=1}^{10} l'_{i,k} W_k^L}{\sum_{j=10}^{13} c'_{i,j} W_j^C} \quad (4.9)$$

where $W_k^L$ and $W_j^C$ are the weight value of load indicator $k$ and carrier indictor $j$ respectively; $l'_{i,k}$ and $c'_{i,j}$ respectively denote the normalized value of load indicator $k$ and carrier indicator $j$ for a certain city. The details of normalized method and weight method (here we use entropy method) can be seen in copious documents such as Shen et al. (2015), Delgado and Romero (2016), Zhang et al. (2006). Herein, $\rho_w$ reflect the state of urban water resources environment carrying capacity instead of its limit as claimed in Sect. 4.4.

### 4.5.3 Data for Case Demonstration and Calculation Results

The demonstration case refers to 35 typical large cities in China. They are the most economically developed and socially important cities, consuming large proportion of water in China. These large cities vary in economic scale, water resources endowment and geographical context.

The data for all the carrier and load indicators in Tables 4.1 and 4.2 are collected for the period of 2008–2017 from the following official statistical documents or reports: China City Statistical Yearbook (2008–2017), China Urban Construction Statistical Yearbook (2009–2018), local water resources statistics bulletin (2008–2017), database of the National Bureau of Statistics of China.

By applying the collected data to the Eqs. (4.5)–(4.9), the results of $\rho_w$ for the 35 large cities can be derived (Table 4.3).

As presented in Table 4.3, according to the capacity index, the top three cities in 2017 are Beijing, Yinchuan and Hohhot, and the bottom three ranked are Haikou, Nanchang and Wuhan. The fluctuation profile of $\rho_w$ varies from city to city. Urumqi has the largest fluctuation interval ranging from 2.603 to 13.091 while Shanghai possesses the smallest one of 3.542–4.115.

## 4.6 Analysis

The analysis on the results will be mainly conducted from the perspective of temporal change of urban water resources carrying capacity. To clearly capture the temporal evolution pattern of $\rho_w$, we can depict graphically the change of $\rho_w$ for 35 cities, and we can classify them into six groups according to the evolving features manifested in the period of 2008–2017 (Fig. 4.6).

As illustrated in Fig. 4.6, different cities exhibit varying changing patterns over the ten years. For the cities with increasing $\rho_w$, Tianjin, Qingdao, Chengdu, Nanjing, Jinan and Xiamen, a total of 6 cities belong to this group (Fig. 4.6a). In the years starting from 2008, $\rho_w$ fluctuated below the average value, and approaching 2017, they all rose to the above-average range. We can see the predominant changing pattern is "decreasing $\rho_w$", with 8 cities out of 35 (Fig. 4.6b), suggesting that during the study period 2008–2017, in these cities, the regulatory measures on water resources carrying capacity have played a significant role in moderating the contradictions between the water resources system and the pressure socioeconomic development. For the cities with irregularly fluctuant $\rho_w$, there are seven cities following this group (Fig. 4.6c), and these cities show similar characteristic of relatively low socioeconomic conditions, indicating the weak capacity to regulate water resources carrying capacity.

**Table 4.3** The value of $\rho_w$ for 35 cities from 2008 to 2017

| City | 2008 | 2009 | 2010 | 2011 | 2012 | 2013 | 2014 | 2015 | 2016 | 2017 | Yearly mean |
|---|---|---|---|---|---|---|---|---|---|---|---|
| Beijing | 5.456 | 6.622 | 5.197 | 4.752 | 4.973 | 5.724 | 8.980 | 7.267 | 8.324 | 6.509 | 6.380 |
| Tianjin | 4.236 | 3.906 | 4.191 | 4.200 | 2.947 | 3.491 | 5.385 | 4.609 | 5.632 | 4.240 | 4.284 |
| Shijiazhuang | 2.439 | 2.810 | 2.950 | 2.311 | 2.415 | 2.096 | 4.002 | 3.118 | 2.750 | 2.333 | 2.722 |
| Taiyuan | 2.605 | 2.009 | 2.626 | 2.157 | 3.740 | 1.683 | 3.391 | 2.841 | 3.577 | 2.160 | 2.679 |
| Hohhot | 5.468 | 7.274 | 2.800 | 5.950 | 1.984 | 1.927 | 7.582 | 4.298 | 4.024 | 3.400 | 4.471 |
| Shenyang | 2.439 | 2.690 | 2.005 | 2.002 | 1.832 | 1.578 | 2.382 | 1.770 | 1.581 | 2.130 | 2.041 |
| Dalian | 2.281 | 1.994 | 1.515 | 1.668 | 1.507 | 1.336 | 2.604 | 1.624 | 2.780 | 2.414 | 1.972 |
| Changchun | 3.085 | 2.707 | 1.758 | 2.110 | 1.769 | 1.815 | 2.823 | 1.845 | 2.149 | 1.914 | 2.198 |
| Ha'erbin | 4.076 | 4.720 | 2.747 | 1.914 | 1.284 | 1.173 | 1.861 | 1.498 | 1.775 | 1.539 | 2.259 |
| Shanghai | 3.542 | 3.920 | 3.929 | 3.975 | 3.794 | 4.115 | 3.699 | 3.609 | 4.057 | 3.885 | 3.853 |
| Nanjing | 2.312 | 2.509 | 2.343 | 2.729 | 3.280 | 3.228 | 3.267 | 2.876 | 3.049 | 3.186 | 2.878 |
| Hangzhou | 2.129 | 1.819 | 1.266 | 1.434 | 1.404 | 1.461 | 2.082 | 1.736 | 2.022 | 1.979 | 1.733 |
| Ningbo | 1.456 | 1.705 | 1.821 | 1.732 | 1.640 | 1.564 | 2.171 | 1.821 | 1.877 | 2.029 | 1.782 |
| Hefei | 2.002 | 2.233 | 1.881 | 1.848 | 1.811 | 1.409 | 2.212 | 1.959 | 2.290 | 2.105 | 1.975 |
| Fuzhou | 1.984 | 2.138 | 1.885 | 1.702 | 1.597 | 1.776 | 1.910 | 1.680 | 1.699 | 2.179 | 1.855 |
| Xiamen | 1.145 | 1.423 | 1.261 | 1.636 | 1.488 | 1.146 | 2.065 | 1.413 | 1.825 | 1.701 | 1.510 |
| Nanchang | 2.144 | 2.309 | 2.063 | 1.645 | 1.361 | 1.333 | 1.679 | 1.533 | 1.786 | 1.284 | 1.714 |
| Jinan | 2.318 | 2.501 | 1.816 | 1.745 | 2.310 | 1.685 | 3.695 | 2.430 | 3.051 | 3.182 | 2.473 |
| Qingdao | 1.733 | 2.405 | 1.892 | 1.724 | 2.209 | 1.923 | 2.848 | 3.175 | 3.301 | 2.532 | 2.374 |
| Zhengzhou | 2.993 | 3.392 | 2.978 | 2.441 | 3.336 | 3.788 | 3.458 | 3.049 | 3.015 | 2.994 | 3.144 |
| Wuhan | 2.265 | 2.488 | 1.819 | 2.006 | 2.227 | 2.094 | 2.681 | 2.357 | 2.415 | 2.615 | 2.297 |
| Changsha | 2.434 | 2.126 | 1.581 | 1.739 | 1.466 | 1.401 | 1.739 | 1.737 | 1.817 | 1.293 | 1.733 |
| Guangzhou | 3.078 | 4.074 | 2.847 | 3.110 | 2.875 | 2.881 | 3.059 | 2.830 | 2.854 | 2.747 | 3.036 |
| Shenzhen | 4.089 | 3.802 | 3.051 | 3.214 | 2.852 | 2.862 | 3.041 | 3.369 | 3.064 | 3.129 | 3.247 |
| Nanning | 2.465 | 3.821 | 4.281 | 3.807 | 2.215 | 3.520 | 2.818 | 2.405 | 2.238 | 2.147 | 2.972 |
| Haikou | 0.867 | 1.269 | 1.665 | 1.112 | 0.854 | 0.864 | 1.384 | 1.042 | 1.275 | 1.051 | 1.138 |
| Chongqing | 1.874 | 2.109 | 2.449 | 2.288 | 2.284 | 2.067 | 2.444 | 2.544 | 2.494 | 2.431 | 2.298 |
| Chengdu | 1.911 | 2.051 | 2.272 | 2.695 | 2.660 | 1.828 | 2.607 | 2.199 | 2.957 | 2.605 | 2.379 |
| Guiyang | 2.735 | 2.682 | 1.745 | 2.021 | 1.559 | 2.198 | 1.763 | 2.424 | 1.853 | 1.639 | 2.062 |
| Kunming | 1.704 | 2.317 | 1.681 | 1.752 | 1.917 | 1.451 | 2.016 | 1.636 | 2.023 | 2.013 | 1.851 |
| Xi'an | 3.242 | 2.827 | 1.791 | 2.095 | 2.476 | 2.705 | 2.960 | 2.586 | 3.716 | 3.277 | 2.768 |
| Lanzhou | 7.637 | 5.066 | 3.710 | 7.925 | 6.332 | 3.598 | 3.397 | 3.978 | 5.073 | 2.202 | 4.892 |
| Xining | 3.128 | 3.316 | 6.016 | 2.517 | 3.194 | 2.413 | 3.607 | 2.633 | 3.696 | 4.154 | 3.467 |
| Yinchuan | 13.091 | 7.616 | 4.006 | 2.603 | 7.170 | 10.504 | 6.096 | 3.606 | 11.163 | 5.783 | 7.164 |
| Urumqi | 3.414 | 4.567 | 2.529 | 3.249 | 5.551 | 2.041 | 3.885 | 3.020 | 3.075 | 1.987 | 3.332 |
| National mean | 3.136 | 3.178 | 2.582 | 2.623 | 2.638 | 2.477 | 3.188 | 2.643 | 3.151 | 2.651 | 2.827 |

The cities with stable $\rho_w$ is characterized by the fact that during the study period, $\rho_w$ is stably vibrated within a narrow interval, and the its changing trend is relatively smooth. Among the 35 sample cities, the cities belonging to the stable type are Shanghai, Ningbo, Fuzhou and Chongqing (Fig. 4.6d), and they have better natural water resources conditions and good ecological status. In addition, the socio-economic developments of these four cities are superior and the artificial water resources facilities such as water supply and sewage treatment are relatively advanced, which can facilitate adapt to the local socio-economic development needs and finally make their water resources carrying capacity remain in a stable state. The fluctuant-stable pattern is a compound of the fluctuant and stable patterns, and it first shows irregularly fluctuations

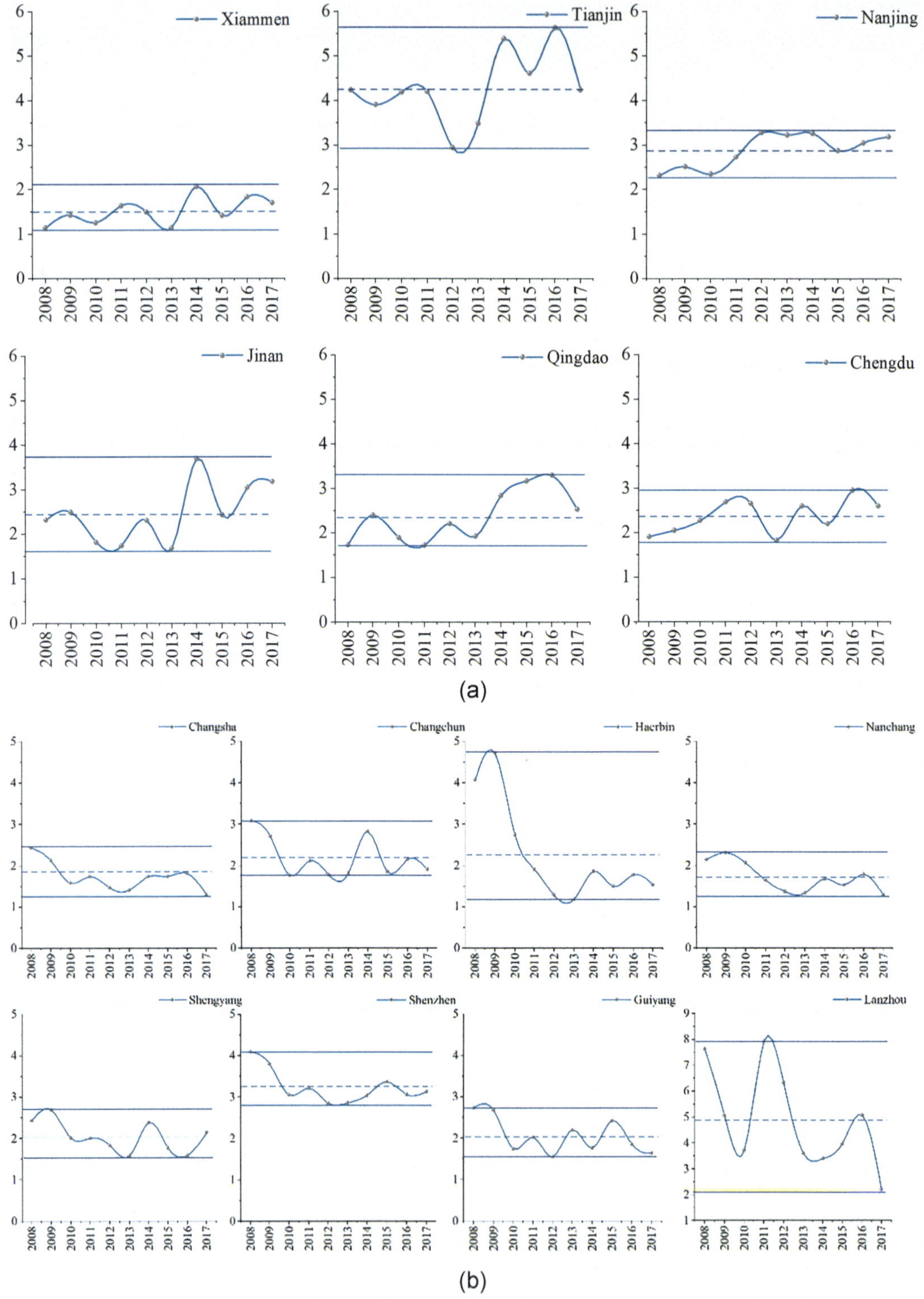


**Fig. 4.6** Temporal evolution pattern of $\rho_w$ for 35 cities. (**a**) Cities with increasing $\rho_w$. (**b**) Cities with decreasing $\rho_w$. (**c**) Cities with irregularly fluctuant $\rho_w$. (**d**) Cities with stable $\rho_w$. (**e**) Cities with fluctuant-stable $\rho_w$. (**f**) Cities with decreasing-increasing $\rho_w$

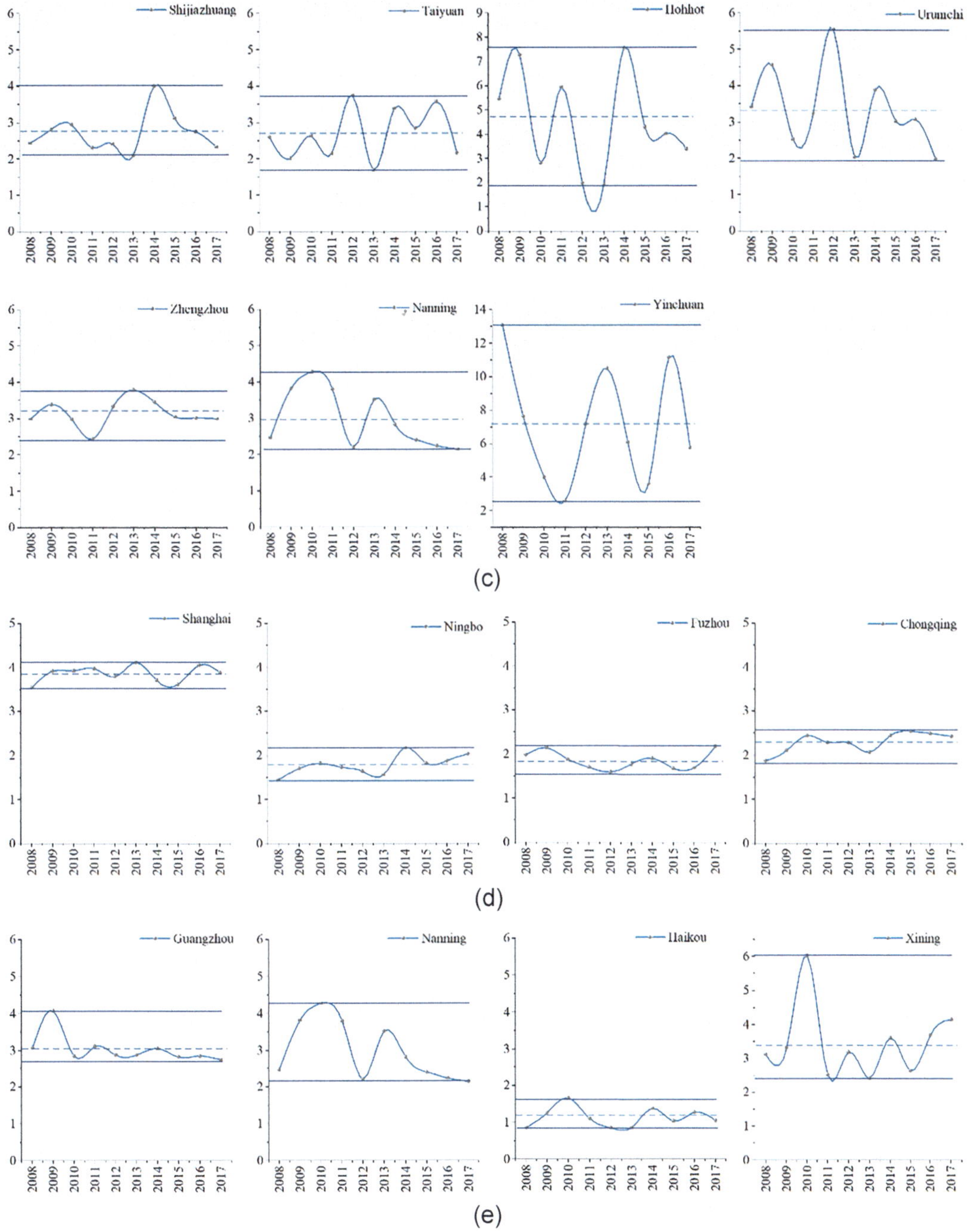


**Fig. 4.6** (continued)

over time, and then tends to stabilize within a certain interval (Fig. 4.6e). Here, four large cities, namely Guangzhou, Nanning, Haikou and Xining, belong to this type. Lastly, as to the decreasing-increasing pattern, it generally shows a trend of first decreasing and then increasing. There are six cities such as Beijing, Hangzhou and Hefei, falling into this category (Fig. 4.6f).

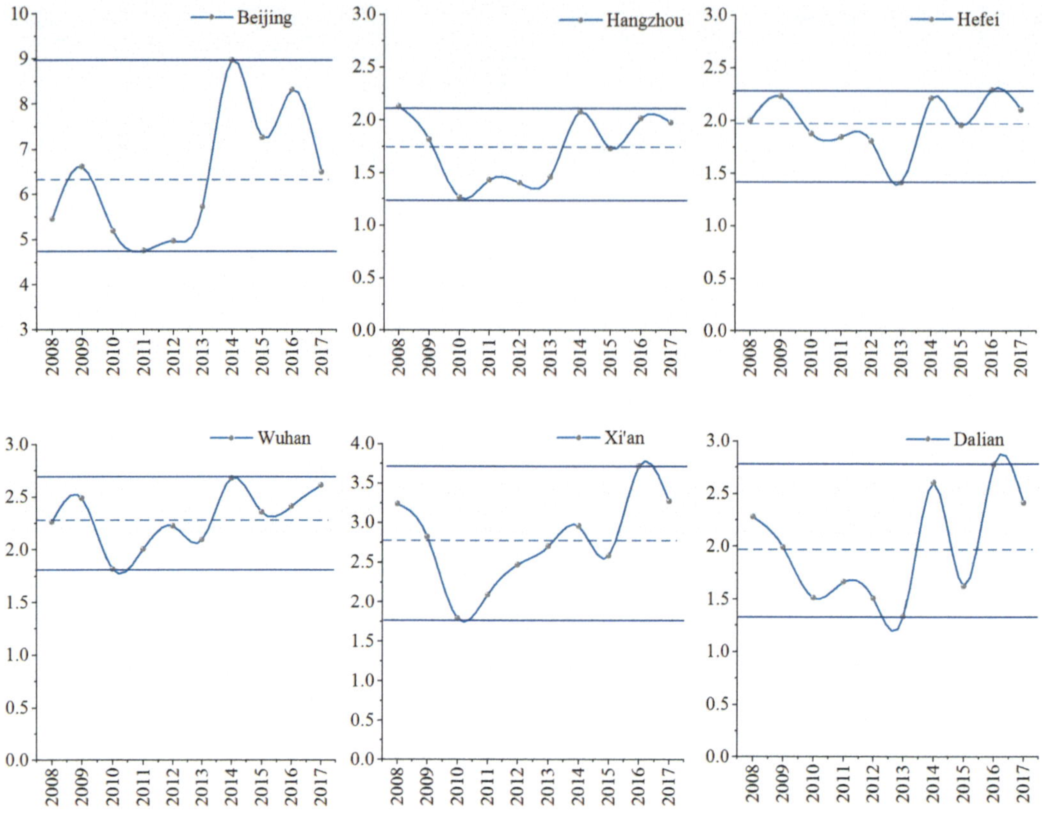


(f)

**Fig. 4.6** (continued)

Taking Beijing as an example, it can be seen that the $\rho_w$ shows a significant decline from 2008 to 2011, reaching a nadir in 2011, and rebounding from 2012, exceeding the average value in 2014 and keeping increasing thereafter.

Further, we will show how the analytical model in Fig. 4.3 can be applied in the context of water resources carrying capacity. As discussed in Sect. 4.4.2, the value of ρ noted in Fig. 4.3 exists virtually but cannot be established in reality. In application, we need to adopt an alternative method to generate the values of ρ. Herein, we deploy the historic information of $\rho_w$ to approximate those typical values. For this purpose, box plot method is used to help calculate the $\rho_0$, $\rho^*$ and $\rho_p$ for water resources carrying capacity. In a specific year, we can obtain the statistical distribution of $\rho_w$ for 35 cities using box plot.

In a box plot, there are three quantile values, namely, lower quartile, median and upper quartile, and they will not be affected by the outliers. Using such method, we can obtain one groups of quantile values for each studied year. Subsequently, a ten-year average value of the three quantile values can be determined. Based on these values, the average value is employed to represent $\rho_0$ (lower quartile), $\rho^*$ (median) and $\rho_p$(upper quartile) (see Fig. 4.7). Using the three values, we can obtain four intervals representing differing state of UWRCC being loaded. For the purpose of practice, we name the four intervals as low-utilization ($I_4$), moderate-utilization ($I_3$), high-utilization ($I_3$), and high-risk-overloaded ($I_4$) in the context of UWRCC. For example, a city belonging to low-utilization interval indicates the capacity provided by the urban water resources system is less utilized by its

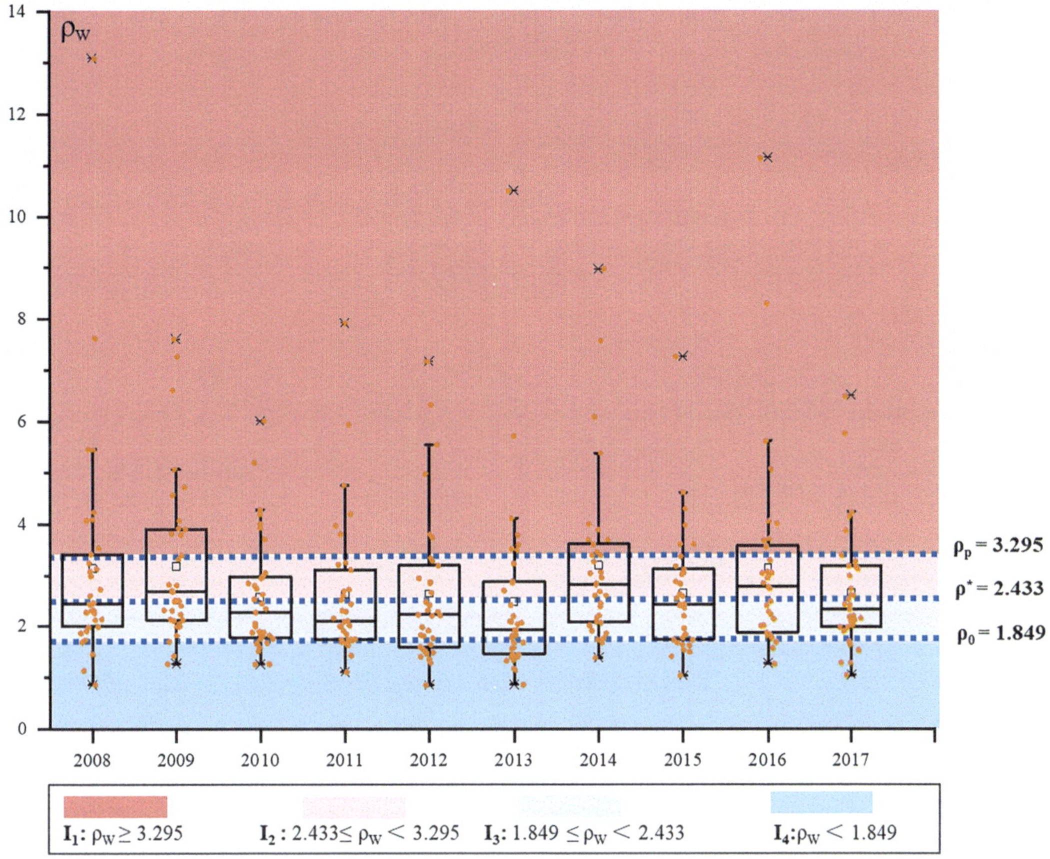


**Fig. 4.7** Box plot of $\rho_w$ and its interval

socioeconomic activities, and a city belonging to high-risk-overloaded interval suggests that the city is prone to falling into the dilemma of UWRCC being overloaded.

Through the intervals defined above, we can observe the change of city falling into different states of UWRCC (Fig. 4.8). As can be seen from this figure, overall, different cities hold differentiated pattern of UWRCC intervals over time, and some cities are dominated by certain intervals. For example, Beijing, Tianjin, Shanghai and Lanzhou have been staying in the interval of high-risk-overloaded, indicating a serious unbalance between the capacity of water resources system and socioeconomic development (consuming water), which should be given with focal attention. Some cities like Hangzhou, Ningbo, Xiamen and Haikou basically lie in low-utilization or moderate utilization intervals, implying that these cities have better potential for steady and sustainable development from the perspective of water resources environment, but the issue of less utilization or waste of UWRCC should also be in hand.

Following Fig. 4.8, we present the overall characteristic of the interval for UWRCC over 2008–2017 (Fig. 4.9). As depicted in Fig. 4.9, the number of cities in each interval fluctuates significantly over the 10-year period. In particular, the number of cities in $I_1$ interval (high-risk-overloaded) declined significantly from 2010 to 2013 compared with the trend in 2008 and 2009. Then, it increased after 2014 and displayed a fluctuant feature between 2014 and 2017. This reflects that with the rapid socio-economic development in China, several cities have entered a state of high risk of water resources carrying capacity being overloaded. However, with the

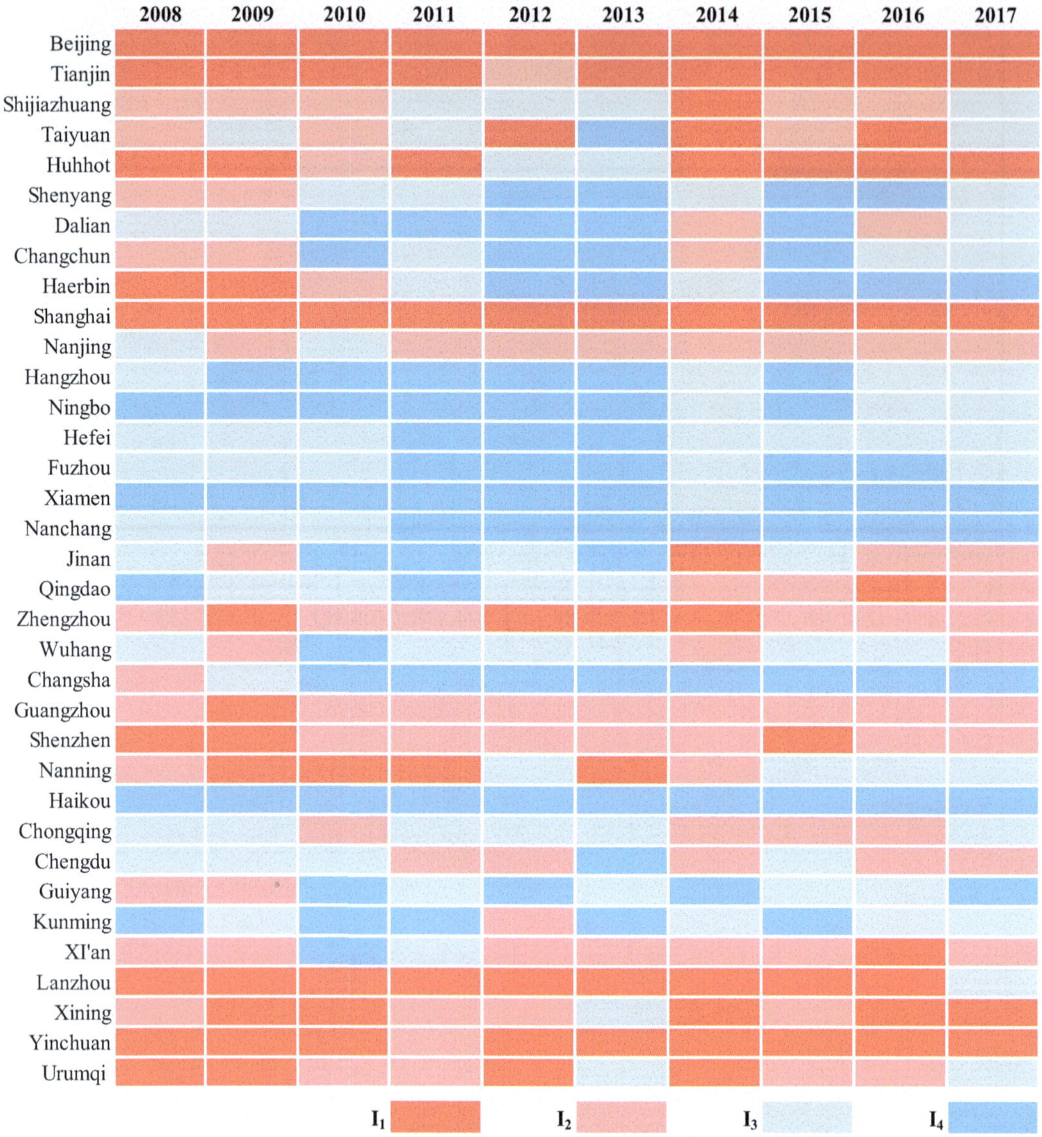


**Fig. 4.8** Distribution of $\rho_w$ interval among 35 cities from 2008 to 2017

practice of new development vision such as high-quality and ecological priority development, cities located in the high-risk overloaded interval began to adjust and regulate their water resources, and it indeed made well effects. As regards the interval of $I_4$ (low-utilization), the number of cities in this interval increased significantly from 2010 to 2013 compared with 2008 and 2009, and began to fluctuated and declined after 2014. This, to a certain extent, mirrors the facts that due to the rapid socio-economic development of most cities in China, UWRCC in some cities (previously underutilized) has been more utilized.

## 4.7 Discussion

The introduced URECC-LC method for evaluating urban resources environment carrying capacity is developed by incorporating the load-and-carrier innovative perspective. From this new perspective, urban resources environment

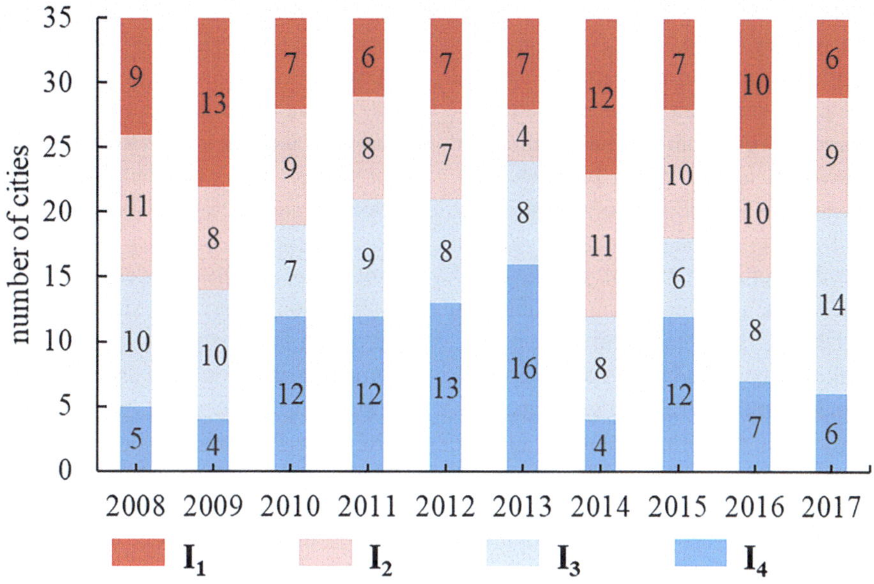


**Fig. 4.9** Number of cities in each $\rho_w$ interval from 2008 to 2017

carrying capacity (URECC) is built up through investigating the interaction between urban loads (UL) and urban carriers (UC) in an urban resources environment system. As a tool for sustainable urban development assessment, URECC measurement emphasizes both the supporting capacity for human pressures provided by urban resources environment and the intensity of human activities. This can verify what extent of current human activities in a city loads in its resources environment (Świąder et al., 2018a, b), and identify the potential risk to prevent city from being overloaded. This indication is similar to the idea of Ecological Footprint (EF) method (see Wackernagel & Rees, 1996; Galli et al., 2020; Świąder et al., 2018a, b): the indicator UL is like Ecological Footprint and the indicator UC is like Biocapacity, and the implications of URECC deficit and URECC surplus are proposed which are analogous to ecological deficit and ecological surplus (see Galli et al., 2020; Świąder et al., 2018a, b). Unlike Ecological Footprint method, this introduced method assesses urban sustainability by evaluating the degree of pressure that human activities place on urban resources environment. This method is a generic evaluation framework for URECC and does not restrict specific factors adopted in the measurement (not like, for instance, EF method must use global general equivalence factor and it is mostly used at national and global level), which is allowed to be applied to assess diverse specific urban resources environment such as land, water, atmosphere, and even social resources like infrastructure resources and public service resources.

The definition of URECC emphasizes on the degree that urban carriers can support urban loads instead of the emphasis given to the limitation of URECC. It is argued that the real threshold of URECC cannot be determined due to the fact that city is an open and resilient system. Instead, a virtual threshold is employed to describe the limitation. This virtual threshold could provide early warning signal for urban managers and planners to monitor the state of urban resources environment properly, which is an important reference to conduct strategic environmental assessment and improve land governance system (Wang et al., 2018a, b).

The measurement of URECC is easy and practical for application, and it presents an explicit indication on the human activities pressures and the capacity of urban resources environment, allowing for the transparency of assessment results. The connotation of URECC is explained in an innovative way. On one hand, it is important to shed light on the state of URECC and its changing pattern, which can assist in monitoring urban resources environment system dynamically for local government and urban planners who have inescapable responsibility to ensure city's development in a sustainable manner (Galli et al., 2020; Peng et al., 2019a, b). On the other hand, resilience is introduced to describe the change of URECC from temporal perspective. In fact, resilience based urban construction and ecological management have been given increasing attention in face of environmental degradation and resources depletion (Ahern, 2013; Romero-Lankao et al., 2016). In applying this new method, a spatial perspective is also employed to compare the spatial differentiations between cities in terms of URECC. Those analysis perspectives enrich the connotations of URECC, and provide deep insights into the value of URECC application.

There are some key issues which should be noted when applying the introduced innovative URECC-LC method. Obviously, the key to use this method for assessing URECC is to identify UL and UC properly. The load factors and carrier factors are very complicated (see Sect. 4.4.1), thus it is essential to ensure that the most representative indicators will be selected. Furthermore, the specific measurement for calculating UL and UC should be as applicable as possible, and the objectivity of that measurement should be also taken into accounts. Additionally, the calculation method for UL and UC should consider the spatial differentiation of the load and carrier factors between different regions. In other words, the specific calculating method for UL and UC should be adjusted when different cities or regions are concerned.

The demonstration case suggests that the introduced URECC-LC method is applicable. The application of the method in 35 large cities in China shows that different cities present diverse changing patterns of water resources carrying capacity. The dynamic change of UWRCC actually indicates the underlying interaction of load and carrier elements in urban water resources system. That is to say, the supporting capacity generated by carriers would dynamically adapt to respond to the pressures induced by water loads, also noted as resilience. Overall, in view of the state interval of UWRCC, the number of major cities in China within the high-risk-overloaded and high-utilization intervals during 2008–2017 shows a declining trend; and the number of cities in the moderate-utilization and low-utilization intervals exhibit an increasing trend. It can be concluded, with the introduction and practice of vision such as high-quality urban development (Chen et al., 2021), the overall UWRCC performance of large cities in China has improved, and China has made progress in alleviating excessive carrying intensity of urban water resources system. On the other hand, the situation of urban water resources carrying capacity being lowly utilized has also been improved. At individual city level, some cities such as Beijing and Tianjin mostly situate at the high risk of UWRCC being overloaded during the studied period, and its UWRCC index show a rising trend, indicating an unsustainable development mode. Thus, the urban managers in these cities should take measures to control the water resources load or enhance the water resources carriers to mitigate the risk of URECC being overloaded. Predominantly, some cities like Dalian and Changchun are in the low-utilization interval of UWRCC and have a decreasing trend in UWRCC over 2008–2017. As such, it is recommended that these cities intensify the proper use of water resources carriers to avoid potential waste and also promote socioeconomic growth and prosperity.

## 4.8 Conclusion

In face of the increasingly serious urban resources environmental problems, cities globally offer both challenges and opportunities for sustainable urban development (Pearson, 2013). The evaluation on URECC establishes a bridge between policy formulation of urban resources environment management and the mission of sustainable development goals (SDGs 11.4, 11.6 and 11.b) at city level. At present, there is no unified theoretical method for evaluating urban resources environment carrying capacity. This study proposes a general theoretical method (URECC-LC) from an innovative perspective, namely the load-and-carrier perspective. Several novel insights have been incorporated when developing this method, and they can be highlighted as: (1) sustainable development based URECC assessment; (2) a system view on urban resources environment; (3) system resilience with timespan; (4) virtual threshold for the limitation of URECC.

The novelty of URECC-LC method lies in that an innovative evaluation framework on URECC is proposed by considering UL and UC collectively, and also the interaction between UL-UC. Practically, this new method allows the comparison of URECC status between different cities or regions as the measurement is an index value without unit. Thus, the introduced generic method can be applied for other administrative units (regions, countries etc.) if the required data for assessing UL and UC is available and sufficient. The assessment results can provide valuable information about the pressures induced by human activities and the carrying state of urban resources environment, based on which, well-informed strategic planning and natural resources environmental management can be implemented. The appropriate evaluation on URECC can also help raise the sustainability awareness and enhance the link between human activities and urban sustainability. Furthermore, it should be noted that indicator based assessment relies strongly on the data availability and data validity (Świąder et al., 2018a, b). Therefore, urban administrators are strongly expected to establish and update databases for diverse types of resources environment so that the URECC status of different resources environments can be assessed accurately and dynamically. It is considered that the method introduced can help guide circular economy from two lens of loads and carriers by identifying resource needs, assessing waste generation, prioritizing circular economy interventions, and engaging stakeholders. By understanding the loads and carriers of a city, circular economy strategies can be developed to promote resource efficiency, waste reduction, and closed-loop material flows. This can help ensure that the urban area can sustainably accommodate its population and support economic activities while minimizing negative environmental impacts.

Theoretically, this piece of research contributes to the advancement of knowledge in sustainability science and the development of a tool that can be applied to various research fields, such as urban and regional planning and design (Ahern, 2013), resilient city construction (Masnavi et al., 2018), urban ecosystem services (Mancini et al., 2018) and others.

## References

Ahern, J. (2013). Urban landscape sustainability and resilience: The promise and challenges of integrating ecology with urban planning and design. *Landscape Ecology, 28*, 1203–1212.

Arrow, K., Bolin, B., Costanza, R., Dasgupta, P., Folke, C., Holling, C. S., Jansson, B.-O., Levin, S., Mäler, K.-G., & Perrings, C. (1995). Economic growth, carrying capacity, and the environment. *Ecological Economics, 15*(2), 91–95.

Aznar-Sánchez, J. A., Belmonte-Ureña, L. J., Velasco-Muñoz, J. F., & Manzano-Agugliaro, F. (2018). Economic analysis of sustainable water use: A review of worldwide research. *Journal of Cleaner Production, 198*, 1120–1132.

Baabou, W., Grunewald, N., Ouellet-Plamondon, C., Gressot, M., & Galli, A. (2017). The Ecological Footprint of Mediterranean cities: Awareness creation and policy implications. *Environmental Science & Policy, 69*, 94–104.

Barrett, G. W., & Odum, E. R. (2000). The twenty-first century: The world at carrying capacity. *Bioscience, 50*(4), 363–368.

Basse, R. M., Charif, O., & Bódis, K. (2016). Spatial and temporal dimensions of land use change in cross border region of Luxembourg. Development of a hybrid approach integrating GIS, cellular automata and decision learning tree models. *Applied Geography, 67*, 94–108.

Bathrellos, G. D., Skilodimou, H. D., Chousianitis, K., Youssef, A. M., & Pradhan, B. (2017). Suitability estimation for urban development using multi-hazard assessment map. *Science of The Total Environment, 575*, 119–134.

Brown, K., Turner, R. K., Hameed, H., & Bateman, I. J. (2000). Environmental carrying capacity and tourism development in the Maldives and Nepal. *Environmental Conservation, 24*(4), 316–325.

Budihardjo, S., Hadi, S. P., Sutikno, S., & Purwanto, P. (2013). The ecological footprint analysis for assessing carrying capacity of industrial zone in Semarang. *Journal of Human Resource and Sustainability Studies, 1*(02), 14–20.

Carey, D. I. (1993). Development based on carrying capacity: A strategy for environmental protection. *Global Environmental Change, 3*(2), 140–148.

Carpenter, S. R., Folke, C., Schefer, M., & Westley, F. (2009). Resilience: accounting for the noncomputable. *Ecology and Society, 14*, 140–154.

Castellani, V., Sala, S., & Pitea, D. (2007). A new method for tourism carrying capacity assessment. *Ecosytems and Sustainable Development, VI*, 365–374.

Cavallaro, M., Asprone, D., Latora, V., Manfredi, G., & Nicosia, V. (2014). Assessment of urban ecosystem resilience through hybrid social–physical complex networks. *Computer-Aided Civil and Infrastructure Engineering, 29*(8), 608–625.

Chapman, R., Howden-Chapman, P., & Capon, A. (2016). Understanding the systemic nature of cities to improve health and climate change mitigation. *Environment International, 94*, 380–387.

Chelleri, L. (2012). From the «resilient city» to urban resilience. A review essay on understanding and integrating the resilience perspective for urban systems. *Documents d'anàlisi geogràfica, 58*(2), 287–306.

Chen, S. (2004). Carrying capacity: An overview. *Chinese Journal of Population Resources and Environment, 2*, 35–40.

Chen, W., Su, H., Yong, Y., & Hu, Z. (2019a). Decision support system for urban major hazard installations management based on 3DGIS. *Physics and Chemistry of the Earth, Parts A/B/C, 110*, 203–210.

Chen, Y., Li, B., Fan, Y., Sun, C., & Fang, G. (2019b). Hydrological and water cycle processes of inland river basins in the arid region of Northwest China. *Journal of Arid Land, 11*(2), 161–179.

Chen, Y., Tian, W., Zhou, Q., & Shi, T. (2021). Spatio-temporal and driving forces of Ecological Carrying Capacity for high-quality development of 286 cities in China. *Journal of Cleaner Production, 293*, 126186.

Cheng, J., Zhou, K., Chen, D., & Fan, J. (2015). Evaluation and analysis of provincial differences in resources and environment carrying capacity in China. *Chinese Geographical Science, 26*(4), 539–549.

Cheng, K., Fu, Q., Meng, J., Li, T. X., & Pei, W. (2018). Analysis of the spatial variation and identification of factors affecting the water resources carrying capacity based on the cloud model. *Water Resources Management, 32*(8), 2767–2781.

Cheng, F., Su, F., Chen, M., Wang, Q., Jiang, H., & Wang, X. (2019). An evolving assessment model for environmental carrying capacity: A case study of coral reef islands. *Journal of Environmental Management, 233*, 543–552.

Cohen, J. E. (1995). Population growth and earth's human carrying capacity. *Science, 269*(5222), 341–346.

Conke, L. S., & Ferreira, T. L. (2015). Urban metabolism: Measuring the city's contribution to sustainable development. *Environmental Pollution, 202*, 146–152.

Costanza, R., d'Arge, R., De Groot, R., Farber, S., Grasso, M., Hannon, B., et al. (1997). The value of the world's ecosystem services and natural capital. *Nature, 387*(6630), 253–260.

Costanza, R., de Groot, R., Braat, L., Kubiszewski, I., Fioramonti, L., Sutton, P., Farber, S., & Grasso, M. (2017). Twenty years of ecosystem services: How far have we come and how far do we still need to go? *Ecosystem Services, 28*, 1–16.

Cui, Y., Feng, P., Jin, J., & Liu, L. (2018). Water resources carrying capacity evaluation and diagnosis based on set pair analysis and improved the entropy weight method. *Entropy, 20*(5).

Cui, G., Zhang, X., Zhang, Z., Cao, Y., & Liu, X. (2019a). Comprehensive land carrying capacities of the cities in the Shandong Peninsula Blue Economic Zone and their spatio-temporal variations. *Sustainability, 11*(2).

Cui, X., Wang, X., & Feng, Y. (2019b). Examining urban metabolism: A material flow perspective on cities and their sustainability. *Journal of Cleaner Production, 214*, 767–781.

Delgado, A., & Romero, I. (2016). Environmental conflict analysis using an integrated grey clustering and entropy-weight method: A case study of a mining project in Peru. *Environmental Modelling & Software, 77*, 108–121.

Delinom, R. M., Assegaf, A., Abidin, H. Z., Taniguchi, M., Suherman, D., Lubis, R. F., & Yulianto, E. (2009). The contribution of human activities to subsurface environment degradation in Greater Jakarta Area, Indonesia. *Science of The Total Environment, 407*(9), 3129–3141.

Dijst, M., Worrell, E., Böcker, L., Brunner, P., Davoudi, S., Geertman, S., Harmsen, R., Helbich, M., Holtslag, A. A. M., Kwan, M.-P., Lenz, B., Lyons, G., Mokhtarian, P. L., Newman, P., Perrels, A., Ribeiro, A. P., Rosales Carreón, J., Thomson, G., Urge-Vorsatz, D., & Zeyringer, M. (2018). Exploring urban metabolism—Towards an interdisciplinary perspective. *Resources, Conservation and Recycling, 132*, 190–203.

Ernstson, H., van der Leeuw, S. E., Redman, C. L., Meffert, D. J., Davis, G., Alfsen, C., & Elmqvist, T. (2010). Urban transitions: on urban resilience and human-dominated ecosystems. *Ambio, 39*, 531–545.

Fan, Y., & Fang, C. (2019). Research on the synergy of urban system operation-Based on the perspective of urban metabolism. *Science of The Total Environment, 662*, 446–454.

Fatemi, M., Rezaei-Moghaddam, K., Wackernagel, M., & Shennan, C. (2018). Sustainability of environmental management in Iran: An ecological footprint analysis. *Iran Agricultural Research, 38*, 53–68.

Feng, Z., Sun, T., Yang, Y., & Yan, H. (2018). The progress of resources and environment carrying capacity: From single-factor carrying capacity research to

comprehensive research. *Journal of Resources and Ecology, 9*(2), 125–134.
Finka, M., & Kluvánková, T. (2015). Managing complexity of urban systems: A polycentric approach. *Land Use Policy, 42*, 602–608.
Fu, C., Zhang, Y., & Yu, X. (2019). How has Beijing's urban weight and composition changed with socioeconomic development? *Science of The Total Environment, 675*, 98–109.
Furmankiewicz, M., Potocki, J., & Kazak, J. (2019). Land-Use Conflicts in the Sudetes, Poland. *IOP Conference Series: Materials Science and Engineering, 471.*
Galli, A., Iha, K., Moreno Pires, S., Mancini, M. S., Alves, A., Zokai, G., Lin, D., Murthy, A., & Wackernagel, M. (2020). Assessing the Ecological Footprint and biocapacity of Portuguese cities: Critical results for environmental awareness and local management. *Cities, 96*, 102442.
González, A., Donnelly, A., Jones, M., Chrysoulakis, N., & Lopes, M. (2013). A decision-support system for sustainable urban metabolism in Europe. *Environmental Impact Assessment Review, 38*, 109–119.
Graymore, M. L. M., Sipe, N. G., & Rickson, R. E. (2010). Sustaining human carrying capacity: A tool for regional sustainability assessment. *Ecological Economics, 69*(3), 459–468.
Guan, X., Wei, H., Lu, S., Dai, Q., & Su, H. (2018). Assessment on the urbanization strategy in China: Achievements, challenges and reflections. *Habitat International, 71*, 97–109.
Guo, Q., Wang, J., Yin, H., & Zhang, G. (2018). A comprehensive evaluation model of regional atmospheric environment carrying capacity: Model development and a case study in China. *Ecological Indicators, 91*, 259–267.
Hong, Y., & Ye, W. (1998). Theoretical analysis of sustainable environmental carrying capacity. *China Population, Resources and Environment, 8*, 3. (in Chinese).
Irankhahi, M., Jozi, S. A., Farshchi, P., Shariat, S. M., & Liaghati, H. (2017). Combination of GISFM and TOPSIS to evaluation of urban environment carrying capacity (Case study: Shemiran City, Iran). *International Journal of Environmental Science and Technology (Tehran), 14*(6), 1317–1332.
Izakovičová, Z., Mederly, P., & Petrovič, F. (2017). Long-term land use changes driven by urbanisation and their environmental effects (Example of Trnava city, Slovakia). *Sustainability, 9.*
Jia, Z., Cai, Y., Chen, Y., & Zeng, W. (2018). Regionalization of water environmental carrying capacity for supporting the sustainable water resources management and development in China. *Resources, Conservation and Recycling, 134*, 282–293.
Joensuu, T., Edelman, H., & Saari, A. (2020). Circular economy practices in the built environment. *Journal of Cleaner Production, 276*, 124215.
Kaur, H., & Garg, P. (2019). Urban sustainability assessment tools: A review. *Journal of Cleaner Production, 210*, 146–158.
Kessler, J. J. (1994). Usefulness of the human carrying capacity concept in assessing ecological sustainability of land-use in semi-arid regions. *Agriculture, Ecosystems and Environment, 48*(3), 273–284.
Lane, M., Dawes, L., & Grace, P. (2014). The essential parameters of a resource-based carrying capacity assessment model: An Australian case study. *Ecological Modelling, 272*, 220–231.
Lane, M. (2010). The carrying capacity imperative: Assessing regional carrying capacity methodologies for sustainable land-use planning. *Land Use Policy, 27*(4), 1038–1045.
Li, R.-m., Yin, Z.-q., Wang, Y., Li, X.-l., Liu, Q., & Gao, M.-m. (2018). Geological resources and environmental carrying capacity evaluation review, theory, and practice in China. *China Geology, 1*(4), 556–565.
Lin, L., Liu, Y., Chen, J., Zhang, T., & Zeng, S. (2011). Comparative analysis of environmental carrying capacity of the Bohai Sea Rim area in China. *Journal of Environmental Monitoring, 13*(11), 3178–3184.
Liu, R. Z., & Borthwick, A. G. (2011). Measurement and assessment of carrying capacity of the environment in Ningbo, China. *Journal of Environmental Management, 92*(8), 2047–2053.
Liu, W., & Zhu, J. (2017). Research progress of resource and environmental carrying capacity: From the perspective of the comprehensive study of geograph. *China Population, Resources and Environment, 27*, 75–86. (in Chinese).
Liu, L., Zhang, Y., Li, X., Xu, J., Zeng, G., Liu, K., Zhang, X., Shen, Q., Hu, Q., & Wang, X. (2008). The spatial relationship analysis of regional development potential and resource and environment carrying capacity in China. In *Geoinformatics 2008 and Joint Conference on GIS and Built Environment: The Built Environment and Its Dynamics.*
Liu, J., Low, S. P., & Wang, L. F. (2017). Critical success factors for eco-city development in China. *International Journal of Construction Management, 18*(6), 497–506.
Liu, Z., Ren, Y., Shen, L., Liao, X., Wei, X., & Wang, J. (2019). Analysis on the effectiveness of indicators for evaluating urban carrying capacity: A popularity-suitability perspective. *Journal of Cleaner Production, 246*, 119019.
Loorbach, D. (2010). Transition management for sustainable development: A prescriptive, complexity-based governance framework. *Governance, 23*(1), 161–183.
Lu, S., Gao, X., Li, W., Jiang, S., & Huang, L. (2018). A study on the spatial and temporal variability of the urban residential water consumption and its influencing factors in the major cities of China. *Habitat International, 78*, 29–40.
Mancini, M. S., Galli, A., Coscieme, L., Niccolucci, V., Lin, D., Pulselli, F. M., Bastianoni, S., & Marchettini, N. (2018). Exploring ecosystem services assessment through Ecological Footprint accounting. *Ecosystem Services, 30*, 228–235.
Marsalek, J. (2014). *Urban water cycle processes and interactions: Urban Water Series-UNESCO-IHP.* CRC Press.
Martire, S., Castellani, V., & Sala, S. (2015). Carrying capacity assessment of forest resources: Enhancing

environmental sustainability in energy production at local scale. *Resources, Conservation and Recycling, 94*, 11–20.

Masnavi, M. R., Gharai, F., & Hajibandeh, M. (2018). Exploring urban resilience thinking for its application in urban planning: a review of literature. *International Journal of Environmental Science and Technology (Tehran), 16*(1), 567–582.

McPhearson, T., Haase, D., Kabisch, N., & Gren, Å. (2016). Advancing understanding of the complex nature of urban systems. *Ecological Indicators, 70*, 566–573.

Meerow, S., Newell, J. P., & Stults, M. (2016). Defining urban resilience: A review. *Landscape and Urban Planning, 147*, 38–49.

Meng, L., Chen, Y., Li, W., & Zhao, R. (2009). Fuzzy evaluation model for water resources carrying capacity in Tarim River Basin, Xinjiang, China. *Chinese Geographical Science, 19*(1), 89–95.

Millennium Ecosystem Assessment (MEA). (2005). *Ecosystems and Human Well-Being: Synthesis*. Island Press.

Monte-Luna, P. D., Brook, B. W., Zetina-Rejón, M. J., & Cruz-Escalona, V. H. (2010). The carrying capacity of ecosystems. *Global Ecology & Biogeography, 13*, 485–495.

Naimi Ait-Aoudia, M., & Berezowska-Azzag, E. (2016). Water resources carrying capacity assessment: The case of Algeria's capital city. *Habitat International, 58*, 51–58.

Nugroho, S. S. (2019). *After Jokowi's first term–Moving the Nation's "Keraton": A javanese perspective* (RSIS Commentaries, No. 174). RSIS Commentaries). Nanyang Technological University.

Oh, K., Jeong, Y., Lee, D., Lee, W., & Choi, J. (2005). Determining development density using the urban carrying capacity assessment system. *Landscape and Urban Planning, 73*(1), 1–15.

Panagopoulos, T., Gonzalez Duque, J. A., & Bostenaru Dan, M. (2016). Urban planning with respect to environmental quality and human well-being. *Environmental Pollution, 208*, 137–144.

Panchal, R., Singh, A., & Diwan, H. (2021). Does circular economy performance lead to sustainable development? – A systematic literature review. *Journal of Environmental Management, 293*, 112811.

Pearson, L. J. (2013). In search of resilient and sustainable cities: Prefatory remarks. *Ecological Economics, 86*, 222–223.

Peng, J., Du, Y., Liu, Y., & Hu, X. (2016). How to assess urban development potential in mountain areas? An approach of ecological carrying capacity in the view of coupled human and natural systems. *Ecological Indicators, 60*, 1017–1030.

Peng, B., Li, Y., Elahi, E., & Wei, G. (2019a). Dynamic evolution of ecological carrying capacity based on the ecological footprint theory: A case study of Jiangsu province. *Ecological Indicators, 99*, 19–26.

Peng, B., Wang, Y., Elahi, E., & Wei, G. (2019b). Behavioral game and simulation analysis of extended producer responsibility system's implementation under environmental regulations. *Environmental Science and Pollution Research, 26*, 17644–17654.

Philpot, T. A. (2014). *Mechanics of materials*. Wiley.

Price, D. (1999). Carrying capacity reconsidered. *Population and Environment, 21*(1), 5–26.

Reed, D., Kapur, K. C., & Christie, R. D. (2009). Methodology for assessing the resilience of networked infrastructure. *IEEE Systems Journal, 3*(2), 174–180.

Rees, W. E. (1996). Revisiting carrying capacity: Area-based indicators of sustainability. *Population and Environment, 17*, 195–215.

Rees, W. E. (2016). Ecological footprints and appropriated carrying capacity: what urban economics leaves out. *Environment and Urbanization, 4*, 121–130.

Ren, C., Guo, P., Li, M., & Li, R. (2016). An innovative method for water resources carrying capacity research–metabolic theory of regional water resources. *Journal of Environmental Management, 167*, 139–146.

Romero-Lankao, P., Gnatz, D., Wilhelmi, O., & Hayden, M. (2016). Urban sustainability and resilience: From theory to practice. *Sustainability, 8*(12).

Seidl, I., & Tisdell, C. A. (1999). Carrying capacity reconsidered: from Malthus' population theory to cultural carrying capacity. *Ecological Economics, 31*(3), 395–408.

Serrao-Neumann, S., Renouf, M. A., Morgan, E., Kenway, S. J., & Choy, D. L. (2019). Urban water metabolism information for planning water sensitive city-regions. *Land Use Policy, 88*, 104144.

Sevegnani, F., Giannetti, B. F., Almeida, C. M. V. B., Agostinho, F., & Brown, M. T. (2018). Accounting for internal stocks in assessing the sustainability of urban systems: The case of ABC Paulista. *Ecological Indicators, 94*, 70–81.

Shapiro, J., & Báldi, A. (2014). Accurate accounting: How to balance ecosystem services and disservices. *Ecosystem Services, 7*, 201–202.

Shen, L., Peng, Y., Zhang, X., & Wu, Y. (2012). An alternative model for evaluating sustainable urbanization. *Cities, 29*(1), 32–39.

Shen, L., Zhou, J., Skitmore, M., & Xia, B. (2015). Application of a hybrid Entropy–McKinsey Matrix method in evaluating sustainable urbanization: A China case study. *Cities, 42*, 186–194.

Shi, Y., Wang, H., & Yin, C. (2013). Evaluation method of urban land population carrying capacity based on GIS—A case of Shanghai, China. *Computers, Environment and Urban Systems, 39*, 27–38.

Shi, L., Yang, S., & Gao, L. (2016). Effects of a compact city on urban resources and environment. *Journal of Urban Planning and Development, 142*(4), 05016002.

Shi, Y., Shi, S., & Wang, H. (2019). Reconsideration of the methodology for estimation of land population carrying capacity in Shanghai metropolis. *Science of Total Environment, 652*, 367–381.

Song, X., Kong, F., & Zhan, C. (2010). Assessment of water resources carrying capacity in Tianjin city of China. *Water Resources Management, 25*(3), 857–873.

Sun, C., Chen, L., & Tian, Y. (2018). Study on the urban state carrying capacity for unbalanced sustainable development regions: Evidence from the Yangtze River Economic Belt. *Ecological Indicators, 89*, 150–158.

Świąder, M. (2018). The implementation of the concept of environmental carrying capacity into spatial management of cities. *Management of Environmental Quality: An International Journal, 29*, 1059–1074.

Świąder, M., Szewrański, S., & Kazak, J. (2018a). Foodshed as an example of preliminary research for conducting environmental carrying capacity analysis. *Sustainability, 10*.

Świąder, M., Szewrański, S., Kazak, J., van Hoof, J., Lin, D., Wackernagel, M., & Alves, A. (2018b). Application of ecological footprint accounting as a part of an integrated assessment of environmental carrying capacity: A case study of the footprint of food of a large city. *Resources, 7*.

Tang, J., & Ye, W. (1998). Study on environmental bearing capacity and its quantification. *China Environmental Science, 18*(3), 227–230. (in Chinese).

Tang, B., Hu, Y., Li, H., Yang, D., & Liu, J. (2015). Research on comprehensive carrying capacity of Beijing–Tianjin–Hebei region based on state-space method. *Natural Hazards, 84*(S1), 113–128.

Tian, Y. N., & Wang, H. Q. (2013). Progress of resources and environmental carrying capacity. *Journal of Clean Energy Technologies, 1*(2), 132–135.

Trubka, R., Newman, P., & Bilsborough, D. (2010). The costs of urban sprawl–Infrastructure and transportation. *Environment design guide*, 1–6.

United Nations. (2018). *World urbanization prospects: The 2018 revision* [Key Facts]. United Nations.

Verhulst, P.-F. (1838). Notice sur la loi que la population suit dans son accroissement. *Correspondance Mathematique et Physique Publiee par A, 10*, 113–126.

Wackernagel, M., & Rees, W. (1996). *Our ecological footprint: Reducing human impact on the earth*. New Society Publishers.

Walker, B., & Meyers, J. (2004). Thresholds in ecological and social–ecological systems: A developing database. *Ecology and Society, 9*(2).

Wang, R., Cheng, J., Zhu, Y., & Lu, P. (2017). Evaluation on the coupling coordination of resources and environment carrying capacity in Chinese mining economic zones. *Resources Policy, 53*, 20–25.

Wang, X. (2010). Research review of the ecological carrying capacity. *Journal of sustainable development, 3*(3), 263–265.

Wang, L., & Liu, H. (2019). Comprehensive evaluation of regional resources and environmental carrying capacity using a PS-DR-DP theoretical model. *Journal of Geographical Sciences, 29*(3), 363–376.

Wang, S., Xu, L., Yang, F., & Wang, H. (2014). Assessment of water ecological carrying capacity under the two policies in Tieling City on the basis of the integrated system dynamics model. *Science of The Total Environment, 472*, 1070–1081.

Wang, W., Sun, Y., & Wu, J. (2018a). Environmental warning system based on the DPSIR model: A practical and concise method for environmental assessment. *Sustainability, 10*.

Wang, Y., Cheng, H., & Huang, L. (2018b). Water resources carrying capacity evaluation of a dense city group: a comprehensive water resources carrying capacity evaluation model of Wuhan urban agglomeration. *Urban Water Journal, 15*(7), 615–625.

Wang, J., Shen, L., Ren, Y., Wei, X., Tan, Y., & Shu, T. (2019). An alternative model for evaluating the balance of carrying capacity between functional urban infrastructures. *Environmental Impact Assessment Review, 79*, 106304.

Wei, Y., Huang, C., Lam, P., Sha, Y., & Feng, Y. (2015a). Using urban-carrying capacity as a benchmark for sustainable urban development: An empirical study of Beijing. *Sustainability, 7*(3), 3244–3268.

Wei, Y., Huang, C., Lam, P. T. I., & Yuan, Z. (2015b). Sustainable urban development: A review on urban carrying capacity assessment. *Habitat International, 46*, 64–71.

Wei, Y., Huang, C., Li, J., & Xie, L. (2016). An evaluation model for urban carrying capacity: A case study of China's mega-cities. *Habitat International, 53*, 87–96.

Widodo, B., Lupyanto, R., Sulistiono, B., Harjito, D. A., Hamidin, J., Hapsari, E., Yasin, M., & Ellinda, C. (2015). Analysis of environmental carrying capacity for the development of sustainable settlement in Yogyakarta urban area. *Procedia Environmental Sciences, 28*, 519–527.

Xu, M., Sbihi, H., Pan, X., & Brauer, M. (2019). Local variation of PM2.5 and NO2 concentrations within metropolitan Beijing. *Atmospheric Environment, 200*, 254–263.

Ye, W., Xu, X., Wang, H., Wang, H., Yang, H., & Yang, Z. (2016). Quantitative assessment of resources and environmental carrying capacity in the northwest temperate continental climate ecotope of China. *Environmental Earth Sciences, 75*(10), 868.

Yu, G., Li, M., Xu, L., Tu, Z., Yu, Q., Yang, D., Xie, X., & Yang, Y. (2018). A theoretical framework of urban systems and their evolution: The GUSE theory and its simulation test. *Sustainable Cities and Society, 41*, 792–801.

Zeng, W., Wang, H., Xue, J., Ye, W., Guan, B., & Mei, F. (1992). Environmental carrying capacity: A key to coordinating population, resources and environment. *Chinese Journal of Population Resources and Environment, 1*, 26–32.

Zhang, Y., Yang, Z., & Li, W. (2006). Analyses of urban ecosystem based on information entropy. *Ecological Modelling, 197*(1–2), 1–12.

Zhang, M., Liu, Y., Wu, J., & Wang, T. (2018). Index system of urban resource and environment carrying capacity based on ecological civilization. *Environmental Impact Assessment Review, 68*, 90–97.

Zhang, F., Wang, Y., Ma, X., Wang, Y., Yang, G., & Zhu, L. (2019). Evaluation of resources and environmental carrying capacity of 36 large cities in China based on a support-pressure coupling mechanism. *Science of The Total Environment, 688*, 838–854.

Zhao, P., Chapman, R., Randal, E., & Howden-Chapman, P. (2013). Understanding resilient urban futures: A systemic modelling approach. *Sustainability, 5*(7), 3202–3223.

Zhou, Y., & Zhou, J. (2017). Urban atmospheric environmental capacity and atmospheric environmental carrying capacity constrained by GDP–PM2.5. *Ecological Indicators, 73*, 637–652.

Zhu, X., Li, Y., Zhang, P., Wei, Y., Zheng, X., & Xie, L. (2019). Temporal–spatial characteristics of urban land use efficiency of China's 35mega cities based on DEA: Decomposing technology and scale efficiency. *Land Use Policy, 88*, 104083.

# 5 Quasi-static and Dynamic Mechanical Properties of Recycled Aggregate Concrete-Filled Steel Tube Columns Exposed to High Temperature

Zhiyu Luo, Hanbing Zhao, and Wengui Li

## 5.1 Introduction

Construction and Demolition waste (C&DW) poses a significant challenge globally. It has been reported that the worldwide C&DW generated already reaches more than 10 billion tonnes every year (Wu et al., 2021a). Besides, most of C&DW is not well disposed of. C&DW accounts for around half of the total waste of landfill (Molla et al., 2021). It not only decreases the available space but also has an adverse impact on the environment. In view of this, many attempts have been made to reuse C&DW (Tang et al., 2019, 2020; Ibrahim et al., 2023). One of the most common approaches is to produce aggregate based on C&DW, called recycled aggregate. Owing to the substantial demand for aggregate in construction, the success of this attempt would be a powerful way to deal with the current massive residual C&DW. Besides, it can alleviate the great consumption of natural aggregate resources, promoting more sustainable development through a closed-loop approach. However, in contrast to natural aggregate, recycled aggregate comprises not only stone but also some cement paste attached to it. In addition, the parent concrete used for aggregate production may have variable properties due to the different sources. For instance, they may previously have different design strengths and suffered from different external loads and damages. As a consequence, concrete made with recycled aggregate generally presents inferior properties and less stable quality than that made with natural aggregate (Luo et al., 2019).

Compared with the method of reinforcing recycled aggregate or RAC (Luo et al., 2019; Li et al., 2016), developing composite components, for instance, RACFST could be a more efficient strategy to apply RAC with sound performance (Li et al., 2017a, b, 2018). It is because that the behaviour of the component is the result of the combined action from RAC and steel tube. On one hand, the contribution from RAC in resisting external load is no longer 100%. On the other hand, steel tube can protect and strengthen the inner RAC under loading. Thus, RACFST has dramatically improved mechanical behavior than the corresponding RAC. Studies also generally indicate that RACFST doesn't show significantly lower mechanical performance such as compressive behavior (Yang & Han, 2006), flexural behavior (Yang & Han, 2006) and seismic

Z. Luo
Centre for Infrastructure Engineering and Safety, School of Civil and Environmental Engineering, The University of New South Wales, Sydney, NSW, Australia

Department of Civil and Environmental Engineering, National University of Singapore, Singapore, Singapore

H. Zhao · W. Li (✉)
Centre for Infrastructure Engineering and Safety, School of Civil and Environmental Engineering, The University of New South Wales, Sydney, NSW, Australia
e-mail: wengui.li@unsw.edu.au



J. Zuo et al. (eds.), *Circular Economy for Buildings and Infrastructure*, Sustainable Development Goals Series, https://doi.org/10.1007/978-3-031-56241-9_5

performance (Tang et al., 2017) than the normal aggregate concrete-filled steel tube columns (NACFST).

When applying RACFST for engineering projects, it could experience very complex loads and environments such as high temperature and dynamic impact. There are scattered studies for the high temperature damaged properties of RACFST (Li et al., 2017a; Wu et al., 2021b; Liu et al., 2017; Yang et al., 2017; Fernandes et al., 2021; Yang & Hou, 2012) and very few investigations for the impact behaviour of RACFST (Guan et al., 2020; Yang et al., 2015; Shakir et al., 2016). Studies show that although is inferior to the performance of NACFST, RACFST exhibits good resistance to high temperature and impact. Compared with single load, the coupled high temperature and dynamic impact is a much severer situation that threatens the safety of concrete structure. The combined action occurs in various cases such as chemical explosions (e.g. Tianjin port explosion (Huang et al., 2020)), terrorist attacks (e.g. 9/11 terrorist attacks (Kiakojouri et al., 2021)) and vehicles/ship crashes and explosions (e.g. iconic mansion in Istanbul (News, 2018)). Despite relatively low possibility, once happened, the extreme loads would lead to serious damage and loss. Typically, the Tianjin port event results in the death of 173 people and economic losses of approximately 70 billion RMB (Kiakojouri et al., 2021). Currently, the performance of RACFST under the coupled high temperature and impact is still seldom reported. More studies are necessary for the systematic understanding of RACFST under coupled effects to ensure the safety of engineering application of RACFST. Besides, the recycled aggregate based composite with outstanding mechanical properties and ductile failure model can be potentially developed for the fast in situ reconstruction of special structures that are highly likely to be exposed to extreme conditions. Therefore, the impact performance of high temperature damaged RACFST and the optimal design of components were studied.

## 5.2 Experimental Program

### 5.2.1 Material Properties and Sample Preparation

River sand and gravels were used as the fine aggregate and natural coarse aggregate (NCA). RCA in this study was obtained by crushing ground concrete with uniform original strength. The grading of NCA and RCA was chosen as 5–20 mm to make the aggregate size smaller than 1/4 of the cross-section of steel tube. The effects of RCA on the properties of concrete were investigated. As listed in Table 5.1, concrete samples prepared contain different RCA proportions of 0%, 50% and 100%. Based on different water absorption of aggregate, additional water was added to compensate for the high-water absorption of recycled aggregate and ensure the same effective water to cement ratios of concrete samples. For concrete samples with 0%, 50% and 100% of recycled aggregate, their cubic compressive strengths (150 mm × 150 mm × 150 mm) were tested as 42.80 MPa, 43.66 MPa and 35.27 MPa, respectively, and their elastic moduli (based on 150 × 300 mm cylinder samples) were obtained as $3.22 \times 10^4$ MPa, $3.16 \times 10^4$ MPa and $2.54 \times 10^4$ MPa, respectively. The steel tubes adopted were Q235 seamless steel tubes, with the average elastic modulus and yield strength determined as 201.5 GPa and 320.1 MPa, respectively.

**Table 5.1** Mix proportion of concrete samples

| Specimens | Water (kg/m$^3$) | Cement (kg/m$^3$) | Sand (kg/m$^3$) | NCA (kg/m$^3$) | RCA (kg/m$^3$) | Additional water (kg/m$^3$) |
|---|---|---|---|---|---|---|
| RAC-0 (NAC) | 205 | 402 | 645 | 1148 | 0 | 0 |
| RAC-50 | 205 | 402 | 645 | 574 | 574 | 28 |
| RAC-100 | 205 | 402 | 645 | 0 | 1148 | 55 |

**Table 5.2** Design of RACFST specimens

| Specimens | Percentage of recycled aggregate | Temperature (°C) | Specimens | Percentage of recycled aggregate | Temperature (°C) |
|---|---|---|---|---|---|
| RAC0FST-20 | 0% | 20 | RAC0FST-500 | 0% | 500 |
| RAC50FST-20 | 50% | 20 | RAC50FST-500 | 50% | 500 |
| RAC100FST-20 | 100% | 20 | RAC100FST-500 | 100% | 500 |
| RAC0FST-200 | 0% | 200 | RAC0FST-700 | 0% | 700 |
| RAC50FST-200 | 50% | 200 | RAC50FST-700 | 50% | 700 |
| RAC100FST-200 | 100% | 200 | RAC100FST-700 | 100% | 700 |

Note: For RACXFST-Y, X means the percentage of recycled aggregate while Y represents the temperature RACFST ever suffered. RAC0FST is equal to NACFST

The RACFST prepared has a diameter of 96 mm and steel tube wall thickness of 2 mm. The concrete in the inner part of RACFST was described in Table 5.2. A length to diameter ratio of 3 was used for the quasi-static compression tests while a length to diameter ratio of 0.5 was used for impact tests to conform to the relevant requirements. The two end faces of samples were polished to be parallel prior to testing.

### 5.2.2 Testing Methods and Procedures

The high temperature damage of RACFST was conducted via a program-controlled electric furnace. The samples were heated from around 20 °C (ambient temperature) to 200 °C, 500 °C and 700 °C with a temperature increasing rate of 10 °C per minute (Huo et al., 2013). After reaching the target temperatures, the temperatures were sustained for 3 h to make the samples heated uniformly. Thereafter, the samples were cooled naturally until to room temperature.

The quasi-static compression and dynamic impact tests of RACFST stub column are shown in Fig. 5.1, conducted via a 3000 kN compression testing machine and a 100 mm-diameter split Hopkinson pressure bar (SHPB), respectively. In the quasi-static test, the longitudinal deformation was tested by two displacement transducers and four evenly distributed strain gauges (half-height section). In the impact test, the stress $\sigma$, strain $\varepsilon$, and strain rate $\dot{\varepsilon}$ were obtained based on incident, reflected and transmission impulses as given in Eqs. (5.1) to (5.3) (Kolsky, 1949; Lindholm, 1964). The impact velocity was 8.79 m/s, achieved by a constant air pressure of 0.9 MPa to launch the striker bar.

$$\sigma = \frac{EA}{A_S}\varepsilon_t \tag{5.1}$$

$$\varepsilon = -\frac{2c_0}{l_0}\int_0^t \varepsilon_r dt \tag{5.2}$$

$$\dot{\varepsilon} = -\frac{2c_0}{l_0}\varepsilon_r \tag{5.3}$$

where $E$ is the elastic modulus and $A$ is the cross-sectional area of the Hopkinson pressure bars; $C_0$ represent the elastic wave velocity in pressure bars; $A_s$ and $l_0$ denote the cross-sectional area and length of the test specimen.

## 5.3 Results and Analysis

### 5.3.1 Failure Patterns

After both quasi-static compression and dynamic impact tests, the failure patterns are not vastly different for RACFST samples with different proportions of recycled aggregate or ever exposed to different high temperatures. After the quasi-static compression test, local outward folding of steel tube was observed as the failure phenomenon of all samples as typically shown in Fig. 5.2a–c. After being subjected to impact load, RACFST samples do not experience obvious change in their shape as typically revealed in Figs. 5.2d–f, but the steel tubes are all yielded according to the measurement. The impact

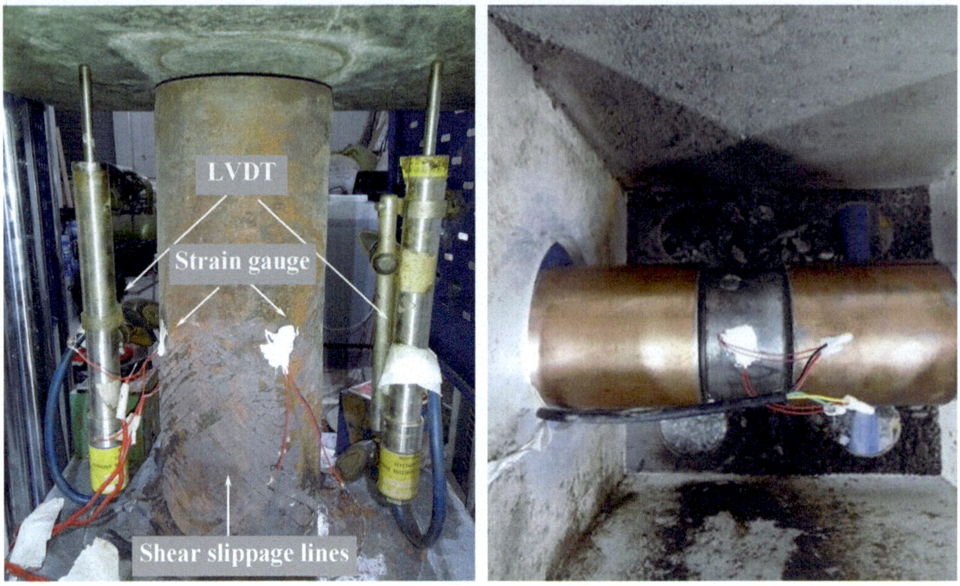


(a) Quasi-static compression (96 × 288 mm) (b) SHPB impact (96 × 48 mm)

**Fig. 5.1** Quasi-static compression and impact tests of RACFST stub column. (**a**) Quasi-static compression (96 × 288 mm). (**b**) SHPB impact (96 × 48 mm). (Reprinted with permission from Li et al., 2017a, 2018)

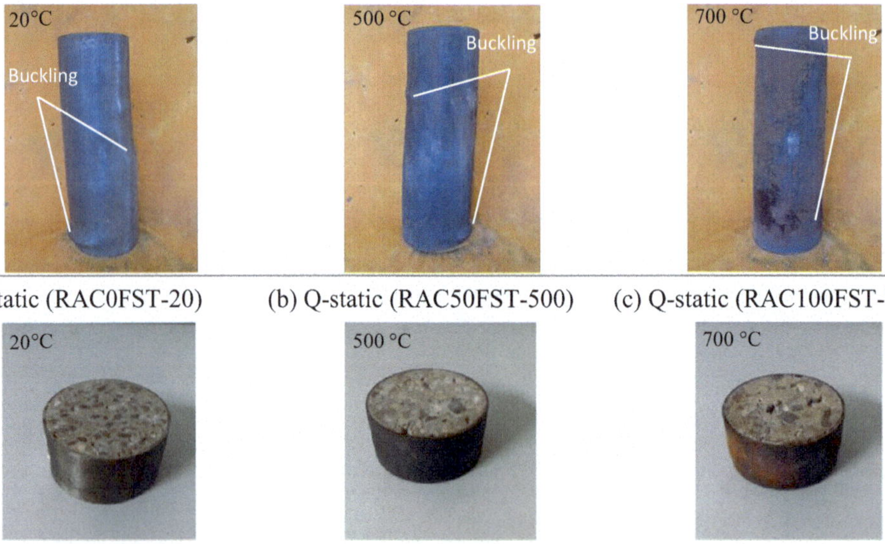


(a) Q-static (RAC0FST-20) (b) Q-static (RAC50FST-500) (c) Q-static (RAC100FST-700)

(d) Impact (RAC0FST-20) (e) Impact (RAC50FST-500) (f) Impact (RAC100FST-700)

**Fig. 5.2** Failure of RACFST after quasi-static compression or impact loads. (**a**) Q-static (RAC0FST-20). (**b**) Q-static (RAC50FST-500). (**c**) Q-static (RAC100FST-700). (**d**) Impact (RAC0FST-20). (**e**) Impact (RAC50FST-500). (**f**) Impact (RAC100FST-700). (Reprinted with permission from Li et al., 2017a, 2018) deformation of RACFST samples is significantly linked with the high temperature suffered but not the recycled aggregate replacement ratio. For RACFST samples ever exposed to a high temperature of 700 °C, their middle part is obviously bulging. Besides, spall of several aggregate particles can be observed.

### 5.3.2 Mechanical Properties and Degradation

The quasi-static compression and dynamic impact tests curves of high temperatures damaged RACFST samples are provided in Fig. 5.3. Even if the results were from three duplicate samples, there were still slight wave oscillations in the impact stress-strain curves due to wave dispersion. Nevertheless, it would not significantly disturb the results. Under both types of loading, the curves present the elastic and elastic-plastic stages. However, the impact stress-strain curves show an unusual near-plateau stage. The main reasons could be the higher structural stability and full play of constraint effect caused by the small length to diameter ratio, and the lower degree of crack propagation in core concrete due to the extremely short loading time. The quasi-static carrying capacity of RACFST was determined by the peak axial load, while the ultimate impact strength was determined by the first peak stress corresponding to the strain at the end of the elastic-plastic stage of the impact stress-strain curves. The elastic modulus was obtained based on secant modulus. Then, the effects of high temperatures exposure and recycled aggregate replacement ratio on the mechanical properties of RACFST tested under different strain rate loading were analysed.

The residual strength index (*RSI*) given in Eq. (5.4) and residual elastic modulus ratio $K_e$ given in Eq. (5.5) are normally used to evaluate the high temperature or fire caused mechanical performance degradation. Similarly, the conception is extended in this study to denote the recycled aggregate replacement induced decrease

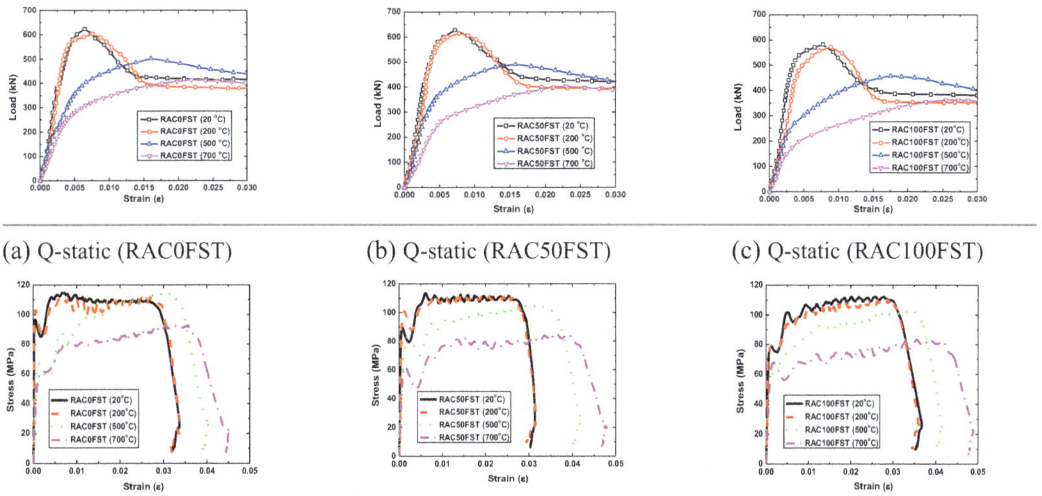


(a) Q-static (RAC0FST) (b) Q-static (RAC50FST) (c) Q-static (RAC100FST)

(d) Impact (RAC0FST) (e) Impact (RAC50FST) (f) Impact (RAC100FST)

**Fig. 5.3** Load/stress versus strain curve of RACFST specimens after being suffered from different high temperatures. (**a**) Q-static (RAC0FST). (**b**) Q-static (RAC50FST). (**c**) Q-static (RAC100FST). (**d**) Impact (RAC0FST). (**e**) Impact (RAC50FST). (**f**) Impact (RAC100FST). (Reprinted with permission from Li et al., 2017a, 2018)

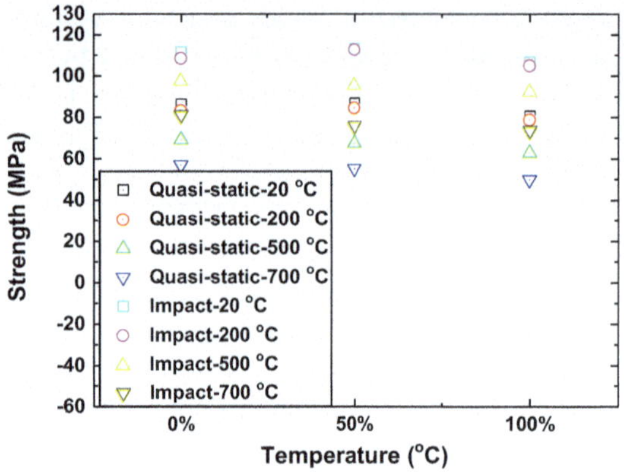


(a) Quasi-static and impact strength (MPa)

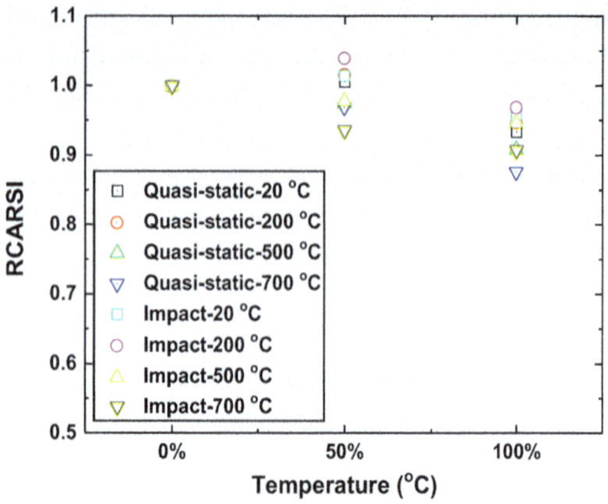


(b) RCA residual strength index

**Fig. 5.4** Effects of different RCA replacement ratios on the strength reduction of heat treated RACFST stub columns. (**a**) Quasi-static and impact strength (MPa). (**b**) RCA residual strength index. (Reprinted with permission from Li et al., 2017a, 2018)

in strength as given in Eq. (5.6), RCA residual strength index (*RCARSI*).

$$RSI = \frac{N_u(T)}{N_u(20)} \tag{5.4}$$

$$K_e = \frac{E_s(T)}{E_s(20)} \tag{5.5}$$

where $N_u(T)$ and $E_s(T)$ in this work refer to the residual strength and residual elastic modulus of RACFST stub column after exposure to high temperature $T$, while $N_u(20)$ and $E_s(20)$ refer to the strength and elastic modulus of RACFST at room temperatures of 20 °C.

$$RCARSI = \frac{N_u(r)}{N_u(0)} \tag{5.6}$$

where $N_u(r)$ is the strength of RACFST stub column with different RCA replacement ratios, and $N_u(0)$ refers to the strength of RACFST one without RCA.

The effects of different RCA replacement ratios and high temperatures on the mechanical performance degradation of heat treated RACFST stub column are presented in Figs. 5.4 and 5.5. As manifested in Fig. 5.4a, both quasi-static and impact strength of samples decreases with the increase of RCA replacement ratios, but just varies in a small range. The results indicate that using recycled aggregate would not significantly compromise the performance of RACFST under the given extreme conditions. This conclusion aligns with the results of previous lateral impact studies on RACFST (Guan et al., 2020; Yang et al., 2015; Shakir et al., 2016). Obviously, the 500 °C and 700 °C high temperatures caused a more distinct recession in mechanical properties than that of RCA replacement ratios. It can be further understood from the quantitative results provided in Figs. 5.4b and 5.5. The *RACRSI* is in the range of 0.88–0.97 when the aggregate of RACFST samples was 100% of RCA, while the *RSI* is in the range of 0.62–0.73 when the temperature that the RACFST samples suffered reached 700 °C. High temperature caused even more clear reduction in elastic modulus as provided in Fig. 5.5b. Compared with using RCA, very high temperature would more seriously damage the material properties. Besides, the properties degradation is not only for concrete (especially higher than 300 °C (Vieira et al., 2011)) but also steel tube (especially higher than 400 °C (Kumar et al., 2013)). Hence, more evident loss of mechanical properties was observed.

The degradation of performance was also found to be different at different loading strain rates. The results in Fig. 5.5 show that the samples under quasi-static load generally experienced a

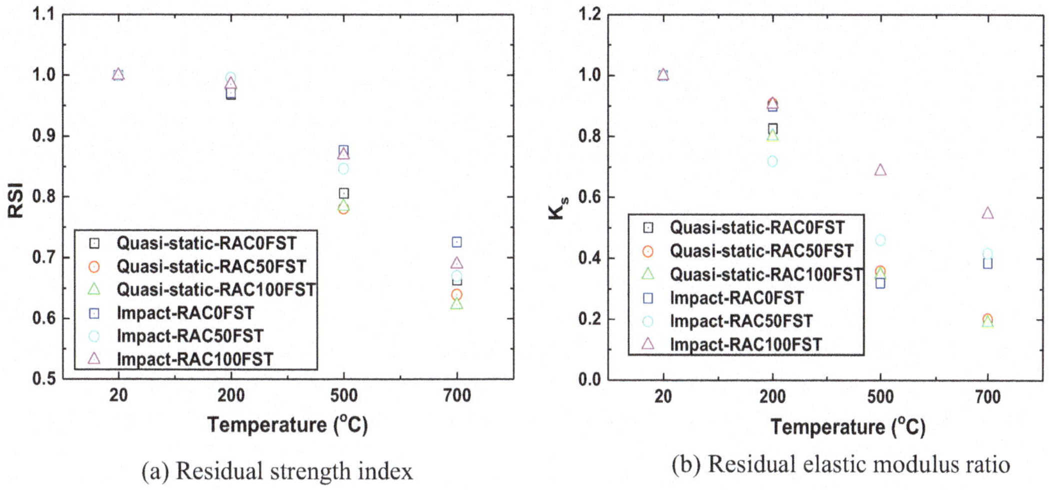


**Fig. 5.5** Effects of different temperatures on the strength degradation of heat treated RACFST stub columns. (**a**) Residual strength index. (**b**) Residual elastic modulus ratio. (Reprinted with permission from Li et al., 2017a, 2018)

larger drop in mechanical properties than the corresponding samples under impact load, especially after being exposed to 500 °C and 700 °C. For instance, after a high temperature of 500 °C, the impact *RSI* value ranges from 84.7% to 87.7%, while the quasi-static *RSI* value ranges from 78.2% to 80.6%; after a high temperature of 700 °C, the impact *RSI* value ranges from 67.0% to 72.6%, while the quasi-static *RSI* value ranges from 62.3% to 66.3%. The strain rate induced strength differences can be also understood through the dynamic increase factor (DIF), the ratio of dynamic strength to the quasi-static strength. As revealed in Fig. 5.6, 200 °C doesn't bring pronounced variation in DIF, whereas apparent increases in DIF were found for samples exposed to the higher temperatures of 500 °C and 700 °C. From 20 to 700 °C, the DIF increases from the range of 1.30–1.33 to 1.37–1.47. The increased DIF could be ascribed to the factor that, 500 °C and 700 °C significantly softened samples and increased loading strain rate, which incurs additional strength increase due to the strain rate effect. It could also be used to explain the small degradation (*RSI*) of impact strength than that of quasi-static strength above. That is, the impact strength is less sensitive to high temperature damage. Results in Fig. 5.6 also show that increasing the temperature from 500 to 700 °C doesn't result in a further increase in DIF. It is possible that very high temperature exposure would increase strain rate but also decrease the strain rate sensitivity of materials. Compared with high temperature exposure, the RCA replacement ratio brings less significant variation in DIF.

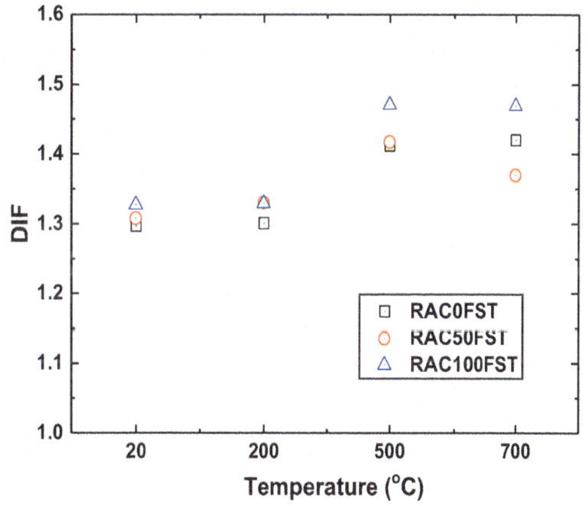


**Fig. 5.6** Dynamic increase factor. (Reprinted with permission from Li et al., 2018)

### 5.3.3 Numerical Model and Parametric Analysis

#### 5.3.3.1 Numerical Model

To further understand the effects of crucial design parameters on the impact performance of high temperature damaged RACFST, finite element analysis (FEA) based on ABAQUS/Explicit module was conducted. The diameter of pressure bars is 100 mm. The length of the striking bar is 1.5 m, incident bar is 6 m, and transmission bar is 4 m. A mesh convergence study was performed to obtain proper mesh size. Based on the experimental records, the striking bar was given an impact velocity of 8.8 m/s and the time step in the simulation was 0.003 s. The model was shown in Fig. 5.7.

The damage plasticity model was adopted to describe the behavior of concrete. The stress-strain constitutive model for core RAC after high temperature was determined as given in Eqs. (5.7) to (5.10) (Yang et al., 2015; Han et al., 2007; Han et al., 2002; Huo, 2011). Previous research indicated that in the case of strong confinement, for instance, the confinement factor $\xi$ greater than 1 for circular CFST, the stress-strain curve of inner concrete would not descend even under large strain. It should be noted that the impact sample has a low length to diameter ratio, which allows the full play of constraint effect. In addition, a short loading time hinders crack development. Thus, the inner concrete was assumed to maintain its peak stress after peak strain. Strain rate effect was generally considered for core concrete. However, the almost intact core concrete in Fig. 5.2 implies that the Stefan effect and the cracking propagation effect would be not significant. The presence of steel tube would also weaken the inertia effect of core concrete (Xiao et al., 2009). Therefore, the CEB empirical equation for unconstrained concrete largely overestimates the strain rate effects in this case. Rather than strain rate effects, the strength enhancement of concrete should be dominantly related to the confinement provided by steel tube, which is considered in the simulation. The damage factor was obtained based on the assumption of strain equivalence and the testing results from Birtel et al. (Birtel, 2006). Besides, the tension constitutive relationships of core RAC was adopted using the fracture energy model (Huo, 2011).

$$y = 2x - x^2, x \leq 1 \tag{5.7}$$

$$y = \frac{x}{\beta(x-1)^2 + x}, x > 1 \tag{5.8}$$

$$\sigma_o(T) = \frac{\sigma_o}{1 + 2.4(T-20)^6 \times 10^{-17}} \tag{5.9}$$

$$\varepsilon_o(T) = \left[1 + \left(1500T + 5T^2\right) \times 10^{-6}\right]\varepsilon_o \tag{5.10}$$

where $x = \varepsilon/\varepsilon_o$, $y = \sigma/\sigma_o$, $\sigma_0 = f'_{c,r}$, $\varepsilon_o = \left(\varepsilon_c + 800\xi_r^{0.2} \times 10^{-6}\right)(1 + r/\theta)$,$\xi_r = f_y A_s/$ $f_{ck,}$ $_r A_c$, $\varepsilon_c = \left(1300 + 12.5f'_{c,r}\right) \times 10^{-6}$,

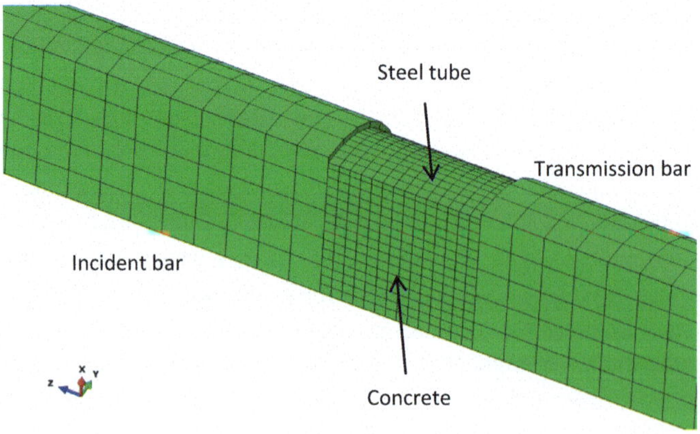


**Fig. 5.7** Model of SHPB impact tests (1/4 model). (Reprinted with permission from Li et al., 2017b)

$\beta = 0.5\left(2.36\times10^{-5}\right)^{\left[0.25+(\xi_r-0.5)^7\right]}\times\left(f'_{c,r}\right)^{0.5}$-$\geq 0.12$, $\theta = 65.715r^2 - 109.43r + 48.989$; $r$ is the replacement ratios of recycled aggregate; $f'_{c,r}$ is the tested cylinder strength of RAC;$f_{ck,r}$ equals to 0.67 $f_{cu}$ for normal strength RAC; $\xi_r$ refers to the confinement factor of RACFST.

For the pressure bars, due to the high strength steel used, they were considered to remain in the elastic stage during impact loading. The Poisson's ratio, mass density and elastic modulus were set as 0.3, 7800 kg/m$^3$ and 210 GPa, respectively. For steel tube in RACFST, the elastic-plastic model was adopted. The stress-strain curve (Han et al., 2002) after high temperature damage were determined as shown in Eqs. (5.11) to (5.14). The degradation of elastic modulus was evaluated accordingly (Lu et al., 2016).

$$\sigma = E_s(T)\varepsilon, \varepsilon \leq \varepsilon_y(T) \quad (5.11)$$

$$\sigma = f_y(T) + E_1(T)\left[\varepsilon - \varepsilon_y(T)\right], \varepsilon > \varepsilon_y(T) \quad (5.12)$$

$$f_y(T) = f_y, T \leq 400 \quad ^\circ\mathrm{C} \quad (5.13)$$

$$f_y(T) = f_y\left[1 + 2.33\times10^{-4}(T-20) - 5.88 \times10^{-7}(T-20)^2\right], T > 400 \quad ^\circ\mathrm{C} \quad (5.14)$$

where $E_s(T)$ refers to the high temperature damaged elastic modulus of the steel; $\varepsilon_y(T) = f_y(T)/E_s(T)$ is the yield strain; $E_1(T) = 0.01E_s(T)$ denotes the modulus in the strengthening stage; $f_y(T)$ refers to degraded yield strength of steel after exposure to temperature $T$.

The strain-rate effect of steel was calculated according to the Cowper-Symonds model, with the dynamic yield function displayed in Eq. (5.15).

$$\sigma_{dy} = \sigma_y\left(1 + (\dot{\varepsilon}/D)^{1/p}\right) \quad (5.15)$$

where the $\sigma_{dy}$ and $\sigma_y$ are the yield strength of steel under dynamic and quasi-static, respectively; $\dot{\varepsilon}$ is the strain rate; The material parameter $D$ is chosen as 40.4 $s^{-1}$ and $P$ is chosen as 5 according to previous research (Deng & Tuan, 2013).

#### 5.3.3.2 Validation and Parametric Analysis

The propagation of stress wave is typically shown as in Fig. 5.8, where the incidence and reflection process can be clearly observed. The impulses in one time of simulation test can be found in Fig. 5.9. Comparisons of impulses obtained from simulation and SHPB experiment are provided in Fig. 5.10 based on representative samples. The impulses from the simulated test are more fluctuant than that from the real experiment because the latter one filtered high frequency oscillation via soft cloth and smoothed data before outputting. Nevertheless, the experimental impulses and simulated impulses are quite similar to each other. Buffer bar was ignored in the model. Consequently, abnormal stain can be observed after transmitted impulses, but they would not affect the stress-strain curves calculated through the two-wave method.

The simulation results for the RACFST sample with 100% of recycled aggregate were presented to promote the further optimal design of RACFST. As given in Fig. 5.11, although there are deviations between experimental curves and the corresponding simulation ones, they don't significantly affect the impact strength (the end of the elastic-plastic stage). The errors for deformation are also acceptable. The model can be used for the following parametric analysis.

The effects of normal design parameters including the strength of concrete (C30, C40 and C50) and steel (Q235, Q345 and Q420), and the steel ratio on the impact strength of RACFST were studied. For RACFST samples with the same external diameter but different steel tube thicknesses of 2 mm, 3 mm and 4 mm, the steel ratio is 8.9%, 13.8% and 19.0%, respectively. Figure 5.12 manifests the improvement in impact strength and deformation of RACFST with the enhanced material strength or steel ratio. It is remarkable that the increase in steel strength and especially steel thickness is better than the increase in concrete strength to help RACFST achieve meliorated impact resistance. For materials themselves, from the grade of C30 to C50 only increased the strength of concrete for 20 MPa, while there are more than 100 MPa

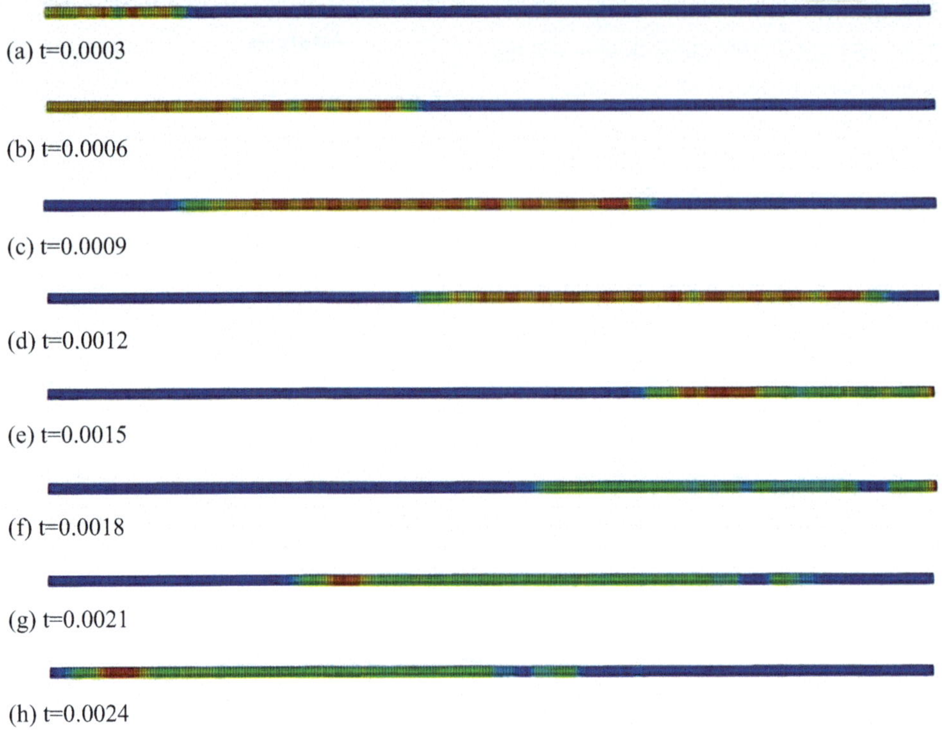


**Fig. 5.8** Propagation of stress wave in incident bar

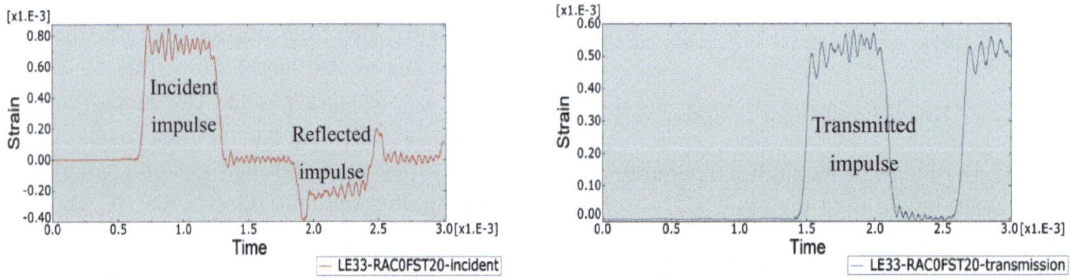


(a) Incident bar (b) Transmission bar

**Fig. 5.9** Typical impulses output from ABAQUS (RAC0FST-20). (**a**) Incident bar. (**b**) Transmission bar

increments when using steel with higher grades. Increasing steel ratio means the increase of the material that has higher strength, temperature resistance and strain rate sensitivity and the improved combination behaviour. These factors to some extend explain the phenomenon presented.

The impact strength enhancement factor $k_{en}$ in Eq. (5.16) and impact deformation decrease factor $\eta_{de}$ in Eq. (5.17) were used to provide a quantitative perspective to understand the effects of design parameters on the impact resistance of RACFST. The corresponding results are presented in Fig. 5.13. Rising the RAC strength grade from C30 to C50 moderately increases the impact strength of RACFST at room temperature but the beneficial effect becomes almost negligible for RACFST samples ever exposure to a high temperature of 700 °C. Unlike concrete, the steel strength and thickness are found to be even more

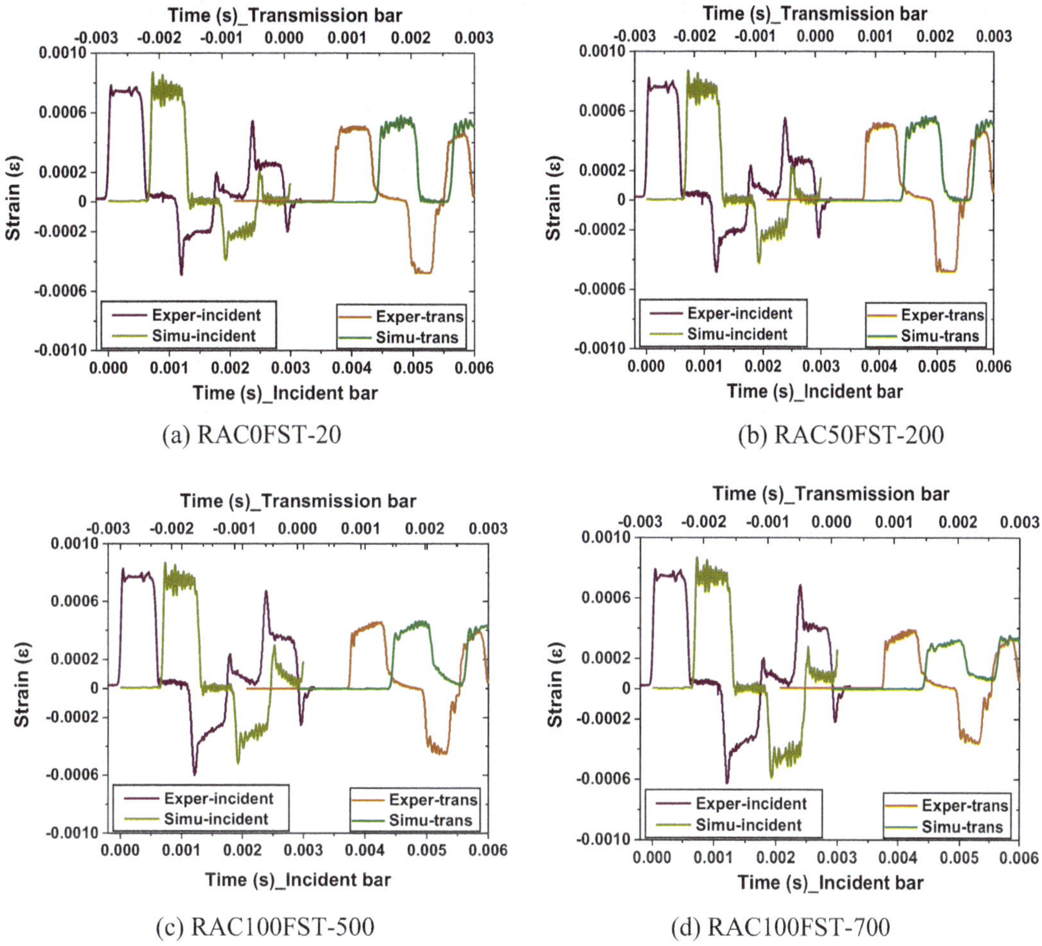


**Fig. 5.10** Comparison of impulse waves acquired from experiment and simulation. (**a**) RAC0FST-20. (**b**) RAC50FST-200. (**c**) RAC100FST-500. (**d**) RAC100FST-700. (Reprinted with permission from Li et al., 2017b)

efficient to help improve impact strength after exposure to high temperatures of 500 °C and 700 °C. For RACFST without being damaged by high temperature, the samples with RAC strength of C50, steel strength of Q420, and steel thickness of 4 mm achieve $k_{en}$ of 1.10, 1.29 and 1.37, respectively. Once the high temperature of 700 °C was applied, the values turn to 1.02, 1.47 and 1.81, respectively. Even just increase the steel tube thickness from 2 to 3 mm, the impact strength is still improved by 1.44 times after 700 °C.

Concrete is more vulnerable than steel to high temperature damage. Thus, for RACFST after high temperature, its steel tube plays an increased role while concrete plays a decreased role in impact resistance. Different types of steel have similar elastic modulus but quite different strengths. Although the difference between different grades of materials is decreased after high temperature, the increased contribution from steel is still able to make the RACFST with higher steel properties/ratios exhibit higher $k_{en}$ after high temperature. In terms of the impact deformation decrease factor, the steel strength and especially steel ratio are still identified as the more influential parameters, but the efficiency decreases with the elevation of high temperature. After 700 °C, $\eta_{de}$ is 0.96 for the sample with RAC strength of C50 and is 0.77 and 0.49 for the sample with steel

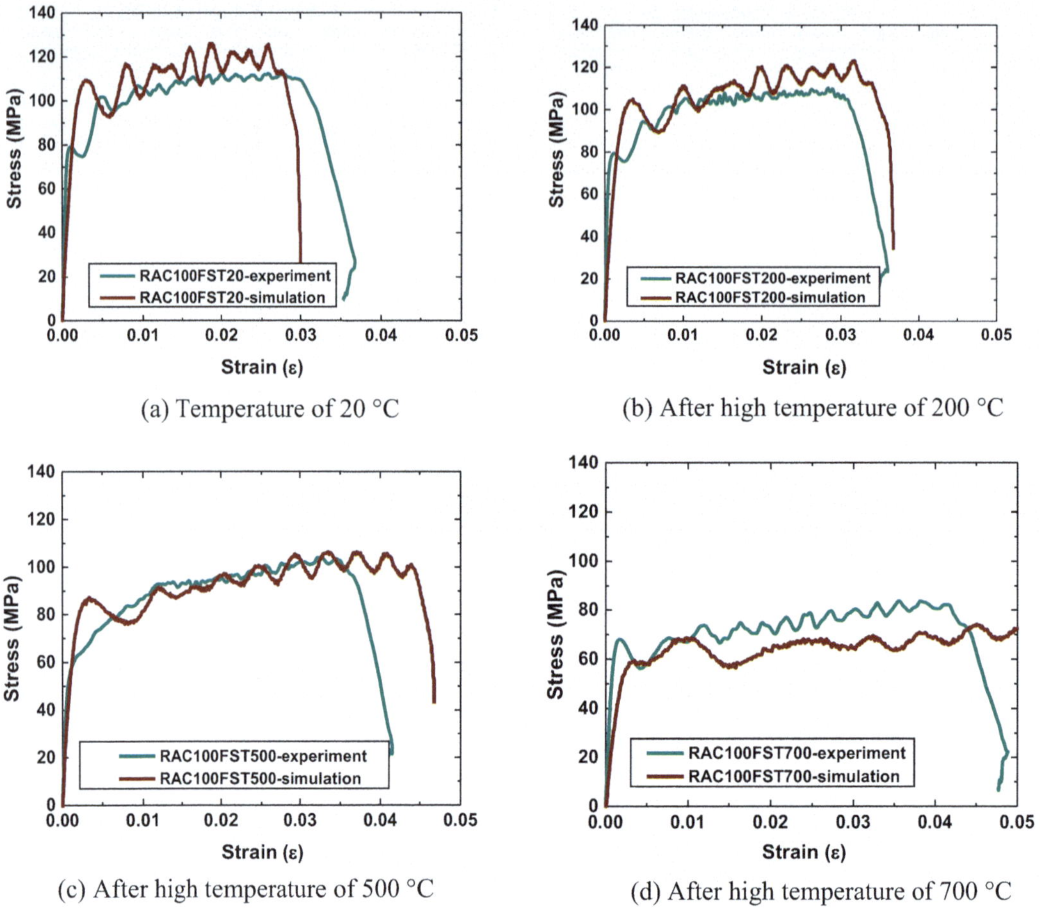


**Fig. 5.11** Comparisons of stress-strain curves calculated from experimental results and simulation results (RAC100FST). (**a**) Temperature of 20 °C. (**b**) After high temperature of 200 °C. (**c**) After high temperature of 500 °C. (**d**) After high temperature of 700 °C. (Reprinted with permission from Li et al., 2017b)

strength of Q420 and steel thickness of 4 mm, respectively.

Among the three material design parameters, the steel ratio is identified as the most efficient one for improving the impact performance of high temperature damaged RACFST because of the stronger higher temperature resistance, mechanical behaviour and strain rate effect of steel than concrete. However, there should be comprehensive consideration for the performance and cost in the design of RACFST.

$$k_{en} = \frac{\sigma}{\sigma_{re}} \tag{5.16}$$

$$\eta_{de} = \frac{\varepsilon}{\varepsilon_{re}} \tag{5.17}$$

where $\sigma_{re}$ is the impact strength of reference RACFST sample and $\sigma$ is the impact strength of samples with different material strength grades or steel ratios. Similarly, $\varepsilon_{re}$ and $\varepsilon$ are the maximum impact strain of the corresponding samples. The reference sample refers to the sample with RAC strength of C30 when investigating the effects of RAC strength grade. Similarly, the reference sample means the sample with steel strength of Q235 or thickness of 2 mm while studying the effects of the other two corresponding parameters.

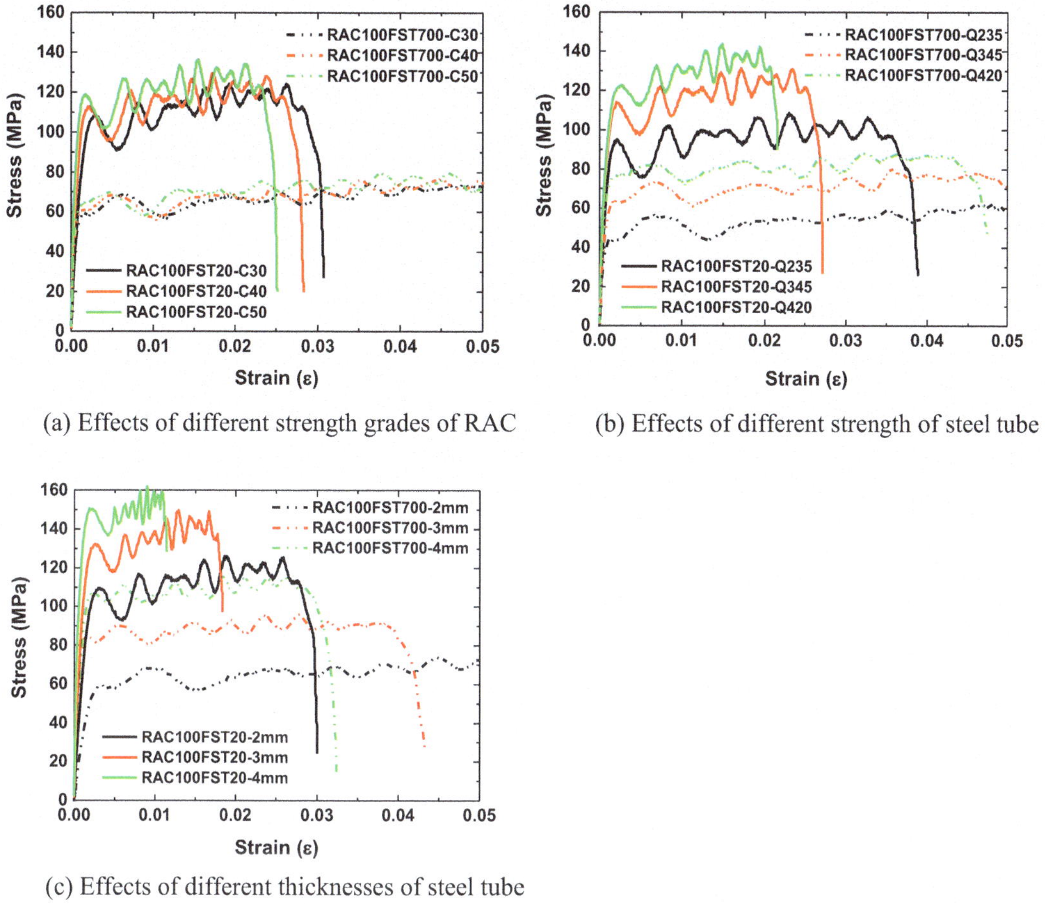


**Fig. 5.12** Effects of design parameters on the simulated impact behavior of RACFST. (**a**) Effects of different strength grades of RAC. (**b**) Effects of different strength of steel tube. (**c**) Effects of different thicknesses of steel tube. (Reprinted with permission from Li et al., 2017b)

## 5.4 Conclusions

The quasi-static to dynamic mechanical performance degradation of recycled aggregate concrete-filled steel tube (RACFST) columns after exposure to high temperature was studied. Within the parameters of this study, the conclusions can be drawn as follows:

The quasi-static compression and impact failure pattern of samples does not vary with the proportion of recycled aggregate or the high temperature experienced. High temperature is relatively more influential than recycled aggregate replacement ratio on the damage and deformation degree of RACFST.

The replacement of natural aggregate by recycled aggregate only brings a moderate decrease in the strengths of high temperature damaged RACFST under different loading rates. The RCA residual strength index is in the range of 0.88~0.97 when natural aggregate is fully replaced by recycled aggregate, with the most severe case (0.88) for quasi-static loaded samples ever suffered from a high temperature of 700 °C.

High temperature exposure more distinctly decreases the mechanical properties of RACFST than aggregate replacement. The residual strength index after a high temperature of 700 °C ranges from 0.62 to 0.73. The degradation is more obvious for quasi-static strength than impact strength,

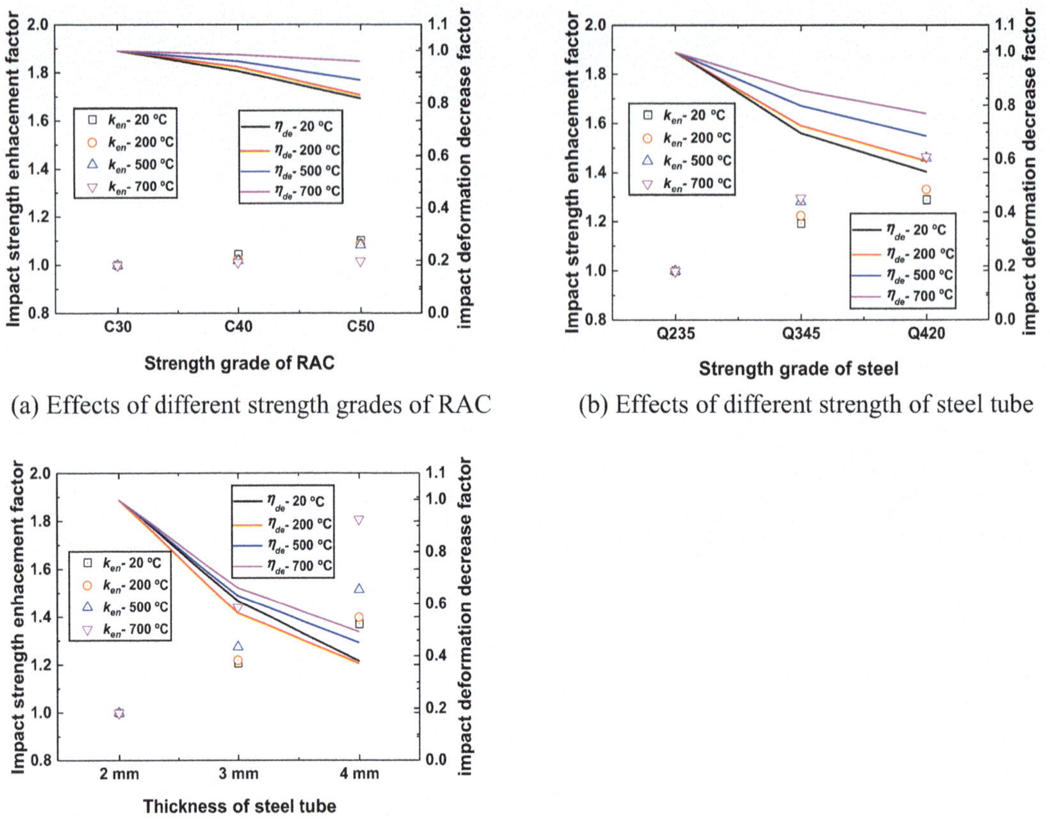


(a) Effects of different strength grades of RAC

(b) Effects of different strength of steel tube

(c) Effects of different thicknesses of steel tube

**Fig. 5.13** Effects of design parameters on the strength and deformation of RACFST in simulated impact. (**a**) Effects of different strength grades of RAC. (**b**) Effects of different strength of steel tube. (**c**) Effects of different thicknesses of steel tube. (Reprinted with permission from Li et al., 2017b)

which is in the range of 0.62–0.66 and 0.67–0.73, respectively.

The dynamic increase factor of RACFST experiences a slight increase after high temperature, which is in the range of 1.30–1.33 for room temperature samples and 1.37–1.47 for 700 °C high temperature damaged samples, respectively. It could be ascribed by the increased impact strain rate and the consequent dynamic strength compensation for high temperature softened samples.

The parametric analysis indicates that the steel related parameters are more efficient than the strength of RAC to help improve the impact strength and deformation of RACFST. Using higher strength steel and a higher steel ratio can even bring more benefits after high temperature exposure due to the less strength degradation of steel than concrete and then the increased role of steel in impact resistance. Compared with the impact strength enhancement factor of 1.29 (Q420) and 1.37 (4 mm) for room temperature samples, the enhancement factor increased to 1.47 (Q420) and 1.81 (4 mm) for samples ever subjected to 700 °C.

The aforementioned findings suggest that the utilization of recycled aggregate in the form of RACFST does not raise significant concerns under extreme conditions of high axial loading rate and high temperature damage. Although the excellent performance of RACFST is demonstrated within an expanded research scope, the current understanding on the performance of RACFST is still not comprehensive. More studies are necessary to promote the practical application of RACFST to efficiently recycle C&DW and develop infrastructure in a sustainable manner.

**Acknowledgements** The authors appreciate the supports from Australian Research Council (FT220100177, IH200100010; LP230100288, DP220101051; DP220100036) and University of Technology Sydney (UTS) Research Academic Program at Tech Lab (UTS RAPT).

## References

Birtel, V. (2006). Mark P parameterised finite element modelling of RC beam shear failure. In *ABAQUS users' conference*.

Deng, Y., & Tuan, C. Y. (2013). Design of concrete-filled circular steel tubes under lateral impact. *ACI Structural Journal, 110*(4), 691.

Fernandes, B., Carré, H., Mindeguia, J.-C., Perlot, C., & La Borderie, C. (2021). Effect of elevated temperatures on concrete made with recycled concrete aggregates – An overview. *Journal of Building Engineering, 44*, 103235.

Guan, Z., Al-Husainy, A., Wang, Q., Jones, S., Su, C., & Liu, L. (2020). Numerical modeling of recycled and normal aggregate CFRP-strengthened concrete-filled steel columns subjected to lateral impact. *Journal of Composites for Construction, 24*(5), 04020048.

Han, L.-H., Yang, Y.-F., Yang, H., & Huo, J.-s. (2002). Residual strength of concrete-filled RHS columns after exposure to the ISO-834 standard fire. *Thin-Walled Structures, 40*(12), 991–1012.

Han, L.-H., Yao, G.-H., & Tao, Z. (2007). Performance of concrete-filled thin-walled steel tubes under pure torsion. *Thin-Walled Structures, 45*(1), 24–36.

Huang, Y., Yip, T. L., & Liang, C. (2020). Risk perception and property value: Evidence from Tianjin port explosion. *Sustainability, 12*(3), 1169.

Huo, R. (2011). *Study on compressive behavior of recycled aggregate concrete-filled steel tubes after exposed to high temperatures*. Dalian University of Technology.

Huo, J., He, Y., Xiao, L., & Chen, B. (2013). Experimental study on dynamic behaviours of concrete after exposure to high temperatures up to 700 C. *Materials and Structures, 46*(1), 255–265.

Ibrahim, M., Alimi, W., Assaggaf, R., Salami, B. A., & Oladapo, E. A. (2023). An overview of factors influencing the properties of concrete incorporating construction and demolition wastes. *Construction and Building Materials, 367*, 130307.

Kiakojouri, F., Sheidaii, M. R., De Biagi, V., & Chiaia, B. (2021). Blast-induced progressive collapse of steel moment-resisting frames: Numerical studies and a framework for updating the alternate load path method. *Engineering Structures, 242*, 112541.

Kolsky, H. (1949). An investigation of the mechanical properties of materials at very high rates of loading. *Proceedings of the Physical Society, 62*(11), 676–700.

Kumar, V., Sharma, U. K., Singh, B., & Bhargava, P. (2013). Effect of temperature on mechanical properties of pre-damaged steel reinforcing bars. *Construction and Building Materials, 46*, 19–27.

Li, W., Luo, Z., Long, C., Wu, C., Duan, W. H., & Shah, S. P. (2016). Effects of nanoparticle on the dynamic behaviors of recycled aggregate concrete under impact loading. *Materials & Design, 112*, 58–66.

Li, W., Luo, Z., Tao, Z., Duan, W. H., & Shah, S. P. (2017a). Mechanical behavior of recycled aggregate concrete-filled steel tube stub columns after exposure to elevated temperatures. *Construction and Building Materials, 146*, 571–581.

Li, W., Luo, Z., Wu, C., Tam, V. W., Duan, W. H., & Shah, S. P. (2017b). Experimental and numerical studies on impact behaviors of recycled aggregate concrete-filled steel tube after exposure to elevated temperature. *Materials & Design, 136*, 103–118.

Li, W., Luo, Z., Wu, C., & Duan, W. H. (2018). Impact performances of steel tube-confined recycled aggregate concrete (STCRAC) after exposure to elevated temperatures. *Cement and Concrete Composites, 86*, 87–97.

Lindholm, U. (1964). Some experiments with the split hopkinson pressure bar∗. *Journal of the Mechanics and Physics of Solids, 12*(5), 317–335.

Liu, W., Cao, W., Zhang, J., Wang, R., & Ren, L. (2017). Mechanical behavior of recycled aggregate concrete-filled steel tubular columns before and after fire. *Materials, 10*(3), 274.

Lu, J., Liu, H., Chen, Z., & Liao, X. (2016). Experimental investigation into the post-fire mechanical properties of hot-rolled and cold-formed steels. *Journal of Constructional Steel Research, 121*, 291–310.

Luo, Z., Li, W., Tam, V. W., Xiao, J., & Shah, S. P. (2019). Current progress on nanotechnology application in recycled aggregate concrete. *Journal of Sustainable Cement-Based Materials, 8*(2), 79–96.

Molla, A. S., Tang, P., Sher, W., & Bekele, D. N. (2021). Chemicals of concern in construction and demolition waste fine residues: A systematic literature review. *Journal of Environmental Management, 299*, 113654.

News, A. (2018). Cargo ship crashes into iconic 18th century mansion in Istanbul.

Shakir, A., Guan, Z., & Jones, S. (2016). Lateral impact response of the concrete filled steel tube columns with and without CFRP strengthening. *Engineering Structures, 116*, 148–162.

Tang, Y.-C., Li, L.-J., Feng, W.-X., Liu, F., & Liao, B. (2017). Seismic performance of recycled aggregate concrete–filled steel tube columns. *Journal of Constructional Steel Research, 133*, 112–124.

Tang, Z., Hu, Y., Tam, V. W., & Li, W. (2019). Uniaxial compressive behaviors of fly ash/slag-based geopolymeric concrete with recycled aggregates. *Cement and Concrete Composites, 104*, 103375.

Tang, Z., Li, W., Tam, V. W., & Xue, C. (2020). Advanced progress in recycling municipal and construction solid wastes for manufacturing sustainable construction materials. *Resources, Conservation and Recycling, 6*, 100036.

Vieira, J., Correia, J., & De Brito, J. (2011). Post-fire residual mechanical properties of concrete made with recycled concrete coarse aggregates. *Cement and Concrete Research, 41*(5), 533–541.

Wu, H., Zuo, J., Zillante, G., Wang, J., & Duan, H. (2021a). Environmental impacts of cross-regional mobility of construction and demolition waste: An Australia study. *Resources, Conservation and Recycling, 174*, 105805.

Wu, Z., Ke, X., & Su, Y. (2021b). Experimental performance of recycled aggregate concrete-filled square steel tubular (RACSST) stub columns after exposure to high temperature and water spraying cooling. *International Journal of Steel Structures, 21*, 787–799.

Xiao, Y., Shan, J., Zheng, Q., Chen, B., & Shen, Y. (2009). Experimental studies on concrete filled steel tubes under high strain rate loading. *Journal of Materials in Civil Engineering, 21*(10), 569–577.

Yang, Y., & Han, L. (2006). Compressive and flexural behaviour of recycled aggregate concrete filled steel tubes (RACFST) under short-term loadings. *Steel & Composite Structures, 6*(3), 257–284.

Yang, Y., & Hou, R. (2012). Experimental behaviour of RACFST stub columns after exposed to high temperatures. *Thin-Walled Structures, 59*, 1–10.

Yang, Y.-F., Zhang, Z.-C., & Fu, F. (2015). Experimental and numerical study on square RACFST members under lateral impact loading. *Journal of Constructional Steel Research, 111*, 43–56.

Yang, Y.-F., Zhang, L., & Dai, X. (2017). Performance of recycled aggregate concrete–filled square steel tubular columns exposed to fire. *Advances in Structural Engineering, 20*(9), 1340–1356.

# Developing an Urban Resource Cadaster for Circular Economy

# 6

Srinivasa Raghavendra Bhuvan Gummidi, Ruichang Mao, Maud Lanau, and Gang Liu

## 6.1 Introduction: Urban Resource Cadaster

The linear production-consumption-disposal economy model is not sustainable, given the compounding energy and resource consumption at the production-use stages and waste accumulation at the end-of-use stage. As an alternative, Walter Stahel, in 1982, suggested a "self-replenishing system" that would loop the end-of-use products/resources back into the use phase (e.g., reuse and repair) or production stage (e.g., recondition and recycle) to extend their lifetime of materials and products (Stahel, 1982). Such a loop-back or circular economy system would negate the environmental impacts caused by the linear production-consumption–disposal economy. The circular economy paradigm has recently become a significant focus of sustainability research as the planet's biophysical boundaries are increasingly transgressed and climate urgency is rising (Steffen et al., 2015).

The potential of applying circular economy principles in the context of the urban built environment is high, given the historical accumulation of resources due to urbanization. For instance, significant resources are stocked in cities in the form of built environment stocks (buildings and infrastructures) and mobile stocks (e.g., electronics and vehicles). The accumulated urban resources can be mined for secondary materials (Jacobs, 2004) that can be looped back into the economy. Especially for developed cities, the circular economy paradigm offers relief by offsetting the rising demand for construction materials. Despite such vast potential, urban mining of the built environment, which represents the prominent use of construction materials and primary waste generation (Agamuthu, 2008), remains less explored.

The barriers associated with urban mining are one of the main reasons behind the lack of proper exploration of the urban mining concept. Some barriers plaguing urban mining are the long lifetimes of urban buildings and infrastructure and the use of bulky and heavy materials. Consequently, the in-situ demolition activities necessitate transportation and storage of recovered

S. R. B. Gummidi
SDU Life Cycle Engineering, Department of Green Technology, University of Southern Denmark, Odense, Denmark

R. Mao
School of Civil Engineering, Tsinghua University, Beijing, China

M. Lanau
Department of Architecture and Civil Engineering, Chalmers University of Technology, Gothenburg, Sweden

G. Liu (✉)
SDU Life Cycle Engineering, Department of Green Technology, University of Southern Denmark, Odense, Denmark

College of Urban and Environmental Sciences, Peking University, Beijing, China
e-mail: gli@igt.sdu.dk; gangliu@pku.edu.cn



J. Zuo et al. (eds.), *Circular Economy for Buildings and Infrastructure*, Sustainable Development Goals Series, https://doi.org/10.1007/978-3-031-56241-9_6

materials, thus requiring planning in advance. However, older buildings (likely to be demolished in the near future) often lack information about their material composition and are not designed for easy disassembly and material recovery (Morlet et al., 2015). Thus, compiling a detailed urban resource cadaster containing information about the material composition, quantity, quality, and spatial distribution of built environment stocks (Kleemann et al., 2017; Oezdemir et al., 2017) is critical to facilitate urban mining and promote the circular economy in the construction industry.

The compiled urban resource cadaster acts as a database for policymakers and administrators for better strategizing sustainable urban planning and circular economy policies. The cadaster can also be used by the construction industry to better plan construction materials demand-supply scenarios and component recovery. Since built environment stocks (i.e., buildings and infrastructure) play multiple roles in our socioeconomic metabolism, compiling the urban resource cadaster can help characterize the extensive reservoir of secondary materials. In addition to understanding the urban mining potential, the urban resource cadaster provides an overview of the essential services provided to citizens. Moreover, the urban resource cadaster helps us understand the drivers of the economy's material cycles and the built environment stocks'associated temporal and spatial lock-ins on energy use and emissions. Although research on built environment stocks is gaining traction, the current literature requires more spatially refined case studies, particularly of urban areas, to analyze further the patterns, drivers, and implications of built environment stocks.

In this book chapter, we describe the process of developing a spatially refined urban resource cadaster by integrating the geo-localized, bottom-up material stock analysis with primary data on building material intensity coefficients. Additionally, we will list the important factors to consider for developing a resource cadaster. Furthermore, we demonstrate the development of a spatially refined urban resource cadaster for three case studies: urban built environment stock cadaster for Odense, Denmark (Lanau & Liu, 2020) and Beijing, China (Mao et al., 2020), and global subway stock cadaster (Mao et al., 2021). Finally, we discuss the potential opportunities and limitations offered by developing a spatially refined urban resource cadaster.

## 6.2 Characterizing Built Environment Stock for Circular Economy

Many interdisciplinary approaches at different spatiotemporal scales have been developed to characterize materials accumulation in built environment stock and explore various policy and resource implications on circular economy in the building and construction sector (See Fig. 6.1). In terms of the granularity of the material-related information they generated, these approaches could be classified into top-down, bottom-up, or both combined.[1]

There are plenty of studies on national or regional level building stocks, which used a top-down approach. It is intuitive to start from nation- or region-based building stock modelling, because relevant data (especially data from housing sector) are well documented in national or regional statistics. Two pioneer case studies (Müller, 2006; Kohler & Yang, 2011) using a stock dynamics approach inaugurated the line of research on building stock's material stocks and flows. The key hypothesis behind the building stock modelling is flows (i.e., resource consumption and waste generation) are driven by stocks which provide essential services (e.g., shelter and working place) to society. This helps connect the biophysical flows between natural environment and socioeconomic systems, which provides an unique understanding of long-term material metabolism of the building stock, such as simultaneously forecasting resource demand and waste generation under different socioeconomic development patterns. The principle of the building stock modelling complies with the principle of mass balance (i.e., within a system, inputs minus outputs should equal to net addition to in-use

[1] This classification is not decisive. The definition of top-down and bottom-up varies in different contexts.

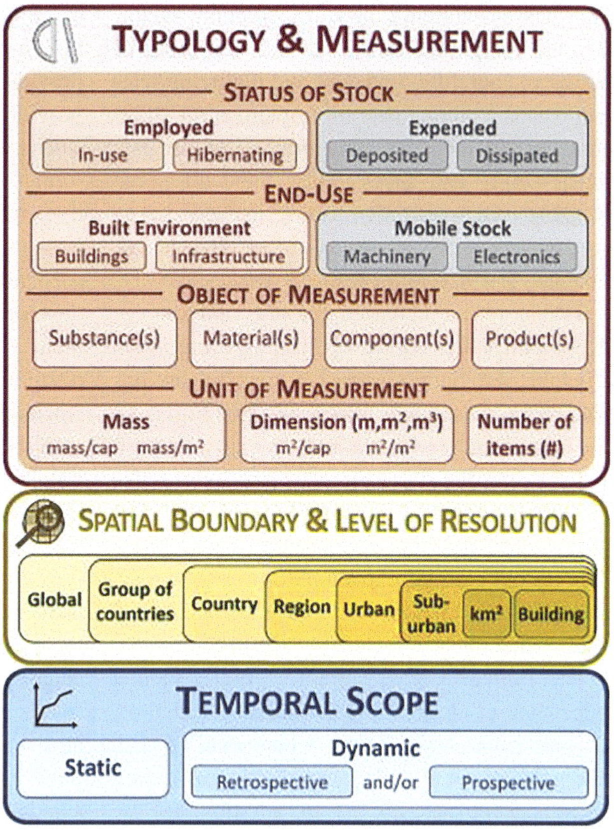


**Fig. 6.1** The three dimensions of the scope of study in any material stock analysis. (Lanau et al., 2019)

stocks), which has been long omitted in economic models.

The last decade witnessed a variety of endeavors using the stock dynamics approach to modelling building stock's metabolism and its implications on various circular economy related topics, such as environmental impacts resulting from material metabolism in building stock (Bergsdal et al., 2007; Shi et al., 2012; Wang et al., 2015), primary steel demand and secondary steel supply, and construction and demolition waste generation. Meanwhile, several important aspects of the building stock have been added to the original modeling framework. Using Norway's dwelling stock as a case, renovation activities of buildings was introduced into the building stock modelling; the heterogeneous nature of different types of building stocks has also been characterized in some case studies (Sandberg et al., 2014, 2017).

Apart from the efforts on extending its conceptual framework, several assumptions (e.g., lifetime and material intensity) used in building stock modelling started to be improved. To make sound conclusions, the lifetime function of a building stock modelling should be chosen with caution, which has been well explored in several case studies (Cao et al., 2019; Miatto et al., 2017). Another key parameter of the building stock modelling is material intensity, which is essential for characterizing the material profile of buildings. The pattern of building functions and their material compositions is time dependent because the ever-changing architectural technologies and lifestyles across the world have led to gradual changes to the functionality and material preference of individual buildings and thus the entire building stock. The complexity and change in material compositions are of great importance for a spectrum of circular economy strategies in building and construction sector, such as, component reuse, material recycling or recovery. Therefore, updating and digitizing relevant data is a prerequisite for implementing those strategies. In order to form joint efforts in data transferability and accessibility, there are some on-going initiatives on collecting material intensity data from various sources (Heeren & Fishman, 2019; Schiller et al., 2019).

Thanks to the increased amounts and sources of geospatial data regarding buildings and infrastructure, the building stock's metabolism has been characterized at a finer scale (See Fig. 6.1). The understanding of building stocks with high spatial resolution started from several pilot cities (i.e., Wakayama, Japan, and Manchester, UK) where data on building stocks are better documented and more available. More examples are scattered across several countries and cities, e.g., Switzerland (Heeren & Hellweg, 2019), Japan (Tanikawa & Hashimoto, 2009), Vienna (Kleemann et al., 2017), Melbourne (Stephan & Athanassiadis, 2017), Taipei (Cheng et al., 2018), Padua (Miatto et al., 2019), and Esch-sur-Alzette (Mastrucci et al., 2017). In addition to the quantity of materials stocked in individual buildings, the high spatial resolution modelling provides additional insights into its location. It provides spatially refined estimates of material stocks, which could be used to identify future 'urban mining' sites. Combined with the stock dynamics approach, it produces information on how much, when, and where the waste streams are, and thus helps establish a sustainable urban development plan and an optimal waste management scheme (Han et al., 2018). The high-resolution characterization of built environment stocks is also a critical step towards a better understanding of the connections between material accumulation and urban fabric (Kohler et al., 2009), which is an underexplored area compared to urban fabric's effects on other urban issues (e.g., urban heat island, urban traffic, and land use). Figure 6.2 illustrates the general workflow to create a spatially refined urban resource cadaster for characterizing urban environment stocks using a bottom-up GIS-based approach and material intensity coefficient data for each building archetype. An archetype describes a standard or representative model that represents

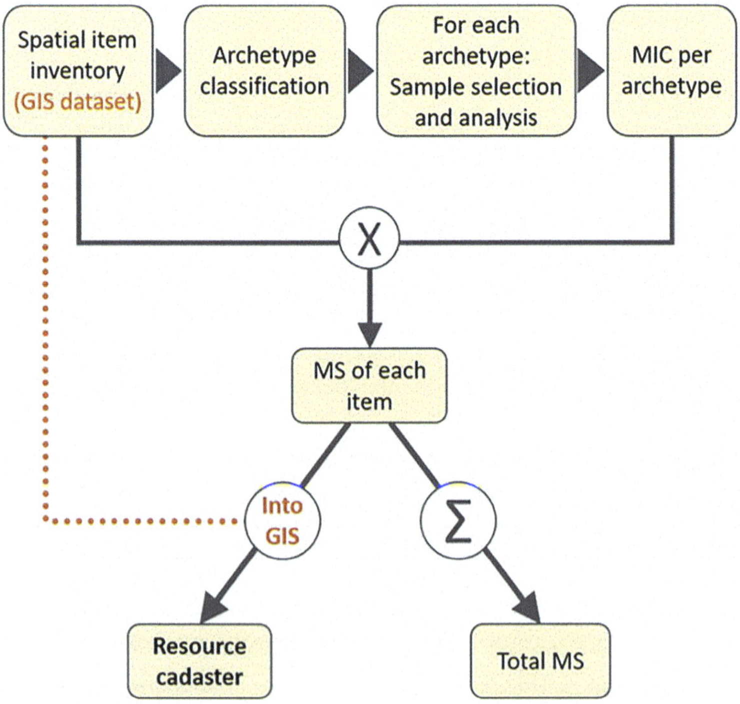


**Fig. 6.2** General methodological framework for modeling an urban resource cadaster through a bottom-up approach. (Lanau & Liu, 2020)

a specific kind of built asset in a cadaster of material resources. For instance, the residential building stock of the Odense case study (Lanau & Liu, 2020) was classified with the age-cohort archetype classification developed by the Danish Building Research Institute (Wittchen & Kragh, 2012).

The urban resource cadaster material stocks are calculated through the following equation:

$$MS_m = \sum_i MS_{m,i} = \sum INV_i \times MIC_{m,i}$$

where $MS_m$ is the total mass of the stock of material 'm' in all items of type 'i'; $INV_i$ is the inventory of items of type 'i', quantified in a dimensional unit; and $MIC_{m,i}$ is the material intensity coefficient of material 'm' (expressed in mass per dimensional unit) in items of type 'i'. The equation is modified according to the dimension of material stock being modelled. For example, inventory data are typically expressed in length for pipe networks and square meters for buildings, so MICs for pipes should be developed in kg/m and in kg/m$^2$ for buildings.

## 6.3 Case Studies on Urban Resource Cadaster

In this section, we demonstrate the development of a spatially refined urban resource cadaster for three case studies: urban built environment stock cadaster for Odense, Denmark (Lanau & Liu, 2020) and Beijing, China (Mao et al., 2020), and global subway stock cadaster (Mao et al., 2021). The case studies involve the development of an urban resource cadaster for a city in a developed country (Odense, Denmark) and a city in a developing country (Beijing, China) and the global historical subway material stock analysis.

### 6.3.1 Odense Case Study

The Odense case study (Lanau & Liu, 2020) analyzes the built environment stock of Odense municipality as of 2018. The case study develops an urban resource cadaster for evaluating the circular economy potential of the third most populous city in Denmark. We modelled both the in-use and hibernating stocks (i.e., no longer in use but not yet disposed of). Furthermore, the stock was modelled according to functionality, i.e., residential buildings (RBs), nonresidential buildings (NRBs), roads (including pedestrian and bike lanes), and pipe networks for water supply, wastewater, and natural gas. Data on other infrastructure could not be retrieved and are thus left outside of the scope of this case study. We have classified the stock further based on the construction period and surveyed sample buildings in each archetype to collect the material composition portfolio. The material composition portfolios for the sampled buildings consist of a total of 46 construction materials, which were aggregated into nine material categories: aggregates (sand, gravel, and stones), concrete (e.g., aerated concrete and asphalt-concrete), mortar and gypsum, clay (e.g., bricks, tiles, and pebbles), timber, metal, plastics, glass and ceramics, and others (e.g., slag and carpet).

The total material stock of Odense municipality is approximately 66.7 megatons (Mt) and mostly comsists of buildings and roads. Considering the spatial coverage and population of Odense municipality, the total material stock equals 329 tons/capita and 219 kilotons/km$^2$. Based on the building use, the residential building stock accounts for 7.47 Mt against 10.15 Mt for non-residential building stock. Figure 6.3 visualizes the developed urban resource cadaster of the secondary materials stocked in the built environment in a high spatial resolution. It can be observed that the city centre has accumulated a significant quantity of resources (ten-fold higher in the city centre than in the outskirts), thus, offering more opportunities for urban mining. We have also observed that the spatial distribution patterns differ according to the material under focus. For example, clay stock is remarkably high in the city centre, as masonry was the preferred construction type in older buildings. On the other hand, concrete stock spreads over the municipality as it is essential for building foundations and roads.

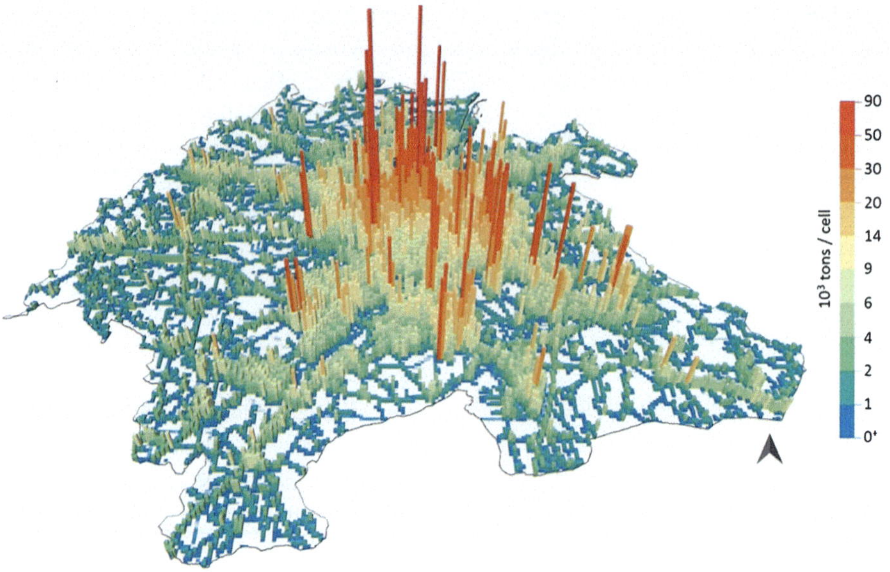


**Fig. 6.3** Spatial distribution of construction materials stocked in the built environment of Odense municipality, as of 2018. (Lanau & Liu, 2020)

### 6.3.2 Beijing Case Study

The Beijing case study (Mao et al., 2020) aims to characterize the quantity, composition, and age of Beijing's urban built environment stocks in a high spatial resolution based on new data science methods and improved material intensity coefficients data from different heterogeneous sources. Specifically, the case study discusses the accumulated stock implications for urban planning and resource, waste, and environmental strategies under a rapid urbanization process. Additionally, this study addresses the knowledge gaps in current built environment stock modelling literature, such as including non-residential buildings, considering the interactions between buildings and infrastructure, and compiling location-context-specific material intensity coefficient data. As a relatively mature city in the world's largest developing country of China, the insights from this case study of Beijing can be helpful both for comparison with cities in industrialized countries and inform urban sustainability transition in other developing cities.

Beijing's material stock comprises of buildings (62.7%), roads (31.1%), railway (3.5%), and metro (2.6%), totaling 3621 Mt (140 ton/cap) of construction materials in 2018. Based on the functionality in building material stocks, residential buildings account for 1188 Mt (52.3%), followed by commercial buildings at 385 Mt (17.0%), public buildings at 208 Mt (9.1%), and educational buildings at 182 Mt (8.0%). In terms of the building age cohort, buildings constructed or reconstructed during the 1990s (826 Mt, 36.4%) and 2000s (1050 Mt, 46.2%) account for nearly 80% of building material stocks in 2018 of Beijing, while buildings built before 1990 account for circa 10%. The potential reasons for this observation are: (a) material intensity coefficients of recently built buildings are higher; (b) the market economy reform and comparably lower quality control increased housing demand, which in turn resulted in shorter lifetime of buildings constructed rapidly in the 1980s (Cao et al., 2019; Fernández, 2007); and (c) Beijing experienced a construction boom before the 2008 Summer Olympics (which was accompanied by a rapid population increase from 13.6 million in 2000 to 21.5 million in 2018 (Statistics, 2008)).

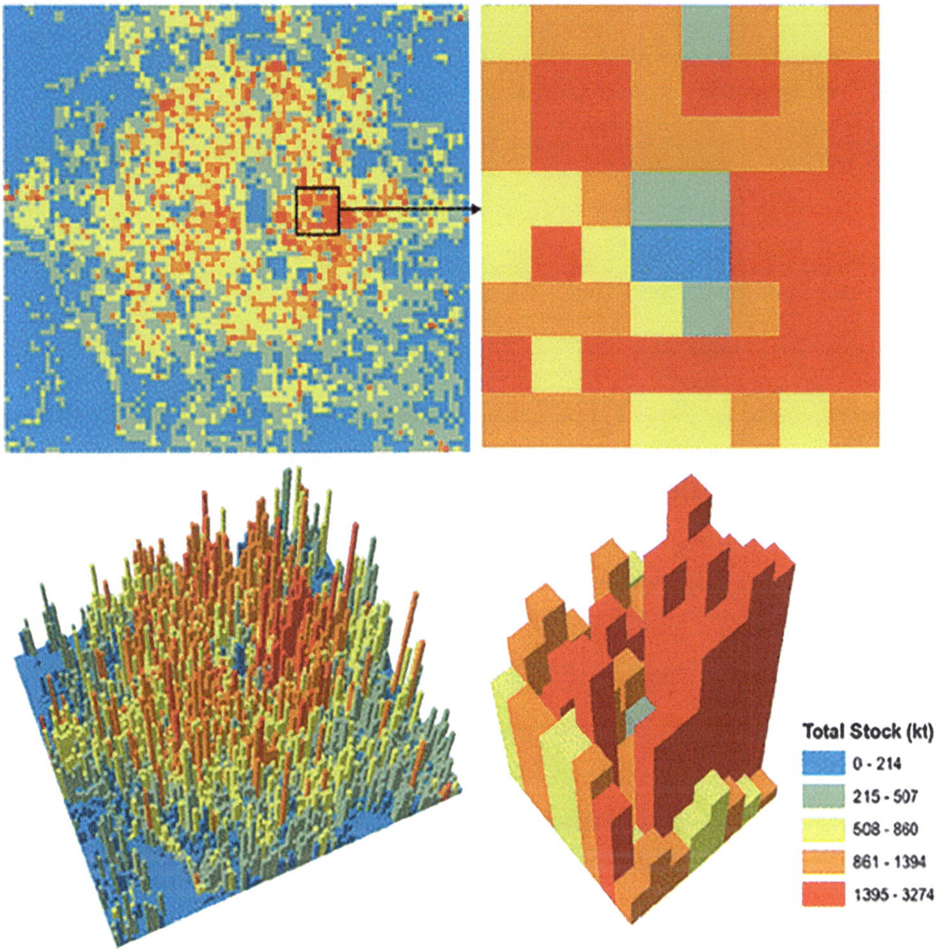


**Fig. 6.4** Built environment stocks of Beijing (left column) and selected area (right column) in 500 m × 500 m grids, both as heat maps (top row) and elevation models (bottom row). (Mao et al., 2020)

The built environment stocks in 500 m × 500 m grids visualized as both grid and elevation maps in Fig. 6.4 indicate that the less dense grids are in both central and outer parts of Beijing. These central grids or clusters with low values represent historical or protected blocks. Urban function wise, commercial grids have the highest material stocks density (7.96 t/m$^2$), followed by residential (4.94 t/m$^2$), public (3.39 t/m$^2$), and educational (2.6 t/m$^2$) grids. In contrast, the building material stock densities of industrial (1.24 t/m$^2$), agricultural (1.19 t/m$^2$), sport (0.9 t/m$^2$), storage (0.85 t/m$^2$), and parking (0.57 t/m$^2$) grids are relatively lower.

### 6.3.3 Global Subway Case Study

In the global subway case study (Mao et al., 2021), we analyzed the historical development patterns of the subway system and quantified the composition, growth, distribution, and associated environmental impacts of global subway material stocks across 219 global cities. Owing to the data availability and the large-scale development of subway systems in China, we have further analyzed the spatiotemporal dynamics of the Chinese subway material stocks development among 37 cities. The collected data is analyzed in the context of sustainable policy implications for

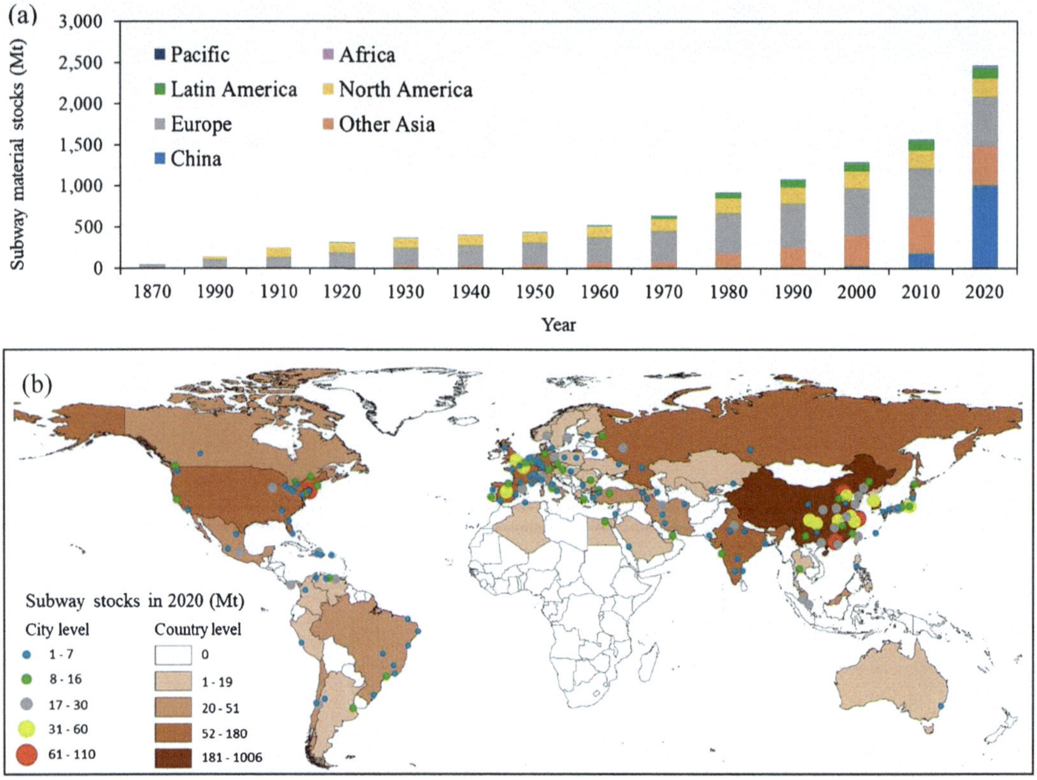


**Fig. 6.5** (**a**) The temporal growth patterns by world region, 1870–2020; and (**b**) the spatial distribution at city and country levels by July 2020. (Mao et al., 2021)

future subway development. The case study addresses knowledge gaps in the existing subway material stock modelling literature as it includes all the large parts of the subway network (i.e., tunnel, station, ancillary road, entrance and exit, and ventilation system), analyzing the spatial development and material stock distribution patterns, and benchmarking subway material stocks results across cities to reveal spatiotemporal variations and growth pathways.

In 2020, the global subway length reached 16,419 km, and the construction material stocks amounted to 2.5 gigatons, equaling to an embodied emission of 560 megatons. China alone constitutes 36.6% (6005 km) of global length, followed by 23.5% in Europe (3866 km), 22.9% in the rest of Asia (3756 km), 10% in North America (1644 km), and 6.2% in Latin America (1021 km). Since the early twentieth century, subway stock witnessed an annual growth of 8 Mt subway stock from the 1900s to 1920s (See Fig. 6.5). However, World War II dampened the stock accumulation to 3 Mt per year from the 1930s to 1940s. Nevertheless, since the 1970s, subway material stocks regained the momentum and increased dramatically. This stock expansion might be attributed to three major reasons: the cheap and efficient nature of commuting with subway systems, construction technological innovations, and rapid urbanization of developing countries. Regarding the stock composition of construction materials and subway components, subway material stocks are dominated by gravel (51%) and stations (49%). In terms of subway stocks at city level, the stock volume ranges from 3.95 Mt to 110.27 Mt, where the most complex urban subway network is found in Shanghai, China.

### 6.3.4 Comparison Between the Case Studies

All three case studies use geographic information systems (GIS) to map and assess the built environment, including the worldwide subway stock cadaster and urban built environment stock cadasters for Beijing, China, and Odense, Denmark. However, these studies differ in spatial data focus, scope and scale, resource type, and data collection methods.

Scope and scale: The Odense case study focuses on a mid-size Danish city for which the material stock of buildings, roads, and underground pipes is modelled. The Beijing case study also focuses on a single city but the spatial scale is much grander as Beijing is a megacity with a significantly larger built environment. The subway case study includes subway systems in cities all over the word; it has a narrower focus in terms of stocks, but a much broader spatial scale. as it covers:

Spatial data focus: The case studies of Odense and Beijing focus on documenting the physical features of buildings and infrastructures, such as age, size, type of construction, and other pertinent attributes related to urban planning and management. In contrast, the global subway case study focuses on describing the physical features and operational specifics of subway systems, including network connectivity, route lengths, station placements, and other pertinent information regarding urban mobility and transportation planning.

Resource Type: The Odense and Beijing case studies are built environment stock cadasters primarily focusing on buildings and their characteristics. In contrast, the global subway case study focuses on subway systems, a specific type of urban transportation infrastructure.

Data sources and collection methodology: Each study used different methodologies and data sources. The Odense and Beijing studies mapped and analyzed the built environment using bottom-up GIS data. For Odense, GIS data was collected from the municipality and infrastructure companies. For Beijing, machine learning methods were used. In contrast, the subway case study identified and mapped subway systems using publicly accessible spatial and non-spatial data sources.

In summary, the first two case studies focus on documenting the physical characteristics of several built environment stocks (i.e., buildings, roads, in two specific cities. On contrary, the third case study focuses on creating a global database of subway systems. All three case studies highlight the importance of spatial analysis and GIS in understanding the built environment and supporting urban planning and management activities. Planners and policymakers can better comprehend the current urban environment and plan for future growth using the urban built environment stock cadasters. On the other hand, the global subway stock cadaster can help with worldwide planning and investment in transportation infrastructure.

## 6.4 Development of the Resource Cadaster

Focus is placed on improving resource management to decrease waste and advance resource efficiency as part of the creation of a material resource cadaster for circular economy applications. Several important factors that can help with the development of a resource cadaster (Li et al., 2022) for enabling a circular economy are listed below:

1. Identification of resource streams. The resource cadaster should identify and track the construction and demolition waste resource's generation, transportation, and disposal.
2. Mapping of resource availability. The resource cadaster should map the availability of resources within urban areas to facilitate different circular economy applications. The mapping can be achieved through archetype-

based or MIC-based bottom-up modelling methods. The mapping can provide insights into resource concentrations across geographical space and can help identify potential resource recovery opportunities.
3. Integration with existing systems. Integrating the resource cadaster with the urban area's current recycling and waste management systems would support implementing circular economy applications. Such integration would require collaboration with various stakeholders in the recycling, waste management, and resource management sectors.
4. Accessibility and usability. The resource cadaster should be accessible to stakeholders, such as locals, companies, and government organizations. A simple and user-friendly user interface should be set up to allow users to obtain information on resources' availability and recycling opportunities.
5. Data-sharing agreements. Setting up data-sharing agreements with stakeholders can help overcome data availability issues for resource cadaster development. These agreements will also ensure that the resource cadaster provides accurate and up-to-date information.
6. Integration with sustainability goals. The resource cadaster should be developed to align with the urban area's sustainability goals, like reducing greenhouse gas emissions, promoting resource efficiency, and minimizing waste.

Ideally, a resource cadaster should be developed considering all six factors. However, owing to different constraints, many of existing resource cadasters lack factors such as an interactive interface for easy accessibility and usability (Lanau & Liu, 2020). A thorough and cooperative strategy is needed to create a resource cadaster for urban sustainability. A resource cadaster can aid in promoting circular economy practices and supporting sustainable urban resource management by mapping resource availability, connecting with current systems, and encouraging data sharing and collaboration.

## 6.5 Opportunities and Challenges of Circular Economy in the Construction Sector

A better understanding of built environment stocks through the compilation of an urban resource cadaster enables us to tackle societal/environmental, and policy problems related to urban sustainability, circular economy, and climate change. Based on the existing literature (Lanau et al., 2019) and our work related to the development of resource cadaster (Lanau & Liu, 2020; Mao et al., 2020, 2021), we summarize the potential contributions of spatially refined built environment stocks to informing a sustainable socio-economic transition.

- First, a spatially refined urban resource cadaster provides a clear overview of the amount, location, and time availability of these material stocks, which are important for facilitating urban mining, smart demolition, and waste management strategies for a variety of stakeholders (e.g., waste and recycling companies, architects, developers, and planners), such as spatially optimized waste management logistics. This information is valuable for upcycling companies that develop their business plan following the amount of material potentially recoverable from demolition. For example, companies specializing in upcycling/reusing old clay bricks (bonded with dissolvable lime mortar) would benefit from information on the amount of clay brick stocked in older archetypes.
- Second, understanding the spatial and contextualized socioeconomic drivers responsible for the historical stock dynamics would help understand the relationship between the embodied emissions and the emissions caused by the operational use of the stock. Such an understanding helps us better predict future resource and energy use (Haberl et al., 2017; Weisz et al., 2015).
- Third, an urban resource cadaster will help us understand the interactions between the different stock entities like service buildings,

transportation facilities, etc. Urban planners require such an overview of interactions between entities to better plan for energy-and resource-efficient cities (Lin et al., 2017).

- Fourth, the resource cadaster can support urban planning decision-makers by addressing spatial considerations. For example, temporary storage for end-of-use stock alleviates the concerns regarding the supply-demand balance. However, the location of such storage facilities should serve smaller areas due to issues related to transporting bulky and heavy construction materials.
- Finally, understanding the past growth of industrialized countries helps to understand the potential pathways for the future development of developing cities/countries.

Despite the advances in the characterization of building stock, several issues still need to be addressed. Information on building stock and its material compositions is still scarce, fragmented, and aggregated, which hampers the transition towards a circular economy in the construction and building sector. Data quality and unavailability hinder the development of urban resource cadasters. Also, inconsistent terminology and non-standardized stock models obstruct the comparison and integration of results between different studies and obstruct the observation of drivers and indicators. Similarly, the common practice of demolition waste recycling worldwide still stays at the material recovery level. For instance, recycled demolition waste is usually used as parts of backfills or road bases. To achieve a higher level of circular economy in the construction and building sector, more refined data on the configuration of building components and materials embedded in these components are urgently needed.

On the other hand, before construction, circular economy thinking (e.g., design for disassembly and reuse of materials or components in new buildings) in the design phase is generally lacking in the building and construction sector. The implementation of circular business models faces numerous barriers at all socio-technical levels (Guldmann & Huulgaard, 2020), such as unclear market demand for a secondary resource (institutional level), concerns about the consistency of flows of returned goods (value chain level), and lack of knowledge and little evidence of financial benefit (organizational level). Costs, technological feasibility, and policy or legal framework conditions for implementing these strategies need to be further evaluated. Additional research on the quality of materials is needed to ensure their safe reuse after demolition, as those might be contaminated (e.g., polychlorinated biphenyls in plastics and asbestos in fiber cement).

## 6.6 Conclusion

A spatially refined urban resource cadaster identifying the location, type, quantity, and quality of materials in the urban built environment stocks has the potential to disrupt the linear production-consumption-disposal economy by informing stakeholders about the circular economy potentials (e.g., reuse, recycling, and refurbishment) when the stocks reach end of life. Given the valuable nature of the spatially refined urban resource cadaster, we provided a detailed description of the steps to characterize the built environment material stocks harnessing the available heterogeneous data sources related to both buildings and infrastructure along with the compilation of material intensity coefficients data by sampling buildings from each defined archetype. Additionally, we have analyzed the different factors that are critical for the development of an ideal urban resource cadaster. We have demonstrated the development of a spatially refined urban resource cadaster using three case studies. We further discussed the different opportunities and challenges highlighting the role of stocks to inform the decision-makers of the potential recyclable secondary resources' whereabouts, enable the exploration of the role of urban form in city weight growth and urban sustainability transition, and facilitate the estimation of material requirements of new cities in the global south based on the growth of industrialized cities.

## References


Agamuthu, P. (2008). Challenges in sustainable management of construction and demolition waste. *Waste Management and Research, 26*(6), 491–492. https://doi.org/10.1177/0734242X08100096

Bergsdal, H., Bohne, R. A., & Brattebø, H. (2007). Projection of construction and demolition waste in Norway. *Journal of Industrial Ecology, 11*(3), 27–39. https://doi.org/10.1162/JIEC.2007.1149

Cao, Z., Liu, G., Duan, H., Xi, F., Liu, G., & Yang, W. (2019). Unravelling the mystery of Chinese building lifetime: A calibration and verification based on dynamic material flow analysis. *Applied Energy, 238*, 442–452. https://doi.org/10.1016/J.APENERGY.2019.01.106

Cheng, K. L., Hsu, S. C., Li, W. M., & Ma, H. W. (2018). Quantifying potential anthropogenic resources of buildings through hot spot analysis. *Resources, Conservation and Recycling, 133*, 10–20. https://doi.org/10.1016/J.RESCONREC.2018.02.003

Fernández, J. E. (2007). Resource consumption of new urban construction in China. *Journal of Industrial Ecology, 11*(2), 99–115. https://doi.org/10.1162/JIE.2007.1199

Guldmann, E., & Huulgaard, R. D. (2020). Barriers to circular business model innovation: A multiple-case study. *Journal of Cleaner Production, 243*, 118160. https://doi.org/10.1016/J.JCLEPRO.2019.118160

Haberl, H., Wiedenhofer, D., Erb, K. H., Görg, C., & Krausmann, F. (2017). The material stock–flow–service nexus: A new approach for tackling the decoupling conundrum. *Sustainability, 9*(7), 1049. https://doi.org/10.3390/SU9071049

Han, J., Chen, W. Q., Zhang, L., & Liu, G. (2018). Uncovering the spatiotemporal dynamics of urban infrastructure development: A high spatial resolution material stock and flow analysis. *Environmental Science & Technology, 52*(21), 12122–12132. https://doi.org/10.1021/acs.est.8b03111

Heeren, N., & Fishman, T. (2019). A database seed for a community-driven material intensity research platform. *Scientific Data, 6*, 23. https://doi.org/10.1038/s41597-019-0021-x

Heeren, N., & Hellweg, S. (2019). Tracking construction material over space and time: Prospective and geo-referenced modeling of building stocks and construction material flows. *Journal of Industrial Ecology, 23*(1), 253–267. https://doi.org/10.1111/jiec.12739

Jacobs, J. (2004). *Dark age ahead: Author of the death and life of great American cities*. Vintage Canada.

Kleemann, F., Lederer, J., Rechberger, H., & Fellner, J. (2017). GIS-based analysis of Vienna's material stock in buildings. *Journal of Industrial Ecology, 21*(2), 368–380. https://doi.org/10.1111/JIEC.12446

Kohler, N., & Yang, W. (2011). Long-term management of building stocks. *Building Research and Information, 35*(4), 351–362. https://doi.org/10.1080/09613210701308962

Kohler, N., Steadman, P., & Hassler, U. (2009). Research on the building stock and its applications. *Building Research and Information, 37*(5–6), 449–454. https://doi.org/10.1080/09613210903189384

Lanau, M., & Liu, G. (2020). Developing an urban resource cadaster for circular economy: A case of Odense, Denmark. *Environmental Science and Technology, 54*(7), 4675–4685. https://doi.org/10.1021/acs.est.9b07749

Lanau, M., Liu, G., Kral, U., Wiedenhofer, D., Keijzer, E., Yu, C., & Ehlert, C. (2019). Taking stock of built environment stock studies: Progress and prospects. *Environmental Science and Technology, 53*(15), 8499–8515. https://doi.org/10.1021/ACS.EST.8B06652

Li, Q., Gummidi, S. R. B., Lanau, M., Yu, B., & Liu, G. (2022). Spatiotemporally explicit mapping of built environment stocks reveals two centuries of urban development in a Fairytale City, Odense, Denmark. *Environmental Science and Technology, 56*(22), 16369–16381. https://doi.org/10.1021/ACS.EST.2C04781/SUPPL_FILE/ES2C04781_SI_002.DOCX

Lin, C., Liu, G., & Müller, D. B. (2017). Characterizing the role of built environment stocks in human development and emission growth. *Resources, Conservation and Recycling, 123*, 67–72. https://doi.org/10.1016/j.resconrec.2016.07.004

Mao, R., Bao, Y., Huang, Z., Liu, Q., & Liu, G. (2020). High-resolution mapping of the urban built environment stocks in Beijing. *Environmental Science & Technology, 54*(9), 5345–5355. https://doi.org/10.1021/acs.est.9b07229

Mao, R., Bao, Y., Duan, H., & Liu, G. (2021). Global urban subway development, construction material stocks, and embodied carbon emissions. *Humanities and Social Sciences Communications, 8*(1), 1–11. https://doi.org/10.1057/s41599-021-00757-2

Mastrucci, A., Marvuglia, A., Popovici, E., Leopold, U., & Benetto, E. (2017). Geospatial characterization of building material stocks for the life cycle assessment of end-of-life scenarios at the urban scale. *Resources, Conservation and Recycling, 123*, 54–66. https://doi.org/10.1016/J.RESCONREC.2016.07.003

Miatto, A., Schandl, H., Wiedenhofer, D., Krausmann, F., & Tanikawa, H. (2017). Modeling material flows and stocks of the road network in the United States 1905–2015. *Resources, Conservation and Recycling, 127*, 168–178. https://doi.org/10.1016/j.resconrec.2017.08.024

Miatto, A., Schandl, H., Forlin, L., Ronzani, F., Borin, P., Giordano, A., & Tanikawa, H. (2019). A spatial analysis of material stock accumulation and demolition waste potential of buildings: A case study of Padua. *Resources, Conservation and Recycling, 142*(July 2018), 245–256. https://doi.org/10.1016/j.resconrec.2018.12.011

Morlet, A., Blériot, J., Hubold, S., Opsomer, R., Linder, M., & Banks, I. (2015). *Potential for Denmark as a circular economy a case study from: Delivering the circular economy – A toolkit for policy makers* (pp. 1–134). Ellen MacArthur Foundation. www.ellenmacarthurfoundation.org/books-and-reports

Müller, B. D. (2006). Stock dynamics for forecasting material flows – Case study for housing in The Netherlands. *Ecological Economics, 59*, 142–156. https://doi.org/10.1016/j.ecolecon.2005.09.025

Oezdemir, O., Krause, K., & Hafner, A. (2017). Creating a resource cadaster – A case study of a district in the Rhine-Ruhr metropolitan area. *Buildings, 7*(2). https://doi.org/10.3390/BUILDINGS7020045

Sandberg, N. H., Sartori, I., & Brattebø, H. (2014). Using a dynamic segmented model to examine future renovation activities in the Norwegian dwelling stock. *Energy and Buildings, 82*, 287–295. https://doi.org/10.1016/J.ENBUILD.2014.07.005

Sandberg, N. H., Sartori, I., Vestrum, M. I., & Brattebø, H. (2017). Using a segmented dynamic dwelling stock model for scenario analysis of future energy demand: The dwelling stock of Norway 2016–2050. *Energy and Buildings, 146*, 220–232. https://doi.org/10.1016/J.ENBUILD.2017.04.016

Schiller, G., Miatto, A., Gruhler, K., Ortlepp, R., Deilmann, C., & Tanikawa, H. (2019). Transferability of material composition indicators for residential buildings: A conceptual approach based on a German-Japanese comparison. *Journal of Industrial Ecology, 23*(4), 796–807. https://doi.org/10.1111/JIEC.12817

Shi, F., Huang, T., Tanikawa, H., Han, J., Hashimoto, S., & Moriguchi, Y. (2012). Toward a low carbon–dematerialization society. *Journal of Industrial Ecology, 16*(4), 493–505. https://doi.org/10.1111/J.1530-9290.2012.00523.X

Stahel, W. R. (1982). The product life factor. *An inquiry into the nature of sustainable societies: The role of the private sector (Series: 1982 Mitchell Prize Papers), NARC*, 74–96.

Statistics, B. M. B. (2008). *Beijing statistical yearbook*. China Statistics Press.

Steffen, W., Richardson, K., Rockström, J., Cornell, S. E., Fetzer, I., Bennett, E. M., Biggs, R., Carpenter, S. R., De Vries, W., De Wit, C. A., Folke, C., Gerten, D., Heinke, J., Mace, G. M., Persson, L. M., Ramanathan, V., Reyers, B., & Sörlin, S. (2015). Planetary boundaries: Guiding human development on a changing planet. *Science, 347*(6223). https://doi.org/10.1126/SCIENCE.1259855

Stephan, A., & Athanassiadis, A. (2017). Quantifying and mapping embodied environmental requirements of urban building stocks. *Building and Environment, 114*, 187–202. https://doi.org/10.1016/j.buildenv.2016.11.043

Tanikawa, H., & Hashimoto, S. (2009). Urban stock over time: Spatial material stock analysis using 4d-GIS. *Building Research and Information, 37*(5–6), 483–502. https://doi.org/10.1080/09613210903169394

Wang, T., Tian, X., Hashimoto, S., & Tanikawa, H. (2015). Concrete transformation of buildings in China and implications for the steel cycle. *Resources, Conservation and Recycling, 103*, 205–215. https://doi.org/10.1016/J.RESCONREC.2015.07.021

Weisz, H., Suh, S., & Graedel, T. E. (2015). Industrial ecology: The role of manufactured capital in sustainability. *Proceedings of the National Academy of Sciences of the United States of America, 112*(20), 6260–6264. https://doi.org/10.1073/PNAS.1506532112/ASSET/F2BAD871-1DB7-4B49-95A8-0F5214146899/ASSETS/PNAS.1506532112.FP.PNG

Wittchen, K., & Kragh, J. (2012). Danish building typologies: Participation in the TABULA project. In *SBi 2012:01*. SBI forlag. http://www.forskningsdatabasen.dk/en/catalog/2389388514

# 7 Modeling Sustainable Housing for Sustainable Development in Cities and Communities: The Perspective of a Developing Economy

Michael Atafo Adabre, Albert P. C. Chan, and Ibrahim Yahaya Wuni

## 7.1 Introduction

Humanity's demand for infrastructure remains unending. Key among the infrastructural needs are housing facilities for adequate shelter and for security. To ensure adequate shelter, policymakers including the United Nations (UN) have stipulated some criteria as requirements for housing development. Among which include price or rental affordability of housing facilities; reduced commuting cost of households (i.e., accessibility); energy efficient housing facilities, environmentally friendly (eco-friendly) facilities; safety of households, reduced lifecycle cost; maintainability of housing facilities and quality housing facilities (i.e., aesthetically pleasing, functionally and technically adequate). These criteria are also known as critical success criteria for sustainable housing (Chan & Adabre, 2019; Adabre et al., 2021a, b) or the sustainable development goals specified for achieving sustainable housing. Thus, sustainable housing is defined as a

> Habitable domicile, positioned within proximity to services, amenities, and places of work (i.e., social infrastructure) and provides safe, sustainable and affordable shelter for a whole range of households (Adabre et al., 2021a, p. 2).

In addition to the importance of sustainable housing for ensuring adequate shelter, it leads to mitigation of greenhouse gas emissions (i.e., carbon dioxide, carbon monoxides) and their negative effects. This is achieved through reduced vehicular movement due to proximity and optimum use of eco-friendly materials for construction. Notwithstanding its essence, sustainable housing development remains elusive globally.

Most developed and developing economies are plagued by the housing crises which are evident in housing deficit and/or deficient housing facilities that do not meet the criteria for sustainable housing (Chan & Adabre, 2019). Yet, the housing situation (i.e., both housing deficit and deficient shelter facilities) is direr in developing economies and, therefore, requires more urgent attention. A socio-technical approach has been recommended in existing studies for sustainable housing (Darko & Chan, 2018; Adabre et al., 2022a). Thus, changes in practices or policies within communities and the adoption of technologies (Adabre et al., 2022a) are two important aspects that have been suggested in

M. A. Adabre (✉) · A. P. C. Chan
Building and Real Estate Department, The Hong Kong Polytechnic University, Hung Hom, Hong Kong
e-mail: 17902405r@connect.polyu.hk; albert.chan@polyu.edu.hk

I. Y. Wuni
Department of Construction Engineering & Management, King Fahd University of Petroleum & Minerals, Dhahran, Saudi Arabia
e-mail: ibrahim.wuni@connect.polyu.hk



J. Zuo et al. (eds.), *Circular Economy for Buildings and Infrastructure*, Sustainable Development Goals Series, https://doi.org/10.1007/978-3-031-56241-9_7

extant literature. Green building technologies and techniques such as energy efficient lighting system, water efficient installations, green roofing and energy efficient ventilation system are essential technologies for sustainable housing (Darko & Chan, 2018). Nonetheless, the adoption of such technologies thrives on policy reformation/changes in practices and behavior by policymakers, practitioners, and other stakeholders (i.e., households and communities). Typical sustainable housing policies such as mixed income housing, high-rise housing development, provision of subsidies and incentives are social factors which influence behavior and practices of households and have ripple effects on adoption of technologies for sustainable housing. Therefore, between the two aspects—socio and technical—the 'socio' has the ascendant for sustainable development in housing.

A typical developing economy that has embraced the socio-technical perspective of sustainable housing is Ghana (Darko et al., 2017). For sustainable housing development towards sustainable cities in Accra and other communities, various technologies were adopted by households. Specifically, incandescent light bulbs were replaced with energy efficient lights such as fluorescent light bulbs. Besides, some old refrigerators that were not energy efficient were replaced with energy efficient refrigerators. Indeed, the adoptions of these and other technologies were aided by reformation in policies and practices. For instance, policies such as supplies of subsidies or incentives were the bases for the adoption of the energy efficient technologies. In addition to these policies, recommendations were proffered for high-rise housing development and direct government initiatives on housing supply. Nonetheless, policies for sustainable housing have still not augmented adequate housing in Ghana particularly in the regional capital—Accra. The nation's capital has been plagued by the deficit and deficient housing crises. The deficit crisis is evinced in high rental charges/high advanced charges while the deficient problem is revealed in the high number of shanty or makeshifts at unapproved sites, which partly leads to the persistent annual flooding in Accra city. Focusing on Ghana particularly Accra, this chapter seeks to model sustainable housing for sustainable development in cities and communities.

The theoretical underpinning of this study is the theory of planned behavior. Through this theory, the study models relationships between sustainable housing goals and key constructs such as barriers, risks and success factors which could influence the attainment of the sustainable development goals in housing. The study's findings seek to apprise policymakers of barriers and risks that could hinder sustainable housing. This will inform policymakers of key challenges that require adequate resource allocation. Findings on the success factors will inform decision makers on essential policies for sustainable housing while cautioning them about counterproductive policies. The remaining of the study is organized as follows: the second section entails the theoretical basis and framework for modeling sustainable housing. Then, the methodology for conducting the study is detailed in section three. This is followed by data analysis and discussion of the results in section four. In section five, the conclusions and recommendations are provided.

## 7.2 Theoretical Basis and Framework of Study

Technological adoption is essential for sustainable housing (Darko et al., 2017; Ohene et al., 2022). Likewise, sustainable housing involves the adoption and implementation of policies to change behavior and practices of various stakeholders. Thus, the espousal of sustainable housing policies entails a shift from prior practices and behavior that do not contribute to sustainable housing to practices that do. Such transition can be modeled using the theory of planned behavior (TPB). The TPB, developed by Ajzen (1991), is the basis for developing a conceptual model for sustainable housing. According to the TPB, adoption of a behavior or practice (i.e., transitioning to sustainable housing) can be explained by three main constructs, namely, perceived behavioral control (to answer the question: what are the barriers to sustainable housing?); subjective norms (to answer the question: what are the uncertainties or risk factors to sustainable housing?) and personal attitude (to answer the question: what are the success factors for sustainable housing?). These three constructs first influence the intention to, or not to, transition to sustainable housing development. Thus, the intention can be modeled using a binary variable such as 'Yes' or 'No'.

Considering the increasing emissions of greenhouse gases and the need to ensure optimum utilization of resources via embracing circular economy principles, the intention to ensure sustainable development including sustainable housing is in the affirmative. An affirmative intention, therefore, leads to the adoption of sustainable housing development as a behavior or practice. After adopting sustainable housing, its assessment is essential to measure its progress. Besides, it is necessary to benchmark and compare sustainable housing development among various cities, communities, regions, and countries. The assessment is conducted by a set of indicators or measured variables referred to as critical success criteria or sustainable housing goals (Gan et al., 2017; Chan & Adabre, 2019; Adabre & Chan, 2020). Chan and Adabre (2019) established six categories of critical success criteria. The first category is 'household satisfaction criteria' (i.e., 'functionality of housing facility', 'end user's satisfaction with the housing facility', 'maintainability of housing facility'). The second is stakeholders' satisfaction criteria (i.e., timely completion of project, project team satisfaction). Third is housing operation cost criteria (i.e., energy efficiency of facility, reduced lifecycle cost of facility and eco-friendliness). Followed by time measurement criteria (i.e., take-up rate of housing facility, waiting time of potential households, and construction cost performance of housing facility). Then, location affordability criteria (i.e., price affordability of housing facility, rent affordability of housing facility and commuting cost) and quality-related criteria (i.e., quality performance of project, aesthetically pleasing view of facility and technical specification of housing facility). Mulliner et al. (2013) established the theoretical positions of the adequacy of these critical success criteria. The attainment of these goals is influenced by varied factors which can be conceptualized broadly using the three TPB constructs.

One of the TPB constructs 'perceived behavioral control' refers to people's perception on the ease or difficulties in adopting a practice of interest. It is often conceptualized as the barrier or impediment that hinders the adoption of a practice. In this study, perceived behavioral control includes the barriers to sustainable housing development as reified in Adabre et al. (2022b). Studies abound on barriers to sustainable housing in both developed and developing countries (Djokoto et al., 2014; Yang & Yang, 2015). In the case of developing countries, Adabre et al. (2021a) established three categories of barriers, with their underlying factors in brackets: the first category is cost-related barrier (i.e., high upfront cost of materials and technologies'; 'high cost of serviced land'; 'high inflation rate'; 'high interest rates' etc.). The second group of barriers is incentive-related (i.e., 'inadequate incentives for private investors'; 'lack of planning control on land development'; 'inadequate subsidies/public funding for sustainable technologies'; 'tight-credit conditions'; 'inadequate infrastructural development'). The third category of barriers is retrofit-related ('low-level or inadequate retrofitting i.e., maintenance operation'; 'poor maintenance culture'; 'inadequate professional skills'; 'inadequate policies or sustainability assessment tools for retrofitting housing facilities'). These are inveterate categories of barriers (Ametepey et al., 2015; Darko et al., 2017; Chan et al., 2018; Alam et al., 2019; Ezennia, 2022).

The second TPB construct is subjective norms which refer to beliefs that an important person or group of people (i.e., community) could adopt to influence a practice. In this study, the subjective norm is adapted as the subjectivities or biases that could influence sustainable housing development. Decisions on sustainable housing could be subjective and biased. This is owing to inadequate information on sustainable housing choices and imperfection in judging information. Furthermore, constraints due to cost and time hinder comprehensive search for sustainable housing choices. In these conditions of uncertainties, decision makers may not aim for optimizing but seek for 'satisfying' during goals attainment. Likewise, in communities or cities with uncertainties in decision making, policymakers and developers, to avoid risks and uncertainties, may opt for 'satisfactory' choices or practices that do not necessarily lead to optimization for sustainable housing. This is a typical case of bounded rationality (Klein & Methlie, 1995; Ortega & Braun, 2011). Therefore, for optimization in decision making for sustainable housing, risks or uncertainties ought to be assessed. Studies have been conducted on various risk factors which could impede sustainable housing. Some risk factors include: 'land acquisition risks', 'inflation volatility', 'fluctuating cost of finance', 'policy instability risks', 'bribery and corruption' and 'fluctuating cost of finance' (i.e., interest rate fluctuations). These risk factors have been categorized into manageable groups, namely, 'political and procurement risks'; 'financial-related risks'; 'design and construction risks' and 'operation and maintenance risks'. These categorizations are essential for ensuring effective resource allocation (Adabre et al., 2021b).

The third TPB construct is attitude. Attitudes are the factors that promote the adoption of a practice or behavior. According to Adabre et al. (2022a), attitudes in the case of the built environment were conceptualized as drivers, promoters, success factors or enablers for circular economy. Similarly, for the adoption of a practice such sustainable housing, success factors are essential to motivate and drive sustainable practices in addition to punishing unsustainable housing practices (Adabre & Chan, 2019; Sony et al., 2021; Moons et al., 2022). For sustainable housing development, Adabre et al. (2021a) established four groups of success factors, namely, (1) 'developers enabling success factors' (i.e., mandatory inclusion of affordable housing unit in developers' project; access to low interest housing loans; effective private sector participation and economic stability), (2) household-enabling success factors (i.e., governments' provision of subsidies to households; adequate maintenance of existing housing facilities; adequate infrastructure supply by government), (3) mixed development success factors (i.e., adequate accessibility to social amenities; sitting or locating housing projects within cities and towns; mixed development of housing projects within cities and towns) and (4) 'land use planning success factors' (i.e., linking commercial development approval to funding for housing'; 'increase tax to discourage long holding periods of vacant land'; 'political will and commitment to low-cost housing by land-use strategy and taxation on property or capital gains for supply of housing facilities).

## 7.3 Research Methodology

This study adopts a positivist philosophical stance and deductive reasoning to test the influence of various constructs (i.e., barriers, risks, and success factors, as depicted in Fig. 7.1) on

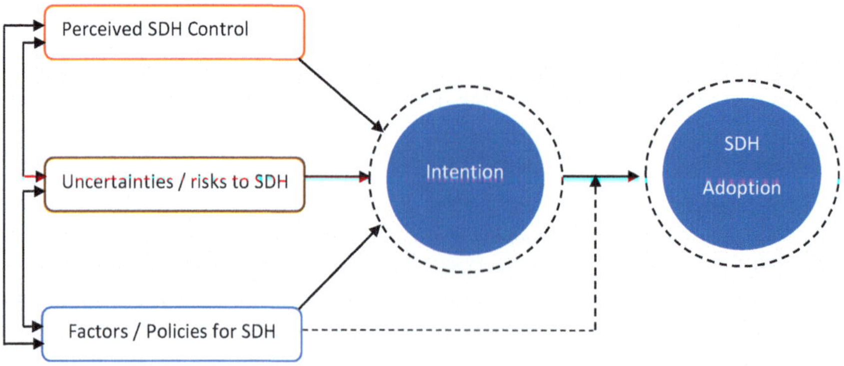


**Fig. 7.1** Conceptual model on relationship among barriers, risk factors, success factors, intention and adoption of Sustainable Development in Housing (SDH). (Adapted from Ajzen, 1991; Adabre et al., 2022a)

sustainable development in housing (i.e., measured by sustainable housing goals). This is achieved through a deterministic modeling of questionnaire data collected through a cross sectional time horizon (Adabre et al., 2021a).

### 7.3.1 Data Collection

A questionnaire was designed for data collection on impacts of barriers, risks, and success factors on goals of sustainable housing. The questionnaire contained five sections: section one entailed questions that solicited the demographic data of respondents, section 2 on goals of sustainable housing, section 3 on barriers, section 4 on risk factors and section 5 on success factors. Prior to the survey, the questionnaire was piloted among four professionals who are abreast of practices and research in the Ghanaian housing market. Feedback from the pilot survey was used to improve and finalize the questionnaire for data collection. The questionnaire was administered to respondents in the formal sector of the Ghanaian housing market. The respondents have participated in housing projects either as contractors/developers or consultants. Some formal institutions that formed the list for data collections include members of the Ghana Real Estate Developers Association (GREDA); State Housing Cooperation (SHC); Tema Development Cooperation (TDC); Public Works Department (PWD); Social Security and National Insurance Trust (SSNIT); Building and Road Research Institute (BRRI) and Architectural and Engineering Service Limited (AESL). Attributed to the lack of a population frame for sampling, non-probability sampling techniques, viz. purposive sampling and snowballing were deployed for the data collection. Respondents were requested to rate the importance of underlying variables of the various constructs (i.e., barriers, risks, and success factors) using a five-point Likert scale (1 = not important, 2 = less important, 3 = neutral, 4 = important and 5 = very important). A three-month duration was dedicated to the data collection. Out of 110 questionnaires administered, 49 questionnaires were received but two were considered invalid because of incompleteness. Thus, 47 questionnaires, forming a 42.7% response rate, were considered valid for data analysis.

### 7.3.2 Data Analysis

Prior studies have utilized various analytical techniques for modeling the impacts of some variables on sustainable housing development. Some techniques have inherent limitations in their assumption (i.e., Roufechaei et al., 2014; Kwak et al., 2018; Guan et al., 2020). For instance, the traditional multiple regression analysis (MRA) has been employed for establishing relationships between energy efficient parameters and the three main dimensions of sustainable development (i.e., the social, economic, and environmental) (Roufechaei et al., 2014). Notwithstanding the essence of the regression technique, it has key weaknesses in its assumptions: (1) the traditional MRA assumes a relationship between one dependent variable and several independent variables. Thus, in a situation where the dependent variables are at least one, then several MRAs must be conducted for each dependent variable. Such analyses could become complicated and impossible when the dependent variables are many i.e., at least five. (2) The MRA further assumes that every variable can be measured directly without errors. However, some variables are constructs and are, therefore, measured by inferring from other variables. For example, sustainable housing is a construct that can be measured by inferring from other underlying variables such as price affordability, rental affordability, energy efficiency, affordable commuting cost and quality of the housing facility. Even in measuring these variables, errors could occur owing to arrangement of the variables in the questionnaire (i.e., random errors) and respondents' fatigue. These underlying shortcomings render the traditional multiple regression technique inappropriate for this study. Regarding another analytical technique interpretive structural modeling (ISM) as deployed by Kwak et al. (2018) and Guan et al. (2020), a major drawback is assessing the statistical significance of established relationships (Adabre et al., 2021b). Therefore, a more rigorous technique is the structural equation modeling.

The structural equation modeling (SEM) could be used to resolve some problems in the MRA and ISM. There are two forms of SEM: partial least squares SEM (PLS-SEM) and covariance-based SEM (CB-SEM). To adopt any of the techniques, certain conditions must be met. For relatively small sample size data (i.e., <200) that are non-normally distributed, the PLS-SEM is more appropriate for data analysis (Aibinu & Al-Lawati, 2010). Since the sample size of this study is relatively low (i.e., 47) and the data are not normally distributed, the PLS-SEM was deployed for data analysis. Besides, because this study seeks to make predictions on impacts of constructs that have relatively high number of observable or underlying variables, the PLS-SEM is more suitable. Before conducting the data analysis with the PLS-SEM, certain statistical requirements must be fulfilled. First, the data must meet a requirement for the central limit theorem. According to this theorem, a sample size of at least 30 is considered appropriate to meet this standard. The second requirement is that the sample size should not be <10 times the number of relationships that a construct can form with other constructs. These requirements were met in this study. For instance, since the study sample size is 47, more than the 30 sample size requirement, the central limit theorem was fulfilled. On the second precondition, the constructs that could form the highest number of inter-constructs relationships are the risk constructs. There are five constructs, and each construct can form a maximum of four relationships between itself and the remaining constructs. Thus, the sample size should not be less that 40 (i.e., 4 by 10, 4 for the number of inter-construct relationship). Recall, the sample size for this study is 47 which is more than the 40-sample size benchmark. Therefore, the data are statistically adequate for the PLS-SEM analysis.

## 7.4 Results of Data Analysis

### 7.4.1 Respondents' Profile

The demographic profile of the respondents showed that 47.9% respondents are in the public sector or department, 35.4% respondents are in the academic/research institutions and 16.7% respondents are private developers or contractors. Regarding profession, most (55.3%) are quantity surveyors, 19.2% are architect, 12.8% are construction managers and then 12.7% constitute planners and engineers. Most respondents (52.2%) have handled at least two housing projects in Ghana of which 55.1% are public projects. Concerning work experience in the Ghanaian housing market, 63.9% have more than 5 years of relevant work experience.

### 7.4.2 Descriptive and Reliability Analysis

The reliability of the data was evaluated using three criteria: the Cronbach Alpha (CA) values, the standard deviation scores, and the mean scores. The CA of the SDGs in housing was 0.878 which is greater than the minimum threshold (i.e., 0.700) for data reliability. Similarly, the CA of the barriers (0.840), risks (0.878) and the success factors (0.897) are all satisfactory. On standard deviations (SD) of underlying variables in the various constructs, most SD values were below one which suggests a relatively high level of consistency among the respondents who ranked the variables of each construct. Furthermore, the mean scores of the underlying variables implied that the variables were considered as important to sustainable housing development since they were all rated above the scale category of *less important* (<2).

### 7.4.3 Results of PLS-SEM Analysis

Three modeling analyses were conducted, namely, (1) modeling the impacts of barriers on sustainable housing development; (2) modeling the impacts of risk factors on sustainable housing development and (3) modeling the impacts of success factors on sustainable housing development. Different underlying variables of sustainable housing were loaded in each modeling analysis as shown in Table 7.1. Nonetheless, common variables that were significantly loaded

**Table 7.1** Constructs and their significantly loaded underlying variables

| Constructs | Underlying variables | Mean | SD | Loadings | AVE | CR | CA |
|---|---|---|---|---|---|---|---|
| Sustainable Development Goals | **Significantly loaded in analysis of barriers** | | | | | | |
| in Housing | Energy efficient housing | 3.915 | 0.880 | 0.715 | 0.504 | 0.876 | 0.812 |
| | Rental cost of housing facility | 4.196 | 0.824 | 0.572 | | | |
| | Commuting cost of household to facility | 3.787 | 0.999 | 0.701 | | | |
| | Technology transfer/innovation (i.e., technological enhancement for adaptability and flexibility in housing and secondary materials usage) | 3.468 | 0.856 | 0.691 | | | |
| | Safety performance (crime prevention) | 4.085 | 0.803 | 0.694 | | | |
| | End user's satisfaction | 4.319 | 0.980 | 0.832 | | | |
| | Stakeholders' satisfaction | 3.957 | 0.833 | 0.741 | | | |
| | **Significantly loaded in analysis of risk factors** | | | | | | |
| | Safety performance | 4.085 | 0.803 | 0.741 | 0.525 | 0.885 | 0.862 |
| | End user's satisfaction with the housing facility | 4.319 | 0.980 | 0.672 | | | |
| | Maintainability of housing facility | 4.283 | 0.851 | 0.795 | | | |
| | Energy efficiency of housing facility | 3.915 | 0.880 | 0.581 | | | |
| | Functionality of housing facility | 4.174 | 0.789 | 0.749 | | | |
| | Technical specification of housing facility | 4.128 | 0.824 | 0.746 | | | |
| | Commuting cost of household to facility | 3.787 | 0.999 | 0.680 | | | |
| | Technology transfer/innovation (i.e., technological enhancement for adaptability and flexibility in housing and secondary materials usage) | 3.468 | 0.856 | 0.750 | | | |
| | **Significantly loaded in analysis of success factors** | | | | | | |
| | Energy efficiency of housing facility | 3.915 | 0.880 | 0.746 | 0.504 | 0.875 | 0.850 |
| | Commuting cost of household to facility | 3.787 | 0.999 | 0.537 | | | |
| | Rental cost of housing facility | 4.196 | 0.824 | 0.689 | | | |
| | Technology transfer/innovation (i.e., technological enhancement for adaptability and flexibility in housing and secondary materials usage) | 3.468 | 0.856 | 0.724 | | | |
| | Safety performance (crime prevention) | 4.085 | 0.803 | 0.682 | | | |
| | End user's satisfaction | 4.319 | 0.980 | 0.811 | | | |
| | Reduced lifecycle cost | 3.933 | 0.918 | 0.749 | | | |
| **Barriers constructs** | **Underlying barriers** | **Mean** | **SD** | **Loadings** | **AVE** | **CR** | **CA** |
| Cost-related barriers | Delays in governmental approval process—CB1 | 3.936 | 0.95 | 0.821 | 0.502 | 0.799 | 0.737 |

(continued)

**Table 7.1** (continued)

| | | | | | | | |
|---|---|---|---|---|---|---|---|
| | High upfront cost of materials and technologies for sustainable housing—CB2 | 4.467 | 0.544 | 0.735 | | | |
| | High cost of serviced land—CB3 | 4.467 | 0.710 | 0.647 | | | |
| | High inflation rate—CB4 | 4.404 | 0.712 | 0.613 | | | |
| Incentive-related barriers | Inadequate incentives for private developers—IB1 | 3.872 | 0.924 | 0.853 | 0.542 | 0.823 | 0.737 |
| | Inadequate access to land among developers—IB2 | 4.043 | 0.908 | 0.742 | | | |
| | Income inequality—IB3 | 3.979 | 0.737 | 0.572 | | | |
| | Lack of planning control on land development—IB4 | 4.239 | 0.728 | 0.749 | | | |
| Retrofit-related barriers | Low-level or inadequate retrofitting (maintenance operation)—RB1 | 3.935 | 0.818 | 0.697 | 0.545 | 0.826 | 0.727 |
| | Inadequate policies or sustainability assessment tools (standard or guidelines) for retrofitting—RB2 | 3.894 | 1.047 | 0.841 | | | |
| | Lack of routine maintenance/poor maintenance culture—RB3 | 4.213 | 0.907 | 0.722 | | | |
| | Policy instability/abandoned or neglected management of public housing projects—RB4 | 4.404 | 0.648 | 0.682 | | | |
| Risk constructs | Underlying Risk variables | MI $=\sqrt{(\text{LO X SI})}$ | | Loading | AVE | CR | CA |
| Political & procurement risks | Bribery and corruption risks—PPR1 | 4.021 | | 0.652 | 0.542 | 0.853 | 0.828 |
| | Inadequate competition risks—PPR2 | 3.433 | | 0.765 | | | |
| | Inadequate project design—PPR3 | 3.741 | | 0.832 | | | |
| | Political continuity risks—PPR4 | 3.908 | | 0.567 | | | |
| | Policy instability risk/inadequate government commitment—PPR5 | 3.779 | | 0.829 | | | |
| Financial-related risks | Delay payment risk—FRR1 | 4.431 | | 0.618 | 0.566 | 0.837 | 0.743 |
| | Fluctuating cost of finance (interest rate fluctuations)—FRR2 | 4.338 | | 0.840 | | | |
| | Inadequate financing institutions for housing—FRR3 | 3.939 | | 0.762 | | | |
| | Inflation rate volatility—FRR4 | 4.174 | | 0.769 | | | |
| Design & Construction Risks | Construction defects/deficiencies—DCR1 | 3.487 | | 0.739 | 0.559 | 0.863 | 0.805 |
| | Cost overrun risk—DCR2 | 4.163 | | 0.848 | | | |
| | Design and construction variation—DCR3 | 3.649 | | 0.677 | | | |
| | Force majeure events—DCR4 | 3.253 | | 0.684 | | | |
| | Time overrun risks—DCR4 | 4.042 | | 0.777 | | | |
| Operation & maintenance risks | Fluctuating demand risk—OMR1 | 3.689 | | 0.700 | 0.559 | 0.833 | 0.792 |

(continued)

**Table 7.1** (continued)

| | | | | | | | |
|---|---|---|---|---|---|---|---|
| | Inadequate infrastructure risks—OMR2 | 3.569 | | 0.585 | | | |
| | Operating cost overrun—OMR3 | 3.316 | | 0.859 | | | |
| | Privatization risk (i.e., selling of rental stock)—OMR4 | 3.574 | | 0.815 | | | |
| Success factors constructs | Underlying success factors | Mean | SD | Loading | AVE | CR | CA |
| Developer-enabling factors | Energy efficient designs and installation—DEF1 | 4.085 | 0.855 | 0.707 | 0.536 | 0.819 | 0.762 |
| | Water efficient design and installation—DEF2 | 4.277 | 0.579 | 0.751 | | | |
| | Use of environmentally friendly materials for construction—DEF3 | 4.370 | 0.671 | 0.867 | | | |
| | Effective private sector participation—DEF4 | 4.064 | 0.845 | 0.574 | | | |
| Household-enabling factors | Monitoring housing conditions/performance for retrofitting—HEF1 | 4.149 | 0.834 | 0.853 | 0.643 | 0.843 | 0.744 |
| | Governments' provision of subsidies to households—HEF2 | 3.979 | 1.073 | 0.727 | | | |
| | Adaptable housing design—HEF3 | 4.044 | 0.833 | 0.820 | | | |
| Mixed-use development factors | Linking commercial development approval to funding for housing—MDF1 | 3.723 | 0.902 | 0.736 | 0.558 | 0.834 | 0.744 |
| | Adequate access to social amenities—MDF2 | 4.340 | 0.668 | 0.836 | | | |
| | Sitting/locating housing projects within cities and towns—MDF3 | 4.362 | 0.705 | 0.732 | | | |
| | Mixed development of housing and commercial buildings—MDF4 | 3.809 | 0.770 | 0.673 | | | |
| Land-use planning factors | Increase tax to discourage long holding periods of vacant land—LPF1 | 3.149 | 1.063 | 0.868 | 0.712 | 0.881 | 0.805 |
| | Taxation on property or capital gains for supplying facilities—LPF2 | 3.362 | 1.112 | 0.855 | | | |
| | Sufficient financial and human resources for planning institutions—LPF3 | 3.575 | 0.773 | 0.806 | | | |

**Note:** *LO* Likelihood of Occurrence, *SI* Severity of Impact

include: energy efficiency of housing facility, commuting cost of households to facility, safety performance (crime prevention), end user's satisfaction and technology transfer (innovation). The underlying variables that are significantly loaded in the groupings of the barriers, risks and success factors constructs are shown in Table 7.1.

#### 7.4.3.1 Results—Impacts of Barriers on Sustainable Housing

Three constructs of barriers, namely, 'cost-related barriers', 'incentive-related barriers' and 'retrofit-related barriers' were adopted and used to model their impacts on sustainable housing development (Adabre & Chan, 2021). The measurement model concerning the relationship between a construct and its underlying variables was first developed and subsequently evaluated based on key

**Table 7.2** Results of hypotheses testing

| Hypotheses | Relationships | Std. Beta | Std. Error | t-value | Decision |
|---|---|---|---|---|---|
| H1 | Cost-related barriers have significant impact on incentive-related barriers | 0.500 | 0.125 | 3.673** | Supported |
| H2 | Incentive-related barriers have significant impact on retrofit-related barriers | 0.651 | 0.122 | 5.316** | Supported |
| H3 | Cost-related barriers have significant impact on retrofit-related barriers | 0.118 | 0.165 | 0.637 | Not supported |
| H4 | Cost-related barriers have significant impact on sustainable housing | 0.273 | 0.186 | 1.507 | Not supported |
| H5 | Incentive-related barriers have significant impact on sustainable housing | −0.574 | 0.161 | 3.443** | Supported |
| H6 | Retrofit-related barriers have significant impact on sustainable housing | 0.824 | 0.155 | 5.087** | Supported |

**Note:** $^{**}p < 0.01$, $^{*}p < 0.05$

statistical requirements. Underlying variables that were not suitable or did not contribute significantly to a construct were identified using the loadings of the variables. Such variables (with loadings <0.500) were deleted, and the analysis reiterated until an appropriate construct was obtained. Each construct was assessed using the average variance extracted (AVE), composite reliability (CR) and Cronbach's Alpha (CA). All constructs of barriers in addition to the sustainable housing constructs met the 0.5 AVE requirement, 0.7 CR condition and 0.7 CA standard which indicate adequate convergent reliability, internal consistency, and indicator reliability, respectively (as shown in Table 7.1). Furthermore, discriminant validity via the Fornell and Lacker criterion, indicators' cross loadings and HTMT were satisfactory. Therefore, the measurement model was deemed adequate for subsequent analyses.

Structural modeling of the data was subsequently conducted to assess the inter-construct relationships among the categories of barriers on one hand and the impacts of the barrier constructs on sustainable housing development. Prior to assessing the statistical significance of the inter-construct relationships, the structural model was evaluated. Multicollinearity was checked using the variance inflation factor (VIF) which had a value of <5.00. This indicates no multicollinearity. Then, statistical significance of the inter-construct relationships was conducted through bootstrapping analysis. This analysis is appropriate since earlier assessment showed that the data followed non-normal distribution. The bootstrapping analysis was conducted to assess the following hypotheses:

**Hypothesis 1:** 'cost-related barriers' have a significant impact on 'incentive-related barriers.'
**Hypothesis 2:** 'incentive-related barriers' have a significant impact on 'retrofit-related barriers.'
**Hypothesis 3:** 'cost-related barriers' have a significant impact on 'retrofit-related barriers.'
**Hypothesis 4:** 'cost-related barriers' have a significant impact on 'sustainable housing.'
**Hypothesis 5:** 'incentive-related barriers' have a significant impact on 'sustainable housing.'
**Hypothesis 6:** 'retrofit-related barriers' have a significant impact on 'sustainable housing.'

Results of the bootstrapping analysis (refer to Table 7.2) are discussed subsequently.

The non-significant impact of cost-related barriers on sustainable housing revealed an interesting finding (indicated by a dotted line as shown in Fig. 7.2). Although prior studies in another developing country such as Malaysia (Durdyev et al., 2018) revealed a significant impact of cost-related barriers on sustainable housing particularly green buildings, the findings from this study showed no such relationship. This is in consonance with prior study by Darko et al. (2017) in which there was no significant relationship between cost and sustainable housing

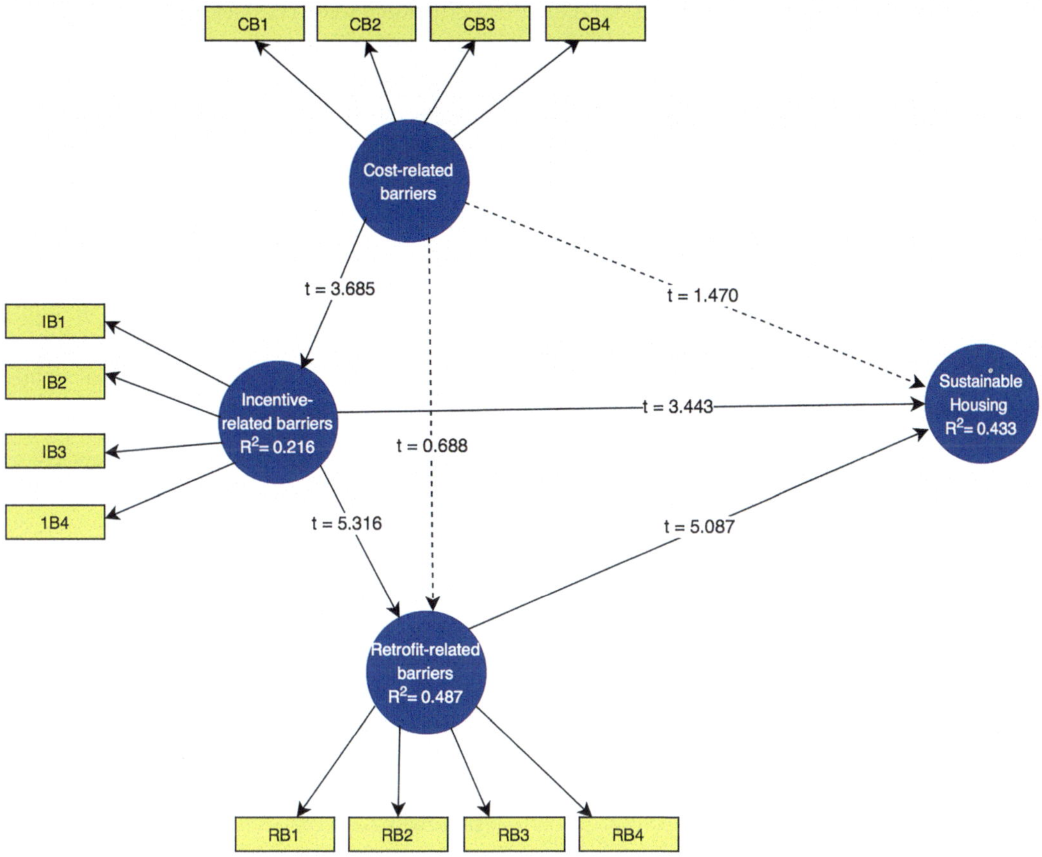


**Fig. 7.2** A model on relationship among barriers and sustainable housing. (Adapted from Adabre & Chan, 2021)

specifically green buildings. This could imply that cost-related barriers are secondary barriers as confirmed in earlier study (Kaygusuz, 2012). Thus, cost-related barriers are not the most important although they could indirectly influence other barriers. This is buttressed by the significant relationship between 'cost-related barriers' and 'incentive-related barriers' (indicated by a solid line as shown in Fig. 7.2). This further suggests that cost as a barrier is attributed to lack of incentives. Therefore, policymakers and decision makers could focus on incentivizing households, suppliers of sustainable technologies and developers through tax-rebates on sustainable technologies, subsidies and interest-free loans. It is worth noting that providing subsidies and interest-free loans to households or potential households could have inflationary impacts (Kumi, 2017). Besides, interest-free loans, if not checked, could be diverted to other uses such as purchase of luxurious goods. To curb this, subsidies and interest-free loans could be directed to suppliers and developers to boost production and supply of housing facilities, efficient energy technologies and electricity production. This strategy could reduce production cost of suppliers and developers and improve supplies which could reduce prices of sustainable technologies and prices of sustainable housing in the market.

Another construct of barriers which requires the attention of decision makers is the 'retrofit-related barriers.' This construct has a significant impact on sustainable housing development (refer to Table 7.2 and Fig. 7.2). This implies that refurbishing existing housing facilities to sustainability standard is hindered by barriers. Therefore, strategies for eliminating barriers to housing retrofit are essential (Ohene et al.,

2022). For example, instead of partial retrofit (i.e., focusing mainly on supplies of energy efficient lighting as evidenced in Ethiopia and Ghana (Costolanski et al., 2013; Kumi, 2017), retrofit could be extended to other household electrical appliances such as electric iron, electric fan, and refrigerators. Indeed, a holistic retrofit is important to prevent 'rebound effect', a situation where positive effects from other retrofitting operations are annihilated by backsliding in other non-retrofitted, inefficient appliances or technologies—the net impact is therefore negligible (Adabre et al., 2021a). Partial retrofit with its concomitant problem of 'rebound effect' has been a major problem in most developing countries as witnessed in Ethiopia (Costolanski et al., 2013). The significant relationship between incentive-related barriers and retrofit-related barriers (indicated by solid line as shown in Fig. 7.2) further reveals an interesting finding. Besides, cost-related barriers did not have a significant impact on retrofit-related barriers (indicated by a dotted line as shown in Fig. 7.2). Thus, most of the barriers to retrofit are attributed to lack of incentives. This also applies to the case of partial retrofit in most developing countries such as Ghana (Kumi, 2017).

#### 7.4.3.2 Results—Impacts of Risk Factors on Sustainable Housing

Measurement models were estimated for five constructs including four risk constructs (i.e., political and procurement risks, financial-related risks, design and construction risks and operation and maintenance risks) and the sustainable housing construct. Assessment of adequacy of the measurement model was conducted by checking the following criteria: loadings of the underlying variables, AVE, CR and CA. Underlying variables that did not meet the loading requirement of $\geq 0.50$ were deleted while the measurement model was estimated iteratively till the loadings of the underlying variables, AVE, CR and CA of the constructs met the required threshold for subsequent analyses (refer to Table 7.1). Furthermore, discriminant analyses via Fornell and Lacker criterion, cross-loading of underlying variables and Heterotrait-monotrait (HTMT) ratio of correlation were conducted and found to be satisfactory for structural modeling.

The structural modeling was carried out to assess the impacts of the risk constructs among themselves and their impacts on the sustainable housing construct. The established model was checked for multicollinearity through the variance inflation factor (VIF). With a VIF $< 5.00$, the structural model was considered appropriate for subsequent analysis since there was no problem of multicollinearity. Bootstrapping analysis was then conducted to assess the statistical significance of the following hypotheses:

**Hypothesis 1:** 'Political and procurement risks' have a significant influence on 'financial-related risks.'

**Hypothesis 2:** 'Political and procurement risks' have a significant influence on 'design and construction risks.'

**Hypothesis 3:** 'Political and procurement risks' have a significant influence on 'operation and maintenance risks.'

**Hypothesis 4:** 'Financial-related risks' have a significant influence on 'design and construction risks.'

**Hypothesis 5:** 'Financial-related risks' have a significant influence on 'operation and maintenance risks.'

**Hypothesis 6:** 'Design and construction risks' have a significant influence on 'operation and maintenance risks.'

**Hypothesis 7:** 'Political and procurement risks' have a significant influence on 'sustainable housing.'

**Hypothesis 8:** 'Financial-related risks' have a significant influence on 'sustainable housing.'

**Hypothesis 9:** 'Design and construction risks' have a significant influence on 'sustainable housing.'

**Hypothesis 10:** 'Operation and maintenance risks' have a significant influence on 'sustainable housing.'

Results of the bootstrapping analysis for testing the hypotheses are shown in Table 7.3 and discussed subsequently.

Among the risk constructs, 'political and procurement risks' have a significant influence on

**Table 7.3** Results of hypothesis testing on risk constructs and sustainable housing

| Hypothesis | Relationship | Std Beta | Std Errors | t-value | Decision |
|---|---|---|---|---|---|
| H1 | Political and procurement risks -> financial-related risks | 0.467 | 0.147 | 3.039** | Supported |
| H2 | Political and procurement risks -> design and construction risks | 0.600 | 0.130 | 4.538** | Supported |
| H3 | Political and procurement risks -> operation and maintenance risks | 0.332 | 0.170 | 2.028* | Supported |
| H4 | Financial-related risks -> design and construction risks | 0.303 | 0.110 | 2.714** | Supported |
| H5 | Financial-related risks -> operation and maintenance risks | 0.117 | 0.123 | 0.850 | Not supported |
| H6 | Design and construction risks -> operation and maintenance risks | 0.349 | 0.239 | 1.359 | Not supported |
| H7 | Political and procurement risks -> sustainable housing | 0.358 | 0.167 | 2.321* | Supported |
| H8 | Financial-related risks -> sustainable housing | 0.181 | 0.150 | 1.257 | Not supported |
| H9 | Design and construction risks -> Sustainable housing | 0.342 | 0.195 | 1.422 | Not supported |
| H10 | Operation and maintenance risks -> sustainable housing | −0.106 | 0.159 | 0.582 | Not supported |

**Note:** $^{**}p < 0.01$, $^{*}p < 0.05$

'financial-related risks' with a t-value of 3.039 at a significant level (p-value) < 0.01. Similarly, with a t-value of 4.538 at a significant level (p-value) < 0.01, 'political and procurement risks' have a significant influence on 'design and construction risks'. Furthermore, 'political and procurement risks' have a significant influence on 'operation and maintenance risks' (t-value of 2.028 at a p-value <0.05) (as indicated by the solid lines in Fig. 7.3). 'Financial-related risks' have a significant influence on 'design and construction risks' at a t-value of 2.714 at a significant level (p-value) < 0.01 (as indicated by the solid lines in Fig. 7.3).

Between the risk constructs and the sustainable housing construct, only the 'political and procurement risks' have a significant influence on 'sustainable housing' (t-value of 2.321 supported at a significance level i.e., p-value <0.05) (as indicated by the solid lines in Fig. 7.3). Thus, to ensure sustainable housing development by the formal sector, decision makers should focus on the 'political and procurement risks'. Indeed, public-private partnerships for housing supplies have often failed attributed to the 'political and procurement risk factors' (Osei-Kyei et al., 2019; Adabre et al., 2022c). Therefore, this category of risks is a primary risk category. It has ripple effects on other risk constructs such as 'financial-related risks'; 'design and construction risks' and 'operation and maintenance risks'. Moreover, it is the only risk category that has a direct impact on sustainable housing development. 'Financial-related risks' have a direct impact on 'design and construction risks'. Thus, financial risk factors could lead to a compromise on design criteria and construction methods. This has often led to relatively low efficient energy designs and installations in most developing economies to avoid high upfront cost of sustainable housing (Darko et al., 2017).

To effectively manage the 'political and procurement risk factors', policymakers should focus on its underlying variables since they account for its ripple effects. Five main underlying variables worth noting include: 'bribery and corruption'; 'inadequate competition'; 'inadequate project design'; 'political continuity risks' and 'policy instability/inadequate government commitment' (refer to Table 7.1). Moreover, considering the impact of 'financial-related risks' on 'design and construction risks', its underlying variables

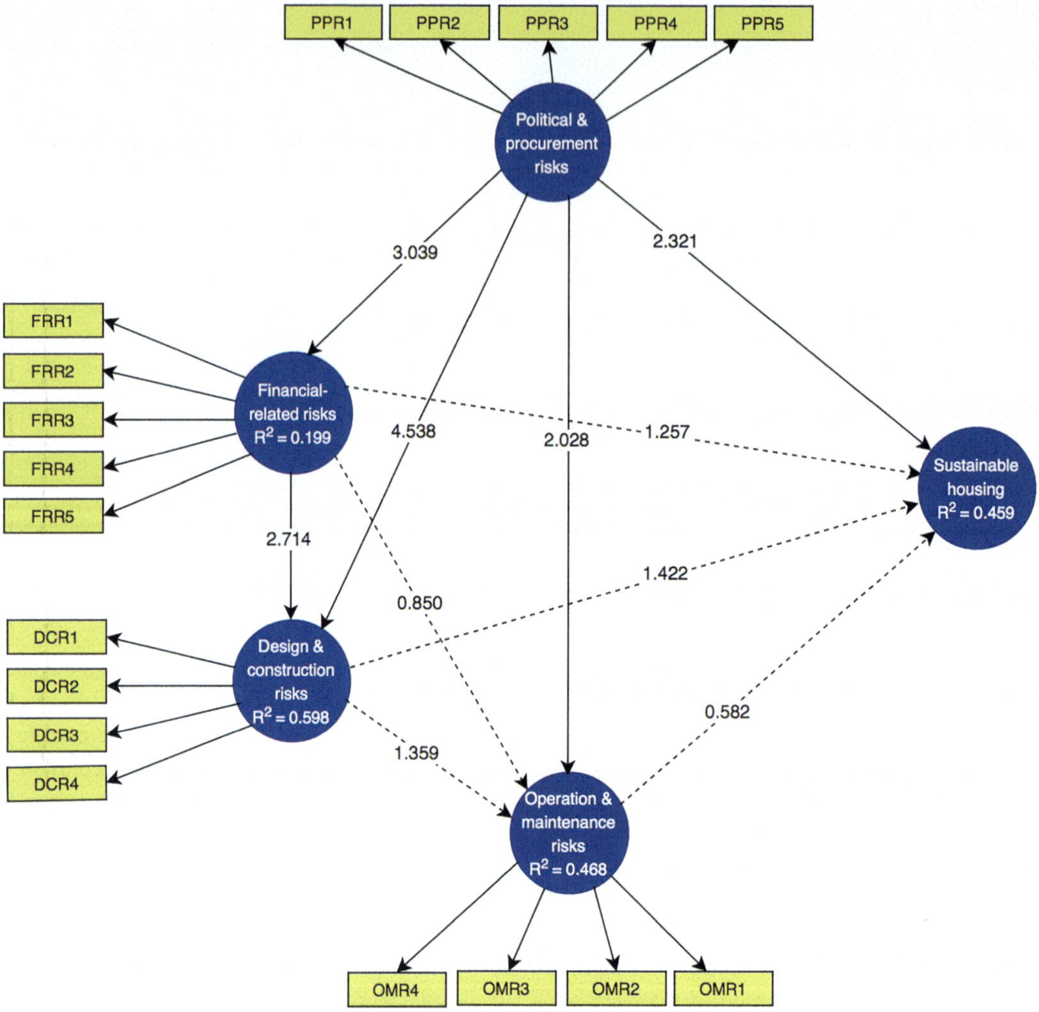


**Fig. 7.3** A model on relationship among risks and sustainable housing. (Adapted from Adabre et al., 2022c)

require utmost attention to manage the 'financial-related risk' construct. Key underlying variables include: 'delay payment'; 'fluctuating cost of finance (interest rate fluctuation)'; 'inadequate financing institutions' and 'inflation rate volatility' (refer to Table 7.1).

#### 7.4.3.3 Results—Impacts of Success Factors on Sustainable Housing

Four constructs of success factors were adopted from extant literature (Adabre et al., 2021a), namely, 'developer-enabling success factors'; 'household enabling success factors'; 'mixed-use development factors' and 'land-use planning factors'. These constructs were used for modeling the influence of success factors on sustainable housing. To do so, the measurement models were first estimated to determine the adequacy of the relationships between each construct and its underlying variables. Suitability of the measurement models was evaluated using the following assessment criteria: loadings of underlying variables and Fornell and Lacker criterion. Upon ascertaining the adequacy of the measurement models, the structural model was estimated. The suitability of the structural model was tested using cross-loading and discriminant analyses. These were found satisfactory and, therefore, test of

**Table 7.4** Results of hypothesis testing on success factors

| Hypotheses | Relationships | Std. Beta | Std. Error | \|t-value\|^ | Decision |
|---|---|---|---|---|---|
| H1 | Developer-enabling success factors -> sustainable housing | 0.380 | 0.144 | 2.64** | Significant |
| H2 | Household-enabling success factors -> sustainable housing | −0.084 | 0.138 | 0.609 | Insignificant |
| H3 | Mixed-use development success factors -> sustainable housing | 0.530 | 0.152 | 3.478** | Significant |
| H4 | Land-use planning success factors -> sustainable housing | 0.040 | 0.126 | 0.117 | Insignificant |

**Note:** **$p < 0.01$, *$p < 0.05$

statistical significance was determined via the bootstrapping analysis. This analysis was conducted to test the following hypotheses:

**Hypothesis 1:** 'Developer-enabling success factors' have a significant effect on sustainable housing.

**Hypothesis 2:** 'Household-enabling success factors' have a significant effect on sustainable housing.

**Hypothesis 3:** 'Mixed-use development success factors' have a significant effect on sustainable housing.

**Hypothesis 4:** 'Land-use planning success factors' have a significant effect on sustainable housing.

Results of the bootstrapping analysis are shown in Table 7.4 and Fig. 7.4.

Among the four constructs of success factors, 'developer-enabling success factors' have a significant effect on sustainable housing development at a t-value of 2.64 with a significant level (p-value) $< 0.01$. Likewise, 'mixed-use development success factors' have a significant effect on sustainable housing (t-value of 3.478, a significant level or p-value $<0.01$) (as indicated by the solid lines in Fig. 7.4). However, 'household-enabling success factors' did not have a significant effect on sustainable housing and likewise 'land-use planning success factors' (as indicated by the dotted lines in Fig. 7.4). The results imply that, regarding the formal sector, incentives to motivate developers could improve sustainable housing supply. Besides, strategies that encourage mixed-use development could reduce commuting distance via proximity. This has sustainability benefits: economic benefits because it reduces commuting cost; environmental benefits since it mitigates emissions owing to short distance travel and social benefits as it improves possibility of interaction among households.

It is surprising that the 'household-enabling success factors' did not yield significant impact on sustainable housing development (as indicated by the dotted lines in Fig. 7.4). Although the government and other policymakers could enhance sustainable housing supply through this construct, it is worth noting that its insignificant effect could be attributed to key reason: some of the household-enabling success factors are counter-productive. For instance, the government of Ghana has offered subsidies to every consumer for the first 50KWh of electricity consumed. This policy aims to improve access to energy (Kumi, 2017). Nonetheless, it is counter-productive since it encourages high consumption of electricity among some households who may not need such subsidies. Besides, some households are still using inefficient energy appliances. Moreover, energy suppliers are not timely reimbursed by the government regarding the subsidies (Eberhard & Shkaratan, 2012). In most developing countries in sub-Saharan Africa, for instance, governments owe electricity producers a high amount of money due to late reimbursement and unpaid subsidies (Kumi, 2017; Gyamfi et al., 2018). The problem is that the electricity producers are not able to expand production, carry out maintenance of plant and equipment. These challenges lead to loss of electricity due

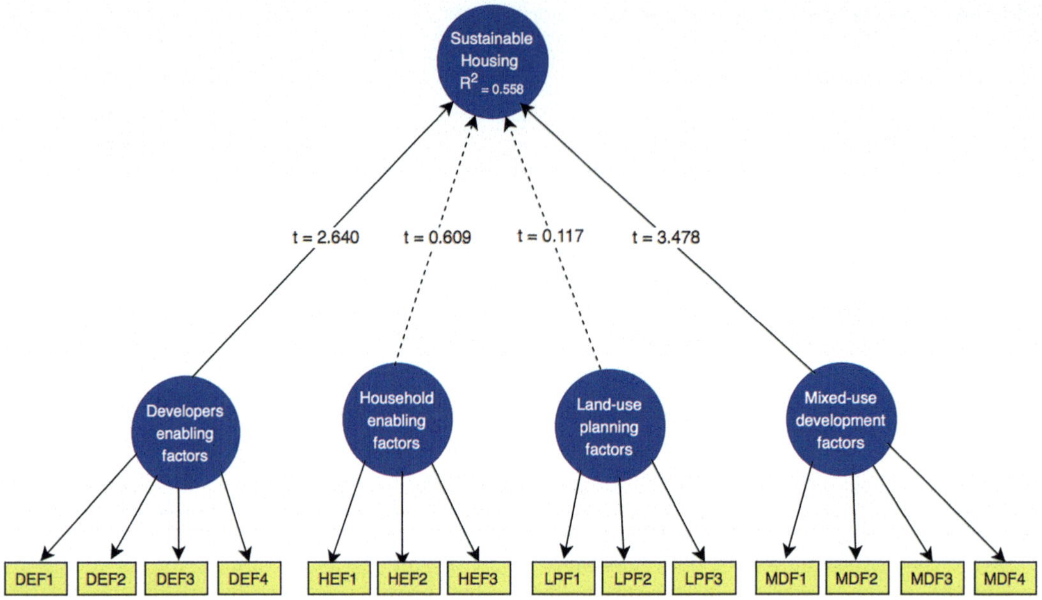


**Fig. 7.4** A model on relationship between success factors and sustainable housing. (Adapted from Adabre et al., 2021a)

to leakages in obsolete equipment. Therefore, instead of providing subsidies on electricity consumption, the subsidies could be directed to subsidizing energy efficient technologies for the households.

The 'land-use planning success factors' did not have a significant effect on sustainable housing (as indicated by the dotted lines in Fig. 7.4). One main reason for this is the nature of land ownership and its effect on housing supply. In most developing countries, for example Ghana, land is owned by the family heads, clans, or chiefs. Because of the different ownerships, even land transaction or purchase of land is problematic: multiple sales of land resulting in multiple owners are common in most sub-Saharan African cities. These lead to litigations and prolongs court proceedings before any development can take place. Dilatory housing supply to match housing demand is partly attributed to the numerous land litigations. Another problem is the different institutions in charge of land development right. Although chiefs, family heads, clans and individuals are landowners, development right on land resides with the government. The two different institutions often conflict each other as the owners mostly usurp the role of the institutions responsible for the development right on land (Siiba et al., 2018). Consequently, housing facilities are constructed at unauthorized places which leads to slums, overcrowding and flooding in major cities of developing countries.

## 7.5 Discussion and Conclusions

Sustainable development is a socio-technical phenomenon (Adabre et al., 2022a). 'Socio-' because it relates to society, communities, cities and changes in practices and policies within the society for the desired sustainable development goals. Technical, since it also involves the adoption of technologies i.e., green building technologies such as LED lighting, solar panels, wind energy and programmable thermostats for the expected goals. Though both aspects—socio-technical—are equally important, progress in the latter (i.e., regards adoption of the technologies) depends on the former (i.e., regards changes in practices or policies and an adjustment in behavior within the community or society). For sustainable housing development in communities and cities, an underlying theory—theory of planned behavior is adopted. Three constructs of the theory, namely, perceived behavioral control (to answer the question: what are the barriers to sustainable housing?); subjective norms (to answer the question: what are the risk factors or uncertainties to sustainable housing?) and personal attitude (to answer the question: what are the success factors for sustainable housing?) were adopted to model their impacts on behavioral adoption/ planned practice (i.e., sustainable housing development in this regard). After establishing this theoretical framing, the partial least squares structural equation modeling (PLS-SEM) was utilized for modeling the impacts of barriers, risk factors and success factors on sustainable housing for communities and cities development.

PLS-SEM analysis of the data revealed the critical success criteria or goals for sustainable development in housing. Some of these criteria include energy efficient housing, rental affordability, commuting cost of households to facility, technology transfer/innovation (i.e., technological enhancement for adaptability, flexibility and secondary materials usage in housing), safety, end user's satisfaction and maintainability of housing facility. Some of these criteria (i.e., innovation for adaptability, flexibility, and secondary materials usage in housing; maintainability of housing for continuous and longer use) are relevant for circular economy in cities and communities (Hosseini et al., 2015; Wuni & Shen, 2022; Oluleye et al., 2023).

The data analysis on the barriers showed that barriers to sustainable housing development could be primary or secondary. Primary barriers have a direct impact on sustainable housing development. And they are, therefore, pressing barriers that require more attention from policymakers. However, secondary barriers do not have a direct impact on sustainable housing. They may influence other barriers. Thus, unlike primary barriers, secondary barriers are less exigent for sustainable housing development. For instance, results of the analysis showed that 'cost-related barriers' are secondary barriers while 'incentive-related barriers' and 'retrofit-related barriers' are primary barriers. The 'cost-related barriers' do not have direct influence on sustainable housing development but directly influence 'incentive-related barriers'. In addition to influencing sustainable housing, 'incentive-related barriers' directly influence 'retrofit-related barriers'. Between 'incentive-related barriers' and 'retrofit-related barriers', 'retrofit-related barriers' have a higher impact on sustainable housing development. This could imply that strategies for mitigating the retrofit-related barriers and for upgrading existing housing facilities to sustainability standards could have more significant impact on sustainable housing development than providing incentives for sustainable housing in new projects. Besides, mitigating the retrofit-related barriers will ensure housing maintenance for continuous and longer use. This is essential for a circular economy in cities and communities.

Analysis regarding the impacts of risk factors on sustainable housing revealed that among the four categories of risks—'political and procurement risks'; 'design and construction risks'; 'operation and maintenance risks' and 'financial-related risks'—political and procurement risks are primary risk factors. Aside directly influencing sustainable housing development, the political and procurement category of risks has ripple effect on the other risk categories as it directly influences them. The 'financial-related risks' which is a secondary risk category has direct influence on 'design and construction risks'. The findings suggest that for sustainable housing development from the formal sector of the Ghanaian housing market, the government, and other policymakers (including international policymakers i.e., UN and World Banks) must focus on the primary risk category—political and procurement risks.

On success factors, the results revealed that two constructs—developer-enabling success factors and mixed-use development success factors—have significant impacts on sustainable housing from the perspective of respondents in the formal sector. However, 'household-enabling success factors' and 'land-use planning success factors' did not have significant impacts on sustainable housing. Moreover, it was revealed that household-enabling success factors could have counter-productive effects on sustainable housing. This is not surprising considering that the allocation of subsidies to households could have inflationary impact. Another reason for this observation is that some practices of households (i.e., non-compliance to planning requirements) have led to detrimental effects (i.e., flooding and slums) in most communities in Accra. The insignificant impact of the 'land-use planning success factors' could be attributed to land litigations and planning challenges owing to differences in ownership right and development right on land.

The study has focused mainly on the formal sector of the Ghanaian housing market. However, the informal sector which is the dominant part of the Ghanaian housing market requires more attention. Future study could model the impacts of the various constructs, viz.: barriers, risks and success factors on sustainable housing development from the perspective of the informal sector (i.e., self-builders).

**Acknowledgements** This chapter is based on my PhD thesis, from which papers of similar background but with different scope and objectives have been published. The authors wish to gratefully acknowledge the Research and Grants Council (RGC) and the Department of Building and Real Estate, the Hong Kong Polytechnic University, for their financial support in conducting this study.

## References

Adabre, M. A., & Chan, A. P. (2019). Critical success factors (CSFs) for sustainable affordable housing. *Building and Environment, 156*, 203–214.

Adabre, M. A., & Chan, A. P. (2020). Towards a sustainability assessment model for affordable housing projects: The perspective of professionals in Ghana. *Engineering, Construction and Architectural Management, 27*(9), 2523–2551.

Adabre, M. A., & Chan, A. P. (2021). Modeling the impact of barriers on sustainable housing in developing countries. *Journal of Urban Planning and Development, 147*(1), 05020032.

Adabre, M. A., Chan, A. P., & Edwards, D. J. (2021a). Modeling relationship between success factors (policies) and critical success criteria (goals) for sustainable housing in developing countries. *International Journal of Construction Management, 23*, 1–11.

Adabre, M. A., Chan, A. P., Edwards, D. J., & Adinyira, E. (2021b). Assessing critical risk factors (CRFs) to sustainable housing: The perspective of a sub-Saharan African country. *Journal of Building Engineering, 41*, 102385.

Adabre, M. A., Chan, A. P., Darko, A., & Hosseini, M. R. (2022a). Facilitating a transition to a circular economy in construction projects: Intermediate theoretical models based on the theory of planned behaviour. *Building Research & Information, 51*(1), 85–104.

Adabre, M. A., Chan, A. P., Edwards, D. J., & Mensah, S. (2022b). Evaluation of symmetries and asymmetries on barriers to sustainable housing in developing countries. *Journal of Building Engineering, 50*, 104174.

Adabre, M. A., Chan, A. P., Edwards, D. J., & Osei-Kyei, R. (2022c). To build or not to build, that is the uncertainty: Fuzzy synthetic evaluation of risks for sustainable housing in developing economies. *Cities, 125*, 103644.

Aibinu, A. A., & Al-Lawati, A. M. (2010). Using PLS-SEM technique to model construction organizations' willingness to participate in e-bidding. *Automation in Construction, 19*(6), 714–724.

Alam, M., Zou, P. X., Stewart, R. A., Bertone, E., Sahin, O., Buntine, C., & Marshall, C. (2019). Government championed strategies to overcome the barriers to public building energy efficiency retrofit projects. *Sustainable Cities and Society, 44*, 56–69.

Ametepey, O., Aigbavboa, C., & Ansah, K. (2015). Barriers to successful implementation of sustainable construction in the Ghanaian construction industry. *Procedia Manufacturing, 3*, 1682–1689.

Ajzen, I. (1991). The theory of planned behavior. *Organizational Behavior and Human Decision Processes, 50*(2), 179–211. https://doi.org/10.1016/0749-5978(91)90020-T

Chan, A. P., & Adabre, M. A. (2019). Bridging the gap between sustainable housing and affordable housing: The required critical success criteria (CSC). *Building and Environment, 151*, 112–125.

Chan, A. P. C., Darko, A., Olanipekun, A. O., & Ameyaw, E. E. (2018). Critical barriers to green building technologies adoption in developing countries: The case of Ghana. *Journal of Cleaner Production, 172*, 1067–1079.

Costolanski, P., Elahi, R., Iimi, A., & Kitchlu, R. (2013). *Impact evaluation of free-of-charge CFL bulb distribution in Ethiopia*. World Bank Policy Research Working Paper (6383).

Darko, A., & Chan, A. P. C. (2018). Strategies to promote green building technologies adoption in developing countries: The case of Ghana. *Building and Environment, 130*, 74–84.

Darko, A., Chan, A. P., Owusu-Manu, D. G., & Ameyaw, E. E. (2017). Drivers for implementing green building technologies: An international survey of experts. *Journal of Cleaner Production, 145*, 386–394.

Djokoto, S. D., Dadzie, J., & Ohemeng-Ababio, E. (2014). Barriers to sustainable construction in the Ghanaian construction industry: Consultants perspectives. *Journal of Sustainable Development, 7*(1), 134.

Durdyev, S., Ismail, S., Ihtiyar, A., Bakar, N. F. S. A., & Darko, A. (2018). A partial least squares structural equation modeling (PLS-SEM) of barriers to sustainable construction in Malaysia. *Journal of Cleaner Production, 204*, 564–572.

Eberhard, A., & Shkaratan, M. (2012). Powering Africa: Meeting the financing and reform challenges. *Energy Policy, 42*, 9–18.

Ezennia, I. S. (2022). Insights of housing providers' on the critical barriers to sustainable affordable housing uptake in Nigeria. *World Development Sustainability, 1*, 100023.

Gan, X., Zuo, J., Wu, P., Wang, J., Chang, R., & Wen, T. (2017). How affordable housing becomes more sustainable? A stakeholder study. *Journal of Cleaner Production, 162*, 427–437.

Guan, L., Abbasi, A., & Ryan, M. J. (2020). Analyzing green building project risk interdependencies using interpretive structural modeling. *Journal of Cleaner Production, 256*, 120372.

Gyamfi, S., Diawuo, F. A., Kumi, E. N., Sika, F., & Modjinou, M. (2018). The energy efficiency situation in Ghana. *Renewable and Sustainable Energy Reviews, 82*, 1415–1423.

Hosseini, M. R., Rameezdeen, R., Chileshe, N., & Lehmann, S. (2015). Reverse logistics in the construction industry. *Waste Management & Research, 33*(6), 499–514.

Kaygusuz, K. (2012). Energy for sustainable development: A case of developing countries. *Renewable and Sustainable Energy Reviews, 16*(2), 1116–1126.

Klein, M. R., & Methlie, L. B. (1995). *Knowledge-based decision support systems: With applications in business*. Wiley.

Kumi, E. N. (2017). *The electricity situation in Ghana: Challenges and opportunities* (p. 30). Center for Global Development.

Kwak, D. W., Rodrigues, V. S., Mason, R., Pettit, S., & Beresford, A. (2018). Risk interaction identification in international supply chain logistics: Developing a holistic model. *International Journal of Operations & Production Management, 38*, 372–389.

Moons, I., De Pelsmacker, P., Pijnenburg, A., Daems, K., & Van de Velde, L. J. (2022). Growers' adoption intention of innovations is crucial to establish a sustainable greenhouse horticultural industry: An empirical study in Flanders and The Netherlands. *Journal of Cleaner Production, 330*, 129752.

Mulliner, E., Smallbone, K., & Maliene, V. (2013). An assessment of sustainable housing affordability using a multiple criteria decision making method. *Omega, 41*(2), 270–279.

Ohene, E., Hsu, S. C., & Chan, A. P. (2022). Feasibility and retrofit guidelines towards net-zero energy buildings in tropical climates: A case of Ghana. *Energy and Buildings, 269*, 112252.

Oluleye, B. I., Chan, D. W., Antwi-Afari, P., & Olawumi, T. O. (2023). Modeling the principal success factors for attaining systemic circularity in the building construction industry: An international survey of circular economy experts. *Sustainable Production and Consumption, 37*, 268–283.

Ortega, D. A., & Braun, P. A. (2011). Information, utility and bounded rationality. In *Artificial general intelligence: 4th international conference, AGI 2011, Mountain View, CA, USA, August 3–6, 2011. Proceedings 4* (pp. 269–274). Springer.

Osei-Kyei, R., Chan, A. P., Yu, Y., Chen, C., & Dansoh, A. (2019). Root causes of conflict and conflict resolution mechanisms in public-private partnerships: Comparative study between Ghana and China. *Cities, 87*, 185–195.

Roufechaei, K. M., Bakar, A. H. A., & Tabassi, A. A. (2014). Energy-efficient design for sustainable housing development. *Journal of Cleaner Production, 65*, 380–388.

Siiba, A., Adams, E. A., & Cobbinah, P. B. (2018). Chieftaincy and sustainable urban land use planning in Yendi, Ghana: Towards congruence. *Cities, 73*, 96–105.

Sony, M., Antony, J., Mc Dermott, O., & Garza-Reyes, J. A. (2021). An empirical examination of benefits, challenges, and critical success factors of industry 4.0 in manufacturing and service sector. *Technology in Society, 67*, 101754.

Wuni, I. Y., & Shen, G. Q. (2022). Developing critical success factors for integrating circular economy into modular construction projects in Hong Kong. *Sustainable Production and Consumption, 29*, 574–587.

Yang, J., & Yang, Z. (2015). Critical factors affecting the implementation of sustainable housing in Australia. *Journal of Housing and the Built Environment, 30*, 275–292.

# Low-Carbon Sustainable Development of 5G Base Stations in China

8

Yangxue Ding, Huabo Duan, Minghui Xie, and Ruichang Mao

## 8.1 Introduction

In recent years, frequent extreme weather events worldwide have highlighted the increasingly prominent environmental issues caused by the overuse of fossil energy. The Intergovernmental Panel on Climate Change's (IPCC) report stated that the earth's average temperature had risen by 0.74 °C over the past 100 years due to greenhouse gas emissions, and if humans continued to rely on fossil fuels, the earth's temperature could rise up to 6.4 °C by the end of the twenty-first century. The greenhouse effect resulting from global warming has caused rising sea levels, frequent extreme weather, accelerated species extinction, and other harmful effects. The United Nations Framework Convention on Climate Change (UNFCCC) defines six greenhouse gases, including carbon dioxide ($CO_2$), nitrous oxide ($N_2O$), methane ($CH_4$), hydrofluorocarbons (HFCs), perfluorocarbons (PFCs), and sulfur hexafluoride (SF6). Among them, $CO_2$ has the most significant impact on atmospheric temperature, accounting for 63% of the greenhouse effect, as detailed in Table 8.1.

The enormity of global carbon emissions is a cause for concern. International Energy Agency (IEA) statistics reveal that global energy carbon emissions exceeded 35,926 million tons $CO_2$eq (Mt $CO_2$eq) in 2019. While the energy carbon emissions decreased in 2020 due to the impact of COVID-19, recent political factors, energy shortages, and increased demand for epidemic relief have resulted in a sharp rise in energy prices (IEA, 2020). Regrettably, despite the rise in energy prices, energy use has not diminished. Nonetheless, Table 8.2 demonstrates that the international community has undertaken multiple initiatives to combat global warming.

China's rapid economic development has led to a significant increase in carbon emissions in recent years. According to the International Energy Agency (IEA), China has now surpassed all other countries in carbon emissions production. In 2019, the energy carbon emissions of China reached 10,619 Mt $CO_2$e, approximately 4.5 times that of India (IEA, 2020). Figure 8.1

Y. Ding
College of Civil & Transportation Engineering, Underground Polis Academy, Shenzhen University, Shenzhen, China

H. Duan (✉)
School of Environmental Science and Engineering, Huazhong University of Science and Technology, Wuhan, China
e-mail: huabo@hust.edu.cn

M. Xie
State Key Laboratory of Environmental Criteria and Risk Assessment, Chinese Research Academy of Environmental Sciences, Beijing, China

R. Mao
DTU Sustain, Department of Environmental & Resource Engineering, Technical University of Denmark, Lyngby, Denmark



J. Zuo et al. (eds.), *Circular Economy for Buildings and Infrastructure*, Sustainable Development Goals Series, https://doi.org/10.1007/978-3-031-56241-9_8

**Table 8.1** Six greenhouse gas warming effects

| Type | Warming effect | Life cycle (Year) |
|---|---|---|
| $CO_2$ | 63% | 50 ~ 200 |
| $CH_4$ | 15% | 12 ~ 17 |
| $N_2O$ | 4% | 120 |
| HFCs | 11% | 13.3 |
| PFCs | | 50,000 |
| $SF_6$ | 7% | – |

**Table 8.2** The major event to mitigate global warming

| Year | Event | Content |
|---|---|---|
| 1988 | The establishment of IPCC | The IPCC is jointly established by the United Nations Environment Programme and the World Meteorological Organization |
| 1998 | United Nation adopted the UNFCC | It forms legally binding climate streaks |
| 1997 | World Climate Conference 3: Phase I of the Kyoto Protocol | It requires developed countries to reduce carbon emissions by 5.2% in 2012 compared to 1990 levels |
| 2012 | World Climate Conference18: Phase II of the Kyoto Protocol | It sets carbon dioxide emissions for developed countries from 2012 to 2020 |
| 2015 | Paris Climate Change Conference: Paris Agreement | To limit global temperature rise to less than 2 °C above pre-industrial levels and, if possible, to less than 1.5 °C |

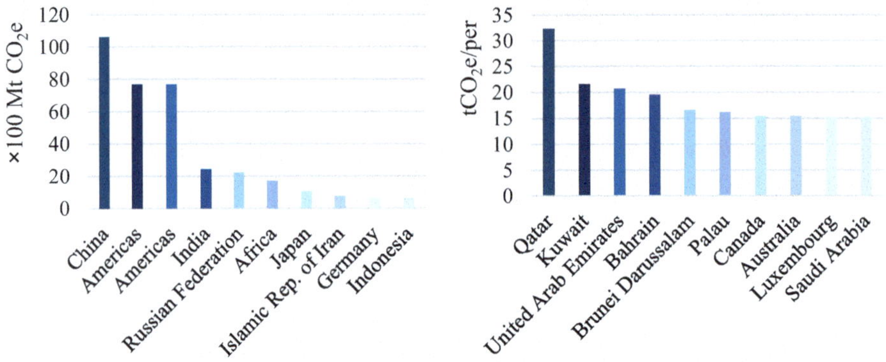


**Fig. 8.1** (**a**) Top 10 countries in energy carbon emissions. (**b**) Top 10 countries in per capita energy carbon emissions

illustrates the energy carbon emissions of the top 10 countries in the world. Despite its high carbon emissions, the per capita energy carbon emission of China is relatively low compared to that of developed countries. In 2019, China's per capita energy carbon emission was 7.41 $tCO_2e$, which is only half of Canada's per capita emission. As a responsible country, China has undertaken the responsibility of global carbon emission reduction. Since joined the United Nations Framework Convention on Climate Change (UNFCC) in 1994, China has actively participated in various Intergovernmental Panel on Climate Change activities, gradually evolving from a participant in climate governance to a leader. Low-carbon city pilot work is being actively carried out in China, and the government has identified 3 batches of 87 pilot cities, taking the lead in exploring the path to carbon peaks and new models of green development. At the 75th United Nations General Assembly, the General Secretary of China made a carbon reduction commitment to the international community, stating that China will strive to reach the peak of carbon emissions by 2030 and achieve carbon neutrality by 2060.

Many countries have made significant investments in digital infrastructure, including 5G base stations which have become a critical component of this infrastructure. However, due to their high radio frequency and limited coverage, the construction and operation of 5G base stations can lead to significant energy consumption and greenhouse gas emissions. To address this challenge, scholars have focused on developing sustainable 5G base stations. For instance, Guo et al. (2022b) utilized LMDI decomposition analysis to estimate carbon emissions from 5G base stations in China, while Ding et al. (2022) conducted the life-cycle assessment (LCA) to calculate carbon emissions of single macro and micro 5G base stations in China. Goncalves et al. (2020) explored carbon neutrality evaluation of 5G base stations from the perspective of network structure and carbon sequestration. Despite the growing attention on sustainable 5G base stations, there remains a gap in research analyzing their environmental impact. Therefore, this chapter aims to provide an overview of green 5G base stations, exploring their construction in China, their environmental impact, and the various factors and measures that influence their carbon reduction, ultimately aiming to find a balance between digitalization and sustainability of infrastructure.

## 8.2 Development of New Infrastructure

### 8.2.1 Development and Research

In recent years, the digitalization of infrastructure has emerged as a crucial element in driving high-quality development in many countries. The US government unveiled the "Domestic Infrastructure Revitalization Plan" in 2021, which outlined the need to modernize aging infrastructure, including bridges, roads, railways, airports, and urban transportation systems, while expanding and enhancing digital infrastructure. The government of China proposed accelerating the adoption of 5G and boosting the development of new infrastructure such as artificial intelligence, the industrial Internet, and the Internet of Things in 2018. The attributes of new infrastructure are shown in Fig. 8.2. The concept of new infrastructure has been interpreted variously. The new infrastructure encompasses seven principal domains: 5G infrastructure, Ultra-high voltage (UHV), intercity high-speed railway and intercity rail transit, new energy vehicle charging piles, big data centers, artificial intelligence, and the Industrial Internet, which span numerous industrial chains. The Bank of China Research Institute estimated that the total investment in China's seven key areas of "new infrastructure" would be approximately 1.2 trillion yuan in 2020. The National Development and Reform Commission provided an official definition of new infrastructure, which consists of three categories: information infrastructure, integration infrastructure, and innovation infrastructure (Liu et al., 2020).

Currently, research on new infrastructure is limited, with a primary focus on policy interpretation and economic impact analysis. In a recent study, Wu et al. (2021) conducted a quantitative evaluation of policies related to new infrastructure in 12 provinces in China, utilizing social network analysis and text mining methods to

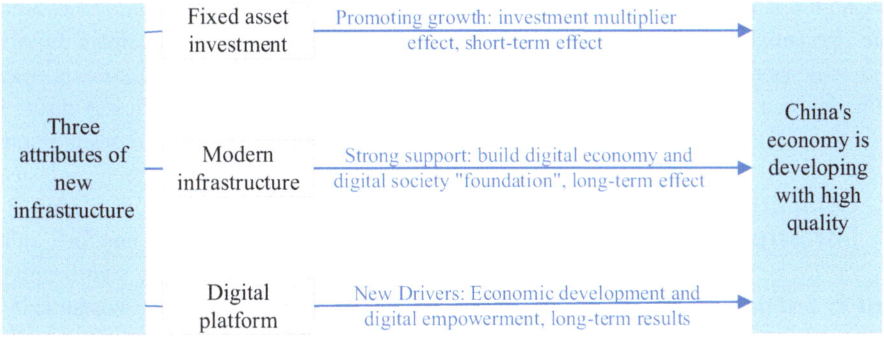


**Fig. 8.2** Three attributes of new infrastructure

reveal the coverage and positioning of new infrastructure. Additionally, they employed the modeling consistency index method to analyze 12 provincial policies, concluding that the Chinese government adopts a bottom-up incremental planning model for new infrastructure policies. Guo et al. (2020) classified the stimulating effects of new infrastructure on the economy into three categories: (i) new infrastructure as a fixed asset investment behavior has a multiplier effect that can drive economic growth and promote inclusive economic growth when compared to traditional infrastructure; (ii) as modern infrastructure, particularly digital infrastructure, new infrastructure can provide fundamental support for the digital transformation of the economy; (iii) as a digital platform, new infrastructure can provide new kinetic energy for China's economic development, promote the integrated development of industries, form a new industrial ecology, and serve the construction of a modern industrial and economic system. Given that society needs to transform to a digital economy, infrastructure transformation is also conducive to the long-term development of the economy (Li, 2020). Du et al. (2022) discussed the theoretical mechanism of the impact of new infrastructure investment on economic growth from three dimensions: economic growth conditions, growth process and growth results, as well as technological innovation, industrial structure, and productivity. Verified by data from provinces in China, the study indicates that new infrastructure investment can significantly improve the quality of economic growth. These studies demonstrate that the construction of new infrastructure is likely to lead to rapid and high-quality economic growth. However, the potential environmental problems associated with new infrastructure construction must not be ignored.

### 8.2.2 Comparison with Traditional Infrastructure

Compared to traditional infrastructure construction, new infrastructure covers a broader range of areas. Traditional infrastructure primarily includes railways, highways, airports, ports, water conservancy, and other physical facilities. In contrast, new construction encompasses 5G base stations, new energy automobile charging stations, large data centers, artificial intelligence, the internet, the internet of things, and other advanced technological areas. The majority of new construction is based on traditional infrastructure systems, but it actively explores emerging areas such as smart logistics, smart cities, smart medical treatment, and distance education. Moreover, new infrastructure is distinguished by its high correlation and wide employment opportunities, as noted by Guo et al. (2020). Therefore, new infrastructure has the potential to stimulate the entire industry chain, with each infrastructure investment area benefiting its upstream and downstream affiliates. For instance, the construction of the internet of things will boost the business expansion of related industries such as modern manufacturing, modern agriculture, logistics, retail, and financial services.

The construction of new infrastructure is often regarded as a more environmentally friendly option compared to traditional infrastructure, due to its reduced material input. However, the production of electronic components required for new infrastructure often involves high-carbon emitting materials such as metals. Additionally, with the fast-paced development of wireless network technology, the replacement cycle of electronic components has become shorter, contributing to an increase in electronic waste. Furthermore, the operation stage of new infrastructure consumes a significant amount of energy. For instance, data centers require power to operate their servers and cooling systems, as well as backup power supplies to ensure uninterrupted service. This energy consumption creates challenges for achieving green and sustainable new infrastructure during construction and operation (Shi et al., 2022). Therefore, addressing these issues has become a significant research area for promoting sustainable new infrastructure.

### 8.2.3 Circular New Infrastructure

Circular new infrastructure is a development model that integrates the concept and principles of circular economy into the construction of new infrastructure, with a focus on achieving sustainable economic, social, and environmental development. This approach emphasizes resource recycling, waste minimization, and energy conservation throughout infrastructure construction and urban development.

Efficient resource utilization is a central aspect of circular new infrastructure. This involves selecting environmentally friendly materials, adopting sustainable design and construction methods, and reducing resource consumption and waste generation during the infrastructure development process. Furthermore, circular new infrastructure actively promotes the application of green energy sources. Renewable energy technologies, such as solar, wind, and geothermal energy, are encouraged to decrease reliance on traditional energy sources, mitigate carbon emissions, and minimize environmental impact. The circular economy model is an essential consideration in the planning, design, and construction of infrastructure projects. This approach promotes the principles of recycling, remanufacturing, and recovery throughout the infrastructure's entire life cycle, minimizing waste generation and facilitating resource reuse.

Through the implementation of circular new infrastructure, society can achieve environmentally friendly, intelligent, and sustainable infrastructure development. This model enhances resource utilization efficiency, reduces environmental pressure, and fosters green economic growth and sustainable social development. The integration of advanced information technology, the Internet of Things, and artificial intelligence plays a crucial role in circular new infrastructure. By leveraging these technologies, infrastructure systems can be intelligently managed, enabling efficient energy utilization, intelligent transportation scheduling, and optimal resource allocation. 5G base stations are the basis for building 5G network implementation technology, so it is very important to realize the cyclic low-carbon development of 5G base stations.

## 8.3 Energy Consumption of 5G Base Stations

### 8.3.1 Development of Wireless Networks

From the first-generation mobile communication system (1G) to the fifth-generation mobile communication system (5G), the services that can be supported have undergone significant changes, and the communication equipment has evolved from cellular phones and Personal Phone Systems (PHS) to smartphones, as shown in Fig. 8.3. During the 1G era, only voice communication was possible through cellular phones, but the signal quality was very poor. The introduction of PHS in 2G significantly improved communication quality and also enabled short message and basic internet services. Since 3G, smartphones have been rapidly developed, enabling multimedia services in addition to basic voice, SMS, and internet services. The fourth generation of mobile communication (4G) combined WLAN technology with 3G communication technology, enabling faster image transmission and clearer image quality, and allowed internet access up to 100 m, 20 times faster than 3G. 4G provided the foundation for high definition (HD), video broadcast, and cloud computing. The main advantage of 5G is its high data transfer rate of up to 10Gbit/s, which is faster than wired internet and 100 times faster than 4G. Additionally, it offers lower network latency, less than 1 millisecond, compared to 30–70 milliseconds for 4G. These features enable 5G networks to be widely used in industrial internet, autonomous driving, telemedicine, and other fields, forming the foundation of new infrastructure. The evolution of these services drives the development of technology and network architecture, while technology and network architecture provide the basis for the evolution of services, leading to diverse services, higher quality, and greater user satisfaction (Yu et al., 2017).

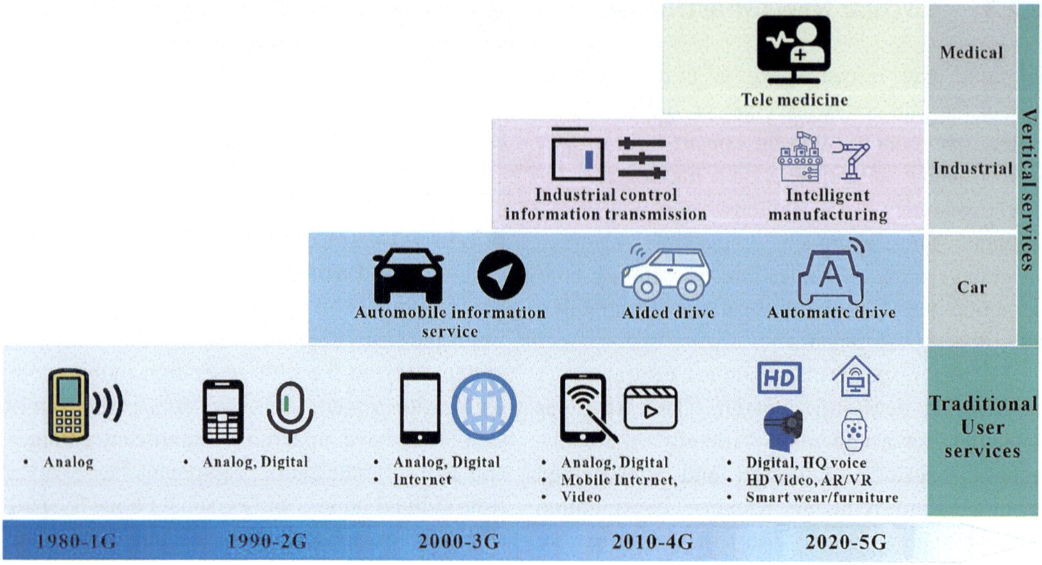


**Fig. 8.3** Main applications from 1G to 5G

The 1G utilized a large area system with base stations that had large transmission power and coverage areas. However, due to the large cell radius and complex landforms, the coverage performance was not satisfactory. The 2G network architecture adopted a cellular network system, with base stations having reduced specifications and transmission power. Each 2G base station covers a hexagonal area that is closely connected, providing better coverage performance than the 1G regional system. Nevertheless, the cellular architecture has coverage blind spots in densely populated or occluded areas. The 3G network introduced a heterogeneous network, coexisting with macro base stations and small cells. Small cells cover and complement areas with coverage holes and high capacity needs, greatly enhancing network performance. The 4G network separated request processing and data service and improved the base station's structure, significantly reducing processing delay. In addition, the 4G network architecture deployed pico, femto, and other heterogeneous components, further improving coverage performance. The 5G network continues to use a heterogeneous network architecture, with macro base stations and small cells, and Wi-Fi is used to supplement coverage and hotspot capacity requirements (Hodara & Skaljo, 2021).

### 8.3.2 Composition and Network Architecture

5G base stations are categorized into micro base stations, macro base stations, and indoor sub-systems based on their transmit power and coverage. As 5G operates at a higher frequency than 4G, its coverage capability is lower and the signal penetration is poor, causing significant signal attenuation. Thus, deploying macro base stations on a large scale is not feasible for 5G networks. Micro base stations, on the other hand, are smaller and more flexible, allowing them to supplement the peripheral communication that cannot be covered by macro stations, thereby improving communication quality and capacity. Therefore, micro stations play a critical role in 5G planning. Additionally, during operation, macro stations consume more energy than micro stations. Hence, this study examines the selection of both macro and micro stations.

The architecture of 5G base station equipment mainly relies on building a Base Band Unit (BBU) and an Active Antenna Unit (AAU) for the radio frequency equipment. The equipment of both 5G macro and micro base stations typically consist of baseband units, radio frequency units, antenna feeder systems, basic components, iron towers and poles, power supply, air conditioning, and computer rooms (Chen et al., 2010; Igor, 2007). The key differences between 5G and 4G base stations are the added radio frequency remote equipment, baseband board, and antenna feeder system. Figure 8.4 illustrates the differences between the equipment of 4G and 5G base stations.

The 5G network architecture is similar to that of the 4G network and is generally composed of a radio access network (NG-RAN) and a core network (5GC). Figure 8.5 shows the 5G network structure provided by 3GPP. 5GC mainly comprises AMF/UPF network elements, where AMF represents the access and mobility management function, and UPF represents the user plane function. NG-RAN is a 5G radio access network, where gNB is a 5G base station, and ng-eNB is a 4G base station, and the upgraded 4G base station can access the 5G core network considering the long-term coexistence of 4G and 5G in the future. Equipment power consumption for the 5G network mainly includes wireless access network equipment and core network equipment. Core network equipment power consumption is similar to that of 4G, while the power consumption of 5G wireless access network equipment, particularly the typical value of gNB operating bandwidth of 100 MHz, transmit power of 200 W, is significantly increased compared with the 4G network, and is the main factor affecting network energy consumption. The 5G base station equipment architecture mainly adopts the BBU + AAU method. The BBU is the baseband part and can be further divided into two logical network elements, CU and DU. The CU handles the

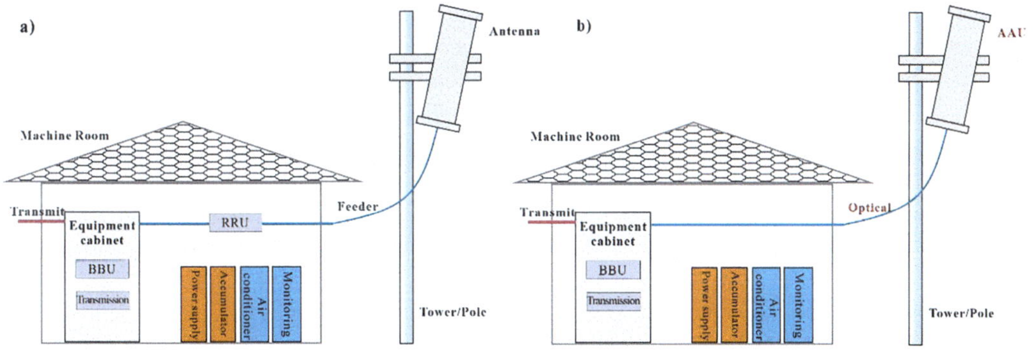


**Fig. 8.4** (**a**) The equipment of the 4G base station, (**b**) The equipment of 5G base station

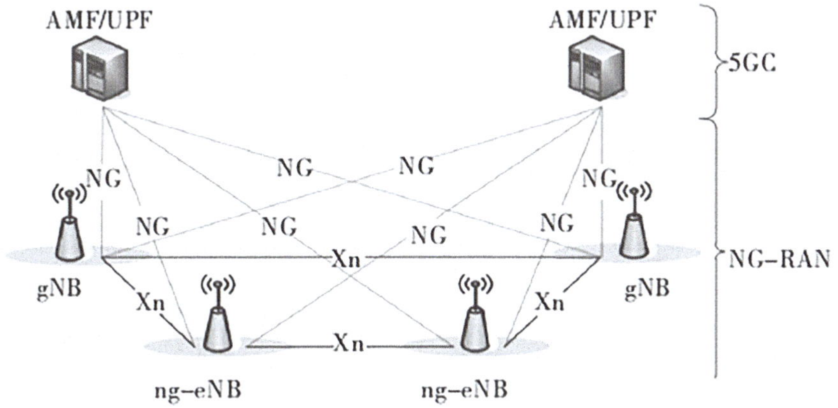


**Fig. 8.5** The 5G network structure

protocol stack functions above the PDCP layer of the wireless network, while the DU handles radio protocol functions below the PDCP layer. The current domestic 5G network construction mainly adopts the CU/DU joint deployment method in principle. The AAU is the radio frequency part, which is an active antenna unit integrated by the radio frequency unit and the antenna. Its operation phase requires more energy consumption, which is 2 ~ 3 times that of 4G base stations. At present, a single 5G base station's full load power is almost 3600 W, while that of a single 4G base station is nearly 1000 W, considering only the power consumption of the baseband unit, radio frequency unit, and air conditioner during the operation phase.

The 5G new wireless network architecture requires a higher bandwidth and lower latency on the forward interface compared to 4G base stations to accommodate 5G technologies such as large-scale MIMO that demand high bandwidth and low latency. It also needs to support data grouping to enable flexible CU/DU deployment in various business scenarios (Halbauer et al., 2018). Moreover, the peak data rate of 5G networks is approximately ten times higher than that of 4G networks (Arshad et al., 2019). Energy efficiency comparison between 5G and 4G networks shows that the former has a higher energy efficiency of about 7.2 W/Mbps.

### 8.3.3 The Construction of 5G Base Stations in China

Since 2019, construction plans have been launched across the country to implement the proposal of new infrastructure. Figure 8.6 depicts the distribution of 5G base stations in China, which shows that the construction of 5G base stations from 2020 to 2021 was mainly concentrated in coastal cities. The number of 5G base stations in economically developed areas and high population density regions has also increased accordingly. Guangdong Province ranks first in the number of base station constructions, with a cumulative construction target of 290,000 seats by 2025. In the Pearl River Delta, 5G broadband city clusters have been built, achieving 5G network coverage in urban areas, counties, and central towns in eastern, western, and northern Guangdong. In 2021, the three major domestic operators, China Mobile, China Telecom, and China Unicom, built more than 1.4 million base stations and 650,000 new ones throughout the year. China has become the first country in the world to build 5G networks on a scale based on the independent networking model, with over 60% of the world's total 5G base stations. Furthermore, fixed broadband will be upgraded to gigabit, with the proportion of fiber users increasing from less than 10% in

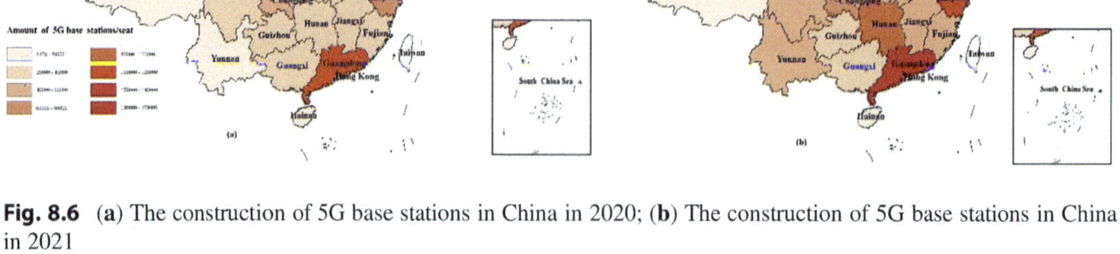


**Fig. 8.6** (**a**) The construction of 5G base stations in China in 2020; (**b**) The construction of 5G base stations in China in 2021

2012 to 94.3% in 2021. All three major operators focus on 5G business, with China Mobile's 5G investment accounting for about 60% of its capital expenditure in 2021, China Telecom's 5G investment accounting for about 44% of its annual cumulative capital expenditure, and China Unicom's 5G investment accounting for about 46% of its cumulative capital expenditure. 5G base stations will coexist with 4G base stations for a long time, and their construction is similar. The number of 4G base stations built in 2021 was 5.9 million, with a gradually slowing growth rate.

### 8.3.4 Energy Consumption

Wireless networks are notorious for their high energy consumption, with the radio access network (RAN) accounting for 70% to 80% of the total electric power consumption (Chih-Lin et al., 2014; Wu et al., 2015). The energy consumption of wireless technology not only results in a significant amount of carbon emissions but also increases operating costs (Panwar et al., 2016). 5G communication represents a significant leap in terms of bandwidth, latency, and other aspects when compared to 4G/3G/2G communication (Shafi et al., 2017). However, the upgrade of wireless networks from 1G to 5G does not necessarily entail the complete replacement of existing equipment. Currently, 2G, 3G, and 4G networks continue to coexist and be developed alongside 5G.

The rapid expansion of new infrastructure has led to a significant increase in energy consumption. Mobile device production, radio access network operations, data center operations, and data transmission contribute to 30%, 29%, and 19% of the carbon footprint, respectively (Belkhir & Elmeligi, 2018). Base stations, which serve as the backbone of wireless networks, consume 60% of the total energy consumed by such networks, and 3G and 4G base stations alone account for 75% to 90% of the total energy consumption (Ahmed et al., 2018). 5G base stations require higher antenna requirements than their 4G counterparts and incorporate new, high-energy components such as millimeter-wave or microwave transceivers, faster data converters, field-programmable gate arrays (FPGAs), integrated MIMO antennas, and high-power/low-noise amplifiers. The mixed deployment of 2G, 3G, and 4G radios with 5G base stations results in a nearly 70% increase in the electric power demand of each site's base station. As 5G networks have higher and higher energy requirements, there may be insufficient AC power, battery capacity, and backup battery capacity, as well as the inability to support high-power long-distance transmission.

## 8.4 Environmental Impact of 5G Base Stations and Influencing Factors

### 8.4.1 Environmental Impact Quantification Methods

Most of existing studies on the environmental impact of wireless networks have mainly concentrated on the overall network, end products, and local applications, such as the Internet of Things. The quantification methodology used in these studies is primarily based on Life Cycle Assessment (LCA) and differentiated using local databases.

Life cycle assessment (LCA) is a standardized method (ISO 14040-14044) used to evaluate the environmental impact of a product or service system from production to disposal. LCA is widely employed to identify opportunities to improve the environmental impact of a product system by analyzing the use of energy and materials and the environmental impact of waste discharge. The International Organization for Standardization has included LCA in the international standard ISO14000 series, and in collaboration with the International Society of Environmental Toxicology and Chemistry, has established the fundamental structure and steps of LCA. The standard outlines four main processes of LCA: defining objectives and scope, conducting inventory analysis, performing impact assessment, and interpreting the results.

Life cycle assessment (LCA) is widely acknowledged as a robust and authoritative methodology for quantifying carbon emissions and evaluating the environmental impact of various products and service systems. With the rise of wireless networks and the consequent environmental concerns, LCA has become a vital tool in environmental assessment. Previous studies have applied the LCA method to assess the carbon emissions of wireless network systems, such as the core network of base stations in the UK (Hatem et al., 2018), the GSM and UMTS networks (Scharnhorst et al., 2006), and all components in 3G networks (Stutz et al., 2006). Mulrow et al. (2022) conducted a literature review to identify the environmental impacts of the Internet of Things (IoT) subsystems and classified them according to the LCA methodology, considering all major stages of the product life cycle. However, it is worth noting that the validity of LCA research results is limited to the research assumptions, as some scholars have pointed out (Guldbrandsson & Bergmark, 2012). By assessing the environmental impact of 5G base stations through the LCA method, the quantification of $CO_2$ emissions is comprehensive, considering the entire life cycle of the product. Moreover, the LCA method has been standardized by the International Standards Organization as ISO 14040-14044, ensuring its reliability and credibility.

### 8.4.2 Carbon Emissions of 5G Base Station

In accordance with the ISO 14040 standard, LCA consists of four key stages, namely purpose and scope determination, inventory analysis, impact assessment, and life cycle interpretation. Figure 8.7 depicts the boundary diagram for the life cycle assessment of 5G base stations. The purpose and scope of the environmental impact study of 5G base stations based on LCA are clearly defined. Firstly, the purpose is to quantify the carbon emissions throughout the entire life cycle of 5G base stations, including the production, transportation, construction, use, and recycling stages. This is intended to clarify the overall environmental impact of 5G base station construction. Secondly, the scope encompasses the production, transportation, and operation processes of various materials, equipment, and components of both single 5G macro and micro base stations.

The civil construction of 5G base stations is typically carried out using the existing infrastructure of 4G base stations, resulting in less material input during the construction phase. The primary focus on carbon emission generation is during the use phase due to power consumption. Previous research has estimated that a single 5G base station will produce approximately 30.2 ~ 33.5 $tCO_2eq$ throughout its life cycle (Ding et al., 2022; Guo et al., 2022a). Consequently, the

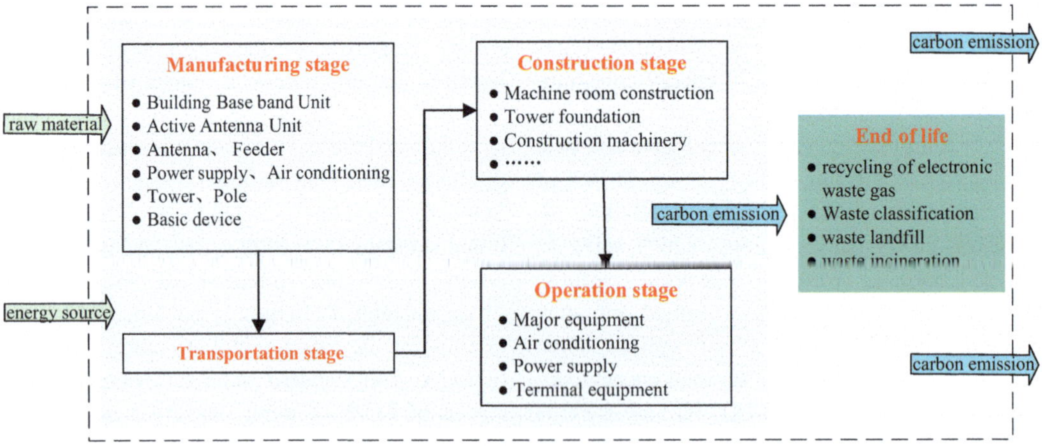


**Fig. 8.7** The system boundary of the $CO_2$ of 5G base station

carbon emissions from 5G base stations in China in 2021 amounted to approximately 49.2 $MtCO_2eq$.

### 8.4.3 Influencing Factors

The carbon emission impact factors of 5G base stations encompass several dimensions. The manufacturing phase is recognized as an important source of carbon emissions, involving the production of manufacturing equipment, assembly, testing, and transportation. Additionally, the construction, operation, and maintenance of 5G base stations will also have an impact on carbon emissions.

Specifically, the construction phase includes base station site selection, site preparation, infrastructure construction, and network equipment deployment. On the other hand, the operation phase is characterized by energy consumption, maintenance, and management. In terms of energy consumption, 5G base stations require continuous operation and stability, which leads to significant electricity consumption (Guo et al., 2022a). This power is mainly supplied by transmission equipment and auxiliary equipment, such as transformers, UPS power supplies, and cooling equipment. Moreover, the base station's operation consumes substantial amounts of fuel, including diesel and LPG, thereby contributing to carbon emissions. All of these energy consumption factors have a bearing on the base station's carbon footprint. Furthermore, the maintenance and management of 5G base stations, which usually are situated in remote areas or high-altitude locations, require considerable transportation and materials, such as maintenance equipment and spare parts. The carbon emissions produced during these processes cannot be disregarded. Overall, the carbon emission impact factors of 5G base stations entail several aspects, including the manufacturing, construction, operation, and maintenance phases. Therefore, to assess the carbon emissions of 5G base stations comprehensively, the impact of these factors must be considered, quantified, and analyzed using scientifically rigorous assessment methods.

## 8.5 Green and Sustainable 5G Base Station

The exponential growth of the communication industry has led to a corresponding increase in power consumption, prompting widespread concern. According to a survey, the annual energy consumption of the communication industry exceeds 30 billion kWh, growing at a rate of approximately 15–20% per year (Fettweis & Zimmermann, 2008). The energy consumption of Information and Communication Technology (ICT) accounts for a progressively larger proportion of the total global energy consumption, accounting for 3% of the total energy consumption. Moreover, the resulting carbon dioxide emissions represent 2% of the global total emissions, which has significant negative impacts on the ecological environment (Oh et al., 2011). The limited site selection of 5G base stations and the increase in their number lead to network interference, and network construction costs increase for operators. With the rapid development of 5G, energy loss in the information and communication industry has become a critical problem. In 2020, the expected energy loss was 1430 tons, accounting for 10% of the global energy loss. Furthermore, data transmission volumes are increasing, resulting in a corresponding increase in energy consumption (Mobile's Green Manifesto, 2012). To address these issues, the concept of green communication emerged to reduce energy waste, pollution, and achieve energy savings and emission reductions (Srivastava et al., 2020). The wireless access network incurs the most significant energy loss, with base station loss being the largest contributor. Consequently, energy-saving and emission reduction measures should primarily focus on reducing the resource demand of base stations (Liu, 2018). Currently, research on energy-saving and emission reduction of base stations is concentrated on network architecture, network deployment, resource scheduling, link-level technology, and other aspects.

### 8.5.1 Network Architecture

The architecture of the 5G network is designed to offer several advantages over its predecessors. The use of a flat IP network results in a system that is both powerful and easy to install, with minimal delays and high levels of application support. Furthermore, it enhances delay over wireless links and ensures good integration with the environment (Agarwal et al., 2021). The flat IP network architecture is designed to reduce the number of nodes in the data channel, which in turn reduces energy loss during transmission and power consumption during use. In addition, the 5G network architecture incorporates nanocore network technology, which includes a flexible wireless access cloud, an open intelligent control cloud, and an efficient and low-cost forwarding cloud. This technology employs cloud storage of data and documents, cloud computing, and software-defined networking (SDN) transmission to significantly reduce material and energy input during 5G network construction and operation, thus leading to a reduction in carbon emissions.

### 8.5.2 Network Deployment

The site layout of 5G network is closely related to the frequency band used for transmission. Presently, 5G network frequency band is mainly deployed in high frequency and millimeter-wave frequency band, which have lower transmission capacity than the previous network frequency band. Consequently, deploying more base stations is necessary for 5G base stations to cover the same area. Macro and micro base stations are currently being deployed for 5G network. The base station is categorized into micro base station, macro base station, and sub-system based on the coverage range. Micro base stations are being deployed to increase coverage. The power consumption of a single macro base station is approximately 5 kW, whereas a Pico Cell requires only about 10 W (Bolla et al., 2012; Deruyck et al., 2014; Hu & Yi, 2014). Deploying multiple micro base stations to cover the blind spots of a macro base station will reduce power consumption during operation, thereby reducing carbon emissions. Chen et al. (2010) proposed that increasing the deployment of micro base stations in areas with high user density could improve the energy efficiency of the entire network. The coexistence of 4G and 5G will last for a long time, and upgrading 4G base stations to 5G base stations will reduce the material input in the construction stage. However, most studies are based on a single case study, which may not be universally applicable. Therefore, subsequent application should be modified based on actual needs.

### 8.5.3 Scheduling of Resources

The deployment of base stations is a critical consideration in meeting users' traffic demand. However, the network flow of different communities changes with time, presenting a state of ups and downs, referred to as the "tidal effect". To ensure the quality of user experience (QOE), operators employ the deployment of additional micro base stations within the coverage area of macro stations. Nevertheless, due to the existence of the tidal effect, when the data demand of a particular cell is low, only a few micro base stations need to be operational to maintain the QOE of the cell. To reduce system energy consumption, a small portion of micro base stations can be placed into a hibernation state. The process of switching users simultaneously necessitates consideration of how to activate the base station from hibernation. Additionally, the power consumed by restarting the base station should be considered to strike a balance between QOE and energy conservation.

In addressing the challenge of balancing quality of experience (QOE) and energy consumption in base station operation, researchers have proposed several approaches. One such approach, as suggested by Tsilimantos et al. (2013), involves closing a specific macro base station, followed by adjusting the transmitting power of the remaining base stations to maintain the coverage area of the macro station system before and after hibernation. Zhou et al. (2009) proposed a method that utilizes network traffic prediction to place links with low utilization rates into dormancy. Another proposed method is the stochastic cellular network performance model based on Poisson-Voronoi-Tessellation (PVT), which was presented by Xiang et al. (2013). However, most current studies assume a uniform distribution of access users and services in a given community, which does not reflect the complexity of real-world scenarios. Therefore, planning and parameter settings should be informed by historical data of specific cases, while ensuring QOE and reserving adjacent base station resources in advance during the planning phase.

### 8.5.4 Link Level Technology

5G networks incorporate several advanced technologies that aim to enhance user experience by improving speed, reducing latency, and minimizing power consumption. Among these technologies are the following: (i) Massive MIMO, which involves the installation of large-scale antenna arrays at base stations to enhance energy and spectrum efficiency, ensure communication security, and improve network stability. Fehske et al. (2011) demonstrated that high-order MIMO can significantly reduce base station power consumption and cost. Additionally, Farhang et al. (2014) proposed that increasing the number of antennas can enable base stations to use low-power amplifiers. (ii) D2D (device-to-device) communication, which facilitates direct data transmission between devices without routing through the base station. This technology reduces base station workload, communication transmission time, and data transmission efficiency. However, power consumption and resource allocation optimization remain problematic in D2D communication. (iii) NOMA (Non-Orthogonal Multiple Access), which enhances system throughput and spectrum efficiency, and supports multiple users on a single resource. NOMA is suitable for future system deployment, especially in scenarios with intense node access and wider coverage, and has significant performance advantages over traditional orthogonal access.

## 8.6 Discussion

This chapter discussed the new infrastructure, with a particular focus on the environment impact of 5G stations. The composition of equipment, network architecture, and methods for evaluating environmental impact are also analyzed. To estimate the carbon emissions of 5G base stations in China in 2021, the LCA method is employed, with carbon emissions serving as the environmental evaluation index. The results indicate that, due to the high carbon emissions resulting from the new infrastructure, the carbon emissions of 5G base stations in China in 2021 amounted to 49.2 $MtCO_2eq$. In order to reduce the carbon emissions of 5G base stations and achieve green 5G, this paper further examines the literature related to existing energy-saving technologies for 5G base stations. The existing energy-saving technologies are categorized based on network architecture, network deployment, scheduling of resources, and link level. While 5G has contributed to rapid and high-quality economic development, the environmental problems caused by the significant electricity consumption in the operation phase cannot be ignored. Therefore, a balance must be struck between digitalization and the sustainability of infrastructure, which should be a priority for local governments. The examination of the impact factors on carbon emissions of 5G base stations, as well as the carbon reduction measures outlined in this chapter, can serve as a valuable reference for local governments seeking to establish sustainable, environmentally-friendly 5G infrastructure.

## References


Agarwal, V., Sharma, C., Shetty, R., et al. (2021). A journey towards a converged 5G architecture & beyond 2021. In *2021 IEEE 4TH 5G WORLD FORUM (5GWF 2021)*.

Ahmed, F., Naeem, M., Ejaz, W., et al. (2018). Resource management in cellular base stations powered by renewable energy sources. *Journal of Network and Computer Applications, 112*, 1–17.

Arshad, Q. K. U. D., Kashif, A., et al. (2019). A review on the evolution of cellular technologies. In *16th International Bhurban Conference on Applied Sciences and Technology (IBCAST)*.

Belkhir, L., & Elmeligi, A. (2018). Assessing ICT global emissions footprint: Trends to 2040 & recommendations. *Journal of Cleaner Production, 177*, 448–463.

Bolla, R., Bruschi, R., Carrega, A., et al. (2012). Cutting the energy bills of internet service providers and telecoms through power management: An impact analysis. *Computer Networks, 56*(10), 2320–2342.

Chen, T., Kim, H., & Yang, Y. (2010). Energy efficiency metrics for green wireless communications. In *2010 International Conference on Wireless Communications & Signal Processing*.

Chih-Lin, I., Rowell, C., Han, S., et al. (2014). Toward green and soft: A 5G perspective. *IEEE Communications Magazine, 52*(2), 66–73.

Deruyck, M., Joseph, W., & Martens, L. (2014). Power consumption model for macrocell and microcell base stations. *Transactions on Emerging Telecommunications Technologies, 25*(3), 320–333.

Ding, Y., Duan, H., Xie, M., et al. (2022). Carbon emissions and mitigation potentials of 5G base station in China. *Resources, Conservation & Recycling, 182*, 106339.

Du, X., Zhang, H., & Han, Y. (2022). How does new infrastructure investment affect economic growth quality? Empirical Evidence from China. *Sustainability, 14*, 3511.

Farhang, A., Marchetti, N., Figueiredo, F., et al. (2014). Massive MIMO and a waveform design of 5th generation wireless communication systems. In *1st International Conference on 5G for Ubiquitous Connectivity, Akaslompolo, Finland* (pp. 70–75).

Fehske, A., Fettweis, G., Malmodin, J., et al. (2011). The global footprint of mobile communications: the ecological and economic perspective. *IEEE Communications Magazine, 8*, 55–62.

Fettweis, G., & Zimmermann, E. (2008). ICT energy consumption-trends and challenges. *Proceedings of the 11th International Symposium on Wireless Personal Multimedia Communications, 2*(4).

Goncalves, L., Sebastiao, P., Souto, N., & Correia, A. (2020). One Step Greener: Reducing 5G and beyond networks' carbon footprint by 2-tiering energy efficiency with CO2 offsetting. *Electronics, 9*(3), 464.

Guldbrandsson, F., & Bergmark, P. (2012). Opportunities and limitations of using life cycle assessment methodology in the ICT sector. In *Electronics Goes Green 2012+* (pp. 1–6).

Guo, C., Wang, J., & Liu, H. (2020). Studies on how new infrastructure empowers high-quality development of China's economy. *Journal of Beijing University of Technology (Social Sciences Edition), 20*(06), 13–21. (in Chinese).

Guo, J., Wang, L., Zhou, W., et al. (2022a). Powering green digitalization: Evidence from 5G network infrastructure in China. *Resources, Conservation & Recycling, 182*(2022), 106286.

Guo, W., Mao, Y., Chen, Y., et al. (2022b). Multi-objective optimization model of micro-grid access to 5G base station under the background of China's carbon peak shaving and carbon neutrality targets. *Frontiers in Energy Research, 10*, 1032993.

Halbauer, H., Weber, A., Wiegner, D., et al. (2018). Energy efficient massive MIMO array configurations. In *2018 IEEE Globecom Workshops (GC Wkshps)* (pp. 1–6).

Hatem, A., Mohamed, M., Taisir, E. H., et al. (2018). Real-time emissions of telecom core networks. In *2018 20th International Conference on Transparent Optical Networks (ICTON)* (pp. 1–4).

Hodara, H., & Skaljo, E. (2021). From 1G to 5G. *Fiber and Integrated Optics, 40*, 85–183.

Hu, R., & Yi, Q. (2014). An energy efficient and spectrum efficient wireless heterogeneous network framework for 5G systems. *IEEE Communications Magazine, 5*, 94–101.

IEA. (2020). *Greenhouse Gas Emissions from Energy Highlights*. Available at: https://www.iea.org/data-and-statistics/data-products/?filter=emissions

Igor, S. (2007). *Evolution of mobile base station architectures*. Available at: https://www.researchgate.net/publication/228643976_Evolution_of_mobile_base_station_architectures

Li, X. (2020). New infrastructure construction and policy orientation for a smart society. *Reformation, 05*, 34–48. (in Chinese).

Liu, Y. (2018). Key technologies of green communication based on 5G mobile communication network. *Information Recording Materials, 19*(04), 83–85.

Liu, Y., Huang, X., & Shi, B. (2020). China's new infrastructure construction: Concepts, current situations and problems. *Journal of Beijing University of Technology (Social Sciences Edition), 20*(06), 1–12. (in Chinese).

Mobile's Green Manifesto 2012 [EB/OL]. [2016-07-09]. Available at: www.Gsma.com/publicpolicy/wp-content/uploads/2012/06/Green-Manifesto-2012.pdf

Mulrow, J., Gali, M., & Grubert, E. (2022). The cyber-consciousness of environmental assessment: how environmental assessments evaluate the impacts of smart, connected, and digital technology. *Environmental Research Letters, 17*(1), 013001.

Oh, E., Krishnamachari, B., Liu, X., et al. (2011). Toward dynamic energy-efficient operation of cellularnetwork infrastructure. *Communications Magazine, IEEE, 49*(6), 56–61.

Panwar, N., Sharma, S., & Singh, A. (2016). A survey on 5G: The next generation of mobile communication. *Physical Communication, 18*, 64–84.

Scharnhorst, W., Hilty, L., & Jolliet, O. (2006). Life cycle assessment of second generation (2G) and third generation (3G) mobile phone networks. *Environment International, 32*, 656–675.

Shafi, M., Molisch, A., Smith, P., et al. (2017). 5G: A tutorial overview of standards, trials, challenges, deployment, and practice. *IEEE Journal on Selected Areas in Communications, 35*(6), 1201–1221.

Shi, W., Zhu, H., Xia, T., & Zhang, J. (2022). The problem of energy consumption appears and there are several barriers to achieve green new infrastructure. *Economic Information, 001521.*

Srivastava, A., Gupta, M. S., & Kaur, G. (2020). Energy efficient transmission trends towards future green cognitive radio networks (5G): Progress, taxonomy and open challenges. *Journal of Network and Computer Applications, 168*(1), 102760.

Stutz, M., Emmenegger, M., Frischknecht, R., et al. (2006). Life cycle assessment of the mobile communication system UMTS: Towards eco-efficient systems. *The International Journal of Life Cycle Assessment, 11*, 265–276.

Tsilimantos, D., Gorce, J. M., & Atman, E. (2013). *32nd EEE International Conference on Computer Communications Turin*, April 14–19, 1097.

Wu, G., Yang, C., Li, S., et al. (2015). Recent advances in energy-efficient networks and their application in 5G systems. *IEEE Wireless Communication, 22*(2), 145–151.

Wu, J., Zhang, Y., & Shi, S. (2021). Crafting a sustainable next generation infrastructure: Evaluation of China's new infrastructure construction policies. *Sustainability, 13*, 6245.

Xiang, L., Ge, X., Wang, C. X., et al. (2013). Energy effeiency evaluation of cellar networks based on spatial distributions of traffic load and power consumption. *IEEE Transactions on Wireless Communications, 12*(3), 961–973.

Yu, H., Lee, H., & Jeon, H. (2017). What is 5G? Emerging 5G mobile services and network requirements. *Sustainability, 9*, 1848.

Zhou, S., Gong, J., Yang, Z., et al. (2009). Green mobile access network with dynamic base station energy saving. *ACM MobiCom, 9*(262), 10–12.

# 9 Affordable Housing and Sustainable Communities in Sub-Saharan Africa: Implementation Drivers, Challenges and Practices

Nicholas Chileshe and Neema Kavishe

## 9.1 Introduction

According to the UN (2020) report, currently 55% of the world's population lives in cities, however, this is estimated to increase further to 70% by 2050. This fact alone is strong enough to indicate the need and urgency of SDG 11. Urbanization is considered as one among the key hindrances towards achieving the SDG 11 because as the urban population grows many countries will struggle to meet the population needs such as housing, education, transportation, sanitation, energy, healthcare and employment to mention a few. Therefore, sustainable development depends largely on the successful management of urban growth by putting in place strategies and integrated policies to improve the community lives both in the urban and rural areas. The 2030 Agenda for Sustainable Development was launched in 2015 aiming at ending poverty in all its forms. A total of 17 Sustainable Development Goals (SDGs) along with 169 targets were defined to mobilize strategies to end all forms of poverty, while ensuring that no one is left behind. These SDGs are intended to define and tackle the world's greatest challenges. However, the progress on addressing and providing solutions to the SDG 11 has been relatively slow. For instance, from the global perspective, there is no any country which has completely managed to achieve the targets set forth in SDG 11. So far, more progress has been made in the developed countries in Europe and North American as opposed to Africa which is still stuck with major challenges to overcome. For instance, looking at the first target and indicator of SDG 11 "ensuring access for all too adequate, safe and affordable housing and basic services and upgrade slums". This continues to be a challenge in the sub-Saharan Africa as it is the only region with little progress made to improve the slum dwellers lives (UN-Habitat, 2018). Despite the challenges, SDG 11 became the driver for collaboration and partnerships between the diverse groups of stakeholders across the globe.

For several decades, this rapid growth in population and urbanisation has contributed to inadequate housing. Across Africa and parts of Asia a severe housing deficiency is being experienced. Drawing from the previous study by Tesfaye (2007), urbanisation and housing demand are directly proportional and, in that case, as the urban population of a city grows, the need for

N. Chileshe (✉)
UniSA STEM, Sustainable Infrastructure and Resource Management (SIRM), University of South Australia, Adelaide, SA, Australia

Faculty of Engineering and the Built Environment, University of Johannesburg, Johannesburg, South Africa
e-mail: nicholas.chileshe@unisa.edu.au

N. Kavishe
School of Built Environment, Engineering and Computing, Leeds Beckett University, Leeds, UK



J. Zuo et al. (eds.), *Circular Economy for Buildings and Infrastructure*, Sustainable Development Goals Series, https://doi.org/10.1007/978-3-031-56241-9_9

more housing also grows, eventually exceeding the availability of formal land for housing, leading to the growth of informal settlement. The informal sector in developing countries has predominantly been absorbing most of this growth. According to Sachs et al. (2020), Côte d'Ivoire, Burkina Faso, and Cambodia are the three countries which have progressed the most in terms of the SDG Index score whereas the following countries Venezuela, Zimbabwe and the Republic of Congo performed the least. More so, according to the UNHABITAT (2021), sub-Saharan Africa is experiencing rapid urbanization as well as a growing slum population, with 4.5 million new residents in informal settlements each year. Further, the average Annual Rate of Change of the Urban Population for Sub- Sahara Africa for 2020–2025 is 3.92% and projected 2045–2050 will be 2.89%.

This chapter will focus more on Sustainable Development Goal (SDG) 11 which is about Sustainable Cities and Communities. According to the United Nations, SDG 11 has 10 Targets and 15 Indicators. These targets stipulate the goals and Indicators signify the metrics by which the world aims to track whether these targets are achieved. The overall objective of SDG 11 is making cities and communities 'inclusive', 'safe', 'resilient' and 'sustainable'. Therefore, based on this objective, the focus is clearly on four main aspects: (1) "Inclusivity" that is to remove inequality associated with access to basic services such as healthcare, education, infrastructure and economic system (2) "Safety" by eradicating extreme poverty which leads to crime and vandalism (3) "resiliency" by gaining the ability to plan and bounce back or recover from disasters and (4) "sustainability" which focuses on the need for resource management.

## 9.2 Conceptualisation and Theoretical Basis: Gidden's Structuration Theory

This chapter is underpinned by the underlying theoretical basis is premised on the following two issues of "Agency" and "Structure", as postulated by Giddens (1984) Structuration Theory. Thus, according to Giddens (1984), structure refers to the rules for acting, thinking, and feeling that are general throughout a society or an organisation, and the available materials and non-material resources that are needed for an action to take place. Baker (2005) also defined Structure as the recurrent patterned arrangements which influence or limit the choices and opportunities available. Therefore, drawing on the arguments put forward by Kavishe et al. (2018), which conceptualised the challenges influencing the delivery of the housing public-private partnership in Tanzania with a number of theoretical perspectives such as Gidden's structuration theory, contingency theory, relational and equity theory, within the context of this study, the "structure" refers to the rules that permeate through the private and public organisations in developing countries with the responsibility for the delivery of affordable housing, and the consideration of the resources that these organisations might have or not have (hence the need for affordable housing) would be required for "action" or delivery of affordable housing to take place. Hence, as pointed out by Galster and Lee (2021), the most successful developed countries (i.e. USA, Hong Kong and Singapore) have shown much stronger government commitment in housing affordability.

In contrast, Baker (2005) defines the "Human agency" as the capacity for human beings to make choices and take action to implement these choices (i.e. whether to undertake affordable housing delivery projects or not). Thus, as alluded by Kavishe et al. (2018), Thus, given the capacity problems facing a number of developing countries in the delivery of affordable housing, and underlying challenges reinforced by the poor PPP performance (World Bank, 2016), it is important to pay attention to both the human agency and affordable housing structures, as they are inseparable and their interactions are significant (Giddens, 1984). Equally, a number of construction and non- construction related studies have employed Giddens (1984) Structuration Theory. For instance, Chileshe et al. (2013) used the concepts of 'agency' and 'structure' in providing a diagnostic model to assess strategic risk in the construction industry. Agyenim-

Boateng et al. (2017), applied Giddens structuration theory in examining the accounting and governance of PPPs; adoption and diffusion of eLearning in UK universities (Hardaker & Singh, 2011); information technology (Cunha, 2013); and structuring of social finance (Rizzi et al., 2018).

## 9.3 Definitions and Background

The focus of this Chapter is on exploring issues around the affordable housing and sustainable communities as defined in SDG11, in sub-Saharan Africa and investigating the influencing drivers, inhibiting challenges and desirable practices associated with the implementation. SDGs are also related to Circular Economy due to their similarities in goals (de Oliveira & Oliveira, 2023). This is captured in Fig. 9.1:

However, before considering the challenges, drivers and practices associated with the delivery and focus on SDG 11 (sustainable cities and communities) with emphasis on developing viable solutions intended to address its first target and indicator, namely that of "ensuring access for all to adequate, safe and affordable housing and basic services and upgrade slums", some of the key terminologies of "affordable housing", "sustainable housing", "slums", "sustainable communities", and the synergies between "affordable housing" and "SDG" will be defined.

### 9.3.1 Definition of Affordable Housing

Different authors have indicated that there is not a typical/universal definition of affordable housing because, the aspect of affordability is subjective and varies from one person to another and also from one country to another. However, some studies have generally defined affordable housing as housing for which an occupant pays not more than 30% of his/her income to cover the costs of rent and utilities (Kalugina, 2016). Figure 9.2 is an illustration of Affordable housing the amenities that would be expected nearer to that.

In addition to what's illustrated in Fig. 9.2, according to the World Bank (2021), the access to affordable housing can have a big impact on a person's well-being. A safe, secure home can improve health, hygiene, and educational opportunities by ensuring access to clean water,

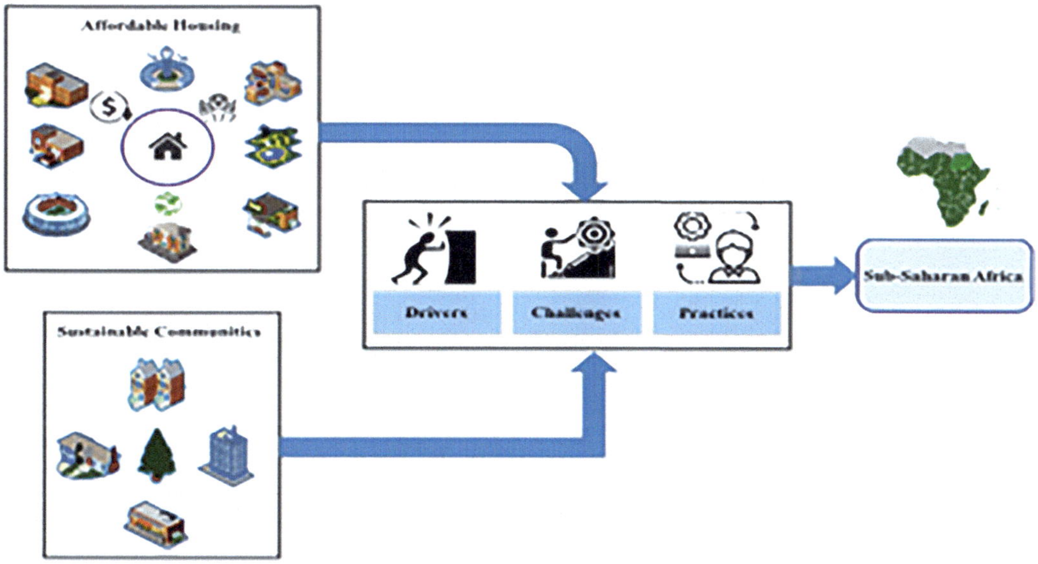


**Fig. 9.1** Affordable Housing and Sustainable Communities in sub-Saharan Africa: Implementation Drivers, Challenges and Practices

**Fig. 9.2** Affordable housing

safe toilets, electricity, and a respite from heat and disease-spreading insects. Similarly, a UK study by Bramley and Karley (2005) defined affordable housing as 'A decent home for every family at a price within their means'; while affordable housing initiatives are viewed as a mechanism for providing provide cheap but decent homes, specifically targeted at the poor citizenry (Obioha, 2021). More recently, the definition has been extended to that of Sustainable Affordable Housing Choice (SAHC) which provides synergies with Sustainable Development (Ezennia & Hoskara, 2019a); and viewed as central to global development strategies (Nubi & Anderson, 2021). Based on these definitions affordable housing is dependent of occupants' income and cost of the house. Thus, the main affordability indicator is housing cost burden against the income earned by the household (Sirmans & Macpherson, 2003).

### 9.3.2 The Need for Affordable Housing

Availability of adequate and affordable housing is critical to promote economically and socially diverse neighbourhoods, in which inhabitants have access to quality services, facilities, amenities, and employment opportunities (UN-Habitat, 2019). Insufficient housing negatively impacts on urban equity and inclusion, urban safety, and livelihood opportunities, and leads to poor health conditions (Ibid).

It is widely researched and reported that the number one housing problem is the lack of affordable housing for very low-income households (Sard, 2001). Globally there has been a noticeable increase on the demand for affordable housing (French & Lalande, 2013) because of rapid population growth and high rate of urbanization in urban areas. This is also evidenced in United Nations (2022) report which claims that half of humanity (3.5 billion people) currently lives in cities and 5 billion people are estimated to be living in cities by 2030. In most urban centres, housing problems are not only associated with the quantity but also with the inferior quality/substandard of housing especially particularly for those in the informal settlements (Makinde, 2014). Despite housing being a basic need, rapid urbanization has caused access to affordable and decent housing be difficult to several low-income earners in most African cities (Aribigbola, 2008; French & Lalande, 2013). The fact that, up until today, many African countries, continue to lack suitable policies for housing development, it has largely contributed to the growth of informal and poor housing conditions (Kavishe, 2018). This is because the demand for housing surpasses its supply. This deficit is experienced in most developing countries such as Nigeria, South Africa, Tanzania, Malaysia, and India just to mention a few. These countries are highlighted due to the varying differences in the delivery of PPP in affordable housing (World Bank, 2016), with Nigeria and South Africa acknowledged as the most experience whereas Tanzania as the least experienced. Further, South Africa is acknowledged as one of the main actors of foreign investment in housing markets in top- tier global cities and address affordable housing challenges associated with these activities (Galster & Lee, 2021). The reference to the 'housing markets' is critical here as this aligns with 'structure' component of Giddens (1984) Structuration Theory. To add some international appeal, two countries outside the SAA are equally selected and these being India and Malaysia. The same report by the World Bank (2016) identified India among the largest emerging markets that have utilised PPPs, with Chou and Pramudawardhani (2015) lending support to this argument by singling out India as where PPPs have been successful in the Asia-Pacific region. To illustrate this few experiences are drawn from few selected countries below.

### 9.3.3 Slums

Whilst the focus of this chapter is on 'Affordable housing', the SDG 11 specifically refers to upgrading of slums through the following goal of: "ensuring access for all, to adequate, safe and affordable housing and basic services and upgrade slums". Figure 9.3a and b are an illustration of Slums with Fig. 9.3b showing Kibera, which is the biggest slum in the world, located in Keny.

Nolan (2015) provides comprehensive overview of the various definitions and evolution of "slums", with emphasis on the UN operational definition of "one or a group of individuals living under the same roof in an urban area, lacking in one or more of the following five

a b

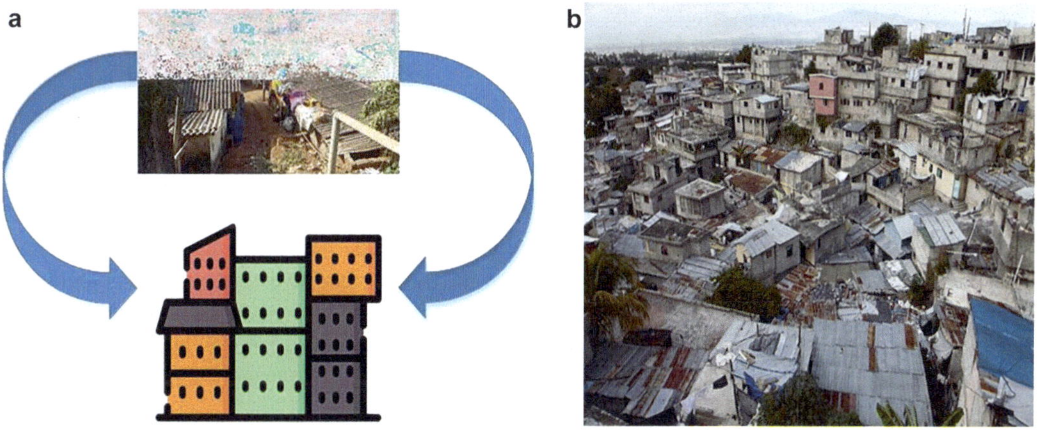

**Fig. 9.3** (**a**) Slums. (**b**) Slums

**Fig. 9.4** Sustainable Communities

amenities" as defined in Appendix A. Further, in sub–Saharan Africa (SSA), approximately 57% of the urban population resided in slums or informal settlements by the year 2018, against the global average of 24% (UNHABITAT, 2021). However, due to the negative connotations associated with the word "Slums", recent studies such as Celentano et al. (2020) have opted for "informal city" or "settlement" when referring to the self-developed urban systems. Other names found in literature which describe negative characteristics of informal settlements includes Colonia's in the state of Texas in the USA (Felipe et al., 2019); 'kampungs' in Indonesia (Alzamil, 2018); and favelas, Squatter settlements, slums, shacks, bidonvilles, villas miseria (Zárate, 2016; Corburn & Sverdilk, 2019).

### 9.3.4 Sustainable Communities

The definition adopted of 'sustainable communities' for this chapter is that as provided for by the UN Nations. Further, 'sustainable communities' as described as those [communities] that are planned, built, or modified to promote sustainable living (Chileshe, 2011). Figure 9.4 is an illustration of Sustainable communities.

Notwithstanding the definitions, over the past few decades, cities in both developed and developing countries have been exposed to severe social and economic pressures, which have led to unequal impact on the urban environment hence increasing the concentration of the most deprived households in the worst urban areas (McDonald et al., 2009). This situation has led to the need for sustainable communities across the globe. For instance, in 2003, the UK government established the Sustainable Communities Plan, explaining a vision of how the communities are to be developed in the next 20 years. The focus of the plan was put to the economic, social and environmental issues in consideration to the needs of future generations (Office of the Deputy Minister, 2003). In the Sustainable Communities Plan, the UK Government has defined sustainable communities as places where people want to live and work, now and in the future. They should meet the unique needs of both existing and future residents, are sensitive to their environment and contribute to a high quality of life. They are safe and inclusive, well planned, built and run and offer equality of opportunities and good services for all (Office of the Deputy Minister, 2003). McDonald et al., (2009) further highlighted that, for communities to be sustainable, they must provide basic services such as hospitals, schools, shops, good public transport, as well as a clean and safe environment. Thus, sustainable communities must offer decent and affordable homes (Malienė & Malys, 2009).

Sustainable communities can be attained through the availability of sustainable affordable housing. According to DeLaTorre (2013); Gan et al. (2017) and Adabre and Chan (2019) sustainable affordable housing is defined as housing which satisfies the demand and needs of the present generation without compromising future generations' ability to meet their housing demands. On the other hand, UN-Habitat (2011) and Arman et al. (2009) describe sustainable affordable housing as housing that is better planned and constructed, more durable, safe, significantly affordable, inexpensive to maintain, and has strong and inter-related economic, social, and environmental components. Appropriate, accessible, and sustainable housing is vital, and it remains a basic need to the people.

### 9.3.5 Sustainable Housing

Based on the discourse around "sustainable communities", it is evident that sustainable housing is necessary, and it is required to continuously be made available, affordable, and durable in the built environment (Ihuah, 2015) as illustrated on Fig. 9.5 below. The following criterion for characterizing sustainable housing are also reported in literature (Malienė & Malys, 2009): (1) Affordability; (2) Availability in the market; (3) Sufficient number of housings; (4) Great variety of green and quality housing; (5) Design, size and comfort; (6) Natural and social environment; (7) Energy efficiency and waste management; (8) Secure and friendly neighbourhood; (9) Accessibility to schools, health, and other services; and (10) Established technical and hygienic requirements.

Having provided the definitions, the interplay between affordable housing and the focus of this chapter, namely SDG11 is illustrated in Fig. 9.6.

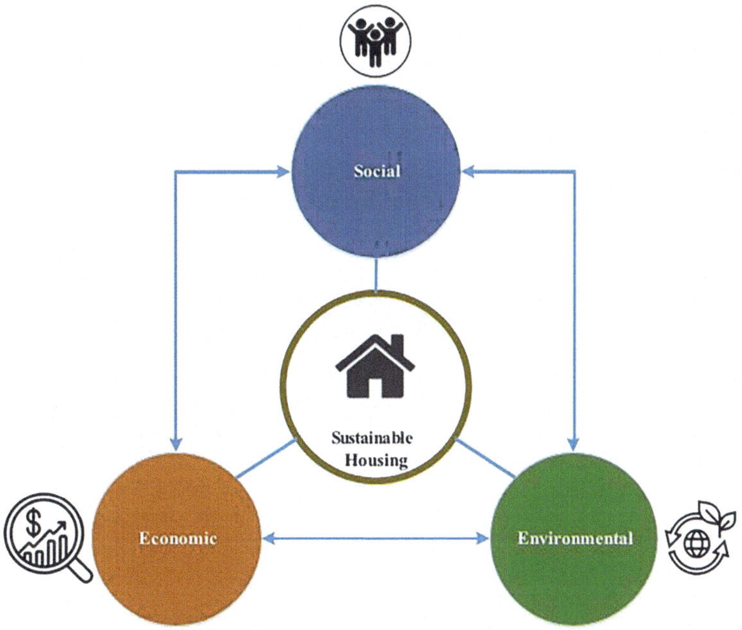


**Fig. 9.5** Sustainable housing

**Fig. 9.6** Interplay between 'Affordable housing' and 'Sustainable Development Goals

## 9.4 Implementation of Affordable Housing in Selected Sub-Saharan African and Developing Countries: Statistics at a Glance

**Nigeria**

According to Egidario et al. (2016), Nigeria, an African country with largest economy, has one of the highest urbanization rates estimated at around 5.8% but is among the countries with the worst urban housing situations across the globe, with 22 million people in Nigeria not having housing they need, and this is the highest number in the world. This problem is overly critical to the urban poor and low-income earners who constitute about 80% of Nigeria's urban population, henceforth, resulting into an increasing need for affordable housing. Further, according to the UNHABITAT (2021), the Average Annual Rate of Change of the Urban Population for Nigeria for 2020–2025 is 3.92% and projected 2045–2050 will be 2.75%. Lateef Olanrewaju et al. (2016) also amplifies the Nigerian housing deficits due to a homeownership of less than 20% and a housing shortage greater than 30 million. Recent statistics by the Voice of America (2022) shows that the housing gaps will worsen by the time Nigeria's population doubles to 400 million, as it is estimated to do by experts in 2050. From the above housing problems, according to the Giddens (1984) Structuration Theory, these low-income earners are limited in the choices towards the ability of securing affordable housing. Further, the demand for affordable housing concerns in Nigeria cuts across several economic, social, and environmental factors (Ezennia & Hoskara, 2019b).

**South Africa**

In South Africa, inadequate provision of affordable human settlements is a huge challenge since its independence. This is further amplified by having a population of 57,779,422 of which only 20.4% urban household that could afford housing prices with finance (Obioha, 2021).

Further, according to the UNHABITAT (2021), the Average Annual Rate of Change of the Urban Population for South Africa for 2020–2025 is 1.72% and projected 2045–2050 will be 0.92%. However, some of the drivers informing most planning- and housing-related policy interventions have advocated for providing the higher-density, more affordable residential housing in well-located areas such as the inner city such as Cape Town, due to its low density and sprawling nature (Massyn et al., 2015). The South Africa government has also intervened in the implementation process of sustainable housing through adoption of subsidy payment as a method of financing affordable housing to ensure that houses are allocated to beneficiaries (Adabre & Chan, 2019). However, the desire to incorporate these drivers as part of the policy to inform the affordable housing issues is some ways premised on the "Human Agency" aspect of the Giddens Structuration Theory which defines "Human agency" as the capacity for human beings to make choices and take action to implement these choices (i.e. whether to undertake PPP training or not; whether to incorporate the influencing drivers or not).

**Tanzania**

According to the World Bank (2021), Tanzania's greatest long-term development challenges has been providing citizens with access to decent and affordable housing. However, this ambition is further from actualization as Tanzania, lacks housing policy, and it has been noted that, since its independence in 1961, it has experienced a significant increase in population from 12.3 million in 1967 to almost 45 million in 2012. Currently, it [Tanzania] has a population of 60 million. Coupled with this is a rise in the urban population from 5.7% in 1967 to 29.1% in 2012, with the overall trend in population growth averaging nearly 3% annually while urbanisation grew by 5% annually (Nnkya, 2014; Wenban-Smith, 2015). Further, according to the UNHABITAT (2021), the Average Annual Rate of Change of the Urban Population for Tanzania for 2020–2025 is 4.89% and projected 2045–2050 will be 3.55%. The result is that the available social facilities and services such as housing have been significantly strained and scarce. To address the housing shortage, Tanzania has developed a viable mortgage finance market for affordable housing using new financing from the highly concessional International Development Association (IDA) (World Bank, 2015a). This will be achieved through the establishment of the Tanzania Mortgage Refinancing Company (TMRC) which is playing a key role in developing the mortgage market. In addition, housing parastatals compete directly with the private sector, but provide units affordable to middle- and upper-income groups (World Bank, 2015b).

**Malaysia**

Malaysia is one such country acknowledged by the World Bank (2016) as having the best markets in Asia relative to affordable housing. However, the current increase in urban population growth which is higher than the supply of affordable housing has led to a severe shortage of housing in Malaysia (Tan et al., 2017). Further, according to the UNHABITAT (2021), the Average Annual Rate of Change of the Urban Population for Nigeria for 2020–2025 is 3.92% and projected 2045–2050 will be 2.75% Malaysia for 2020–2025 is 2.33% and projected 2045–2050 will be 1.54%. This status is also supported by a number of studies around affordable housing which have been investigated in this context and have ranged from housing preference and housing affordability in Malaysian housing markets (Soon & Tan, 2019), with urbanization acknowledged as having increased the demand of houses in urban areas; to the identification of the differences concerning the importance of the success factors between the public and private sectors (Ismail, 2013). Further, despite having the best markets, Malaysia still has challenges within the housing policy and its implementation problems due to the loose connection between the federal and state government among the (Kamal et al., 2020).

**India**

As with most developing countries, India faces challenges associated with the provision of affordable housing, with the push driven by rapid urbanization pressure and poverty (Malik et al., 2022a); and thus, access to housing remains high on the agenda of the Indian government (Sandhu, 2013). Further, according to the UNHABITAT (2021), the Average Annual Rate of Change of the Urban Population for India for 2020–2025 is 2.33% and projected 2045–2050 will be 1.54%. Coupled with those statistics, India is also known as the second most populated country globally (Khaire & Muniappa, 2022). Therefore, several studies have also explored the issue of affordable housing within the Indian context. These have ranged from the provision of affordable housing to low- and middle-income groups (Ghumare et al., 2020); Affordable housing development (Sengupta, 2013); to low-income resettlement archetypes (Sarkar & Bardhan, 2020) and critical success factors (Agarwal et al., 2022). However, most of affordable housing research have focused on demand-supply constraints, stakeholders' perspectives, and definitional issues around housing affordability (Khaire & Muniappa, 2022).

## 9.5 Drivers Influencing the Implementation of Affordable Housing in Developing Countries

In understanding the drivers influencing the implementation of affordable housing, the approach undertaken in our Chapter is conceptualise 'drivers' as factors influencing positive outcomes on the affordable housing. Hence, the terminology is used interchangeably with 'enablers' 'facilitators' 'critical success criteria', and 'critical success factors'. For instance, according to King et al. (2017), migration and population growth are two major drivers of insufficiency in adequate, secure, and affordable housing, especially for those newly arrived at the city without appropriate identification and savings.

Since the supply of adequate and affordable houses remains a major challenge in most developing countries due to continuously growing population and urbanization rates, most Governments have made efforts to alleviate the housing shortage as one among the response/strategy towards achieving Sustainable Development Goal 11 (Make cities inclusive, safe, resilient, and sustainable). The following are the drivers for implementing affordable housing in developing countries as identified by different authors Izar and Limbumba (2021); French and Lalande (2013) and SHELTER (2013).

1. The need for slum clearance (urban and regional planning)
2. Improving the standard of living
3. Enabling sustainable communities and regions
4. Supporting the provision of adequate shelter
5. Facilitating access to affordable housing for all
6. A strategy towards sustainable development
7. Creating additional affordable housing stock
8. To cater to the future growth of low-income segments of the population
9. Reduce urban vulnerability and poverty
10. Contributing to economic growth and
11. Improve the health conditions of low-income families
12. Helps prevent homelessness

Despite the identified drivers, that motivates the development of affordable housing still, little has been done to develop strategy and mechanism to achieve the development of affordable housing. Even the existing strategy of Public Private Partnership (PPP) does not accurately reflect what is required to influence and broaden affordable housing provision in many developing countries due to several different existing challenges. According to Mohammed et al. (2017). Effective public-private partnership has the potential to guarantee the supply of reasonably priced and affordable housing provision, yet this has not been achieved due to several challenges. The following subsection identifies all the challenges hindering the implementation of affordable housing. In Nigeria, the main drivers of urbanization have been identified as including natural population increase, economic and religious activities, state policies on reclassification of settlements and decentralization of governance structure (Egidario et al., 2016). There is also a push for promotion of local circular economy and sustainable resource management (Talamo et al., 2022).

**Critical Success Factors**

Adabre and Chan (2019) study aimed at reviewing the literature on the critical success criteria (CSC) for affordable housing and based on the outcomes and developing a framework for measuring success in affordable housing projects identified 13 CSCs. These were further categorized into the following four underlying components: developer's enabling CSFs; household-demand enabling CSFs; mixed land use CSFs; land use planning CSFs. Using Tanzania as the case study, Kavishe and Chileshe (2019) study aimed at identifying the critical success factors (CSFs) and key areas for improvement for public–private partnerships (PPP) in affordable housing schemes (AHS) projects established 17 CSFs.

## 9.6 Challenges Hindering the Implementation of Affordable Housing

Having discussed the drivers influencing the implementing of affordable housing in developing countries, this sub section now focuses on the 'challenges'(for eradication) hindering the implementation of affordable housing as a solution to addressing some of the objectives of SDGs 11. However, the 'concept' of challenges is also understood as any factors that might have a negative effect on the implementation process. Therefore, the 'challenges' are used interchangeably with 'barriers' and 'risks' to the implementation process. In addition to the rapid population and urbanization, there are numerous challenges hindering the implementation/development of affordable houses. For example, in Tanzania several public organizations such as National Housing Cooperation (NHC), National Social Security Fund (NSSF), Public Pension Fund (PPF) and Tanzania Building Agency (TBA) have initiated a few affordable housing projects across the country however, the selling price is still judged unaffordable to the majority of low-income group, and these houses end up being bought by the high-income group, defeating the original aim. The main reasons for such a challenge are associated with excessive cost of building materials, lack of cheap building technology to facilitate construction of affordable houses, lack of government support to subsidize the development of such houses. Several challenges in providing affordable housing are identified in literature. For instance, King et al. (2017) identified the following three key challenges to providing adequate, secure, and affordable housing in the global South: (1) the growth of informal or substandard settlements, (2) the overemphasis on home ownership, and (3) inappropriate policies or laws that push the poor out of the city. These Africa's housing challenges have also been explored from the dimensions of governance, finance, urban management, urban health and community development (Oyalowo & Lawanson, 2021). These [housing] challenges have also been viewed from the dimensions of governance, finance, urban management, urban health, and community development (Oyalowo & Lawanson, 2021).

Other challenges affecting the delivery of affordable housing projects in SSA are prevalent in literature. Damoah et al. (2020) study aimed at exploring the challenges facing affordable housing in Ghana identified a number of challenges and factors such as those (challenges) as related to politics and governance, international forces, administration, financial resources, material resources, management practices, technical, local external forces and natural forces, and factors related to politics and governance, administration, financial resources, materials resources, local external forces and natural forces classified as external challenges, those that relate to management and technical are internal to the performing organization. Considerations should be made in applying the circular design principles for the materials used (Oorschot & Asselbergs, 2021).

### 9.6.1 Risks Associated with the Implementation of Affordable Housing

There quite several studies that have investigated the risks associated with the implementation of affordable housing with the developing countries. For instance, a study by Adabre et al. (2021) aimed at exploring linear relationships among clusters of these risk factors and the sustainable development goals in the Ghanaian housing market established a comprehensive list of 21 risk factors from literature review, and these were subsequently categorized into the following four main clusters/constructs, namely: (1) political and procurement risks; (2) financial-related risks; (3) design and construction risks and (4) operation and maintenance risks. Furthermore, using Ghana as a case study, Adabre et al. (2022a) explored the risk factors that hinder sustainable housing in developing economies and categorized the 30 risk factors into the following five thematic groups: 'financing- related', 'procurement-related', 'design and construction', 'operation and maintenance' and 'political-related' risk factors.

### 9.6.2 Barriers Associated with the Implementation of Affordable Housing

Adabre et al. (2022b) study focused on the Ghanaian housing market explored the concurrent evaluation of sustainability challenges from both perspectives towards identifying symmetries and asymmetries on sustainable housing barriers and developed the following four categorizations of barriers from extant literature: (1) economic; (2) social; (3) environmental; and (4) institutional barriers. Four categorizations of barriers were developed from extant literature, viz.: 'economic', 'social', 'environmental' and 'institutional' barriers. The earlier studies by Adabre and Chan (2021) which assessed the influence of barriers on sustainable housing in Ghana identified the following three main groupings of barriers: (1) cost-related; (2) incentive-related; and (3) retrofit-related. Likewise, Batra (2021) study aimed at exploring PPP for the linkages between the urban functioning, governance, and management for housing provision, addressing the pressing need of today's housing demand and delivery identified 31 barriers categorized into the following 4 themes: Structural, contextual, environmental (institutional and strategic), and financial.

## 9.7 Best Practices for Actualizing Affordable Housing

In the context of this chapter, 'practices' are based on the definition as provided by Rommes and Reymen (2018) using the design science (DS) perspective as involving the application or instantiation of values, constructs, models and or principles. Thus, this entails the application of any principles and frameworks associated with the implementations of affordable housing. A simpler definition of practices would that be of viewing the preceding 'values, concepts, and principles' as the 'what' components or foundation building blocks of 'Affordable housing' whereas the 'practices' would be the 'how' components to aid the translation of the building blocks. Chileshe (2007) also used such an approach providing a practical approach for understanding the quality management (QM) terminology. Currently, all governments worldwide have tried diverse ways to address the problem of sustainable, adequate, and yet affordable housing for the past four decades, but the progress is slow, and the challenge of housing shortage and informal settlements is still growing especially in the developing countries. The United Nations 2030 Agenda has also weighted in the agenda for best practices as an avenue for addressing what is considered as sustainability interlinked challenges for humanity though its well-established seventeen specific Sustainable Development Goals (SDGs) (Jiménez-Aceituno et al., 2020). Therefore, the next sub sections identify and discussed the following different approaches adopted by different governments/ countries as best practices for actualising Affording Housing: (1) Multi sectoral approach collaboration; (2) Technology; (3) Smart home; (4) Good governance; (5) Government intervention; (6) Public Private Partnerships. These are captured in Fig. 9.7.

### 9.7.1 Multi Sectoral Approach/Collaboration

The delivery of affordable housing, and actual problem is too big for an individual sector to solve it alone, because the overall population needs support to obtain decent housing. Malik et al. (2022b) sums this up by acknowledging that housing development calls for collaboration through a multipronged and concerted effort from all stakeholders to acquire resources within institutional arrangements set by state and local governments. The need for multi sectoral approach and collaboration is also evident in several developing countries. For instance, within the context of Malaysia, Cheah and Lee (2022) study used the network perspective to explore how political, business, and social actors collectively address the housing needs of the urban poor and reported that the adoption of a collaborative strategy using strategic networks contributes to meeting the growing demands of urbanites' housing needs. The 'social actors' referred to align with Giddens (1984) Structuration Theory relative here also to the structure. Likewise, with the sub-Saharan African countries, the previous Tanzanian Vice-President Dr. Mohammed also affirmed this.

Gharib Bilal while addressing the housing shortage by stressing that, "housing issues can only be achieved through Public Private Partnership (PPP) and not through the government alone". It sounded relevant and valid since several

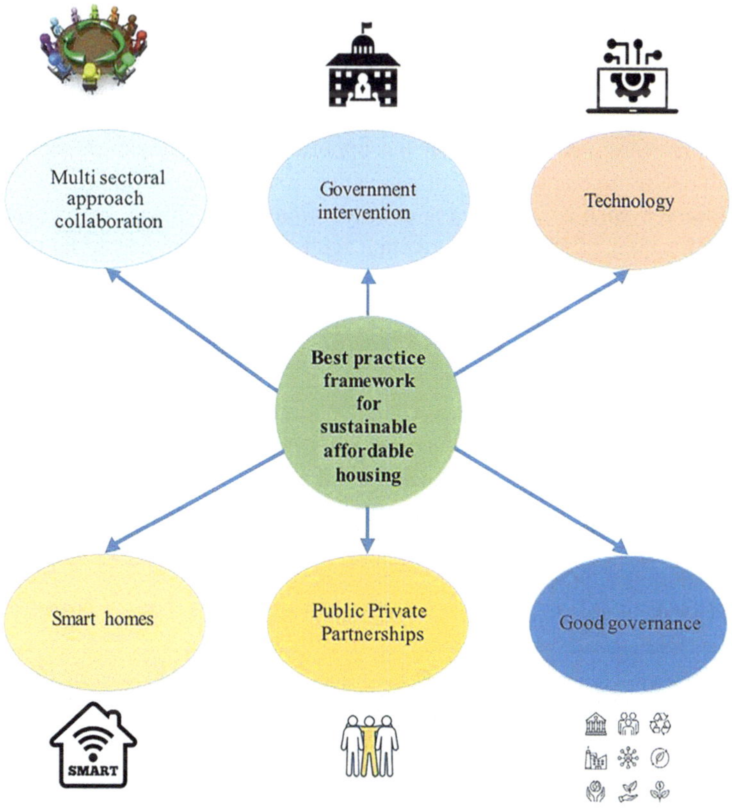


**Fig. 9.7** Best practice framework for sustainable affordable housing

strategies have been employed to solve the urban housing problems to date but with little success. Stakeholder management is also identified as a possible solution to reduce disputes experienced because of legal and regulatory issues in public–private partnership (PPP) projects within the South African context (Dithebe et al., 2021). Further, in addition to the collaboration, stakeholders are also expected to have social responsibilities towards supporting government initiatives for affordable housing development (Othman & Abdellatif, 2011). In Nigeria, as a way of bringing government and industry together to address the lack of affordable housing, advocacy campaign for affordable houses is also organised (VOA, 2022). The sectoral approach and collaboration should also be extended to resident participation and inclusion as this is interlinked to the best practice of PPPs and considered the best strategy for PPP to evolve as a future guarantor of the sustainable city (Fell & Mattsson, 2021). There is also potential for applying the principles of the 'passive house', which is modular and circular, as well as a smart building skin and smart core with open floor space in between in designing these smart homes (Oorschot & Asselbergs, 2021). Further, according to Marchesi et al. (2020), affordable housing can leverage on the opportunities offered by circular economy in innovative design. However, such an approach is dependent on the combination of technical innovation with social innovation.

### 9.7.2 Technology

Application of technology is another best practice that actualising Affordable Housing. For instance, in addressing the stakeholder collaboration, Neumann (2019) suggests some best practice for structuring a SDG model, facilitating participatory stakeholder workshops, identifying levers for successful action, and establishing a culture of reflection. This could be done using technology through usage of the software iMODELER. In South Africa, emergent technologies such as 3D printing have been suggested to government as a potential strategy for alleviating the housing shortages through reaping of the significant benefits in terms of cost, time, quality, productivity, and stakeholders' satisfaction (Aghimien et al., 2020). Further, due to the era of the 4Ir, digital technologies such as 3D technologies are suggested as the means and solutions to many of the world's problems (Hoosain et al., 2020). Likewise, from the construction industry perspective, the application of technology can enhance the effective delivery of sustainable affordable housing (Moghayedi et al., 2022); 3-D printing technology for concrete homes (Galster & Lee, 2021); and smart technologies can be embedded into the affordable house to achieve community happiness (Mohamad et al., 2022). Further, in developing countries such as India, the adoption new construction technologies could be a solution in dealing with such a huge construction demand (Pawar et al., 2022). Others include prefabricated construction technology (PCT) (Jain & Bhandari, 2022).

### 9.7.3 Smart Homes

The 'technology' best practice as discussed above could also be extended to smart homes as good strategies or 'best-practice' to foster a sustainable adoption of smart housing in the sub-Saharan African region (Tetteh & Amponsah, 2020). Further, according to Moghayedi et al. (2021), housing technology can be used as a vehicle for achieving sustainable, innovative, affordable housing. However, the success of this practice is very much dependent of the levels of technological awareness. Furthermore, as pointed out in the seminal study of Mehta and Bridwell (2005), any application of technology would need to be tailored for local use under local conditions.

### 9.7.4 Good Governance

Good governance has always been at the core of implementation initiatives in both developing and developing countries. However, most developing countries particularly those in sub- Saharan African countries lacks the capacity for effective translation and implementation of policies. This is unlike developed counties such as the USA and in Asia that have implemented large-scale public housing programmes (Galster & Lee, 2021). A case in example is Tanzania's quest in undertaking the transformation of Dar es Salaam to be a smart city which have been affected by poor implementation of master plans (Todd et al., 2019). In Ghana, the challenges related to 'politics and governance' were ranked top out of thirty-four as facing affordable housing delivery that uses private-public partnership (PPP) scheme as the delivery mode (Damoah et al., 2020).

### 9.7.5 Government Interventions

Relative to government interventions, best practices of alleviating the sustainable affordable housing involves the government's provision of more lands to the developers or the government directly building affordable housing on their lands as suggested in Malaysia (Lateef Olanrewaju & Idrus, 2020). Other interventions identified in literature undertaken in countries such as Nigeria, Tanzania and Malaysia include land provision (discount) to housing producer (Galster & Lee, 2021); and cost of land reduction (VOA, 2022). Some seminal studies such as Sard (2001) had previously suggested using vouchers to obtain housing as a mechanism for increasing families' success at having additional funding. There equally emerging trends such as participatory and assisted self- help housing among low-income people as strategies to address the affordable housing deficit (Egidario et al., 2016). However, some of the government interventions haven't been successful. According to the World Bank (2015b), in Tanzania, the government subsidy programs have done little to promote wide-scale housing affordability. Likewise, in some developed countries such as Denmark, the exposure of the non-profit housing systems to state intervention risks has weakened its governance and financial power (Noring et al., 2022).

### 9.7.6 Public Private Partnerships

The role of the PPP in alleviating the affordable housing challenges whilst evident in literature is also mixed with supporting studies (Mohammed et al., 2017; Nzau & Trillo, 2020; Kavishe et al., 2019; Chileshe & Kavishe, 2021; Damoah et al., 2020; World Bank, 2015b, 2021); and unsupportive (Fell & Mattsson, 2021). For instance, in Ghana, effective public-private partnership has the potential to guarantee the supply of reasonably priced and affordable housing provision (Mohammed et al., 2017). Within the sub-Saharan African context, and more specifically Kenya, a study by Nzau and Trillo (2020) which sought answers to the following question of: "How can urban public policy facilitate slum regeneration, increase affordable housing, and enhance social inclusion in cities of developing countries?" established that the residents' housing needs could be met by including at least 27.9% affordable housing in new developments, entirely borne by the private sector. In Tanzania, the World Bank (2021) has demonstrated how private capital, even in less-developed financial markets, can be mobilized successfully through the right public investments. However, despite the benefits of using the PPP is addressing the affordable housing delivery challenges, this mechanism is not with any challenges of its own. As shown by Kavishe et al. (2019) study aimed at identifying and ranking policy and regulatory framework factors and inherent pitfalls in the delivery of Tanzanian public–private partnerships (PPPs) affordable housing schemes, the significant challenges were found to be: current PPP policy and guidelines that needed further improvement; having a PPP policy and clear regulatory framework of which Tanzania had. In contrast, "poor planning skills and analytical capacity", "high cost of building materials" and "inadequate access to housing finance" were found to be the critical pitfalls. Likewise, Fell and Mattsson (2021) provides evidence evidence that PPPs are unjust and exclude local actors from collaborations. Further, PPPs are prone to barriers, gaps, and challenges that are required to be dealt with at the policy, governance, and industry level (Batra, 2021). From the above descriptions of the challenges facing PPPs, it's clear how the Giddens Structuration Theory (Giddens, 1984) manifests itself as the identified challenges of the "PPP policy, guidelines, legal framework and procurement regulations" are akin to the "structures" and which act as a set of rules, with power to manage and informing over the actions of the members of a society or organisation (i.e., public and private organisations, collaborating stakeholders).

## Appendix 1. Criteria for UN Operational Definition of Slums

1. Durable housing (a permanent structure providing protection from extreme climatic conditions);
2. Sufficient living area (no more than three people sharing a room);
3. Access to improved water (water that is sufficient, affordable, and can be obtained without extreme effort);
4. Access to improved sanitation facilities (a private toilet, or a public one shared with a reasonable number of people); and
5. Secure tenure (de facto or de jure secure tenure status and protection against forced eviction)

## References


Adabre, M. A., & Chan, A. P. (2019). Critical success factors (CSFs) for sustainable affordable housing. *Building and Environment, 156*, 203–214.

Adabre, M. A., & Chan, A. P. C. (2021). Modeling the impact of barriers on sustainable housing in developing countries. *Journal of Urban Planning and Development, 147*(1), 639.

Adabre, M. A., Chan, A. P. C., Edwards, D. J., & Adinyira, E. (2021). Assessing critical risk factors (CRFs) to sustainable housing: The perspective of a sub-Saharan African country. *Journal of Building Engineering, 41*, 102385.

Adabre, M. A., Chan, A. P. C., Edwards, D. J., & Osei-Kyei, R. (2022a). To build or not to build, that is the uncertainty: Fuzzy synthetic evaluation of risks for sustainable housing in developing economies. *Cities, 125*, 103644.

Adabre, M. A., Chan, A. P. C., Edwards, D. J., & Mensah, S. (2022b). Evaluation of symmetries and asymmetries on barriers to sustainable housing in developing countries. *Journal of Building Engineering, 50*, 104174.

Agarwal, S., Singh, T. P., Bajaj, D., & Pant, V. (2022). Affordable housing in urban India: a review of critical success factors (CSFs) addressing housing adequacy with affordability for the urban poor. *Housing, Care and Support, 25*(1), 61–79.

Aghimien, D., Aigbavboa, C., Aghimien, L., Thwala, W. D., & Ndlovu, L. (2020). Making a case for 3D printing for housing delivery in South Africa. *International Journal of Housing Markets and Analysis, 13*(4), 565–581.

Agyenim-Boateng, C., Stafford, A., & Stapleton, P. (2017). The role of structure in manipulating PPP accountability. *Accounting, Auditing & Accountability Journal, 30*(1), 119–144.

Alzamil, W. S. (2018). Evaluating urban status of informal settlements in Indonesia: a comparative analysis of three case studies in north Jakarta, Journal of. *Sustainable Development, 11, No. 148.*

Aribigbola, A. (2008). Housing policy formulation in developing countries: evidence of programme implementation from Akure, Ondo State, Nigeria. *Journal of Human Ecology, 23*(2), 125–133.

Arman, M., Zuo, J., Wilson, L., Zillante, G., & Pullen, S. (2009). Challenges of responding to sustainability with implications for affordable housing. *Ecological Economics, 68*(12), 3034–3041.

Baker, C. (2005). *Cultural studies: Theory and practice*. Sage.

Batra, R. (2021). A thematic analysis to identify barriers, gaps, and challenges for the implementation of public-private-partnerships in housing. *Habitat International, 118*, 102454.

Bramley, G., & Karley, N. K. (2005). How much extra affordable housing is needed in England? *Housing Studies, 20*(5), 685–715.

Celentano, G., Goswein, V., Magyar, J., & Habert, G. (2020). The informal city as a socio-technical system: Construction management and money distribution in the informal and upgraded communities of Bangkok. *Journal of Cleaner Production, 256*, 120142.

Cheah, C. W., & Lee, C. K. C. (2022). Housing the urban poor through strategic networks: A cross- case analysis. *Habitat International, 124*, 102579.

Chileshe, N. (2007). Quality management concepts, principles, tools and philosophies: A valid methodology for deployment within UK construction-related SMEs. *Journal of Engineering, Design and Technology, 5*(1), 49–67.

Chileshe, N. (2011). Delivering sustainability through construction and project management: principles, tools and practices. In P. Roetman & C. Daniels

(Eds.), *Creating sustainable communities in a changing world* (pp. 111–118). Crawford House Publishing.
Chileshe, N., & Kavishe, N. (2021). Readiness assessment of public–private partnerships (PPPs) adoption in developing countries: The case of Tanzania. *Built Environment Project and Asset Management, 11*(1), 71–87.
Chileshe, N., Wilson, L. J., Zuo, J., Zillante, G., & Pullen, S. (2013). Strategic risk assessment for pursuing sustainable business in the construction industry: Diagnostic models. In G. Wells (Ed.), *Sustainable business: Theory and practice of business under sustainability principles, ch. 9* (pp. 155–177). Edward Elgar.
Chou, J. S., & Pramudawardhani, D. (2015). Cross-country comparisons of key drivers, critical success factors and risk allocation for public-private partnership projects. *International Journal of Project Management, 33*(5), 1136–1150.
Corburn, J., & Sverdilk, A. (2019). Informal settlements and human health. In M. Nieuwenhuijsen & H. Khreis (Eds.), *Integrating human health into urban*. Springer.
Cunha, J. (2013). A distributed model of the structuration of information technology in organizations. *Management Research, 11*(3), 330–351.
Damoah, I. S., Ayakwah, A., Aryee, K. J., & Twum, P. (2020). The rise of PPPs in public sector affordable housing project delivery in Ghana: challenges and policy direction. *International Journal of Construction Management, 20*(6), 690–703. https://doi.org/10.1080/15623599.2020.1763897
de Oliveira, C. T., & Oliveira, G. G. A. (2023). What Circular economy indicators really measure? An overview of circular economy principles and sustainable development goals. *Resources, Conservation and Recycling, 190*, 106850.
DeLaTorre, A. K. (2013). *Sustainable, affordable housing for older adults: A case study of factors that affect development in Portland, Oregon*. Portland State University.
Dithebe, K., Thwala, W. D. D., Aigbavboa, C. O., Edwards, D. J., Hayhow, S., & Talebi, S. (2021). Stakeholder management in the alleviation of legal and regulatory disputes in public-private partnership projects in South Africa. *Journal of Engineering, Design and Technology*. https://doi.org/10.1108/JEDT-01-2021-0037
Egidario, B. A., Patrick, A. E., & Eziyi, O. I. (2016). Urbanization and housing for low-income earners in Nigeria: A review of features, challenges and prospects. *Mediterranean Journal of Social Sciences, 7*(3 S1), 347.
Ezennia, I. S., & Hoskara, S. O. (2019a). Methodological weaknesses in the measurement approaches and concept of housing affordability used in housing research: A qualitative study. *PLoS ONE, 14*, 221246.
Ezennia, I. S., & Hoskara, S. O. (2019b). Exploring the severity of factors influencing sustainable affordable housing choice: Evidence from Abuja, Nigeria. *Sustainability, 11*(20), 5792. https://doi.org/10.3390/su11205792
Felipe, J., Mehta, A., & Rhee, C. (2019). Manufacturing matters... but it's the jobs that count. *Cambridge Journal of Economics, 43*, 139–168.
Fell, T., & Mattsson, J. (2021). The role of public-private partnerships in housing as a potential contributor to sustainable cities and communities: A systematic review. *Sustainability (Switzerland), 13*(14), 7783.
French, M. A., & Lalande, C. (2013). Green cities require green housing: Advancing the economic and environmental sustainability of housing and slum upgrading in cities in developing countries. In *The Economy of Green Cities* (pp. 257–284). Springer.
Galster, G., & Lee, K. O. (2021). Housing affordability: a framing, synthesis of research and policy, and future directions. *International Journal of Urban Sciences, 25*(sup1), 7–58.
Gan, X., Zuo, J., Wu, P., Wang, J., Chang, R., & Wen, T. (2017). How affordable housing becomes more sustainable? A stakeholder study. *Journal of Cleaner Production, 162*, 427–437.
Ghumare, P. N., Chauhan, K. A., & Yadav, S. K. M. (2020). Housing attributes affecting buyers in India: Analysis of perceptions in the context of EWS/LIG consumers view. *International Journal of Housing Markets and Analysis, 13*(4), 533–552.
Giddens, A. (1984). *The constitution of society: Outline of the theory of structuration*. University of California Press.
Hardaker, G., & Singh, G. (2011). The adoption and diffusion of eLearning in UK universities: A comparative case study using Giddens's Theory of Structuration. *Campus-Wide Information Systems, 28*(4), 221–233.
Hoosain, M. S., Paul, B. S., & Ramakrishna, S. (2020). The impact of 4ir digital technologies and circular thinking on the United Nations sustainable development goals. *Sustainability (Switzerland), 12*(23), 10143.
Ihuah, P. W. (2015). Building materials costs increases and sustainability in real estate development in Nigeria. *African Journal of Economic and Sustainable Development, 4*(3), 218–233.
Ismail, S. (2013). Critical success factors of public private partnership (PPP) implementation in Malaysia. *Asia-Pacific Journal of Business Administration, 5*(1), 6–19.
Izar, P., & Limbumba, T. M. (2021). A matter of value: Assessing the scope and effects of Tanzania's national housing corporation's development strategy on Dar es Salaam's urban neighbourhoods. *International Journal of Urban Sciences, 25*(1), 195–217.
Jain, S., & Bhandari, H. (2022). Affordable housing with prefabricated construction technology in India: An approach to sustainable supply. *ECS Transactions, 107*(1), 8513–8520.
Jiménez-Aceituno, A., Peterson, G. D., Norström, A. V., Wong, G. Y., & Downing, A. S. (2020). Local lens for SDG implementation: Lessons from bottom-up

approaches in Africa. *Sustainability Science, 15*(3), 729–743.
Kalugina, A. (2016). Affordable housing policies: An overview. *Cornell Real Estate Review, 14*(1), 76–83.
Kamal, E. M., Lai, K. S., & Yusof, N. (2020). The low-middle income housing challenges in Malaysia. *Planning Malaysia, 18*(1), 102–117.
Kavishe, N. (2018). Improving the delivery of PPP housing projects in developing countries. PhD Thesis, University of Birmingham, Birmingham UK.
Kavishe, N., & Chileshe, N. (2019). Critical success factors in public-private partnerships (PPPs) on affordable housing schemes delivery in Tanzania: A qualitative study. *Journal of Facilities Management, 17*(2), 188–207.
Kavishe, N., Jefferson, I., & Chileshe, N. (2018). An analysis of the delivery challenges influencing public-private partnership in housing projects: The case of Tanzania. *Engineering, Construction and Architectural Management, 25*(2), 202–240.
Kavishe, N., Chileshe, N., & Jefferson, I. (2019). Public–private partnerships in Tanzanian affordable housing schemes: Policy and regulatory issues, pitfalls and solutions. *Built Environment Project and Asset Management, 9*(2), 233–247.
Khaire, M., & Muniappa, N. (2022). Housing policy research in India: a scoping review and thematic analysis. *International Journal of Housing Markets and Analysis, 15*(2), 339–356.
King, R., Orloff, M., Virsilas, T., & Pande, T. (2017). *Confronting the urban housing crisis in the Global South: Adequate, secure, and affordable housing, working paper.* World Resources Institute. Available online at: www.citiesforall.org, Date accessed, 22nd May 2022.
Lateef Olanrewaju, A., & Idrus, A. (2020). What is determining affordable housing shortages in the Greater Kuala Lumpur, Malaysia? *Property Management, 38*(1), 52–81.
Lateef Olanrewaju, A., Anavhe, P., & Hai, T. K. (2016). A framework for affordable housing governance for the Nigerian property market, Creative Construction Conference 2016, CCC 2016, 25–28 June 2016. *Procedia Engineering, 164*, 307–314.
Makinde, O. O. (2014). Housing delivery system, need and demand. *Environment, Development and Sustainability, 16*(1), 49–69.
Maliene, V., & Malys, N. (2009). High-quality housing-A key issue in delivering sustainable communities. *Building and Environment, 44*(2), 426–430.
Malik, J., Bardhan, R., Hong, T., & Piette, M. A. (2022a). Developing occupant archetypes within urban low-income housing: A case study in Mumbai, India. *Building Simulation, 15*(9), 1661–1683.
Malik, S., Roosli, R., & Yusof, N. (2022b). Institutional stakeholder collaborations (ISCs): A conceptual framework for housing research. *Journal of Housing and the Built Environment, 37*(1), 213–239.
Marchesi, M., Tweed, C., & Gerber, D. (2020). Applying circular economy principles to urban housing. *IOP Conference Series: Earth and Environmental Science, 588*(5), 052065.
Massyn, M. W., McGaffin, R., Viruly, F., & Hopkins, N. (2015). The challenge of developing higher density, affordable housing in the inner city of cape town. *International Journal of Housing Markets and Analysis, 8*(3), 412–428.
McDonald, S., Malys, N., & Maliene, V. (2009). Urban regeneration for sustainable communities: A case study. *Technological and Economic Development of Economy, 15*(1), 49–59.
Mehta, R., & Bridwell, L. (2005). Innovative construction technology for affordable mass housing in Tanzania, East Africa. *Construction Management and Economics, 23*(1), 69–79.
Moghayedi, A., Awuzie, B., Omotayo, T., Le Jeune, K., Massyn, M., Ekpo, C. O., Braune, M., & Byron, P. (2021). A critical success factor framework for implementing sustainable innovative and affordable housing: A systematic review and bibliometric analysis. *Buildings, 11*(8), 317.
Moghayedi, A., Awuzie, B., Omotayo, T., Le Jeune, K., & Massyn, M. (2022). Appraising the nexus between influencers and sustainability-oriented innovation adoption in affordable housing projects. *Sustainable Development.* https://doi.org/10.1002/sd.2306
Mohamad, Z. Z., Yang, F. C., Charles Ramendran, S. P. R., Rehman, M., Nee, A. Y. H., & Yin, Y. C. (2022). Embedding eco-friendly and smart technology features in affordable housing for community happiness in Malaysia. *GeoJournal, 87*(1), 167–181.
Mohammed, I., Nahiduzzaman, K. M., & Aldosary, A. (2017). Pro-poor urban housing provision in Ghana: Implementation challenges and prospect. *Open House International, 42*(4), 98–107.
Neumann, K. (2019). Sustainable cities and communities - Best practices for structuring a SDG model. *IOP Conference Series: Earth and Environmental Science, 323*(1), 012094.
Nnkya, T. J. (2014). Financing affordable housing in Tanzania: policy, initiatives, challenges, and opportunities. *6th Global Housing Finance Conference*, Washington, DC.
Nolan, L. B. (2015). Slum definitions in urban India: Implications for the measurement of health inequalities. *Population and Development Review, 41*(1), 59–84. https://doi.org/10.1111/j.1728-4457.2015.00026.x
Noring, L., Struthers, D., & Grydehøj, A. (2022). Governing and financing affordable housing at the intersection of the market and the state:Denmark's private non-profit housing system. *Urban Research and Practice, 15*(2), 258–274.

Nubi, T., & Anderson, I. (2021). Africa's housing sector as a pathway to achieving the SDGs. In *Advances in 21st century human settlements* (pp. 1–12).

Nzau, B., & Trillo, C. (2020). Affordable housing provision in informal settlements through land value capture and inclusionary housing. *Sustainability, 12*, 5975. https://doi.org/10.3390/su12155975

Obioha, E. E. (2021). Mission unaccomplished: Impediments to affordable housing drive in addressing homelessness in sub-Saharan Africa. *Social Sciences, 10*(8), 310. https://doi.org/10.3390/socsci10080310

Office of the Deputy Prime Minister. (2003). *Sustainable communities: Building for the future*. ODPM.

Oorschot, L., & Asselbergs, T. (2021). New housing concepts: Modular, circular, biobased, reproducible, and affordable. *Sustainability (Switzerland), 13*(24), 13772.

Othman, A., & Abdellatif, M. (2011). Partnership for integrating the corporate social responsibility of project stakeholders towards affordable housing development: A South African perspective. *Journal of Engineering, Design and Technology, 9*(3), 273–295.

Oyalowo, B., & Lawanson, T. (2021). Housing and the SDGs in African Cities: Towards a sustainable future. In *Advances in 21st century human settlements* (pp. 321–332).

Pawar, P., Minde, P., & Kulkarni, M. (2022). Analysis of challenges and opportunities of prefabricated sandwich panel system: A solution for affordable housing in India. *Materials Today Proceedings*. https://doi.org/10.1016/j.matpr.2022.05.19

Rizzi, F., Pellegrini, C., & Battaglia, M. (2018). The structuring of social finance: Emerging approaches for supporting environmentally and socially impactful projects. *Journal of Cleaner Production, 170*, 805–817.

Romme, A. G. L., & Reymen, I. M. M. J. (2018). Entrepreneurship at the interface of design and science: Toward an inclusive framework. *Journal of Business Venturing Insights, 10*, e00094.

Sachs, J., Schmidt-Traub, G., Kroll, C., Lafortune, G., Fuller, G., & Woelm, F. (2020). *The sustainable development goals and COVID-19. Sustainable Development Report 2020*. Cambridge University Press.

Sandhu, K. (2013). Formal housing finance outreach and the urban poor in India. *International Journal of Housing Markets and Analysis, 6*(3), 269–283.

Sard, B. (2001). Housing vouchers should be a major component of future housing policy for the lowest income families. *Cityscape: A Journal of Policy Development and Research, 5*(2), 89–110.

Sarkar, A., & Bardhan, R. (2020). Socio-physical liveability through socio-spatiality in low- income resettlement archetypes– A case of slum rehabilitation housing in Mumbai, India. *Cities, 105*, 1–18.

Sengupta, U. (2013). Affordable housing development in India: a real deal for low-income people? *International Development Planning Review, 35*(3), 261–281.

SHELTER. (2013). Affordable Housing. *Hudco Publication, 14*(1), 74–79.

Sirmans, G. S., & Macpherson, D. A. (2003). The state of affordable housing. *Journal of Real Estate Literature, 11*(2), 133–155.

Soon, A., & Tan, C. (2019). An analysis on housing affordability in Malaysian housing markets and the home buyers' preference. *International Journal of Housing Markets and Analysis, 13*(3), 375–392. https://doi.org/10.1108/IJHMA-01-2019-0009

Talamo, C. M. L., Atta, N., Valle, A. D., & Campioli, A. (2022). Ecosystem perspective for sustainable settlements in East Africa. In *Research for Development* (pp. 3–24).

Tan, T. H., Samihah, H. K., & Phang, S. N. (2017). Building affordable housing in urban Malaysia: Economic and institutional challenges to housing developers. *Open House International, 42*(4), 28–35.

Tesfaye, A. (2007). Problems and prospects of housing development in Ethiopia. *Property Management, 25*(1), 27–53.

Tetteh, N., & Amponsah, O. (2020). Sustainable adoption of smart homes from the sub-Saharan African perspective. *Sustainable Cities and Society, 63*, 102434.

Todd, G., Msuya, I., Levira, F., & Moshi, I. (2019). City Profile: Dar es Salaam, Tanzania. *Environment and Urbanization ASIA, 10*(2), 193–215.

UNHABITAT. (2021). *Housing, Slums and Informal Settlements, Urban Indicators Database*, Available online at <www.unhabitat.org>, Date accessed, 30th May January 2022.

UN-Habitat. (2018). Tracking progress towards inclusive, safe, resilient and sustainable cities and human settlements. SDG 11 synthesis report-high level political forum 2018.

UN-Habitat. (2019). The strategic plan 2020–2023.

UN-Habitat. (2011). *Condominium Housing in Ethiopia: The Integrated Housing Development Programme*. Nairobi, Kenya.

United Nations. (2022). *Sustainable development goals*. Available at: https://www.un.org/sustainabledevelopment/cities/. Accessed on 21/05/22.

Voice of America. (2022). *Millions in Nigeria Struggle for Affordable Housing Amid Real Estate Boom*, Available from https://www.voanews.com/a/millions-in-nigeria-struggle-for-affordable-housing-amid-real-estate-boom-/6458742.html, Date accessed – 27th May 2022.

Wenban-Smith, H. (2015). Population growth, internal migration, and urbanization in Tanzania, 1967–2012: Phase 2 (Final Report). Available at: http://www.theigc.org/wp-content/uploads/2015/09/Wenban-Smith-2015-Working-paper.pdf

World Bank. (2015a). *World Bank approves new financing for more affordable housing in Tanzania as its cities expand*, available at < https://www.worldbank.org/en/news/press-release/2015/02/24/world-bank-approves-new-financing-for-more-affordable-housing-in-tanzania-as-its-cities-expand> (Date accessed – 27th May 2022).

World Bank. (2015b). *Stocktaking of the housing sector in Sub-Saharan Africa, Summary Report*, available at: https://documents1.worldbank.org/curated/en/916331468184793126/pdf/101153-v1-revised-PUBLIC.pdf, Date accessed – 27th May 2022.

World Bank. (2016). *Tanzania economic update the road less traveled unleashing public private partnerships in Tanzania*, Africa Region Macroeconomics and Fiscal Management Global Practice, available at: <www.worldbank.org/tanianla/economlcupdate>, (accessed 10 March 2018).

World Bank. (2021). *Making Housing Affordable and Accessible in Tanzania*, available at: https://www.worldbank.org/en/news/feature/2021/07/07/making-housing-affordable-and- accessible-in-tanzania, (Accessed 27th May 2022).

Zarate, L. (2016), They are not "informal settlements"–They are habitats made by the people. *The Nature of Cities, 26*.

# Towards City 5.0: Building Resilient Communities Through the Interplay of Circular Economy, Resilient and Sustainable Supply Chains, and Reverse Logistics

10

Nicholas Chileshe, Larissa Statsenko, Ruchini Jayasinghe, and Ethan Nikookar

## 10.1 Introduction

According to the United Nations Economic and Social Council (2020), a smart, sustainable city is defined as "an innovative city that uses information and communication technologies (I.C.T.s) and other means to improve quality of life, the efficiency of urban operation and services, and competitiveness, while ensuring that it meets the needs of present and future generations with respect to economic, social, environmental as well as cultural aspects". Clarification of this definition is further viewed from the lenses of 'innovation' and 'smartness', with Ahvenniemi et al. (2017) highlighting the differences between 'sustainable' and 'smart cities', and thus recommending the usage of a more accurate term "smart sustainable cities" as opposed to "smart cities". Recently coined concept of 'City 5.0' and 'Industry 5.0' shifts the focus from technology development to more people centric sustainable and resilient communities. Whilst this is not the focus of this book chapter, we adopt the definition as that of 'sustainable and resilient cities and communities' with emphasis on 'resilience', and in line with SDG11, which is aimed at making cities and human settlements inclusive, safe, resilient, and sustainable.

The smart aspect is still accommodated by how the usage of 'technology', 'innovation' and digitization could enable the attainment of integrated circular business models and the application of risk management techniques. Examples of technology-based solutions for smart cities include the application of blockchain-enabled frameworks for energy-efficient smart parking (Singh et al., 2022) and; the application of machine learning as a precursor for dealing with effective treatment of industrial sewage in sustainable cities (Miao et al., 2021). From the transportation perspective, Chehri and Mouftah (2019) point to the application of intelligent Transportation Systems (ITS), such as connected and automated vehicles (C.A.V.) in smart and sustainable cities. Therefore, in summation, it could be

N. Chileshe (✉)
UniSA STEM, University of South Australia, Adelaide, Australia

Faculty of Engineering and the Built Environment, University of Johannesburg, Johannesburg, South Africa
e-mail: nicholas.chileshe@unisa.edu.au

L. Statsenko
UniSA STEM, University of South Australia, Adelaide, Australia

R. Jayasinghe
UniSA Online, University of South Australia, Adelaide, Australia

E. Nikookar
School of Business, University of Wollongong, Wollongong, NSW, Australia



J. Zuo et al. (eds.), *Circular Economy for Buildings and Infrastructure*, Sustainable Development Goals Series, https://doi.org/10.1007/978-3-031-56241-9_10

argued that for the health of residents, and the prevention of further environmental deterioration, effective treatment of industrial sewage has become very necessary. Sewage treatment in sustainable cities attracts more researchers both from academic and industrial communities.

Resilient and Sustainable Supply Chains (R.S. S.C.) are one of the key enablers for building sustainable cities, determined by U.N. Sustainable Development Goal 11 to "make cities and human settlements inclusive, safe, resilient and sustainable". R.S.S.C.N.D. is critical for being prepared for unforeseen 'black swan events in the world of increased interconnectivity, escalating political tensions among countries and biological threats, which pose additional threats and constraints for local and global communities and cities.

Sustainable and resilient supply chains can ensure the continuity of business operations in the face of disasters and unforeseen threats, minimize or at least control their negative impacts on communities, generate employment, and positively influence society and communities (López-Castro & Solano-Charris, 2021). Referring to U.N. Sustainable Development Goals, sustainable and resilient supply chains are critical for them. For example, to achieve 'Goal 2: Zero hunger', efficient supply chains are needed to ensure a continuous supply in the face of adverse events (López-Castro & Solano-Charris, 2021; Maiyar & Thakkar, 2020; de Souza et al., 2019). Achievement of 'Goal 3: Good health and well-being' requires an adequate infrastructure, technology, and supplies—the pharmaceuticals and medical S.C. (Zahiri et al., 2017). 'Goal 7: Affordable and clean energy' is focused on expanding the infrastructure and updating the technology to provide clean, efficient, and reliable energy. This is only possible with resilient and sustainable supply chains in the resource and energy sectors (Fattahi et al., 2020; Ahranjani et al., 2020). 'Goal 12: Responsible consumption and production' is reliant upon sustainable and resilient closed-loop and reverse supply chains, which reduce environmental impact, and enable value recovery, second use, and proper final disposal of products (Lotfi et al., 2021; Mishra & Singh, 2020).

The recent example of the COVID-19 pandemic has highlighted vulnerabilities of globally connected supply chains utilizing just-in-time, low-inventory disciplines. To name a few, the food and agriculture sectors, consumer staple goods, manufacturing, and pharmaceutical supply in case of a global disruption proved to be critical in ensuring that cities and communities recover fast and with the least impact. The pandemic has left organizations and governments around the world puzzled about the strategies and approaches for tackling the issues at hand, but most importantly, looking into the future of engineering sustainable and resilient supply chains to make local communities and cities less vulnerable in the face of disruptions.

SRSC allows overcoming common challenges such as geographic dispersion of stakeholders, lack of infrastructure, adverse climate conditions, natural and man-made disasters and so on.

## 10.2 Supply Chain Disruptions

According to Talluri et al. (2013, p. 254), a supply chain disruption refers to the situation 'when the supply chain is radically and unexpectedly transformed through nonavailability of certain production, warehousing, distribution, or transportation options...'. Supply chain disruptions are rare and unpredictable; nevertheless, they result in long-lasting supply chain discontinuity (Greening & Rutherford, 2011; Scheibe & Blackhurst, 2018). Natural catastrophes and man-made crises are frequently cited causes of supply chain disruptions (Durach et al., 2017). Examples of recent causes of significant supply chain disruptions include: the COVID-19 pandemic in 2021, the 6.4 magnitude earthquake that shook Taiwan in February 2018; the Laredo border closing that affected Mexico in May 2017; Hurricane Irma hitting the U.S. in September 2017; Typhoon Napartak striking Taiwan in July 2016; and the Tianjin explosion that destroyed Port Tianjin in August 2015 (Resilince, 2016, 2017, 2018, 2019). According to Resilince (2016, 2017, 2018, 2019), it took 22–56 weeks for supply chains affected by these disruptions to

**Table 10.1** Examples of recent significant supply chain disruptions

| | Type | Event | Date | Location | Average Site Time to Recover |
|---|---|---|---|---|---|
| 1 | Natural catastrophe | Covid-19 pandemic | March 2020 | Globally | 56 |
| 2 | Natural catastrophe | 6.4 M earthquake in Taiwan | Feb-2018 | Taiwan | 22 |
| 3 | Natural catastrophe | Hurricane Irma | Sep-2017 | U.S.A. | 33 |
| 4 | Natural catastrophe | Typhoon Napartak | July-2016 | Taiwan | 30 |
| 5 | Man-made crisis | Laredo border closing | May-2017 | Mexico | 24 |
| 6 | Man-made crisis | Tianjin explosion | Aug-2015 | China | 37 |

recover and restore their operations (see Table 10.1 for details).

Supply chain disruptions can spread rapidly through the supply chain, a phenomenon known as the 'ripple effect' (Sokolov et al., 2015). The main reason for the ripple effect is the tight couplings in the current supply chains (Ivanov et al., 2013). For example, COVID-19 has caused an unprecedented runaway chain of product shortages and delivery delays, particularly the shortage of consumer staple goods like hand sanitizers that subsequently impacted the life of many people extensively worldwide in a very short timeframe (Nikookar & Yanadori, 2022).

A supply chain disruption differs from everyday business incidents in terms of its severe negative impact on the financial performance of the parties involved (Craighead et al., 2007). Because a supply chain disruption may materialize almost anywhere throughout a supply chain without warning, its consequences are unpredictable (Xu et al., 2014). Supply chain disruptions can not only negatively impact the financial performance of firms by jeopardizing the smooth flow of materials and products (Greening & Rutherford, 2011; Kleindorfer & Saad, 2005; Zsidisin & Wagner, 2010), but their impact on vulnerable populations will also be particularly dire. Supply chain disruptions do not just create higher prices and shortages among high-end consumer products, such as cars. They also affect more-basic commodities such as generic drugs or energy, increasing the cost of living and the provision of basic needs (Bradley, 2021).

The negative impact of supply chain disruptions is not limited to costs. Supply chain disruptions also have non-financial and long-lasting impacts on organizational reputation and customer satisfaction (Ivanov et al., 2013). Lee and Whang (2005) highlight the effect of supply chain disruptions on increased uncertainties regarding delivery time and product quantity and quality, which consequently lead to lower service levels and customer dissatisfaction. Wagner and Bode (2008) argue that supply chain disruptions negatively impact supply chain performance with respect to delivery dependability, order fulfilment, delivery speed, and customer satisfaction.

The impact of a supply chain disruption is not evenly distributed across supply chains. When a disruption occurs in a supply chain, some firms recover quickly, whereas others may take considerable time to resume operations. In some cases, supply chains may collapse completely, resulting in firms going out of business permanently. The severity of the impact varies depending on firms' reactions to supply chain disruption. For example, a fire at one of Philips' electronic production facilities in Albuquerque, New Mexico (U.S.A.) in 2000 caused Nokia and Ericsson to contend with the loss of the supply of radio-frequency chips, a key component in producing cell phones (Rice & Caniato, 2003). While both companies were similarly affected by the disruption, their reactions were remarkably different. Ericsson did not realize the seriousness of the disruption and took no counteractive measures. The fire caused Ericsson to miss its sales plan, which resulted in a loss of market share and ultimately led to the shutdown of its cell phone production facility. In contrast, Nokia's supply chain manager immediately sensed the seriousness of the situation and collaborated with the supplier to solve the problem. They also changed the design of their products to accommodate the radio-frequency chip shortage. Nokia successfully served its customers globally and became the dominant market supplier (Rice & Caniato, 2003). This example illustrates that the way a supply chain disruption is addressed has important implications for its recovery from the disruption.

The adverse impacts of supply chain disruptions require firms to adopt systematic approaches for dealing with them. Enterprise risk management (ERM) is a dominant paradigm popularized in the 1990s to cope with supply chain disruptions that firms might face (Pettit et al., 2019). According to the ERM paradigm, a supply chain member, independently or in collaboration with other members, applies risk management tools and processes to mitigate the negative impacts associated with supply chain disruptions (Imbiri et al., 2021, 2023; Lindroth & Norrman, 2001). This is achieved through a systematic process constituting disruption identification, assessment, treatment, and monitoring. However, the considerable reliance of ERM on subjective information to identify disruption has been heavily criticized by researchers (Pettit et al., 2013, 2019). Much of the existing approaches in ERM assume that all potential disruptions are predictable, but recent disruptions caused by natural or man-made disasters refute this assumption. ERM may handle situations where disruptions are predictable to some extent, for example, disruptions due to a predictable labour strike in a supplier site. However, it fails to address unpredictable disruptions such as terrorist attacks or natural disasters (Greening & Rutherford, 2011; Ivanov et al., 2013).

As well as the subjective nature of disruption identification, Fiksel et al. (2015) assert that ERM is based on historical data that may not be available in some cases, such as data about natural disasters. Moreover, ERM evaluates potential disruptions independently and fails to recognize the interaction between predicted disruption and some environmental factors which may intensify the consequences. As a result, ERM may underestimate the severity or probability of disruptions (Fiksel et al., 2015). According to this argument, it is evident that ERM approaches, and solutions are not sufficient to deal with supply chain disruptions (Barroso et al., 2011; Pettit et al., 2013, 2019). Firms need a new approach that enables them to be proactively prepared for major disruptions emerging in the new business environment (Scholten et al., 2014).

## 10.3 Sustainable and Resilient Supply Chains

Sustainable supply chains are defined as "a supply chain that not only simultaneously makes a profit and achieves its potential, but it is one that also is responsible to its consumers, suppliers, societies and environments by innovative strategic, tactics and management technologies" (Kim et al., 2014, p. 8). The sustainability in SC emphasizes the need for environmental, social, and economic criteria to be satisfied (Sánchez-Flores et al., 2020). Sustainable supply chains are managed considering the integration of economic, environmental, and social considerations, which facilitates stakeholder involvement and improves the long-term profitability, competitiveness, and resilience of the organization (Ahi & Searcy, 2013).

Sustainable supply chain management usually involves designing efficient materials and information flows, equipping traditional supply chains with capabilities to act upon the challenges and the ability to anticipate uncertainties while remaining competitive and connected with the customers in the long run (Sharma et al., 2020). For example, (Jouzdani & Govindan, 2021) discussed, among others, sustainable network design to ensure the achievement of the objectives of zero hunger, affordable and clean energy, decent work and economic growth, and climate actions (Tsolakis et al., 2020), explored the design of blockchain-centric supply chains to achieve S.D.G.s (Genc, 2021), discussed closed-loop supply chain relations considering industry investment, innovation, affordability, clean product, and responsible consumption/production.

The resilience aspect of supply chain design, on the other hand, also plays a critical role in building sustainable communities since it facilitates and enables speedy recovery from unforeseen events, as discussed in Sect. 10.1.

Those might include disasters, wars, political sanctions, pandemics and other threats. Supply chain resilience is the adaptive capacity to tolerate disturbances and manage environmental variability by rebuilding itself to either the original state or a more robust state than before (Pires Ribeiro & Barbosa-Povoa, 2018; Lam & Bai, 2016). The key underlying principles of resilient supply chains have been established in the works of Christopher and Peck (2004), Sheffi and Rice (2005) and re-evaluated by Walker (2020) and Wieland and Durach (2021). Originally, Sheffi and Rice (2005) defined supply chain resilience as "the ability to bounce back from disruptions" (p. 41). "The ability of a system to return to its original state or move to a new, more desirable state after being disturbed" (Christopher & Peck, 2004, p. 4). More recent definitions reinterpret the resilience concept referring to social–ecological phenomena (Wieland & Durach, 2021) by including SC adaptability and transformability in the face of change, but also SC viability as a system ability to maintain itself and recover in the presence of disturbances over a long-term horizon' (Ivanov & Dolgui, 2020).

In the face of disruption, resilient supply chain antecedents enabling one to quickly recover from a shock include agility, collaboration, supply chain risk management culture and supply chain re-engineering (Christopher & Peck, 2004), flexibility, redundancy, collaboration visibility, velocity properties (Christopher & Peck, 2004; Jüttner & Maklan, 2011; Brandon-Jones et al., 2014), robustness (Wieland & Wallenburg, 2013), risk awareness and knowledge management (Scholten et al., 2014), resource reconfiguration, disruption orientation, learning and risk management culture (e.g. Scholten & Schilder, 2015; Chowdhury & Quaddas, 2016).

Agility is defined as "the ability of a supply chain to rapidly respond to change by adapting its initial stable configuration" (Wieland & Wallenburg, 2013, p. 302), often attributed to visibility and velocity properties (Kamalahmadi & Parast, 2016). While visibility is the knowledge of the operating assets and environment in the supply chain along with upstream and downstream inventories, demand, and supply status (Kamalahmadi & Parast, 2016), velocity relates to the speed of the supply chain under the resilience concept when there are re-designing procedures in action (Wieland & Wallenburg, 2013).

Supply chain re-engineering is the re-design of traditional supply chains to integrate built-in resilience into the supply chain (Christopher & Peck, 2004; Kamalahmadi & Parast, 2016; MacCarthy et al., 2016). The supply chain re-engineering is underpinned by supply chain visibility, including the identification of potential vulnerabilities of critical suppliers; supply base strategies, including risk awareness of the suppliers and supply chain design principles based on the trade between redundancy and flexibility. Redundancy emphasizes having multiple sourcing and suppliers and maintaining backups and safety stocks (Lee, 2004). Flexibility is another central capability at the process level referring to the ability of production, sourcing, and transportation systems in the supply chain to adapt to a dynamic environment (Sheffi, 2005).

The difficulty in designing sustainable and resilient supply chains lies in the contradictory objectives that entail, for example, focusing on ethical considerations and long-term benefits for the next generations—sustainability, and for example, properties or agility and flexibility that define resilience. To build those into the supply chain design, it is often required to have a diversified supplier base rather than building long-term trust based long lasting relationships with suppliers (Rajesh, 2021; Statsenko et al., 2018a, b). However, corporate social responsibility and recent pressures on multinationals for localizing supply chains force industries and communities to adopt more responsible sourcing by reporting on incorporating local suppliers and ensuring that procurement contributes to the long-term development of the region, making the entire supply chain less vulnerable and self-sufficient over the long term.

Despite its academic and practical relevance, until recently, the SCM literature often studied resilience and sustainability separately in SC. One of the contributions of this work is fulfilling this gap by providing an overview of principles for Sustainable and Resilient Supply Chains (SRSC).

Integration of resilience and sustainability concepts in supply chain management. Although the concepts of 'resilient supply chains and 'sustainable supply chains have been discussed in recent literature, designing a supply chain that incorporates both properties are a complex task from management and modelling perspectives (Corrales-Estrada et al., 2021; Fahimnia & Jabbarzadeh, 2016; Rajesh, 2021). The difficulties arise due to sometimes conflictive objectives and success criteria of 'resilience' and 'sustainability'(Rajesh, 2021).

A few studies have attempted to investigate the trade-off between these two concepts (Fahimnia & Jabbarzadeh, 2016; Jabbarzadeh et al., 2018; Rajesh, 2021). The integration must be considered from the strategic design level and tactical and operational planning level (López-Castro & Solano-Charris, 2021). The concept of sustainable supply chains has evolved from applying lean principles and focusing on efficiency gained through the waste reduction in a product value chain. In contrast, resilience in supply chains is largely based on efficient responses to disruptions enabled by agility and flexibility (Ambulkar et al., 2015; Rajesh, 2018). For instance, from the strategic perspective, once the supply chains are environmentally, economically, and socially sustainable, they become highly sensitive to disruptions and can be easily compromised. When the sustainable supply chain gains efficiency through applications of lean principles and waste reduction, this may result in decreased flexibility since this often reduces available time and inventory buffers (Rajesh, 2021). Furthermore, from an operational perspective, process streamlining in sustainable supply chains to reduce the amount of waste, and to support resource conservation, reduces the capabilities for rapid reconfiguration, flexibility, and redundancy (Luthra et al., 2017; Saenz & Revilla, 2014). Scholars acknowledge the necessity of sustainable supply chains to incorporate resilience aspect to be able to handle unforeseen disruptions (Bai et al., 2019; Fahimnia & Jabbarzadeh, 2016). López-Castro and Solano-Charris (2021) incorporated sustainable and resilience criteria into a supply chain design and declared their contribution towards sustainable development goals (SDG). These approaches depict a proper positioning of the partition line to the concepts within the same supply chain.

The COVID-19 pandemic has disrupted both global supply chains and has changed regular consumption patterns. It has induced the demand for the consumer staples, such as sanitizers, gloves, masks, immune-boosting medicines etc. It has also increased the complexity of plastic waste management as those essential goods and often food is sold in a plastic package (Vanapalli et al., 2021). To reduce the effects of disruptions, supply chains should practice resource efficiency by sourcing green and recycled materials (Majumdar et al., 2021; El-Baz & Ruel, 2021). On the other hand, when there is a rapid response to local needs due to the pandemic, it encourages searching for novel techno-savvy approaches in supply chain and logistics (Sarkis et al., 2010). During the post-covid era, organizations are in the process of creating a 'new normal' mode of operations, restructuring processes and value chains, and they must consider novel approaches that incorporate sustainability and resilience principles (Rai et al., 2021a, b). For instance, the pandemic is considered a wake-up call for building sustainability into food supply chains (Alsetoohy et al., 2021). The social distancing and close contact minimization facilitated a swift change to online and contactless delivery, which indirectly led to more lean and efficient logistic solutions. The pandemic had a positive impact on the sustainable local food supply chain by generating more business opportunities for companies that can meet environmental, economic, and social considerations successfully (Alsetoohy et al., 2021). Rajesh (2021) suggested that organizations should plan for optimum level of efficiency, while maintaining flexibility to generate a trade-off between sustainability and resilience. Maintaining the lean-agile paradigms and sustainable principles throughout the entire supply chain is key. During the recovery phase, sustainability aspects should also be considered in supply chains and help better preparedness and

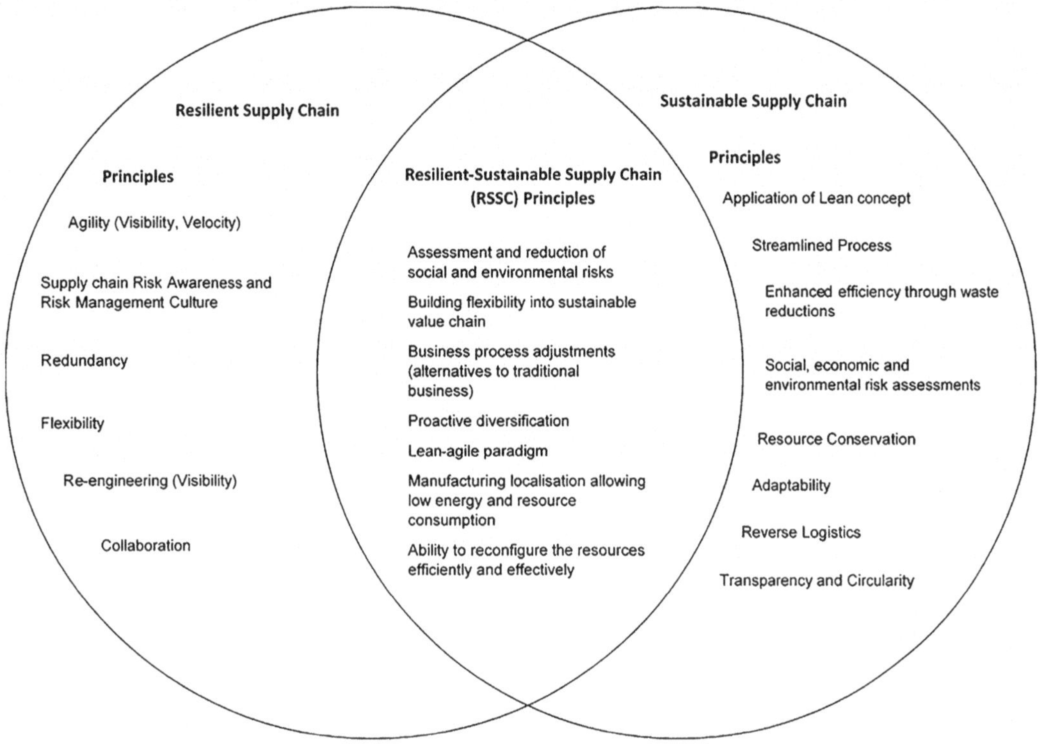


**Fig. 10.1** Sustainability and resilience principles in SRCH

response to unknown risks (Rajesh, 2018; Paul et al., 2021; Rai et al., 2021a, b).

Figure 10.1 illustrates the conceptual framework that brings together principles of resilience and sustainability in the context of supply chain management, also highlighting their benefits.

## 10.4 Role of Circular Economies and Reverse Logistics

Before considering the key conceptual and empirical issues around how cities and communities can overcome risks and uncertainties in their quest to become sustainable and resilient, it is necessary to define the terms 'reverse logistics', 'circular economy'and the interplay between 'reverse logistics'and 'circular economy'.

Reverse logistics: There are quite several definitions of R.L. in literature. However, the commonly used is that provided by Tibben-Lembke and Rogers (2002), which defines it as: "The process of planning, implementing and controlling the efficient, cost-effective flow of raw materials, in-process inventory, finished goods and related information from the point of consumption to the point of origin for the purpose of recapturing value of proper disposal". Other definitions are provided by Guide and Van Wassenhove (2009, p. 10) as follows "...the design, control, and operation of a system to maximize value creation over the entire life cycle of a product with the dynamic recovery of value from different types and volumes of returns over time." Within the construction industry context, RL could be viewed as the management of the flow of salvaged building materials from established buildings back into the construction market. From the 'sector' perspective, Vegter et al. (2020) defines reverse flow through distinctions between 'open' and 'closed' loops, with the reverse flow in the same sector referred to as a 'closed-loop supply chain', and in another sector as 'an open-loop supply chain'. However, the simplified definition version of RL is described as simplified as the flow of goods from end-users to producers (Vegter et al., 2020).

Circular economy: There are different interpretations and definitions of the circular economy available in the literature. However, the majority have evolved from the works of Ellen MacArthur Foundation (2015), which view the circular economy as a systems solution framework that tackles global challenges like climate change, biodiversity loss, waste, and pollution. Accordingly, this is anchored on the following three principles driven by design: (1) Eliminate waste and pollution; (2) Circulate products and materials (at their highest value); and (3) Regenerate nature. In contrast, the European Commission (2015) defines C.E. as "one in which the value of products, materials and resources is maintained for as long as possible, therefore diminishing waste and resource use". CE can also be defined and researched at three different levels: Meso, Macro and Micro. According to Pomponi and Moncaster (2017), the buildings can be seen as meso-level, whereas the macro-level is urban agglomerates which include cities, neighbourhoods, and built environments. In contrast, the micro level deals with building components. From a solutions perspective, Lahane et al. (2020) view the CE as acting as alternative solution to the dominant linear economic model of production and consumption.

Relationship between the interplay of RL and CE and the SDGs. The combination of the concepts of 'Reverse logistics'and 'Circular Economy', which is termed 'circularity'have emerged as remedial strategies for waste recovery (Chileshe et al., 2016). Further, according to Lahane et al. (2020), the incorporation of the CE into supply chain management is known as circular supply chain management (CSCM); and according to Geissdoerfer et al. (2018), these are also often referred to as open loop supply chains, closed-loop supply chains as well as circular supply chains.

Each concept makes a unique contribution to the attainment of Sustainable Development Goals (SDGs). For instance, Lewandowski (2016), the creation of reverse logistics systems to close a material loop is also needed and one of the key factors in circular business implementation. The synergies between R.L. and C.E. are further amplified by Julianelli et al. (2020), who proposed a framework aimed at illustrating how each component of the circular supply chain (i.e., Circular Inputs, Product Design, Process Design and Circular Flows) is influenced by the critical success factors (CSFs) as they influenced all components of the circular supply chain and, consequently, reverse logistics. Therefore, the CE could be enabled by RL, as acknowledged by Julianelli et al. (2020), that reverse logistics achieves a wider scope, incorporating other circular and sustainable processes, which are enhanced by critical success factors and allows absorbing and creating value from circular supply chain components.

The synergies between SDGs and the concepts of CE and R. can be found in the SGD12, which is aimed at ensuring sustainable consumption and production patterns. Here, the concept of circular economy (CE) has similarities to a recent popular global business trend, which considerably minimizes waste and environmental pollution (Moktadir et al., 2020). Further, according to Bjørnbet et al. (2021), this is one of the strongest links between SDG12, and particularly the sub-target of 12.5 around substantial waste reduction, through prevention, reduction, recycling, and reuse, as this is the main or core aspect of C.E. From the economic perspective, the interplay between C.E. and S.D.G. are nested within the aspirations of SGD8, which is "Promote sustained, inclusive and sustainable economic growth, full and productive employment and decent work for all". According to Grdic et al. (2020), the circular economic model enables economic growth and prosperity in accordance with environmental protection and sustainable development because of the current unsustainable linear means of production in the economy.

Similarly, economic drivers have been found to be dominant in the implementation of R.L. (Chileshe et al., 2018). Other studies have also explored the relationships between CE and sustainable development (Sauvé et al., 2016; Ghisellini et al. 2018).

## 10.5 The Framework for Sustainable Cities, Incorporating Reverse Logistics, Circular Economy, and Sustainable and Resilient Supply Chains

The sustainability and resilience of communities and cities largely depend on the health of their supply chains. Current principles of just-in-time, low cost, low inventory for global supply chains operation have proved inefficient and risky, including disruptions induced by pandemics, geo-political issues between powerful countries and reliance on critical raw materials. These issues can be tackled by promoting circularity, sustainability, and resilience concepts within the global supply chains for a smooth transition towards sustainable and resilient economies. These approaches facilitate supply chain resilience and strengthen governments' efforts to meet sustainability targets (Schneider-Petsinger, 2021).

Figure 10.2 demonstrates the interplay between those concepts, CE, RL, and SGD, in a nutshell. Along with the above discussion, Fig. 10.2 depicts that the interplay between R.L. and C.E. is not only the RL concept is closely related to CE practice but also the environmentally friendly and RL practices and facilities themselves are essential pre-requisites for CE implementation and gradually support sustainable development (Moktadir et al., 2020). While R.L. and its CSF directly influence the circularity and circular supply chain within the CE context (Julianelli et al., 2020), the inner circle/inner point of the diagram represents the "Reverse Logistics"concept while considering it as an integral part of the broader "CE concept". According to the above discussion related to the RL and CE interplay, CE and RL are linked

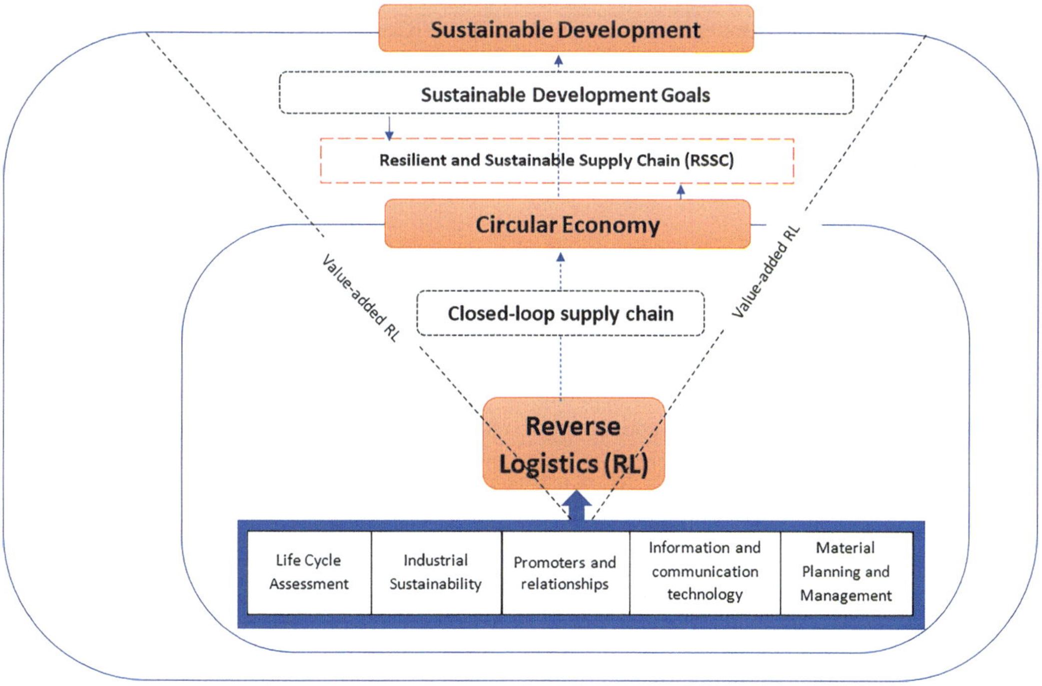


**Fig. 10.2** Interplay between CE, RL and SDGs

through the characteristics of the circular supply chain, which is facilitated through the "value-added" RL which considers developing policies for circular supply chains such as maintenance policy and enabling redistribution aspects such as reuse/redistribute, refurbish/remanufacture, recycle, rethink and re-design and a broader sustainability perspective within the CE's technical cycle. While the sustainable development goals are discussed at a macro level (global perspective), it is evident that CE is the main influencing factor for SDGs and its contribution to the economy and overall sustainable development in the community (Bjørnbet et al., 2021). Further, the diagram also illustrates the overall CE concept as a tool to attain a considerable number of SDGs. Once such a strong link can be found with Goal 12-responsible production and consumption, the outer circle in this diagram could be "sustainable development" as the ultimate objective.

Recommendations for policy and practice. To achieve a trade-off between sustainability and resilience in supply chains, there is a need to plan for an optimum level of efficiency while maintaining flexibility (Rajesh, 2021). Maintaining the lean-agile paradigms and sustainable principles throughout the entire upstream and downstream supply chain is important. Resilience in sustainable supply chains can also be achieved through diversification. Taking an example of agro-industrial supply chains, the implementation of competitive business strategies enables to sustain business processes by adjusting and proactive diversify to reduce reliance on traditional markets in case of a sharp shift in demand because of the disruption (Stadnyk et al., 2021). Despite the effects of the pandemic, disruptions often push businesses for product and supply chain diversification, adjustments, and alternatives to traditional ways of doing business (Weersink et al., 2021).

Information technology-related disruptions, including security breaches or technical computer system disruptions, are also common, resulting in increased waiting times decreeing the performance (Gölgeci et al., 2020; Klumpp & Loske, 2021). In such cases, supply chain resilience can be enhanced through the application of Artificial Intelligence (AI) and Data Analytics, IoT technologies etc. As a result of this supply chain resilience-based technological orientation, cost-effectiveness and economic benefits in the competitive world can be obtained, enabling timely, efficient recovery to maintain business continuity (Pettit, 2018; Sikken, 2011). The advent of technology in the digital age has led to the popularization of the term Industry 4.0, which serves as a reference for research and development in various technological sectors. This has inspired people to continue advancing technology for the betterment of human life. The notion of City 5.0 explains the impact of the fourth industrial revolution on people's lives, highlighting the revolution that not only utilizes technology but also considers the human aspect (Rosemann et al., 2021). Integration of technology and human values will enable more efficient practices facilitating circular economy principles from the consumer perspective.

Supply chain risk management culture with both organizational and people-oriented soft processes to create resilience and sustainability mindsets that guide procurement decision-making (Gorod et al., 2021). The supply chain risk management culture pertains to two aspects such as leadership and innovation. Leadership is providing support and guidance according to the company policies and guidelines for the purpose of changing organization culture (Kamalahmadi & Parast, 2016). Innovation stands as a key element for the long-term survival and growth of a supply chain, which guides the organization in adapting and responding to the changes (Sharifirad & Ataei, 2012).

The role of individuals, particularly supply chain decision-makers, in improving the resilience and sustainability of supply chains is undeniable. For example, the results of a study by Nikookar and Yanadori (2022) indicate that s supply chain managers' work experience significantly improves supply chain resilience. Work experience can also contribute to the sustainability of the supply chain, as managers with greater experience are more likely to be aware of sustainable practices in the field.

Keeping other supply chain resilience or sustainability antecedent's constant. Supply chains that feel their current level of resilience or sustainability are low may be interested in hiring supply chain managers with more industry-specific work experience.

## References


Ahi, P., & Searcy, C. (2013). A comparative literature analysis of definitions for green and sustainable supply chain management. *Journal of Cleaner Production, 52*, 329–341.

Ahranjani, M. P., Ghaderi, S. F., Azadeh, A., & Babazadeh, R. (2020). Robust design of a sustainable and resilient bioethanol supply chain under operational and disruption risks. *Clean Technologies and Environmental Policy, 22*(1), 119–151.

Ahvenniemi, H., Huovila, A., Pinto-Seppä, I., & Airaksinen, M. (2017). What are the differences between sustainable and smart cities? *Cities, 60*, 234–245.

Alsetoohy, O., Ayoun, B., & Abou-Kamar, M. (2021). Covid-19 pandemic is a wake-up call for sustainable local food supply chains: Evidence from green restaurants in the U.S.A. *Sustainability (Switzerland), 13*(16).

Ambulkar, S., Blackhurst, J., & Grawe, S. (2015). Firm's resilience to supply chain disruptions: Scale development and empirical examination. *Journal of Operations Management, 33*, 111–122.

Bai, C., Kusi-Sarpong, S., Badri Ahmadi, H., & Sarkis, J. (2019). Social sustainable supplier evaluation and selection: A group decision-support approach. *International Journal of Production Research*, 1–22.

Barroso, A. P., Machado, V. H., & Machado, V. C. (2011). Supply chain resilience using the mapping approach, In Li P. (Ed.), Supply chain management, InTech, 161–184. Available at: http://dx.doi.org/10.5772/15006

Bjørnbet, M. M., Skaar, C., Fet, A. M., & Schulte, K. Ø. (2021). Circular economy in manufacturing companies: A review of case study literature. *Journal of Cleaner Production, 294*, 126268.

Bradley, M. (2021). *Supply chain disruptions: The risks and consequences.* Rand viewed 22 April 2022, https://www.rand.org/blog/2021/11/supply-chain-disruptions-the-risks-and-consequences.html#:~:text=The%20impact%20on%20vulnerable%20populations,the%20provision%20of%20basic%20needs

Brandon-Jones, E., Squire, B., Autry, C.W., & Petersen, K.J. (2014). A contingent resource-based perspective of supply chain resilience and robustness. *Journal of Supply Chain Management, 50*(3), 55–73.

Chehri, A., & Mouftah, H. T. (2019). Autonomous vehicles in the sustainable cities, the beginning of a green adventure. *Sustainable Cities and Society, 51*.

Chileshe, N., Rameezdeen, R., & Hosseini, M. R. (2016). Drivers for adopting reverse logistics in the construction industry: A qualitative study. *Engineering Construction and Architectural Management, 23*(2), 134–157.

Chileshe, N., Rameezdeen, R., Hosseini, M. R., Martek, I., Li, H. X., & Panjehbashi-Aghdam, P. (2018). Factors driving the implementation of reverse logistics: A quantified model for the construction industry. *Waste Management, 79*, 48–57.

Chowdhury, M. M. H., & Quaddas, M. (2016). Supply chain readiness, response and recovery for resilience. *Supply Chain Management: An International Journal, 21*(6), 709–731.

Christopher, M., & Peck, H. (2004). Building the resilient supply chain. *The International Journal of Logistics Management, 15*, 1–14.

Corrales-Estrada, A. M., Gómez-Santos, L. L., Bernal-Torres, C. A., & Rodriguez-López, J. E. (2021). Sustainability and resilience organizational capabilities to enhance business continuity management: A literature review. *Sustainability, 13*, 81–96.

Craighead, C., Blackhurst, J., Rungtusanatham, J., & Handfield, R. (2007). The severity of supply chain disruptions: Design characteristics and mitigation capabilities. *Decision Sciences, 38*(1), 131–156.

de Souza, V., Bloemhof-Ruwaard, J., & Borsato, M. (2019). Towards regenerative supply networks: A design framework proposal. *Journal of Cleaner Production, 221*, 145–156.

Durach, C. F., Glasen, P. C., & Straube, F. (2017). Disruption causes and disruption management in supply chains with Chinese suppliers: Managing cultural differences. *International Journal of Physical Distribution & Logistics Management, 47*(9), 843–863.

El Baz, J., & Ruel, S. (2021). Can supply chain risk management practices mitigate the disruption impacts on supply chains' resilience and robustness? Evidence from an empirical survey in a COVID-19 outbreak era. *International Journal of Production Economics, 233*, 107972.

Ellen MacArthur Foundation. (2015). *Growth within: A circular economy vision for a competitive Europe*, Vol. 100.

European Commission. (2015), *Closing the loop—An E.U. action plan for the circular economy.* In European Commission, n/a.

Fahimnia, B., & Jabbarzadeh, A. (2016). Marrying supply chain sustainability and resilience: A match made in heaven. *Transportation Research Part E: Logistics and Transportation Review, 91*, 306–324.

Fattahi, M., Govindan, K., & Farhadkhani, M. (2020). Sustainable supply chain planning for biomass-based power generation with environmental risk and supply uncertainty considerations: A real-life case study. *International Journal of Production Research*, 1–25.

Fiksel, J., Polyviou, M., Croxton, K. L., & Pettit, T. J. (2015). From risk to resilience: Learning to deal with disruption, M.I.T. *Sloan Management Review, 56*(2), 79–86.

Geissdoerfer, M., Morioka, S. N., de Carvalho, M. M., & Evans, S. (2018). Business models and supply chains for the circular economy. *Journal of Cleaner Production, 190*, 712–721.

Genc, T. S. (2021). Implementing the United Nations sustainable development goals to supply chains with behavioral consumers. *Annals of Operations Research*, 1–32.

Ghisellini, P., Ripa, M., & Ulgiati, S. (2018). Exploring environmental and economic costs and benefits of a circular economy approach to the construction and demolition sector. A literature review. *Journal of Cleaner Production, 178*, 618–643.

Golgeci, I., Yildiz, H. E., & Andersson, U. R. (2020). The rising tensions between efficiency and resilience in global value chains in the post-COVID-19 world. *Transnational Corporations Journal, 27*(2).

Gorod, A., Hallo, L., Statsenko, L., Nguyen, T., & Chileshe, N. (2021). Integrating hierarchical and network centric management approaches in construction megaprojects using a holonic methodology. *Engineering, Construction and Architectural Management, 28*(3), 627–661.

Grdic, Z. S., Nizic, M. K., & Rudan, E. (2020). Circular economy concept in the context of economic development in E.U. countries. *Sustainability (Switzerland), 12*(7) art. no. 3060.

Greening, P., & Rutherford, C. (2011). Disruptions and supply networks: A multi-level, multi-theoretical relational perspective. *The International Journal of Logistics Management, 22*(1), 104–126.

Guide, V. D. R., & Van Wassenhove, L. N. (2009). The evolution of closed-loop supply chain research. *Operational Research, 57*(1), 10–18.

Imbiri, S., Rameezdeen, R., Chileshe, N., & Statsenko, L. (2021). A novel taxonomy for risks in agribusiness supply chains: A systematic literature review. *Sustainability, 13*(16), 9217.

Imbiri, S., Rameezdeen, R., Chileshe, N., & Statsenko, L. (2023). Risk propagation and resilience in the agribusiness supply chain: A systematic literature review. *Journal of Agribusiness in Developing and Emerging Economies*. (ahead-of-print).

Ivanov, D., & Dolgui, A. (2020). Viability of intertwined supply networks: Extending the supply chain resilience angles towards survivability. A position paper motivated by COVID-19 outbreak. *International Journal of Production Research, 58*(10), 2904–2915.

Ivanov, D., Sokolov, B., & Dolgui, A. (2013). The ripple effect in supply chains: Trade-off 'efficiency-flexibility-resilience' in disruption management. *International Journal of Production Research, 52*(7), 2154–2172.

Jabbarzadeh, A., Fahimnia, B., & Sabouhi, F. (2018). Resilient and sustainable supply chain design: Sustainability analysis under disruption risks. *International Journal of Production Research, 56*(17), 5945–5968.

Jouzdani, J., & Govindan, K. (2021). On the sustainable perishable food supply chain network design: A dairy products case to achieve sustainable development goals. *Journal of Cleaner Production, 278*.

Julianelli, V., Caiado, R. G. G., Scavarda, L. F., & Cruz, S. P. D. M. F. (2020). Interplay between reverse logistics and circular economy: Critical success factors-based taxonomy and framework. *Resources, Conservation and Recycling, 158*, 104784.

Jüttner, U., & Maklan, S. (2011). Supply chain resilience in the global financial crisis: An empirical study. *Supply Chain Management: An International Journal, 16*(4), 246–259.

Kamalahmadi, M., & Parast, M. M. (2016). A review of the literature on the principles of enterprise and supply chain resilience: Major findings and directions for future research. *International Journal of Production Economics, 171*, 116–133.

Kim, K., Jeong, B., & Jung, H. (2014). Supply chain surplus: Comparing conventional and sustainable supply chains. *Flexible Services and Manufacturing Journal, 26*(1–2), 5–23.

Kleindorfer, P. R., & Saad, G. H. (2005). Managing disruption risks in supply chains. *Production and Operations Management, 14*(1), 53–68.

Klumpp, M., & Loske, D. (2021). Sustainability and resilience revisited: Impact of information technology disruptions on empirical retail logistics efficiency. *Sustainability, 13*(10), 5650.

Lahane, S., Kant, R., & Shankar, R. (2020). Circular supply chain management: A state-of-art review and future opportunities. *Journal of Cleaner Production, 258*.

Lam, J. S. L., & Bai, X. (2016). A quality function deployment approach to improve maritime supply chain resilience. *Transportation Research Part E: Logistics and Transportation Review, 92*, 16–27.

Lee, H. L. (2004). The triple-a supply chain. *Harvard Business Review, 82*(10), 102–113.

Lee, H. L., & Whang, S. (2005). Higher supply chain security with lower cost: Lessons from total quality management. *International Journal of Production Economics, 96*(3), 289–300.

Lewandowski, M. (2016). Designing the business models for circular economy—Towards the conceptual framework. *Sustainability, 8*(1), 43–58.

Lindroth, R., & Norrman, A. (2001). Supply chain risk management: Purchasers' vs. planners view on sharing capacity investment risks in the telecom industry. In *Proceedings of the I.P.S.E.R.A. 11th international conference*, Twente University 25–27 March, pp. 25–27.

López-Castro, L. F., & Solano-Charris, E. L. (2021). Integrating resilience and sustainability criteria in the supply chain network design: A systematic literature review. *Sustainability, 13*(19), 10925.

Lotfi, R., Mehrjerdi, Y. Z., Pishvaee, M. S., Sadeghieh, A., & Weber, G. W. (2021). A robust optimization model for sustainable and resilient closed-loop supply chain network design considering conditional value at risk. *Numerical Algebra, Control & Optimization, 11*(2), 221.

Luthra, S., Govindan, K., Kannan, D., Mangla, S. K., & Garg, C. P. (2017). An integrated framework for sustainable supplier selection and evaluation in supply chains. *Journal of Cleaner Production, 140*, 1686–1698.

MacCarthy, B. L., Blome, C., Olhager, J., Srai, J. S., & Zhao, X. (2016). Supply chain evolution–theory, concepts and science. *International Journal of Operations & Production Management, 36*(12), 1696–1718.

Maiyar, L. M., & Thakkar, J. J. (2020). Robust optimization of sustainable food grain transportation with uncertain supply and intentional disruptions. *International Journal of Production Research, 58*, 5651–5675.

Majumdar, A., Sinha, S. K., Shaw, M., & Mathiyazhagan, K. (2021). Analyzing the vulnerability of green clothing supply chains in South and Southeast Asia using fuzzy analytic hierarchy process. *International Journal of Production Research, 59*(3), 752–771.

Miao, S., Zhou, C., AlQahtani, S. A., Alrashoud, M., Ghoneim, A., & Lv, Z. (2021). Applying machine learning in intelligent sewage treatment: A case study of chemical plant in sustainable cities. *Sustainable Cities and Society, 72*, 103009.

Mishra, S., & Singh, S. P. (2020). A stochastic disaster-resilient and sustainable reverse logistics model in big data environment. *Annals of Operations Research*, 1–32.

Moktadir, M. A., Ahmadi, H. B., Sultana, R., Zohra, F.-T., Liou, J. J. H., & Rezaei, J. (2020). Circular economy practices in the leather industry: A practical step towards sustainable development. *Journal of Cleaner Production, 251*, 119737.

Nikookar, E., & Yanadori, Y. (2022). Preparing supply chain for the next disruption beyond COVID-19: Managerial antecedents of supply chain resilience. *International Journal of Operations & Production Management, 42*(1), 59–90.

Paul, S. K., Moktadir, M. A., & Ahsan, K. (2021). Key supply chain strategies for the post-COVID-19 era: Implications for resilience and sustainability. *International Journal of Logistics Management., 34*, 1165–1187.

Pettit, T. J., Croxton, K. L., & Fiksel, J. (2013). Ensuring supply chain resilience: Development and implementation of an assessment tool. *Journal of Business Logistics, 34*(1), 46–76.

Pettit, T. J. (2018). *Supply chain resilience: A case of balancing the supply chain for long-term sustainability*. Council of Supply Chain Management Professionals Cases.

Pettit, T. J., Fiksel, J., & Croxton, K. L. (2019). The evolution of resilience in supply chain management: a retrospective on ensuring supply chain resilience. *Journal of Business Logistics, 40*(1), 56–65.

Pires Ribeiro, J., & Barbosa-Povoa, A. (2018). Supply chain resilience: Definitions and quantitative modelling approaches—A literature review. *Computers & Industrial Engineering, 115*, 109–122.

Pomponi, F., & Moncaster, A. (2017). Circular economy for the built environment: A research framework. *Journal of Cleaner Production, 143*, 710–718.

Rai, S. S., Rai, S., & Singh, N. K. (2021a). Organizational resilience and social-economic sustainability: COVID-19 perspective. *Environment Development and Sustainability, 23*(8), 12006–12023.

Rai, S. S., Rai, S., & Singh, N. K. (2021b). Organizational resilience and social-economic sustainability: COVID-19 perspective. *Environment, Development and Sustainability*, 1–18.

Rajesh, R. (2018). On sustainability, resilience, and the sustainable–resilient supply networks. *Sustainable Production and Consumption, 15*, 74–88.

Rajesh, R. (2021). Optimal trade-offs in decision-making for sustainability and resilience in manufacturing supply chains. *Journal of Cleaner Production, 313*.

Resilince. (2016). *EventWatch® 2015 annual report*. Viewed 26 February 2019. https://info.resilinc.com/eventwatch-2015-annual-report

Resilince. (2017). *EventWatch® 2016 annual report*. Viewed 26 February 2019. https://info.resilinc.com/eventwatch-2016-annual-report

Resilince. (2018). *EventWatch® 2017 annual report*. Viewed 26 February 2019. https://info.resilinc.com/eventwatch-2017-annual-report

Resilince. (2019). *EventWatch® 2018 annual report*. Viewed 4 March 2019. https://info.resilinc.com/eventwatch-2018-annual-report

Rice, J., & Caniato, F. (2003). Building a secure and resilient supply network. *Supply Chain Management Review, 7*(5), 22–30.

Rosemann, M., Becker, J., & Chasin, F. (2021). City 5.0. *Business & Information Systems Engineering, 63*, 71–77.

Saenz, M. J., & Revilla, E. (2014). Creating more resilient supply chains. *MIT Sloan Management Review, 55*(4), 22–24.

Sánchez-Flores, R. B., Cruz-Sotelo, S. E., Ojeda-Benitez, S., & Ramírez-Barreto, M. E. (2020). Sustainable supply chain management – A literature review on emerging economies. *Sustainability (Switzerland), 12*(17), 1–27.

Sarkis, J., Helms, M. M., & Hervani, A. A. (2010). Reverse logistics and social sustainability. *Corporate Social Responsibility and Environmental Management, 17*, 337–354.

Sauvé, S., Bernard, S., & Sloan, P. (2016). Environmental sciences, sustainable development and circular economy: Alternative concepts for trans-disciplinary research. *Environmental Development, 17*, 48–56.

Scheibe, K. P., & Blackhurst, J. (2018). Supply chain disruption propagation: A systemic risk and normal

accident theory perspective. *International Journal of Production Research, 56*(1–2), 43–59.
Schneider-Petsinger, M. (2021). *U.S. and European strategies for resilient supply chains.* Viewed 20 April 2022. https://www.chathamhouse.org/2021/09/us-and-european-strategies-resilient-supply-chains
Scholten, K., & Schilder, S. (2015). The role of collaboration in supply chain resilience. *Supply Chain Management: An International Journal, 20*(4), 471–484.
Scholten, K., Scott, P. S., & Fynes, B. (2014). Mitigation processes–Antecedents for building supply chain resilience. *Supply Chain Management: An International Journal, 19*(2), 211–228.
Sharifirad, M. S., & Ataei, V. (2012). Organizational culture and innovation culture: Exploring the relationships between constructs. *Leadership & Organization Development Journal, 33*(5), 494–517.
Sharma, M., Luthra, S., Joshi, S., & Kumar, A. (2020). Developing a framework for enhancing survivability of sustainable supply chains during and post-COVID-19 pandemic. *International Journal of Logistics Research and Applications*, 1–21.
Sheffi, Y. (2005). Preparing for the big one [supply chain management]. *Manufacturing Engineer, 84*(5), 12–15.
Sheffi, Y., & Rice, J. B. (2005). A supply chain view of the resilient enterprise. *M.I.T. Sloan Management Review, 47*(1).
Sikken, B. J. (2011). *Accelerating the transition towards sustainable investing-strategic options for investors, corporations, and other key stakeholders.* Corporate Finance: Valuation. World Economic Forum.
Singh, S. K., Pan, Y., & Park, J. H. (2022). Blockchain-enabled secure framework for energy-efficient smart parking in sustainable city environment. *Sustainable Cities and Society, 76*, 103364.
Sokolov, B., Ivanov, D., Dolgui, A., & Pavlov, A. (2015). Structural quantification of the ripple effect in the supply chain. *International Journal of Production Research, 54*(1), 152–169.
Stadnyk, V., Krasovska, G., Pchelianska, G., & Holovchuk, Y. (2021). *Determinants of "green entrepreneurship" competitive strategies implementation in the agro-industrial sector of Ukraine.* I.O.P. Conference Series: Earth and Environmental Science, Vol. 628, No. 1.
Statsenko, L., Gorod, A., & Ireland, V. (2018a). A supply network governance framework: A case study of the south Australian mining industry. *Journal of Global Operations and Strategic Sourcing, 11*(1), 55–78.
Statsenko, L., Gorod, A., & Ireland, V. (2018b). A complex adaptive systems governance framework for regional supply networks. *Supply Chain Management, 23*(4), 293–312.
Talluri, S., Kull, T., Yildiz, H., & Yoon, J. (2013). Assessing the efficiency of risk mitigation strategies in supply chains. *Journal of Business Logistics, 34*(4), 253–269.
Tibben-Lembke, R. S., & Rogers, D. S. (2002). Differences between forward and reverse logistics in a retail environment. *Supply Chain Management: An International Journal, 7*(5), 271–282.
Tsolakis, N., Niedenzu, D., Simonetto, M., Dora, M., & Kumar, M. (2020). Supply network design to address United Nations sustainable development goals: A case study of blockchain implementation in Thai fish industry. *Journal of Business Research, 131*, 495–519.
United Nations Economic and Social Council. (2020). United Nations Economic and Social Council. The UNECE-ITU smart sustainable cities indicators.
Vanapalli, K. R., Sharma, H. B., Ranjan, V. P., Samal, B., Bhattacharya, J., Dubey, B. K., & Goel, S. (2021). Challenges and strategies for effective plastic waste management during and post COVID-19 pandemic. *Science of The Total Environment, 750*, 141514.
Vegter, D., Hillegersberg, J., & Matthias Olthaar, M. (2020). Supply chains in circular business models: Processes and performance objectives. *Resources, Conservation and Recycling, 162*, 105046.
Wagner, S. M., & Bode, C. (2008). An empirical examination of supply chain performance along several dimensions of risk. *Journal of Business Logistics, 29*(1), 307325.
Walker, B. (2020). Resilience: What it is and is not. *Ecology and Society, 25*(2).
Weersink, A., von Massow, M., Bannon, N., Ifft, J., Maples, J., McEwan, K., McKendree, M. G. S., Nicholson, C., Novakovic, A., Rangarajan, A., Richards, T., Rickard, B., Rude, J., Schipanski, M., Schnitkey, G., Schulz, L., Schuurman, D., Schwartzkopf-Genswein, K., Stephenson, M., & Wood, K. (2021). COVID-19 and the agri-food system in the United States and Canada. *Agricultural Systems, 188*.
Wieland, A., & Durach, C. F. (2021). Two perspectives on supply chain resilience. *Journal of Business Logistics, 42*(3), 315–322.
Wieland, A., & Wallenburg, M. C. (2013). The influence of relational competencies on supply chain resilience: A relational view. *International Journal of Physical Distribution & Logistics Management, 43*(4), 300–320.
Xu, M., Wang, X., & Zhao, L. (2014). Predicted supply chain resilience based on structural evolution against random supply disruptions. *International Journal of Systems Science: Operations & Logistics, 1*(2), 105–117.
Zahiri, B., Zhuang, J., & Mohammadi, M. (2017). Toward an integrated sustainable-resilient supply chain: A pharmaceutical case study. *Transportation Research Part E: Logistics and Transportation Review, 103*, 109–142.
Zsidisin, G. A., & Wagner, S. M. (2010). Do perceptions become reality? The moderating role of supply chain resiliency on disruption occurrence. *Journal of Business Logistics, 31*(2), 1–20.

# 11 The Impact of Transportation and Emissions Allocation on the Global Warming Potential of Concretes Containing Alternative Binders and Fillers

Vahid Shobeiri, Bree Bennett, Tianyu Xie, and Phillip Visintin

## 11.1 Introduction

The building and construction industry has a significant impact on the natural environment, being responsible for approximately 33% of global carbon equivalent emissions, the consumption of approximately 40% of natural resources and the production of approximately 25% of global waste streams (Du et al., 2022). For example, the global production of concrete consumes more than four billion tons of Ordinary Portland Cement (OPC), eight billion tons of natural aggregates and two billion tons of freshwater annually (Sharma et al., 2022). This demand for concrete is expected to further rise as population growth and increased standards of living causes a higher demand for public infrastructure and high-density housing. As a result, it has been estimated that the worldwide production of concrete will increase by more than 50% through the year 2033 (ARTBA; Hossain et al., 2022).

With rising awareness of the environmental impacts of the building and construction industry, it has become increasingly important to develop a more sustainable approach for concrete production (Manjunatha et al., 2022). In response researchers, practitioners and policymakers have sought to establish sustainable development strategies based on the principles of a circular economy (CE) (European Commission, 2020). The CE provides a closed production model as an alternative to today's predominantly linear *"consume–use–waste"* business models leading to an economic growth without an increased pressure on the environment (Marsh et al., 2022). As shown in Fig. 11.1, the CE aims to reduce resource use, emissions, waste and energy leakage by extending the useful life of products and minimizing the waste produced (Shehata et al., 2022). Therefore, circular products can positively contribute to the environment and economy through a reduction in consumption of materials, energy, and resource (Ding et al., 2023).

When considering concrete binders, the use of industrial wastes such as fly ash (FA) and ground granulated blast-furnace (GGBFS) as either supplementary cementitious materials (SCMs) (Busari et al., 2018; Pang et al., 2022) or as primary binders in geopolymer concrete (GPC) (Amran et al., 2020; Thomas et al., 2022) is an attractive approach. When utilising these materials as an SCM it is possible to significantly reduce the need for OPC, and when used as a binder in GPC it is possible to entirely avoid the need to OPC. At the same time as limiting the need for the natural resources required to produce OPC and avoiding the significant emissions

V. Shobeiri · B. Bennett · T. Xie · P. Visintin (✉)
School of Architecture and Civil Engineering, The University of Adelaide, Adelaide, SA, Australia
e-mail: phillip.visintin@adelaide.edu.au



J. Zuo et al. (eds.), *Circular Economy for Buildings and Infrastructure*, Sustainable Development Goals Series, https://doi.org/10.1007/978-3-031-56241-9_11

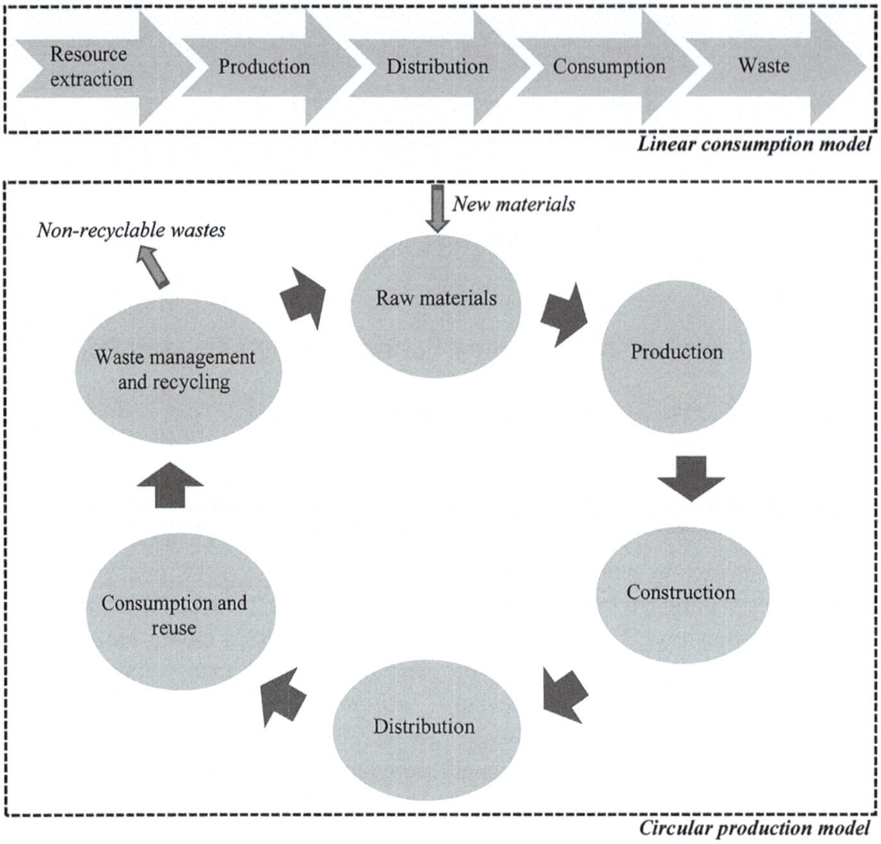


**Fig. 11.1** The linear and closed production models

associated with production, the use of industrial by-products such as GGBFS and FA provides an alternative diversion away from landfill.

When considering concrete fillers, materials such as discarded plastic, rubber, glass, and construction and demolition (C&D) wastes have all been proposed as alternatives to natural aggregates. These materials are produced in large quantities, for example each year 2–3 billion tons of C&D waste, 250 million tons of plastic waste and 10 million tons of glass waste are produced worldwide (Vishnu & Singh, 2021; Boora & Dharma, 2022). At present the collection and management of these wastes creates a significant environmental burden, and so utilisation as an alternative to natural aggregates may not only provide an alternative waste management pathway but also minimise the need to mine finite natural aggregate resources (Nasrudin et al., 2022).

To date the concrete technology research community has largely assumed that the utilisation of industrial and municipal waste products will lead to environmental benefits. Recent research has however showed that detailed life-cycle assessment (LCA) covering impacts from extraction to end-of-life is required, and that without this modelling it is possible to produce concrete mix designs that contain wastes and by-products that have significantly higher global warming potential than the materials that they are intended to replace (Shobeiri et al., 2021). This research has shown that of critical importance is the impact of allocating emissions associated with the production of by-products such as FA and GGBFS and that when considering alternative fillers it is essential to consider the emissions associated with both their preparation and transportation (Zhang et al., 2023).

The aim of this chapter is therefore to show the need to couple a detailed LCA with concrete mix design even when considering the utilisation of waste and by-products and to explore the significant variation in results depending on the assumptions made. This is done via two scenario analyses.

In the first scenario 843 geopolymer concrete (GPC) mix designs are considered and through a detailed LCA it is shown that depending on local material availability and the modelling assumptions applied geopolymer concretes can be either significantly less, or significantly more carbon intensive than the OPC concretes that they are intended to replace.

In the second scenario, through a comparison of emissions factors associated with the preparation of natural and alternative aggregates critical transportation distances are identified at which from an emissions perspective it is better to utilise naturally available resources rather than utilising wastes as aggregates.

Importantly the results obtained through these scenario analyses are not intended to promote or discourage the use of any specific materials, but rather they are intended to show the need to conduct concrete mix designs in conjunction with LCA. The findings of this chapter can help identify the optimum concrete types in terms of alternative binders and fillers based on a detailed life-cycle assessment, coupling growth to positive environmental and social outcomes and provides a viable solution to reducing carbon emissions from concrete industry as it encourages the elimination of waste and continued conservations of natural resources.

## 11.2 Life Cycle Assessment Model

LCA is a technique that can be used to assess the environmental burdens of a product throughout its entire life cycle from cradle to grave (Munir et al., 2023). By including the total impact in terms of energy, resource and emissions during the life cycle of a product, LCA can provide an accurate picture of its environment performance helping decision-makers to select between the alternatives (Alhassan et al., 2023; Sharma & Manepatil, 2009).

Within the framework of ISO 14040, a LCA is developed here to quantify the Carbon Dioxide equivalent ($CO_{2\text{-eq}}$) emissions of 1 $m^3$ of concrete with the traditional and alternative binders (OPC based or geopolymer concretes) and fillers (natural aggregates, recycled concrete, plastic, glass or rubber aggregates). Note that the effects of alternative fillers are only studied for OPC due to the availability of data related to GPC with alternative fillers. Here a focus is taken specifically on global warming potential (GWP) to quantify the environmental burdens contributing to greenhouse gas emissions of traditional concretes when replaced with alternative concretes. This is important because the GWP as a measure of sustainability can provide guidance on whether waste materials which are used in concrete for the sake of land and natural resource conservation can have higher or lower emissions than the materials that they are intended to replace.

In this analysis, the GWP is presented in terms of $CO_{2\text{-eq}}$ emissions associated to fuel pre-combustion, fuel combustion and electricity generation for all stages of concrete production from raw material extraction through to production and manufacture (see Fig. 11.2). The system boundary for the OPC concrete includes the emissions arising from fuel pre-combustion, fuel combustion and electricity use for (i) mining, crushing, drilling and preparation of cement raw materials; (ii) cement pyro processing; (iii) cement clinker cooling; (iv) cement finish milling and grinding; (v) cement conveying; (vi) natural aggregate production; (vii) superplasticizer production and (viii) concrete batching. A detailed description of the inventory of each of these processes can found in Gursel (2014).

The system boundary for GPC mixes also includes the emissions arising from fuel pre-combustion, fuel combustion and electricity use for: (i) FA and GGBFS manufacture; and (ii) alkali activator production (sodium silicates and sodium hydroxide) based on Wernet et al. (2016) and Fawer et al. (1999).

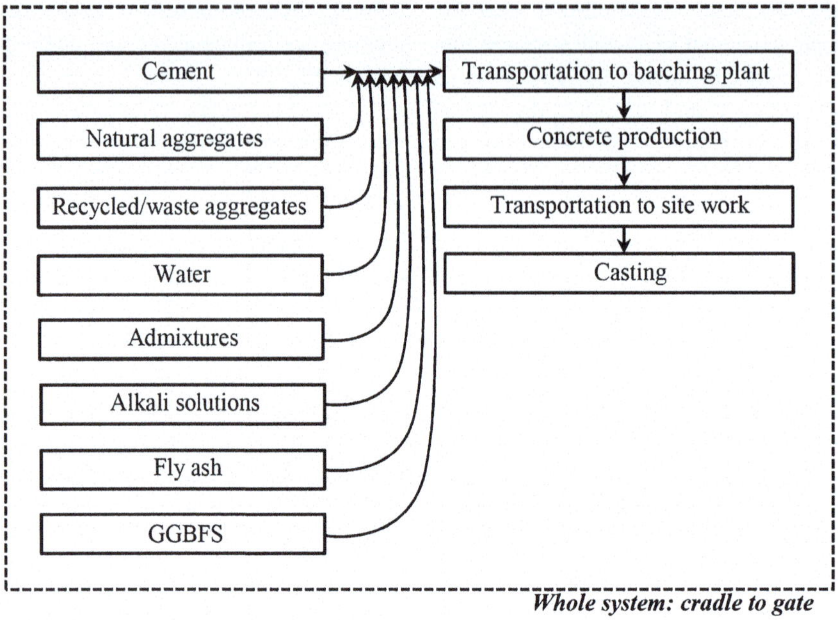


**Fig. 11.2** Scope of life cycle assessment system

The emissions associated with alternative aggregates including recycled concrete, glass waste, plastic waste and rubber waste are taken from Tahanpour Javadabadi (2019), Roh et al. (2020) and Gravina and Xie (2022), and the energy requirement for the production of concrete is assumed to be provided by the average US grid mix in the Ecoinvent database (Ecoinvent 2.2).

For implementation, the emissions associated with transportation are limited to three common transportation classes: road (class 8b (model 2005)), rail (diesel locomotive) and water. Unit emissions associated with transportation per km (kg $CO_{2\text{-eq}}$/km-kg) are taken from Facanha and Horvath (2007), Weber and Matthews (2008) and Nealer et al. (2012).

Importantly, because FA and GGBFS are industrial by-products, according to European Union Directive (EU, 2008), a proportion of the emissions associated with the production of FA through the burning of coal, or the production of GGBFS through the smelting of metals may be allocated to the by-products. This is in contrast to what would be done if FA and GGBFS were considered industrial wastes, where all emissions associated with their primary production could be ignored. Here the approach of Chen et al. (2010) is applied to quantify the emissions of FA and GGBFS based on either no allocation, financial allocation or mass allocation. The intention here is not to promote the definition of materials as a waste or by-product but rather to show the impact of making these different assumptions.

### 11.2.1 Quantification of Critical Transport Distances

A wide variety of waste materials have been used for concrete production including industrial by-products (e.g., FA and GGBFS) and municipal solid wastes (e.g., plastic, rubber and glass) (Tavakoli et al., 2018; Kishore & Gupta, 2020). The utilization of these waste materials is assumed to lead to the production of more sustainable concrete (Helmy et al., 2023). However, such assumptions may not be true because wastes and by-products most suitable for concrete production are often not produced locally and need to be transported considerable distances (Sandanayake et al., 2018).

Moreover, such assumptions may not be supported when FA and GGBFS are classified as by-products rather than waste products (EU, 2008; Zhao & Yang, 2023). Therefore, to better understand the significant impact that the transport and allocation emissions of the alternative binders and fillers can have, their critical transport distance (with several allocation scenarios) will be determined. Where the critical transport distance is defined as the tipping point at which the increased emissions associated with sourcing an alternative binder or filler is greater than any reduction in emissions associated with the use of the alternative material. This analysis is done in two stages (i) considering binders, that is, the point at which the production of a geopolymer concrete which utilizes high volumes of alternative binders (FA and GGBFS) has higher emissions than the manufacture of an OPC based concrete; and (ii) where the sourcing of an alternative aggregate (recycled concrete, plastic rubber or glass) has higher emissions than sourcing a natural aggregate.

#### 11.2.1.1 Critical Transport Distances for Geopolymer Concrete

When considering alternative binders, the maximum transportation distances in km for FA and GGBFS (here referred to as critical distance ($D_{crit-bind}$)) over which geopolymer concrete can be more emissions intensive than an equivalent OPC concrete is found using the following equation:

$$D_{crit-bind} = \frac{[CO_{2-eq}]_{OPC} - [CO_{2-eq}]_{GPC}}{TE_{GPC}} \tag{11.1}$$

where $TE_{GPC} = (m_{FA} + m_{GGBFS}) \times C_{trans}$ is the total transportation emissions factor (kg $CO_{2\text{-eq}}$/km) for alternative binders which can be calculated by multiplying the total mass of FA and GGBFS by the emissions associated with transportation per km-kg ($C_{trans}$) (kg $CO_2$/kg- km). It should be noted that $C_{trans}$ can change based on the transportation mode (road, rail and sea) of FA and GGBFS. $[CO_{2-eq}]_{GPC}$ and $[CO_{2-eq}]_{OPC}$ are respectively the total $CO_{2\text{-eq}}$ emissions per 1 $m^3$ of geopolymer concrete and an equivalent OPC concrete of the same strength which can be calculated as follows:

$$CO_{2-eq} = \sum_i m_i \left(C_{prod_i} + C_{trans_i} \times d_i\right) + \sum_j m_j \left(C_{prod_j} + F_j\right) \tag{11.2}$$

where $m_i$, $d_i$ and $C_{prod_i}$ are respectively the mass (kg), average transportation distance (km) and production emissions per mass (kg $CO_2$/kg) of cement, aggregates, water, admixture and alkali activator solutions which are applicable based on the concrete type (OPC or geopolymer), and $m_j$, $C_{prod_j}$ and $F_j$ are respectively the mass (kg), the production emissions and the allocation emissions per kg of FA and GGBFS which are applicable to geopolymer concrete.

When calculating the critical transportation distance for geopolymer concrete the following additional assumptions are applied:

- To make a fair comparison between GPC and OPC concrete, they are assessed using a functional unit combining concrete volume and concrete compressive strength (1 $m^3$ of geopolymer concrete versus 1$m^3$ of OPC concrete with the same compressive strength). To make this comparison the Feret equation (Ouellet-Plamondon & Habert, 2015) is used to generate OPC concrete mixes with the same compressive strength as the GPC mixes
- To account for curing emissions, the curing energy for GPC mixes requiring heat curing is calculated based on the heat transfer theory as follows:

$$Energy_{curing} = P \times t_{curing} + m \times c \times \left(T_{curing} - 25\right) \tag{11.3}$$

where $P$ is the power of oven; $m$ and $c$ are the mass and specific heat of concrete. $T_{curing}$ and $t_{curing}$ respectively represent the curing temperature and time, and the ambient temperature is considered to be 25 °C.

- To accurately capture the emissions from the transportation of all concrete ingredients, the average transportation distances taken from Grant (2015) are considered for all concrete components other than FA and GGBFS. It is assumed that aggregates, water and admixtures are provided from the same suppliers for both GPC mixes and OPC concretes
- While the road transportation is considered for other concrete ingredients as they are usually provided locally, rail and sea transportation mode options are also considered for FA and GGBFS as they usually need to be transported longer distances. It is also assumed that FA and GGBFS can have the same transportation mode.
- To examine the effects of allocation scenario on the critical distance of FA and GGBFS, their allocation emissions on either mass or financial basis are calculated as follows (Chen et al., 2010):

$$C_m = \frac{m_{by-product}}{m_{by-product} + m_{main-product}} \quad (11.4)$$

$$C_f = \frac{U_p.m_{by-product}}{U_p.m_{by-product} + U_p.m_{main-product}} \quad (11.5)$$

where $C_m$ and $C_f$ are respectively the mass and financial allocation coefficients; $m_{by-product}$ and $m_{main-product}$ are the mass of by-product and main-product; and $U_p.\ m_{by-product}$ and $U_p.\ m_{main-product}$ are the price per unit volume of by-product and main-product. While the price of FA, GGBFS, electricity and steel are taken from Teh et al. (2017), the produced mass of FA and GGBFS during the process of coal burning and smelting iron are taken from Chen et al. (2010). Having the allocation coefficients, the allocation emissions per unit mass of FA and GGBFS can be calculated according to their produced mass and the emissions associated with their main-products (electricity and steel). In this study, the economic allocation emissions are calculated as 0.23 and 1.30 kg $CO_{2-eq}$/kg of GGBFS and FA respectively and the mass allocation emissions are obtained as 0.68 and 2.27 kg $CO_{2-eq}$/kg of GGBFS and FA respectively.

To have a clear understanding of the trade-off between concrete mix ingredients, concrete technology and the generated emissions, a large database including 843 GPC mixes is collected from the previous studies and for which details of the database can be found in Xie et al. (2020). Figure 11.3 illustrates the distribution of compressive strength and GPC mix designs developed with FA, GGBFS, and blended GGBFS and FA from which it can be seen that the developed database covers a wide range of mix designs with a

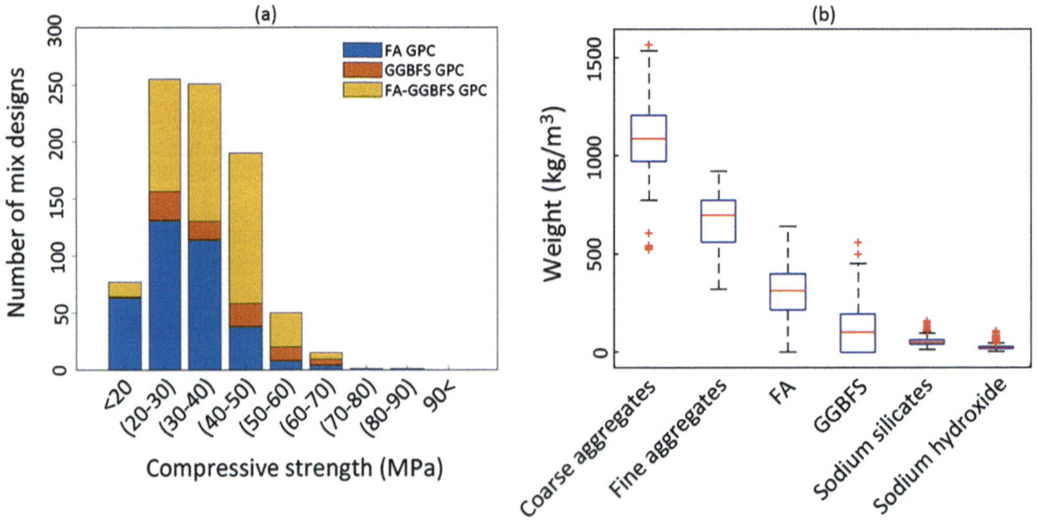


**Fig. 11.3** Summary of (**a**) compressive strength; (**b**) GPC mix components in the database

compressive strength range between 10 MPa and 90 MPa.

#### 11.2.1.2 Critical Transport Distances for Alternative Aggregates

When considering alternative fillers, the critical distance at which the use of natural aggregates can become more emissions intensive than the use of alternative aggregates in concrete can be found by:

$$D_{crit-agg} = \frac{[CO_{2-eq}]_{Altr.Agg} - [CO_{2-eq}]_{Nat.Agg}}{C_{trans}} \tag{11.6}$$

where $D_{crit-agg}$ is the critical transport distance of natural aggregates in km, $C_{trans}$ is the emissions associated with transportation of aggregates per kg-km which can vary based on the transportation mode, and $[CO_2-eq]_{Altr.\ Agg}$ and $[CO_2-eq]_{Nat.\ Agg}$ are respectively the emissions associated with the production of alternative and natural aggregates which are taken from Gursel (2014), Tahanpour Javadabadi (2019), Roh et al. (2020) and Gravina and Xie (2022).

Importantly, in Eq. 11.6 only the emissions associated with the production of aggregates are considered, by ignoring the other components of the concrete in this way it is implicitly assumed that the use of alternative aggregates will not change the functionality of the concrete and hence comparison is done only on a volumetric basis. Although this is a significant simplification that does allow the impact of the aggregate alone to be considered. It is however recommended that further research be undertaken to allow for the variation in binder that is required to maintain mechanical and durability properties that may be negatively impacted by the addition of alternative fillers (Sofi, 2018; Steyn et al., 2021). Further, because of the diverse functions of the sources that generated these municipal solid wastes, it is difficult to identify the corresponding primary products. Therefore, in this section, a cut-off is applied assuming no environmental burden allocated to these alternative fillers.

## 11.3 Results and Discussion

### 11.3.1 Critical Transport Distances for Geopolymer Concrete

Figures 11.4, 11.5, and 11.6 show the critical distance of GPC produced with either FA or GGBFS and a blended of FA and GGBFS binder, and where each figure focuses on a different transportation mode (road, rail and sea) and within each subplot the outcomes are presented for three different allocation scenarios (from left to right cut-off, no, economic and mass allocation).

Where a blended FA and GGBFS binder is used it is assumed that each binder type has the same transportation mode, and hence the critical distances are the total transport distances of FA and GGBFS.

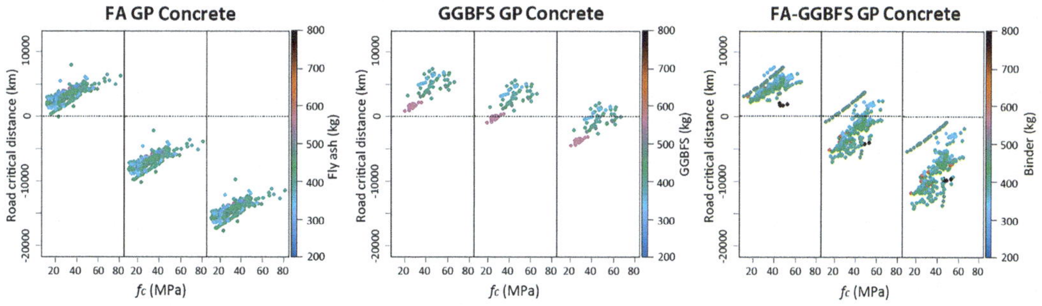


**Fig. 11.4** Road critical distances of alternative binders for GPC mixes with different allocation scenarios (from left to right cut-off, none, economic and mass allocation)

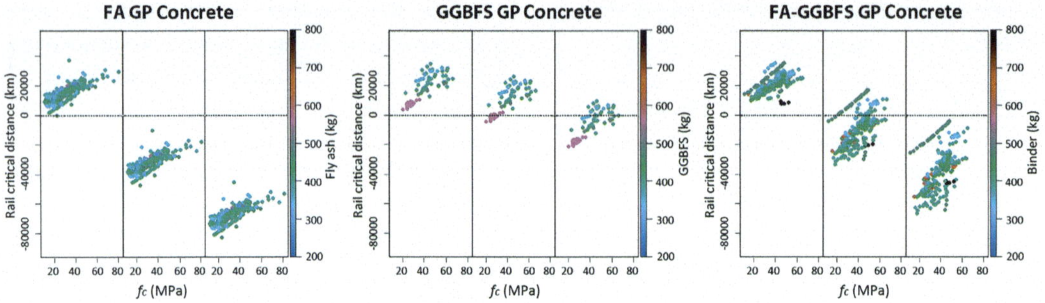


**Fig. 11.5** Rail critical distances of alternative binders for GP concrete with different allocation scenarios (from left to right cut-off, none, economic and mass allocation)

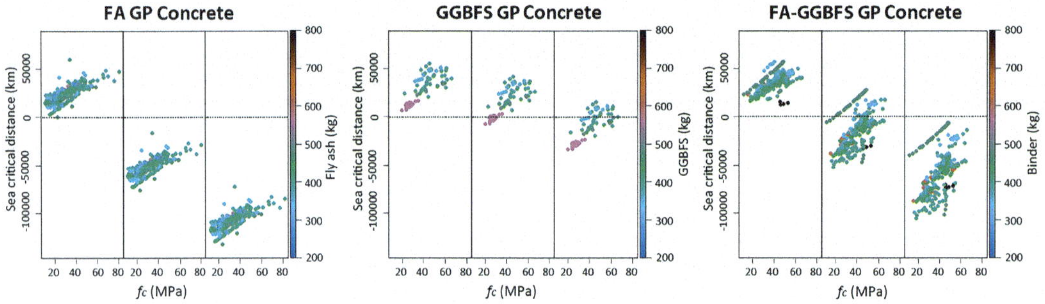


**Fig. 11.6** Sea critical distances of alternative binders for GPC mixes with different allocation scenarios (from left to right cut-off, none, economic and mass allocation)

When interpreting the results, a zero critical distance indicates that the GPC and the equivalent strength OPC concrete have identical emissions. If the critical distance is negative, this implies that GPC is more emissions intensive than the equivalent strength OPC concrete regardless of how locally the materials can be sourced the OPC concrete will always be less emissions intensive than GPC. For example, a GPC mix with a total mass binder of 500 kg and a road critical distance of −5000 km can be 5000 (km) × 500 (kg) × 0.000128 (kg $CO_2$/kg-km) = 320 kg $CO_2$ per 1 $m^3$ of concrete (where 0.000128 kg $CO_2$/kg-km is the unit road transportation emissions) more emissions intensive than its equivalent OPC.

It is clear from Figs. 11.4, 11.5, and 11.6 that the critical distances generally increase with an increase in the compressive strength regardless of transportation mode and allocation scenario. For instance, the critical distances of GPC with different binder types can generally increase up to 600% when the compressive strength increases more than 200%. The reason for this is that as the compressive strength of OPC concrete greatly depends on the cement content, the $CO_{2\text{-eq}}$ emissions from the OPC concrete increase at a faster rate than that from GPC as compressive strength increases (Busch et al., 2022). However, it can be seen that the critical distances for sea and rail transportation are longer than that for road transportation. The reason for this is that rail and sea transport are generally more efficient than road, with the unit $CO_{2\text{-eq}}$ emissions per kg-km for sea and rail transportation are about 87% and 78% respectively less than that for road transportation (Gursel, 2014). Because of the high mass of binders within concrete this change in transportation emissions allows that the critical distances of mixes with the same strength to increase by up to 5 and 7 times when the transportation mode changes from road to rail and sea respectively. For instance, while the road critical distances of FA for FA GPC mixes with the strengths of 15 MPa and 50 MPa are about 500 km 4300 km respectively, the sea critical distances for the same mixes and strengths are about 3700 km and 32,000 km respectively.

It also appears from Figs. 11.4, 11.5, and 11.6 that the critical distances of mixes with the same strength can vary by a change in the mass of binders regardless of the transportation mode and allocation scenario. For instance, the critical distance of GGBFS in the GGBFS GPC mixes with the same strength can increase more than 2 times when the mass of GGBFS decreases more than 40%. The reason for this is that less GGBFS can be generally transported longer distances with the same emissions when more GGBFS is transported shorter distances. Interestingly, it can also be seen that the critical distances of mixes with the same strength and binder content can also change regardless of the transportation mode and allocation scenario. The reason for this can be attributed to the fact that multiple different mix designs can achieve the same compressive strength. For instance, the critical distances of FA in the FA GPC mixes with similar strength and FA content can increase up to 50% when the mass of other ingredients such as alkali activator solution decreases. Similarly, the critical distances of FA and GGBFS in FA GGBFS GPC mixes with the same strength and binder content can be twice with a change in the mass of fillers, sodium silicates-to-sodium hydroxide ratio, FA-to-GGBFS ratio and curing temperature. Note that the discussion on the concrete mix ingredients in the database is out of the scope of this chapter and the reader is referred to Fig. 11.3a for a summary of GPC components, and the original database for more information on the curing regime (Xie et al., 2020).

It is also clear that the critical distances are highly sensitive to the product status of FA and GGBFS to be either a by-product or a waste product. This is important because classifying as a by-product rather than a waste product can lead to an increase in the FA and GGBFS emissions factors of up to 50 and 3 times on an economic basis and up to 88 and 8 times on a mass basis, respectively. It is seen that while in the case of no allocation the critical distances of FA and GGBFS for all types of GPC are positive, the critical distances can either be positive or negative depending on the allocation type (either economic or mass) and the binder and activator content and constituents of GPC. For instance, while the critical distances for all FA GPC mixes with both economic and mass allocation are negative, the critical distances for two other types of GPC mixes can either be negative or positive depending on allocation emissions and binder content. The reason for this can be attributed to the fact that the impact of both economic and mass allocation for FA is higher than that for GGBFS (more than 17 and 11 times respectively) and therefore GPC mixes relying more on the FA suffer more from the allocation impacts. This is especially true for the GGBFS GPC mixes where there is no FA in the mix and therefore allocation has the least impact on this type of concrete compared to other GPC mix types. For instance, while about 20% and 70% of GGBFS GPC mixes with respectively economic and mass allocation have a negative critical distance, about 85% and 100% of FA GGBFS GPC mixes with the same allocation scenarios have a negative critical distance as this type of GPC also includes FA in its mix binder.

### 11.3.2 Critical Transport Distances for Aggregates

Figure 11.7 shows the critical distances of natural aggregates compared to alternative aggregates with different transportation modes (road, rail and sea), where the critical distance for recycled, plastic and glass aggregates are read from the left-hand axis and for sea transport from the right-hand axis. When reading Fig. 11.7 natural aggregates can be less emissions intensive than alternative aggregates if they are provided from a quarry within the critical distance from the construction site.

It can be seen in Fig. 11.7 that, as expected, the critical distances increase with a change in the transportation mode from road to rail and sea. However, this rise is less apparent for recycled concrete aggregates as their production emissions are much lower than those from other alternative

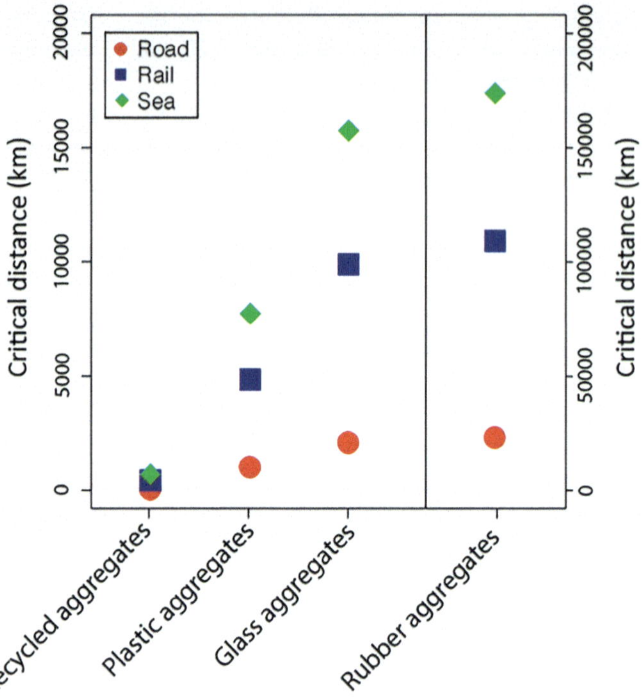


**Fig. 11.7** Critical distances of natural aggregates compared to locally provided alternative aggregates

aggregates. In fact, the production emissions of recycled concrete aggregates are about 2 times more than that from natural aggregates (Roh et al., 2020) leading to the critical distances of 100 km (road), 400 km (rail) and 700 km (sea). However, the production emissions of plastic, glass and rubber aggregates are respectively about 8, 17 and 200 times more than that from recycled concrete aggregates (Tahanpour Javadabadi, 2019; Gravina & Xie, 2022). These high production emissions have resulted in the critical distances of 1000 km (road), 5000 km (rail) and 8000 km (sea) for the plastic aggregates; 2000 km (road), 10,000 km (rail) and 15,000 km (sea) for the glass aggregates; and 23,000 km (road), 110,000 km (rail) and 170,000 km (sea) for rubber aggregates.

The results of this analysis show that from the perspective of GWP, recycled concrete aggregates should be a preferred substitute to natural aggregates over other alternative aggregates including plastic, glass and rubber aggregates. However, their possible negative impacts on the concrete properties should also be taken into account when it comes to the structural performance (Shi et al., 2019; Guo et al., 2020). It is also important to remember here that the analysis presented only considers GWP and environmental benefit, and natural resource utilisation and landfill are not included.

It is also important to emphasise here that this analysis is conducted here to simply explore the GWP and the benefits of recycled aggregate use from the perspective of avoiding the use of natural materials is not considered.

## 11.4 Conclusion

The population growth and industrial development have led to significant increases in the production of industrial and commercial wastes making their management a challenging task. For instance, the large amount of FA and GGBFS generated from power stations and iron-making process adds extra cost for their proper disposal. Additionally, the large amount of C&D wastes generated from construction industry makes their disposal a significant issue for environment. Hence, to address the waste management issues, it is important to reuse and recycle the waste materials for the purpose of circular economy. In this case, several industrial and commercial wastes including FA, GGBFS, recycled concrete, plastic, glass and rubber wastes can be used for the concrete production. This chapter presents a transportation analysis for the possible use of such waste materials in concrete through a detailed life-cycle assessment. The following findings have been found from the results of this analysis:

- Having similar emissions from GPC and OPC concrete of the same strength, FA and GGBFS can be sourced from longer distances as the compressive strength increases. The reason for this is that while the compressive strength of the OPC concrete highly depends on the cement, the compressive strength of GPC mixes depends on several factors such as binder content and type, and alkali activator solutions
- In the case that GPC and OPC concretes of the same strength have similar emissions, FA and GGBFS can be transported longer distances of up to 5 and 7 times by rail and sea rather than road. This is because the sea and rail are more effective transportation means than road
- A change in binder content, FA-to-GGBFS ratio, sodium silicates-to-sodium hydroxide ratio and curing regime can change the maximum transportation distances of FA and GGBFS in which the GPC and OPC concrete of the same strength can have similar emissions. It was shown that this transportation distance can increase up to 100% when the sodium silicates-to-sodium hydroxide ratio, curing temperature and FA-to-GGBFS ratio change significantly for the mixes of the same strength
- The maximum transportation distances of FA and GGBFS in which the GPC and OPC concrete of the same strength can have similar emissions are highly affected by the allocation scenarios. It was shown that this transportation distance can either be positive or negative depending on the allocation type (no, economic and mass) and binder type (FA or GGBFS)
- The maximum transportation distances of natural aggregates in which they can have similar emissions to alternative aggregates for the recycled concrete aggregates are much less than plastic, glass and rubber aggregates. This indicates the emissions advantages of using recycled concrete aggregates over other alternatives as a replacement for natural aggregates in concrete production.

More importantly, the results found in this chapter can aid to identify the optimum type of concrete in terms of alternative binders and fillers considering their GWP through a detailed life-cycle assessment.

## References


Alhassan, M., Alkhawaldeh, A., Betoush, N., Alkhawaldeh, M., Huseien, G. F., Amaireh, L., & Elrefae, A. (2023). Life Cycle Assessment of the Sustainability of alkali-activated binders. *Biomimetics, 8*(1), 58.

Amran, Y. M., Alyousef, R., Alabduljabbar, H., & El-Zeadani, M. (2020). Clean production and properties of geopolymer concrete; A review. *Journal of Cleaner Production, 251*, 119679.

Boora, A., & Dharma, A. (2022). State of art: Review for sustainable application of waste material in rigid pavement. In *Advances in construction materials and sustainable environment* (pp. 127–141).

Busari, A. A., Akinmusuru, J. O., & Dahunsi, B. I. (2018). Review of sustainability in self-compacting concrete: The use of waste and mineral additives as supplementary cementitious materials and aggregates. *Portugaliae Electrochimica Acta, 36*(3), 147–162.

Busch, P., Kendall, A., Murphy, C. W., & Miller, S. A. (2022). Literature review on policies to mitigate GHG emissions for cement and concrete. *Resources, Conservation and Recycling, 182*, 106278.

Chen, C., Habert, G., Bouzidi, Y., Jullien, A., & Ventura, A. (2010). LCA allocation procedure used as an incitative method for waste recycling: An application to mineral additions in concrete. *Resources, Conservation and Recycling, 54*(12), 1231–1240.

Ding, L., Wang, T., & Chan, P. (2023). Forward and reverse logistics for circular economy in construction: A systematic literature review. *Journal of Cleaner Production, 388*, 135981.

Du, Q., Deng, Y., Zhou, J., Wu, J., & Pang, Q. (2022). Spatial spillover effect of carbon emission efficiency in the construction industry of China. *Environmental Science and Pollution Research, 29*(2), 2466–2479.

Ecoinvent, Ecoinvent Database Version 2.2. (2011). Swiss Center for Life Cycle Inventories, Hamburg, Germany.

EU. (2008). *Directive 2008/98/EC of 19 November 2008 on waste and repealing certain Directives*. https://eur-lex.europa.eu/legal-content/EN/TXT/?uri=CELEX:32008L0098

European Commission. (2020). *A new circular economy action plan for a cleaner and more competitive Europe*. https://eur-lex.europa.eu/legal-content/EN/TXT/?uri=COM:2020:98:FIN

Facanha, C., & Horvath, A. (2007). Evaluation of life-cycle air emission factors of freight transportation. *Environmental Science & Technology, 41*(20), 7138–7144.

Fawer, M., Concannon, M., & Rieber, W. (1999). Life cycle inventories for the production of sodium silicates. *International Journal of Life Cycle Assessment, 4*(4), 207.

Grant, T. F. (2015). *Life cycle inventory of cement & concrete produced in Australia*. Life Cycle Strategies Pty Ltd.

Gravina, R. J., & Xie, T. (2022). Toward the development of sustainable concrete with Crumb Rubber: Design-oriented Models, Life-Cycle-Assessment and a site application. *Construction and Building Materials, 315*, 125565.

Guo, Z., Jiang, T., Zhang, J., Kong, X., Chen, C., & Lehman, D. E. (2020). Mechanical and durability properties of sustainable self-compacting concrete with recycled concrete aggregate and fly ash, slag and silica fume. *Construction and Building Materials, 231*, 117115.

Gursel, A. P. (2014). *Life-cycle assessment of concrete: Decision-support tool and case study application*. University of California.

Helmy, S. H., Tahwia, A. M., Mahdy, M. G., & Abd Elrahman, M. (2023). Development and characterization of sustainable concrete incorporating a high volume of industrial waste materials. *Construction and Building Materials, 365*, 130160.

Hossain, M., Sultana, R., Patwary, M. M., Khunga, N., Sharma, P., & Shaker, S. J. (2022). Self-healing concrete for sustainable buildings. A review. *Environmental Chemistry Letters*, 1–9.

Kishore, K., & Gupta, N. (2020). Application of domestic & industrial waste materials in concrete: A review. *Materials Today Proceedings, 26*, 2926–2931.

Manjunatha, M., Seth, D., Balaji, K. V. G. D., & Bharath, A. (2022). Engineering properties and environmental impact assessment of green concrete prepared with PVC waste powder: A step towards sustainable approach. *Case Studies in Construction Materials, 17*, e01404.

Marsh, A. T., Velenturf, A. P., & Bernal, S. A. (2022). Circular Economy strategies for concrete: Implementation and integration. *Journal of Cleaner Production, 132*, 486.

Munir, Q., Abdulkareem, M., Horttanainen, M., & Kärki, T. (2023). A comparative cradle-to-gate life cycle assessment of geopolymer concrete produced from industrial side streams in comparison with traditional concrete. *Science of the Total Environment, 865*, 161230.

Nasrudin, N. N., Ariffin, N. F., Hasim, A. M., & Zaimi, M. N. S. (2022). A review: Utilization of waste materials in concrete. In *Materials science forum* (Vol. 1056, pp. 61–68). Trans Tech Publications Ltd..

Nealer, R., Matthews, H. S., & Hendrickson, C. (2012). Assessing the energy and greenhouse gas emissions mitigation effectiveness of potential US modal freight policies. *Transportation Research Part A: Policy and Practice, 46*(3), 588–601.

Ouellet-Plamondon, C., & Habert, G. (2015). Life cycle assessment (LCA) of alkali-activated cements and concretes. In *Handbook of alkali-activated cements, mortars and concretes* (pp. 663–686). Woodhead Publishing.

Pang, L., Liu, Z., Wang, D., & An, M. (2022). Review on the application of supplementary cementitious materials in self-compacting concrete. *Crystals, 12*(2), 180.

Roh, S., Kim, R., Park, W. J., & Ban, H. (2020). Environmental evaluation of concrete containing recycled and by-product aggregates based on life cycle assessment. *Applied Sciences, 10*(21), 7503.

Sandanayake, M., Gunasekara, C., Law, D., Zhang, G., & Setunge, S. (2018). Greenhouse gas emissions of different fly ash based geopolymer concretes in building construction. *Journal of Cleaner Production, 204*, 399–408.

Sharma, S., & Manepatil, S. (2009). Eco design: Selection of material for important parts of a tri-cycle through LCA. ICORD 09: Proceedings of the 2nd International Conference on Research into Design, Bangalore, India.

Sharma, N., Sharma, P., & Parashar, A. K. (2022). Use of waste glass and demolished brick as coarse aggregate in production of sustainable concrete. *Materials Today Proceedings, 62*, 4030–4035.

Shehata, N., Mohamed, O. A., Sayed, E. T., Abdelkareem, M. A., & Olabi, A. G. (2022). Geopolymer concrete as green building materials: Recent applications, sustainable development and circular economy potentials. *Science of the Total Environment, 836*, 155577.

Shi, X., Mukhopadhyay, A., Zollinger, D., & Grasley, Z. (2019). Economic input-output life cycle assessment of concrete pavement containing recycled concrete aggregate. *Journal of Cleaner Production, 225*, 414–425.

Shobeiri, V., Bennett, B., Xie, T., & Visintin, P. (2021). A comprehensive assessment of the global warming potential of geopolymer concrete. *Journal of Cleaner Production, 297*, 126669.

Sofi, A. (2018). Effect of waste tyre rubber on mechanical and durability properties of concrete–A review. *Ain Shams Engineering Journal, 9*(4), 2691–2700.

Steyn, Z. C., Babafemi, A. J., Fataar, H., & Combrinck, R. (2021). Concrete containing waste recycled glass, plastic and rubber as sand replacement. *Construction and Building Materials, 269*, 121242.

Tahanpour Javadabadi, M. (2019). Comparative Life Cycle Assessment of Incorporating Recycled PET Aggregates into Concrete (Master's thesis, NTNU).

Tavakoli, D., Hashempour, M., & Heidari, A. (2018). Use of waste materials in concrete: A review. *Pertanika Journal of Science & Technology, 26*(2), 499–522.

Teh, S. H., Wiedmann, T., Castel, A., & de Burgh, J. (2017). Hybrid life cycle assessment of greenhouse gas emissions from cement, concrete and geopolymer concrete in Australia. *Journal of Cleaner Production, 152*, 312–320.

Thomas, B. S., Yang, J., Bahurudeen, A., Chinnu, S. N., Abdalla, J. A., Hawileh, R. A., et al. (2022). Geopolymer concrete incorporating recycled aggregates: A comprehensive review. *Cleaner Materials, 3*, 100056.

Vishnu, T. B., & Singh, K. (2021). A study on the suitability of solid waste materials in pavement construction: A review. *International Journal of Pavement Research and Technology, 14*(5), 625–637.

Weber, C. L., & Matthews, H. S. (2008). *Food-miles and the relative climate impacts of food choices in the United States.*

Wernet, G., Bauer, C., Steubing, B., Reinhard, J., Moreno-Ruiz, E., & Weidema, B. (2016). The Ecoinvent database version 3 (part I): Overview and methodology. *International Journal of Life Cycle Assessment, 21*(9), 1218e1230.

Xie, T., Visintin, P., Zhao, X., & Gravina, R. (2020). Mix design and mechanical properties of geopolymer and alkali activated concrete: Review of the state-of-the-art and the development of a new unified approach. *Construction and Building Materials, 256*, 119380.

Zhang, C., Hu, M., van der Meide, M., Di Maio, F., Yang, X., Gao, X., et al. (2023). Life cycle assessment of material footprint in recycling: A case of concrete recycling. *Waste Management, 155*, 311–319.

Zhao, W., & Yang, Q. (2023). Life cycle assessment and multi-index performance evaluation of semi-flexible pavement after composite modification by using fly ash, rubber particles, warm mixing asphalt and recycled asphalt pavement. *Construction and Building Materials, 364*, 129945.

# 12 Costs and Pricing of Green Buildings

Yongsheng Jiang, Dong Zhao, Zihao Xu, Yunjia Zhang, Zhongyi Men, and Tao Hu

## 12.1 Introduction

Reducing global warming, extreme weather events, improving resource and energy usage efficiency, and promoting circular economy development through reduction, reuse, and recycling of resources are essential goals. However, the construction industry is one of the industries that consume a significant amount of energy and contribute to social and environmental problems. According to the International Energy Agency (IEA, 2019), the construction industry accounts for 36% of energy consumption and 39% of $CO_2$ emissions. Additionally, construction and demolition activities generate high levels of noise, air pollutants, and waste (Zhang et al., 2018). Green buildings have emerged as an effective solution to reduce energy consumption and air pollution (Zhao et al., 2016; McCoy et al., 2018). Green buildings refer to sustainable buildings that reduce or eliminate negative impacts and create positive impacts on the environment (Son et al., 2011). Several countries, including China, have introduced standards such as the "Green Building Assessment Standard" GREEN BUILDING/T50378–2019 to certify green buildings based on land, energy, water, and material conservation indicators. Encouraging the promotion and upgrading of green buildings is critical to promote the development of circular economy in the construction industry. However, the growth of green buildings faces several challenges, with incremental costs being the most significant (Jiang et al., 2022).

Green buildings have higher costs compared to non-green buildings, including higher construction (hard cost) and financing (soft cost) costs (Mo et al., 2017). The incremental cost is considered the primary obstacle to the adoption of green buildings (Darko & Chan, 2017; Feng et al., 2021; Hwang & Tan, 2012; Kats, 2003; Yudelson, 2010). Green buildings cost more on materials and technologies in the design and construction phases than traditional buildings. Therefore, the incremental costs hinder developers from building green buildings (Yeganeh et al., 2021).

Green buildings have several positive externalities, including reduced pollution on water, soil, and air during the construction stage (Bianchini & Hewage, 2012; Zuo & Zhao, 2014), reduced noise and construction waste (Olubunmi et al., 2016), and increased cost savings, health and well-being, and productivity during operation

Y. Jiang · Z. Xu · Y. Zhang · Z. Men · T. Hu
School of Management Engineering, Shandong Jianzhu University, Jinan, Shandong, China

D. Zhao (✉)
School of Planning, Design and Construction, Michigan State University, East Lansing, MI, USA

Department of Civil and Environmental Engineering, Michigan State University, East Lansing, MI, USA
e-mail: dz@msu.edu



J. Zuo et al. (eds.), *Circular Economy for Buildings and Infrastructure*, Sustainable Development Goals Series, https://doi.org/10.1007/978-3-031-56241-9_12

(Fuerst & McAllister, 2011; Heerwagen, 2000). These benefits are incorporated into the value of green buildings, known as the price premium. The majority of studies confirm significant sale and rent premiums for green buildings (Bruegge et al., 2016; Hyland et al., 2013; Zhang et al., 2017). Price premiums can compensate developers for the incremental costs and encourage them to develop green buildings.

The price premiums associated with green buildings can be a barrier for homebuyers who may need to increase their expenses to purchase green buildings, limiting their willingness to pay (Deng & Wu, 2014; Jiang & Payne, 2022). Homebuyers play a critical role in the movement of green buildings (Zhao et al., 2017, 2019), as their willingness to pay determines the demand and supply in the building market (Zhang et al., 2016). If the price premiums exceed their willingness to pay (WTP), homebuyers may not choose green buildings. Unfortunately, the current WTP is not high due to a mismatch in information and cost-benefit analysis (Kahn & Kok, 2014).

Government policies play a crucial role in guiding and supporting the development of green buildings (Yang et al., 2021). For instance, they may provide monetary incentives to encourage green buildings development and promote the green buildings market. However, effective pricing policies need to balance the incremental cost, price premium, and WTP, as stakeholders have varying interests (Menassa & Baer, 2014; Feng et al., 2021). Figure 12.1 depicts the critical drivers and barriers faced by developers, homebuyers, and the government in the market.

Figure 12.1 reveals that different stakeholders in green buildings have diverse interest claims. Developers may be willing to build green buildings when the price premium and government subsidies exceed the incremental costs. Homebuyers may be willing to pay a price premium for green buildings if they can derive greater benefits. The government may be constrained by fiscal feasibility but has a mandate to promote green buildings development. To promote green buildings development and realize a circular economy in the construction industry, it is crucial to balance the relationships between incremental cost, price premium, and WTP.

Previous research on the costs and pricing of green buildings has mainly been conducted from the perspective of a single stakeholder. For example, some studies have focused on the life cycle cost-benefits of green buildings from the perspective of developers, while others have examined the influencing factors of homebuyers' WTP. However, little research has been conducted on the conflicts between developers, homebuyers, and the government. Hence, the purpose of this chapter is to analyze the incremental cost, price premium, and WTP of green buildings and to assist the government in resolving conflicts between these stakeholders through policy tools. Ultimately, this will promote the development of green buildings and foster the realization of a circular economy in the construction industry.

## 12.2 Research Method

This study utilized the Web of Science database to conduct a systematic review of literature on green buildings, green housing, and green apartments. The research questions were designed to identify articles that focused on incremental cost, price premium, WTP, and public policies related to green buildings. We used four theme words: (1) TS = (green building, green housing, or green apartment and incremental cost or cost premium); (2) TS = (green building, green housing, or green apartment and price premium); (3) TS = (green building, green housing, or green apartment and WTP or purchase intention); and (4) TS = (green building, green housing, or green apartment and public policy).

After the initial search, we conducted a further screening of the articles by examining the title, abstract, and keywords to eliminate irrelevant studies. Finally, we identified the literature related to the research objective and conducted a systematic review of the chosen literature. Our focus was on the barriers to the development of green buildings, such as incremental costs, price premiums, willingness to pay, and public policies, as well as the interaction between these factors.

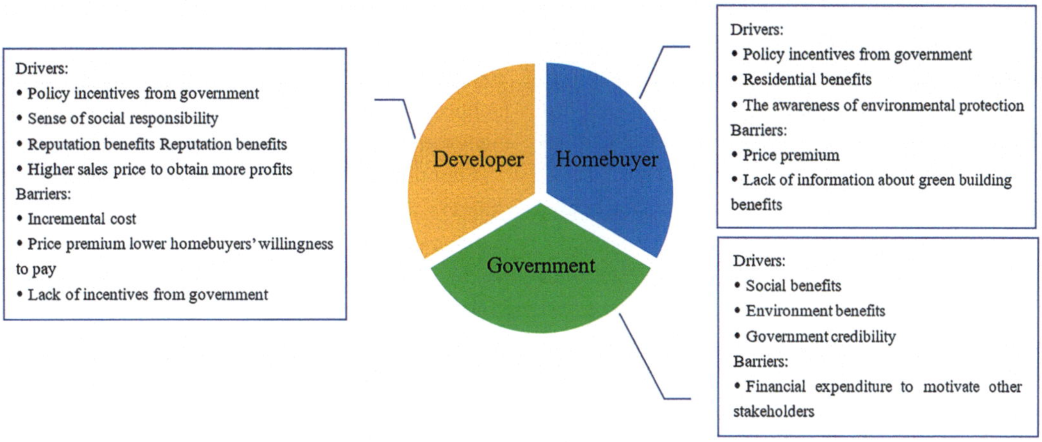


**Fig. 12.1** Drivers and barriers in the green building market

Through this study, we aim to provide effective decision-making references for stakeholders and policymakers in the green buildings market to promote the sustainable development of green buildings. We believe that our review of the literature will shed light on the key factors and challenges that need to be considered to promote the adoption of green buildings.

## 12.3 Incremental Cost

The incremental cost is a significant obstacle to the development of green buildings. The hidden cost of green buildings can offset the innovation in construction, which is why a comprehensive understanding of the incremental cost is essential to facilitate the development of green buildings (Sun et al., 2019). There are three aspects to consider when discussing the incremental cost of green buildings: (1) the composition of incremental cost; (2) the measurement of incremental cost; and (3) the benefit compensation for incremental cost.

### 12.3.1 Composition of Incremental Costs

The incremental cost of green buildings can be divided into soft costs and hard costs (Abidin & Azizi, 2021). Many scholars have focused on the study of hard costs, which mainly refers to the actual and direct costs of green building projects generated by the use of energy-saving equipment, green materials, and technologies. For example, Wan et al. pointed out that green materials account for 65–75% of the incremental cost (Wan, 2018), while Hwang et al. found that the incremental cost mainly comes from the use of green technologies and materials (Hwang & Tan, 2012). The incremental costs of various types of green technologies also differ, and previous studies have noted that passive technologies have a lower incremental cost than active technologies (Weerasinghe et al., 2021).

On the other hand, some studies show that soft costs are the main source of incremental cost, which can be summarized as design cost, authority requirement, and development provision (Abidin & Azizi, 2021). According to Li et al., the incremental cost will increase with a higher green certification level (Li et al., 2020). Another source of soft cost is that developers may increase the cost of marketing green buildings for a profit (Abidin & Azizi, 2021).

Moreover, some studies show that the incremental cost of green buildings mainly occurs in the pre-operation stage. Green buildings, as an innovation, require more costs in the design and construction process (Devine & McCollum, 2019; Mollaoglu et al., 2016). For example,

Zhang et al. pointed out that the additional cost of green elements in the construction stage is higher than that of conventional property projects (Zhang et al., 2018). Hwang et al. believed that the incremental cost caused by cost overruns and project delays is twice as high as that of traditional projects (Hwang & Leong, 2013; Hwang et al., 2017).

Overall, understanding the composition of incremental cost is essential for promoting the development of green buildings. By identifying the sources of incremental cost, stakeholders and policymakers can take effective measures to reduce the incremental cost of green buildings, thereby promoting the sustainable development of green buildings.

### 12.3.2 Measurement of Incremental Costs

There are two main views on how to estimate the incremental cost of green buildings. One perspective suggests that green buildings have higher incremental costs than traditional buildings. Many studies have estimated the incremental cost of green buildings, and their findings show that the average incremental cost of green buildings is around 6–10% (AlAwam & Alshamrani, 2021). The incremental cost of green buildings increases as their green certification level rises. For instance, according to the study of Vyas and Jha, the incremental costs of the 3-star and 5-star green buildings are 3.10% and 9.37%, respectively (Vyas & Jha, 2018). Furthermore, the incremental cost of green buildings differs by country and region. In the United States, the average green premium cost is 1.84% (Kats, 2003), whereas in Europe, the specific extra investment of residential energy-efficient buildings was found to be 8% of the entire building cost (Schnieders & Hermelink, 2006).

On the other hand, some scholars argue that high-performance green buildings do not necessarily need more cost due to adopting sustainable materials and systems (Rehm & Ade, 2013), and some green characteristics can even be attained in green buildings with a very low or zero cost premium (Dwaikat & Ali, 2016; Rehm & Ade, 2013). Even if the cost of adopting green technologies, green materials, and energy efficiency measures is higher, the incremental cost will continue to fall with mass adoption and faster delivery (Abidin & Azizi, 2021). Furthermore, some case studies demonstrate that, while the initial cost of green industrial buildings is 29% higher than that of traditional buildings, their operation and maintenance costs can be reduced by 23% and 15%, respectively, which partially offsets the high incremental cost in the early stage (Weerasinghe et al., 2021).

### 12.3.3 Benefit Compensation for Incremental Costs

Some scholars have studied incremental costs and benefits together. Price premium is frequently mentioned by researchers to compensate for the incremental cost (Fuerst & McAllister, 2011; Kahn & Kok, 2014; Portnov et al., 2018). Developers will only build if the price premium is sufficient to cover the incremental cost (Deng & Wu, 2014). Portonv et al. concluded that green building accrue 1–12.5% additional costs to developers, and 10–31% of the additional costs are passed through to homebuyers (Portnov et al., 2018). According to some quantitative studies, the price or rent premium brought by green buildings to developers frequently exceeds the incremental cost they paid (Kahn & Kok, 2014; Zhang et al., 2018).

Green buildings can provide environmental and social benefits in addition to the economic benefits mentioned above (Li et al., 2020). The increased environmental benefit accounted for 63% of the economic benefits brought by green buildings (Liu et al., 2014). According to Miraj et al., green buildings can achieve 58.65% energy and water conservation across the whole life cycle, although they cost 9.22% more (Miraj et al., 2021). Mao et al. also estimated the incremental costs and benefits and found that the incremental return on green energy technologies is greater than 25% of the incremental cost (Mao

& Yang, 2012). Gabay et al. conducted a cost-benefit analysis of a green office building case in Israel, and the results showed that the benefits of the building were completely worth the initial investment (Gabay et al., 2014). In addition to energy savings and reduced environmental impact, green buildings can also provide a better indoor environment for occupants, which can have positive effects on their health, well-being, and productivity (Gao et al., 2020; Li et al., 2020). For instance, a study by the Harvard School of Public Health found that occupants in green-certified buildings had a 26% higher cognitive function test score than those in non-green buildings (Allen et al., 2016). Furthermore, green buildings can help to mitigate climate change and improve air quality, which can have significant public health benefits (Franco et al., 2021). Overall, the benefits of green buildings can outweigh the incremental costs, making them a worthwhile investment for both developers and occupants.

## 12.4 Price Premium

Price premium refers to the additional cost that homebuyers or tenants pay for purchasing or renting a green building compared to a traditional building. The price premium compensates for the incremental cost of incorporating sustainable features and materials in the construction of green building. The relevant studies focus on the following aspects: (1) source of price premium; (2) impact of price premium; and (3) measure of price premium.

### 12.4.1 Source of Price Premium

There are three viewpoints regarding the source of the price premium for green buildings. The first view is that the higher cost of constructing green buildings leads to their expensive pricing (Council, 2013). Empirical research indicates that green buildings with new technologies and materials command a significant premium in rent and market value (Eichholtz et al., 2010, 2013). The second view is that the benefits of green features and energy efficiency are incorporated into the green buildings' sale price (Kahn & Kok, 2014). The benefits are enjoyed by various stakeholders, including lower operating costs (Council, 2013; Sabapathy et al., 2010; Zhang, 2015), improved occupant comfort, health, and productivity (Heerwagen, 2000; Zhang et al., 2018). Kahn and Kok observed that green attributes of homes can be capitalized on in sales deals (Kahn & Kok, 2014). Hopkins found that some LEED-certified campus buildings produced positive financial effects (Hopkins, 2015). The third view is that occupants who prioritize quality of life and health are willing to pay a premium (Kahn & Kok, 2014).

### 12.4.2 Impact of Price Premium

In addition to reputation, developers also consider the economic returns of their investments and whether the incremental cost is justified (Zhang et al., 2018). Copiello and Donati highlighted the importance of government tax breaks, as the incremental cost of green buildings can be offset by a price premium (Copiello & Donati, 2021). However, for homebuyers, the need to pay more for green buildings has always been a significant barrier due to financial viability concerns (Deng & Wu, 2014). According to Copiello and Donati, the price premium may discourage potential homebuyers from entering the green buildings market (Copiello & Donati, 2021). Homebuyers are typically price sensitive, and an excessive premium may reduce their willingness to pursue green buildings, even if the premium is offset by life cycle energy savings (Dell'Anna & Marta, 2021; Portnov et al., 2018).

### 12.4.3 Measure of Price Premium

The concept of price premium has been a key focus in green building research for many years, and it is often measured using a hedonic regression model (Heinzle et al., 2013). The premium can vary significantly depending on the location, building type, and certification standards. For instance, the premium for Gold and Platinum LEED certified buildings in Italy is 7% and 11%, respectively (Mangialardo et al., 2018), while A-rated properties in Ireland receive a 9% premium (Hyland et al., 2013). Similarly, energy-efficient residential buildings in Finland generate a 3.3% price premium (Fuerst et al., 2016), and green building housing in China may be 6.9% more expensive than their counterparts.

Several studies have also demonstrated the existence of a green premium in various types of green buildings. For instance, green offices in Tokyo receive a 2.6–5.4% rent premium (Onishi et al., 2021), while green hotels in Beijing receive a 6.5% room rate premium (Zhang et al., 2020). Moreover, on average, green residential buildings result in a 5% increase in property value (Shewmake & Viscusi, 2015), and green offices in Hong Kong with HKBEAM certification have a rental premium of about 10.9% (Mesthrige & Chan, 2019).

Studies have also shown that the green premium is likely to increase with green certification stringency (Chegut et al., 2016; Deng et al., 2012; Zhang et al., 2018). Green buildings also have a significant premium at the resale stage, which can make up for the original homebuyers' economic viability. Homebuyers may realize an economic return if the premium at the resale stage is higher than that at the presale stage. For instance, Deng and Wu found that GM-rated homes in Singapore bring in an 11.5% resale premium, compared with a 4% green price premium during presale (Deng & Wu, 2014). Additionally, Jiang et al. found that certified green buildings have a 6% price premium compared to non-certified housing at the resale stage (Jiang et al., 2021).

In summary, the premium associated with green buildings varies depending on the certification standard, location, and building type. Nonetheless, research shows that green buildings have a significant price premium both during presale and resale, and the premium can be even higher for dwellings certified at higher levels. Furthermore, the green premium is likely to increase with certification stringency.

## 12.5 Willingness to Pay

Higher prices have limited homebuyers' WTP for green buildings due to the price premium generated by green buildings. Furthermore, homebuyers are unlikely to quantify the benefits of green buildings in the short term. It is difficult for them to increase their green building budget (Chau et al., 2010; Jiang & Payne, 2022). Homebuyers' WTP for green buildings has become a vital link between developers and homebuyers. It is critical to increase homebuyers' WTP. Therefore, this section identifies the factors affecting homebuyers' WTP for green buildings at the current stage, measures the level of homebuyers' WTP, analyzes the gap between price premiums and WTP, and provides a theoretical reference for the government to introduce incentive policies for homebuyers.

### 12.5.1 Factors Affecting Homebuyers' WTP

Homebuyers' WTP for green buildings is influenced by two types of factors: green building features and homebuyer features. Green building features such as low utilities, low maintenance costs, and recycling water and garbage can improve homebuyers' WTP to varying degrees (Abraham & Gundimeda, 2020; Al-Surf et al., 2021; Chau et al., 2010; Ofek & Portnov, 2020; Portnov et al., 2018; Tapsuwan et al., 2018). Studies have shown that factors such as thermal and lighting comfort, low noise levels, and good indoor air quality can also increase homebuyers' WTP (Abraham & Gundimeda, 2020; Al-Surf et al., 2021; Chau et al., 2010; Golbazi et al., 2020; Robinson et al., 2016; Tapsuwan et al., 2018). Moreover, homebuyers have different preferences for the benefits of green buildings. For instance, waste classification and air quality technologies are preferred by homebuyers working in green buildings, while energy conservation is preferred by people with green living experience (Chau et al., 2010; Xie et al., 2017).

However, there are also some barriers that may reduce homebuyers' WTP for green buildings, such as long payback times, agency problems, and visibility (Bonde, 2012). Homebuyers are more willing to pay for green buildings if they understand them well. Ofek et al. found that people who are more familiar with green buildings have a higher WTP (Ofek & Portnov, 2020). If homebuyers realize that their buildings have green certification, they are more willing to pay a rental premium (Robinson et al., 2016). Additionally, homebuyers' WTP is also influenced by regional and demographic preferences (Robinson et al., 2016), such as gender (Tapsuwan et al., 2018), education level (Abraham & Gundimeda, 2020; Tapsuwan et al., 2018), socioeconomic status (Hu et al., 2014), and age (Tapsuwan et al., 2018). Sang et al. found that green building education and public welfare activities can improve homebuyers' WTP for green buildings (Sang et al., 2020), and tenants with high ecological friendliness are more willing to rent green buildings (Xie et al., 2017). Durdyev et al. categorized the factors affecting homebuyers' WTP into the following categories: attitude, encouragement from surrounding people, perceptual behavior control, awareness and knowledge of green building, marketing and promotion, concern for the environment, government incentives, and purchase purpose (Durdyev & Tokbolat, 2022).

### 12.5.2 Measure of WTP

Homebuyers are only willing to pay a reasonable premium for reducing life cycle costs and improving living comfort (Heerwagen, 2000; Kats, 2003). A national survey conducted by Portnov et al. in Israel indicated that homebuyers' WTP for green buildings is in the range of about 7–10% (Portnov et al., 2018). Xie et al. studied the WTP for pro-environmental behavior and found that the majority of respondents were willing to pay less than a 5% premium for a green building (Xie et al., 2017). The study also suggests that homebuyers.

## 12.6 Public Policy

Due to the additional costs and lack of knowledge or uncertainty about potential benefits, many developers and homeowners may be hesitant to enter the green building market. Therefore, it is important for stakeholders to have green building incentives adopted by governments in their public policies around the world to promote green building (Olubunmi et al., 2016; Wang et al., 2021; Zou et al., 2017).

For example, many government incentives in China are given to developers who adopt energy-efficient technologies (Kong & He, 2021). The findings of the above studies suggest that the government should implement green building incentives for developers to compensate for the additional costs and for homebuyers to compensate for the price premium. This can help reduce the impact of the factors affecting the green building market and promote the development of the green building market.

Effective policy measures need to be identified to act as a lubricant in the entire sales chain of green buildings. These measures can minimize the negative impact of the factors affecting the green building market and promote the development of the green building market. Therefore, this section will search the policy literature in the aforementioned research fields to identify such effective policy measures.

### 12.6.1 Green Building Policies for Developers

Many studies have focused on the impact of green building policies on developers, particularly in reducing the incremental cost of adopting green buildings. Zhang et al. found that tax breaks and subsidies can effectively lower this cost (Zhang et al., 2018). Green building incentives policies are also available to encourage developers to adopt green buildings. The results show that some green building incentives, such as monetary subsidies, tax concessions, and exemptions for developers, are effective in reducing their incremental cost (Olubunmi et al., 2016). Non-financial incentives, such as accelerated licensing or technical assistance, fee relief, and development applications, are also viable options for the government to provide support for developers facing financial constraints (Perkins & McDonagh, 2012). Zou et al. further suggest that non-financial incentives for developers are more effective than financial incentives (Zou et al., 2017).

In addition, mandatory green buildings policies are also available to coerce developers into adopting green buildings. These policies can be used in combination with green building incentives to accelerate the development of green buildings (Rainwater & Martin, 2008; Shi et al., 2014). Energy consumption and carbon emissions from traditional construction have become increasingly important issues in sustainable development, making mandatory measures necessary for countries to solve environmental and energy problems (Wu & Xu, 2013).

### 12.6.2 Green Building Policies for Homebuyers

One of the challenges in promoting green buildings among homebuyers is their low WTP due to information asymmetry and a cost-benefit mismatch (Kahn & Kok, 2014). Activities such as advertising, education, and promotion can help to clear up misconceptions about green buildings and boost their WTP (Abraham & Gundimeda, 2020; Al-Surf et al., 2021; Ofek & Portnov, 2020). Studies also show that homebuyers are willing to pay more for certified green buildings (Jang et al., 2018; Xie et al., 2017). Furthermore, developers should consider adding homebuyers' preferred benefits to green buildings to make them more attractive (Chau et al., 2010).

Various green building incentives are being used to encourage homebuyers to purchase green buildings. For instance, green building incentives to homebuyers, such as loan interest rates and tax breaks, can encourage them to purchase green buildings (Cohen et al., 2019). However, short-term economic incentives may have a detrimental impact on homebuyers' WTP because they may perceive them as promotions of inferior products. Long-term economic incentives, on the other hand, have been shown to increase homebuyers' WTP (Al-Surf et al., 2021; Durdyev & Tokbolat, 2022; Rajaee et al., 2019). The study shows that green building incentives for homebuyers are superior to those for developers in promoting the development of green buildings (He & Chen, 2021).

## 12.7 Conclusion

The development and adoption of green buildings is crucial for achieving sustainable development goals and reducing negative environmental impacts. However, the incremental cost and lack of knowledge about potential benefits may deter developers and homebuyers from entering the green building market. Therefore, government incentives in the form of monetary subsidies, tax concessions and exemptions, accelerated licensing, technical assistance, fee relief, and development applications are necessary to promote green building adoption. Mandatory policies, such as energy consumption and carbon emission regulations, can also encourage developers to adopt green buildings. For homebuyers, education, promotion, and certification can increase willingness to pay for green buildings. Incentives, such as loan interest rates and tax breaks, are also effective in encouraging homebuyers to purchase green buildings in the long-term. Green building incentives for homebuyers are superior to those for developers in promoting the development of green buildings.

**Acknowledgements** This study was funded by of Shandong Social Science Foundation (# 20CJJJ23).

## References

Abidin, N. Z., & Azizi, N. Z. M. (2021). Soft cost elements: Exploring management components of project costs in green building projects. *Environmental Impact Assessment Review, 87*, 106545.

Abraham, P. S., & Gundimeda, H. (2020). Greening offices: Willingness to pay for green-certified office spaces in Bengaluru, India. *Environment, Development and Sustainability, 22*, 1839–1857.

AlAwam, Y. S., & Alshamrani, O. S. (2021). Initial cost assessment stochastic model for green buildings based on LEED score. *Energy and Buildings, 245*(111), 045.

Allen, J. G., MacNaughton, P., Satish, U., Santanam, S., Vallarino, J., & Spengler, J. D. (2016). Associations of cognitive function scores with carbon dioxide, ventilation, and volatile organic compound exposures in office workers: A controlled exposure study of green and conventional office environments. *Environmental Health Perspectives, 124*(6), 805–812.

Al-Surf, M., Balabel, A., Alwetaishi, M., Abdelhafiz, A., Issa, U., Sharaky, I., Shamseldin, A., & Al-Harthi, M. (2021). Stakeholder's perspective on green building rating systems in Saudi Arabia: The case of LEED, Mostadam, and the SDGs. *Sustainability, 13*(15), 8463.

Bianchini, F., & Hewage, K. (2012). How "green" are the green roofs? Lifecycle analysis of green roof materials. *Building and Environment, 48*, 57–65.

Bonde, M. (2012). Difficulties in changing existing leases–one explanation of the "energy paradox"? *Journal of Corporate Real Estate, 14*(1), 63–76.

Bruegge, C., Carrión-Flores, C., & Pope, J. C. (2016). Does the housing market value energy efficient homes? Evidence from the energy star program. *Regional Science and Urban Economics, 57*, 63–76.

Chau, C. K., Tse, M. S., & Chung, K. Y. (2010). A choice experiment to estimate the effect of green experience on preferences and willingness-to-pay for green building attributes. *Building and Environment, 45*(11), 2553–2561.

Chegut, A., Eichholtz, P., & Holtermans, R. (2016). Energy efficiency and economic value in affordable housing. *Energy Policy, 97*, 39–49.

Cohen, C., Pearlmutter, D., & Schwartz, M. (2019). Promoting green building in Israel: A game theory-based analysis. *Building and Environment, 163*, 106227.

Copiello, S., & Donati, E. (2021). Is investing in energy efficiency worth it? Evidence for substantial price premiums but limited profitability in the housing sector. *Energy and Buildings, 251*, 111371.

Council, W. G. B. (2013). *The business case for green building: A review of the costs and benefits for developers, investors and occupants*.

Darko, A., & Chan, A. P. (2017). Review of barriers to green building adoption. *Sustainable Development, 25*(3), 167–179.

Dell'Anna, F., & Bottero, M. (2021). Green premium in buildings: Evidence from the real estate market of Singapore. *Journal of Cleaner Production, 286*, 125327.

Deng, Y., & Wu, J. (2014). Economic returns to residential green building investment: The developers' perspective. *Regional Science and Urban Economics, 47*, 35–44.

Deng, Y., Li, Z., & Quigley, J. M. (2012). Economic returns to energy-efficient investments in the housing market: Evidence from Singapore. *Regional Science and Urban Economics, 42*(3), 506–515.

Devine, A., & McCollum, M. (2019). Understanding social system drivers of green building innovation adoption in emerging market countries: The role of foreign direct investment. *Cities, 92*, 303–317.

Durdyev, S., & Tokbolat, S. (2022). A quantified model for assessment of drivers of acquiring green buildings by potential clients. *Environment, Development and Sustainability, 24*(3), 3815–3831.

Dwaikat, L. N., & Ali, K. N. (2016). Green buildings cost premium: A review of empirical evidence. *Energy and Buildings, 110*, 396–403.

Eichholtz, P., Kok, N., & Quigley, J. M. (2010). Doing well by doing good? Green office buildings. *American Economic Review, 100*(5), 2492–2509.

Eichholtz, P., Kok, N., & Quigley, J. M. (2013). The economics of green building. *Review of Economics and Statistics, 95*(1), 50–63.

Feng, Q., Wang, Y., Chen, C., Dong, Z., & Shi, X. (2021). Effect of homebuyer comment on green housing purchase intention—mediation role of psychological distance. *Frontiers in Psychology, 12*, 568451.

Franco, M. A. J. Q., Pawar, P., & Wu, X. (2021). Green building policies in cities: A comparative assessment and analysis. *Energy and Buildings, 231*, 110561.

Fuerst, F., & McAllister, P. (2011). Green noise or green value? Measuring the effects of environmental certification on office values. *Real Estate Economics, 39*(1), 45–69.

Fuerst, F., Oikarinen, E., & Harjunen, O. (2016). Green signalling effects in the market for energy-efficient residential buildings. *Applied Energy, 180*, 560–571.

Gabay, H., Meir, I. A., Schwartz, M., & Werzberger, E. (2014). Cost-benefit analysis of green buildings: An Israeli office buildings case study. *Energy and Buildings, 76*, 558–564.

Gao, Y., Yang, G., & Xie, Q. (2020). Spatial-temporal evolution and driving factors of green building development in China. *Sustainability, 12*(7), 2773.

Golbazi, M., El Danaf, A., & Aktas, C. B. (2020). Willingness to pay for green buildings: A survey on students' perception in higher education. *Energy and Buildings, 216*, 109956.

He, L., & Chen, L. (2021). The incentive effects of different government subsidy policies on green buildings. *Renewable and Sustainable Energy Reviews, 135*, 110123.

Heerwagen, J. (2000). Green buildings, organizational success and occupant productivity. *Building Research and Information, 28*(5–6), 353–367.

Heinzle, S. L., Yip, A. B. Y., & Xing, M. L. Y. (2013). The influence of green building certification schemes on real estate investor behaviour: Evidence from Singapore. *Urban Studies, 50*(10), 1970–1987.

Hopkins, E. A. (2015). LEED certification of campus buildings: A cost-benefit approach. *Journal of Sustainable Real Estate, 7*(1), 99–111.

Hu, H., Geertman, S., & Hooimeijer, P. (2014). The willingness to pay for green apartments: The case of Nanjing, China. *Urban Studies, 51*(16), 3459–3478.

Hwang, B. G., & Leong, L. P. (2013). Comparison of schedule delay and causal factors between traditional and green construction projects. *Technological and Economic Development of Economy, 19*(2), 310–330.

Hwang, B. G., & Tan, J. S. (2012). Green building project management: Obstacles and solutions for sustainable development. *Sustainable Development, 20*(5), 335–349.

Hwang, B. G., Zhu, L., Wang, Y., & Cheong, X. (2017). Green building construction projects in Singapore: Cost premiums and cost performance. *Project Management Journal, 48*(4), 67–79.

Hyland, M., Lyons, R. C., & Lyons, S. (2013). The value of domestic building energy efficiency—evidence from Ireland. *Energy Economics, 40*, 943–952.

IEA. (2019). *Global status report for buildings and construction*.

Jang, D. C., Kim, B., & Kim, S. H. (2018). The effect of green building certification on potential tenants' willingness to rent space in a building. *Journal of Cleaner Production, 194*, 645–655.

Jiang, H., & Payne, S. (2022). Examining regime complexity in China's green housing transition: A housing developers' perspective. *Building Research and Information, 50*(3), 291–307.

Jiang, Y., Xing, Y., Zhao, D., & Jiao, R. (2021). Resale of green housing compensates for its premium pricing: An empirical study of China. *Journal of Green Building, 16*(4), 45–61.

Jiang, Y., Hu, T., Zhao, D., Liu, B., Zhang, H., Zhang, Y., & Xu, Z. (2022). Decision model to optimize long-term subsidy strategy for green building promotion. *Sustainable Cities and Society, 86*, 104126.

Kahn, M. E., & Kok, N. (2014). The capitalization of green labels in the California housing market. *Regional Science and Urban Economics, 47*, 25–34.

Kats, G. (2003). *Green building costs and financial benefits*.

Kong, F., & He, L. (2021). Impacts of supply-sided and demand-sided policies on innovation in green building technologies: A case study of China. *Journal of Cleaner Production, 294*, 126279.

Li, S., Lu, Y., Kua, H. W., & Chang, R. (2020). The economics of green buildings: A life cycle cost analysis of non-residential buildings in tropic climates. *Journal of Cleaner Production, 252*, 119771.

Liu, Y., Guo, X., & Hu, F. (2014). Cost-benefit analysis on green building energy efficiency technology application: A case in China. *Energy and Buildings, 82*, 37–46.

Mangialardo, A., Micelli, E., & Saccani, F. (2018). Does sustainability affect real estate market values? Empirical evidence from the office buildings market in Milan (Italy). *Sustainability, 11*(1), 12.

Mao, Y. H., & Yang, G. H. (2012). Sustainable development drivers for green buildings: Incremental costs-benefits analysis of green buildings. In *Advanced materials research* (Vol. 374, pp. 76–81). Trans Tech Publications Ltd..

McCoy, A. P., Zhao, D., Ladipo, T., Agee, P., & Mo, Y. (2018). Comparison of green home energy performance between simulation and observation: A case of Virginia, United States. *Journal of Green Building, 13*(3), 70–88.

Menassa, C. C., & Baer, B. (2014). A framework to assess the role of stakeholders in sustainable building retrofit decisions. *Sustainable Cities and Society, 10*, 207–221.

Mesthrige, J. W., & Chan, H. T. (2019). Environmental certification schemes and property values: Evidence from the Hong Kong prime commercial office market. *International Journal of Strategic Property Management, 23*(2), 81–95.

Miraj, P., Berawi, M. A., & Utami, S. R. (2021). Economic feasibility of green office building: Combining life cycle cost analysis and cost–benefit evaluation. *Building Research and Information, 49*(6), 624–638.

Mo, Y., Zhao, D., McCoy, A., Du, J., & Agee, P. (2017). Latent relationship between construction cost and energy efficiency in multifamily green buildings. In *Computing in Civil Engineering 2017* (pp. 273–280).

Mollaoglu, S., Chergia, C., Ergen, E., & Syal, M. (2016). Diffusion of green building guidelines as innovation in developing countries. *Construction Innovation*.

Ofek, S., & Portnov, B. A. (2020). Differential effect of knowledge on stakeholders' willingness to pay green building price premium: Implications for cleaner production. *Journal of Cleaner Production, 251*, 119575.

Olubunmi, O. A., Xia, P. B., & Skitmore, M. (2016). Green building incentives: A review. *Renewable and Sustainable Energy Reviews, 59*, 1611–1621.

Onishi, J., Deng, Y., & Shimizu, C. (2021). Green premium in the Tokyo office rent market. *Sustainability, 13*(21), 12227.

Perkins, M. A., & McDonagh, J. (2012). *New Zealand local government initiatives and incentives for sustainable design in commercial buildings*.

Portnov, B. A., Trop, T., Svechkina, A., Ofek, S., Akron, S., & Ghermandi, A. (2018). Factors affecting homebuyers' willingness to pay green building price premium: Evidence from a nationwide survey in Israel. *Building and Environment, 137*, 280–291.

Rainwater, B., & Martin, C. (2008). *Local leaders in sustainability: Green counties*. American Institute of Architects.

Rajaee, M., Hoseini, S. M., & Malekmohammadi, I. (2019). Proposing a socio-psychological model for adopting green building technologies: A case study from Iran. *Sustainable Cities and Society, 45*, 657–668.

Rehm, M., & Ade, R. (2013). Construction costs comparison between 'green'and conventional office buildings. *Building Research and Information, 41*(2), 198–208.
Robinson, S., Simons, R., Lee, E., & Kern, A. (2016). Demand for green buildings: Office tenants' stated willingness-to-pay for green features. *Journal of Real Estate Research, 38*(3), 423–452.
Sabapathy, A., Ragavan, S. K. V., Vijendra, M., & Nataraja, A. G. (2010). Energy efficiency benchmarks and the performance of LEED rated buildings for Information Technology facilities in Bangalore, India. *Energy and Buildings, 42*(11), 2206–2212.
Sang, P., Yao, H., Zhang, L., Wang, S., Wang, Y., & Liu, J. (2020). Influencing factors of consumers' willingness to purchase green housing: A survey from Shandong Province, China. *Environment, Development and Sustainability, 22*, 4267–4287.
Schnieders, J., & Hermelink, A. (2006). CEPHEUS results: Measurements and occupants' satisfaction provide evidence for Passive Houses being an option for sustainable building. *Energy Policy, 34*(2), 151–171.
Shewmake, S., & Viscusi, W. K. (2015). Producer and consumer responses to green housing labels. *Economic Inquiry, 53*(1), 681–699.
Shi, Q., Lai, X., Xie, X., & Zuo, J. (2014). Assessment of green building policies–A fuzzy impact matrix approach. *Renewable and Sustainable Energy Reviews, 36*, 203–211.
Son, H., Kim, C., Chong, W. K., & Chou, J. S. (2011). Implementing sustainable development in the construction industry: Constructors' perspectives in the US and Korea. *Sustainable Development, 19*(5), 337–347.
Sun, C., Li, Z., & Li, X. (2019). Research on green building incremental cost optimization. In *IOP Conference Series. Earth and environmental science (Vol. 267, No. 5)*. IOP Publishing.
Tapsuwan, S., Mathot, C., Walker, I., & Barnett, G. (2018). Preferences for sustainable, liveable and resilient neighbourhoods and homes: A case of Canberra, Australia. *Sustainable Cities and Society, 37*, 133–145.
Vyas, G. S., & Jha, K. N. (2018). What does it cost to convert a non-rated building into a green building? *Sustainable Cities and Society, 36*, 107–115.
Wan, K. S. (2018). Cost budget and cost control analysis of new green building projects. Paper presented at the 5th International Conference on Business, Economics and Management (BUSEM), Xian, PEOPLES R CHINA.
Wang, W., Tian, Z., Xi, W., Tan, Y. R., & Deng, Y. (2021). The influencing factors of China's green building development: An analysis using RBF-WINGS method. *Building and Environment, 188*, 107425.
Weerasinghe, A. S., Ramachandra, T., & Rotimi, J. O. (2021). Comparative life-cycle cost (LCC) study of green and traditional industrial buildings in Sri Lanka. *Energy and Buildings, 234*, 110732.
Wu, Y. N., & Xu, R. H. (2013). Green building development in China-based on heat pump demonstration projects. *Renewable Energy, 53*, 211–219.
Xie, X., Lu, Y., & Gou, Z. (2017). Green building pro-environment behaviors: Are green users also green buyers? *Sustainability, 9*(10), 1703.
Yang, R., Tang, W., & Zhang, J. (2021). Technology improvement strategy for green products under competition: The role of government subsidy. *European Journal of Operational Research, 289*(2), 553–568.
Yeganeh, A., McCoy, A. P., Reichard, G., Schenk, T., & Hankey, S. (2021). Green building and policy innovation in the US Low-Income Housing Tax Credit programme. *Building Research and Information, 49*(5), 543–560.
Yudelson, J. (2010). *The green building revolution*. Island Press.
Zhang, X. L. (2015). Green real estate development in China: State of art and prospect agenda-A review. *Renewable & Sustainable Energy Reviews, 47*, 1–13.
Zhang, L., Sun, C., Liu, H. Y., & Zheng, S. Q. (2016). The role of public information in increasing homebuyers' willingness-to-pay for green housing: Evidence from Beijing. *Ecological Economics, 129*, 40–49.
Zhang, L., Liu, H., & Wu, J. (2017). The price premium for green-labelled housing: Evidence from China. *Urban Studies, 54*(15), 3524–3541.
Zhang, L., Wu, J., & Liu, H. (2018). Turning green into gold: A review on the economics of green buildings. *Journal of Cleaner Production, 172*, 2234–2245.
Zhang, L., Wu, J., Liu, H. Y., & Zhang, X. L. (2020). The value of going green in the hotel industry: Evidence from Beijing. *Real Estate Economics, 48*(1), 174–199.
Zhao, D., McCoy, A., & Du, J. (2016). An empirical study on the energy consumption in residential buildings after adopting green building standards. *Procedia Engineering, 145*, 766–773.
Zhao, D., McCoy, A. P., Du, J., Agee, P., & Lu, Y. (2017). Interaction effects of building technology and resident behavior on energy consumption in residential buildings. *Energy and Buildings, 134*, 223–233.
Zhao, D., Miotto, A. B., Syal, M., & Chen, J. (2019). Framework for Benchmarking green building movement: A case of Brazil. *Sustainable Cities and Society, 48*, 101545.
Zou, Y. H., Zhao, W. X., & Zhong, R. J. (2017). The spatial distribution of green buildings in China: Regional imbalance, economic fundamentals, and policy incentives. *Applied Geography, 88*, 38–47.
Zuo, J., & Zhao, Z. Y. (2014). Green building research–current status and future agenda: A review. *Renewable and Sustainable Energy Reviews, 30*, 271–281.

# C&D Waste Recycling and Management: Chinese Practices

# 13

Mingxue Ma, Vivian W. Y. Tam, and Khoa N. Le

## 13.1 Introduction

Construction industry has consumed a substantial amount of natural resources and is responsible for a quarter of the solid waste generated globally (Benachio et al., 2020). Massive construction and demolition (C&D) waste is generated owing to the rapid development of construction industry (Yu et al., 2022). Of all the countries in the world, China produces the largest amount of C&D waste, with 2.4 billion tonnes generated in 2015 (Duan et al., 2019). The quantity of C&D waste in China is projected to reach 7.3 billion tonnes by 2030 (Nanchang Municipal Ecology and Environment Bureau, 2020). Currently, most C&D waste in China is disposed of in landfills, which has inevitably occupied land resources and destroyed natural habitats (Li et al., 2022).

In recent years, the concept of circular economy has received increasing attentions from industry and academia. One goal of circular economy is to achieve sustainable development and resource efficiency through reusing, recycling and recovering materials in production, distribution and consumption processes (Campitelli et al., 2022). A transition from linear economy (make-use-dispose) to circular patterns of resource use and management is inevitable, because circular economy can direct the circular flow of materials, reduce waste and contribute to sustainable production (Hossain et al., 2020). Since large-scale C&D activities are undergoing in China, production of C&D waste is unavoidable. However, landfilling contradicts the core principles of circular economy, because it leads to a loss of resources, and materials that end up in this destination cannot to be reused in other products (Pires & Martinho, 2019). Effective and proper treatment of C&D waste is imperative. It is generally agreed that the principle of "reduce, reuse, and recycle" can contribute to a sustainable future (Huang et al., 2018). Specifically, approximately 80% of C&D waste in China could be processed into secondary construction materials, contributing to resource conservation and reducing reliance on landfilling (Ding & Xiao, 2014; Liu et al., 2020). Recycling the waste is in accordance with the idea of circular economy which seeks to minimize waste and maximize the use of materials within the economic cycle (Campitelli et al., 2022).

There is a significant quantity of C&D waste that requires management in China (Yu et al., 2022). In addition, the quantity of C&D waste continues to increase, leading to a growing need to recycle the waste (Yu et al., 2022). In recent couple of years, China has placed increased emphasis on the recycling of C&D waste, while

M. Ma · V. W. Y. Tam (✉) · K. N. Le
School of Engineering, Design and Built Environment, Western Sydney University, Penrith, NSW, Australia
e-mail: M.Ma@westernsydney.edu.au; k.le@westernsydney.edu.au



J. Zuo et al. (eds.), *Circular Economy for Buildings and Infrastructure*, Sustainable Development Goals Series, https://doi.org/10.1007/978-3-031-56241-9_13

the performance varies among different Chinese cities (Li et al., 2020a). According to Nanchang Municipal Ecology and Environment Bureau (2020), the total resource utilization rate for C&D waste is still below 10% in China. The best-performing cities, such as Shanghai and Shenzhen, have achieved recovery rates of >15% (Ghisellini et al., 2018). However, this number is still significantly lower than that in some developed countries, such as 96% in Japan (Ministry of Land Infrastructure Transport and Tourism (Japan), 2018), 88% in Germany (Deloitte, 2015) and 70% of the United States (Duan et al., 2019). Moreover, the majority of Chinese cities still lags behind (Li et al., 2020a). Although it is difficult to attribute this to ineffective policies, because the effectiveness of policies is affected by multiple reasons (Yu et al., 2022), the relatively low recovery rate of C&D waste indicates that certain challenges might prevent the successful implementation of policies in practice. Therefore, identification of challenges in the process of promoting C&D waste recycling is essential.

At present, recycling of C&D waste in China is still in its primary stage (Li et al., 2020a). Around 70 recycling lines have been set up in China, and each can handle over one million tonnes of C&D waste per year. Despite the presence of hundreds of small-scale recycling plants, the total resource utilization rate for C&D waste is below 10% (Nanchang Municipal Ecology and Environment Bureau, 2020). Studies on C&D waste management in China increase rapidly. A comprehensive collection of past studies conducted empirical investigation of C&D waste management in specific Chinese cities, including Shenzhen (Ding et al., 2022; Bao & Lu, 2020; Yu et al., 2018; Yuan, 2017; Wu & Yuan, 2016), Hong Kong (Bao et al., 2020; Wu et al., 2020; Au et al., 2018), Beijing (Shi et al., 2019), Shanghai (Jin et al., 2017), and Chongqing (Yong et al., 2016; Wang et al., 2013; Xu et al., 2012). In addition, possible challenges hindering implementation of C&D waste recycling in different countries or regions were investigated. Major challenges include insufficient regulations (Huang et al., 2018; Yuan et al., 2011; Al-Otaibi et al., 2022), involvement of multiple government departments (Yuan, 2017), uncertainty in quantity estimation (Islam et al., 2019; Lu et al., 2017), unorganized collection (Wang et al., 2020; Zhao et al., 2011), immature recycling technology (Huang et al., 2018), underdeveloped market for products recycled from C&D waste (Shooshtarian et al., 2022; Huang et al., 2018; Yuan, 2017), insufficient attention from industry (Yuan, 2017; Huang et al., 2018; Shooshtarian et al., 2022), extra cost for recycling treatment (Tam et al., 2009; Huang et al., 2018), and a lack of higher value use of recycled aggregate (Lockrey et al., 2016). Although C&D waste management in China has received increasing attentions from researchers and a few empirical studies were conducted, most of past studies focused on C&D waste management in one specific city. Previous studies were limitedly conducting investigations in the scope of different Chinese cities. In addition, many researchers primarily addressed one problem and explored its detailed solution. For instance, Wang et al. (2020) proposed a prototype robotic system which could automatically sort and recycle C&D waste on site. Islam et al. (2019) estimated C&D waste generation rate and quantified waste generation through regression analysis. Guo et al. (2022) studied different types of government subsidies to effectively promote C&D waste recycling. There are limited number of articles that comprehensively investigate the challenges in promoting C&D waste recycling across multiple cities of China. It has significant rooms for analysis.

This study employs site visits to recycling centres in five different Chinese cities (i.e., Shanghai, Suzhou, Chongqing, Changsha, and Shenzhen) and interviews with project managers and on-site workers. It aims to identify the challenges in the current C&D waste management in these five selected cities and provide recommendations to improve recycling rate. The results of this study are expected to aid policy makers in formulation of proper C&D waste management, encourage recycling of C&D waste, reduce landfill, and promote circular economy.

## 13.2 Methodology

For the aim of this study, a mixed-method approach which combines site visits, and semi-structured interviews is adopted. Five Chinese cities from different provinces were chosen. These five cities are Shanghai (Direct-administered municipalities), Suzhou (Jiangsu Province), Chongqing (Direct-administered municipalities), Changsha (Hunan Province), and Shenzhen (Guangdong Province).

### 13.2.1 Case Description

In 2018, the Ministry of Housing and Urban-Rural Development of the People's Republic of China (2018) selected 35 cities as pilot cities to mandatorily initiate their C&D waste recycling plans, because these cities are at the forefront of development and economic levels. Additionally, as construction activities are the major economic activities, these cities have a more urgent need to manage waste compared with other Chinese cities. The successful implementation of C&D waste recycling in these cities is expected to serve as a model for the promotion of C&D waste recycling in non-pilot cities across whole China.

Because of uneven development in economy and construction, there exists significant variety in current status of municipal C&D waste management. In some developed cities (such as Shenzhen), recycling C&D waste into recycled aggregates or bricks has gradually become a major approach to deal with the waste, whereas most C&D waste in underdeveloped cities (such as Guiyang) ends up in landfills (Yuan, 2017). In this study, five Chinese cities were chosen, because they have implemented relevant regulations and actively promoted recycling. All of the five selected cities were among the 35 pilot cities: Shanghai (Direct-administered municipalities), Suzhou (Jiangsu Province), Chongqing (Direct-administered municipalities), Changsha (Hunan Province), and Shenzhen (Guangdong Province).

#### 13.2.1.1 Eastern China

Shanghai, as a commercial, financial, and political centre, is one of the most advanced cities in China with regard to C&D waste management, with a recycling rate of 20% (Huang et al., 2018). In the past decades, Shanghai Municipal People's Government (2017) enacted and updated recycling regulations, but further improvement is still needed with a view to obtaining long-term benefits (Yu et al., 2022). The operation of the case project in Shanghai starts from 2016 and is run by a private company. This recycling plant is one of the largest recycling plants in Shanghai which occupies one million $m^2$. Approximately 4000 tonnes of C&D waste are treated per day and the annual recycling capacity is estimated to be one million tonnes.

Suzhou is one of the largest economies in the Yangtze River delta. Local government in Suzhou is concerned with the C&D waste management in response to the city's accelerated economic urbanisation and extensive C&D activities (Bao et al., 2019). The case project in Suzhou is operated under a Build-operate-transfer (BOT) initiative which occupies an area of 78,000 $m^2$. This facility could recycle 4000 tonnes per day and is operated by a private company, whose business scope includes C&D waste transportation. In 2019, this company was awarded two BOT contracts for C&D waste transportation and recycling.

#### 13.2.1.2 Western China

Chongqing is one of the most populous cities in western China (Zhang, 2009). The average quantity of C&D waste collected in urban areas of Chongqing is estimated to reach 4.03 million tonnes (Lin et al., 2020). However, landfilling is still a broadly used method to dispose C&D waste, because it is impractical to eliminate all the C&D waste through recycling. The case project in Chongqing occupies 65,333 $m^2$, processing 2000 tonnes of C&D waste per day. This recycling plant is located in rural areas, making it difficult to find a stable source of C&D waste for recycling.

#### 13.2.1.3 Southern-Central China

Current C&D waste management in Changsha might fail to follow the increasing pace of construction activities, as there were only two recycling enterprises in 2014 (Huang et al., 2018). The Bureau of Housing and Urban-Rural Development of Changsha (2019) formulated polices for C&D waste recycling, presenting an urgent need to address the issue of waste generation. Besides, Changsha Municipal Government (2018) introduced a subsidy scheme to recycling companies of approximately 3 CNY (or US $0.43) for per cubic meter of C&D waste recycled. The case project in Changsha is a pilot project, which was initiated by local government and managed by a state-owned company. This project is featured for on-site recycling, with 420–480 tonnes C&D waste directly recycled on the demolition site.

Shenzhen, one of the economically developed cities in southern China, has begun to recycle C&D waste into recycled aggregate and bricks, whereas some underdeveloped cities have not taken measures to treat the waste (Yuan, 2017). Management on C&D Waste in Shenzhen (2007) forms the foundation for C&D waste recycling. Over the past few years, Shenzhen has played a leading role toward achieving sustainable C&D waste management in China, partly owing to its many recycling enterprises (AECOM Asia Company Limited, 2018). The case project in Shenzhen starts with an initial investment of 65 million CNY (or US $9,390,213). It occupies 5000 $m^2$ and two crushing lines to convert C&D waste into recycled aggregate. Approximately 7000 tonnes of C&D waste could be recycled per day. In 2018, approximately 1,230,000 tonnes of recycled aggregate were produced, and this company achieved total revenues over 60 million CNY (or US $8,667,889).

### 13.2.2 Site Visits

Despite the introduction of municipal policies in the five selected cities, the industry practices of C&D waste management have not been officially documented. There exists a policy-practice gap between government regulations and industry practices (Wafula et al., 2014). Therefore, site visits to recycling plants and interviews with industry practitioners appear to be appropriate methods to explore industry practice of C&D waste management in its real-life context. Site visits were conducted to C&D waste recycling plants in the five selected cities to explore the current practices of C&D recycling to the real context. These five recycling plants were investigated from December 2019 to January 2020. Before site visits and interviews, ethical approval for this project was granted by Human Research Ethics Committee (HREC). Approval Number is H13470. Researchers firstly obtained permission from C&D waste recycling companies for site visits and interviews and specified selection criteria for interviewees. The average duration of a site visit was approximately one hour. At least one project manager and 1–2 on-site workers participated in site visits and interviews at each site. Throughout the site visit process, direct observation of recycling process, production environment, area of recycling plants, professional equipment, and major products was involved. A series questions related to recycling process and products were asked during the visits, and answers would be obtained directly from accompanying professionals.

### 13.2.3 Interviews

Semi-structured interviews were conducted face-to-face with a total of 12 professionals, including 5 project managers and 7 on-site workers. The selection of interviewees is based on three criteria: (1) project managers should have adequate industry experience, i.e., at least one project; (2) project managers should have a comprehensive understanding of regulatory environment, local regulations and market; (3) on-site workers should have adequate industry practice, i.e., at least 3-year experience in related fields, who could supplement additional information of recycling technologies, facilities, production process, and products. These 12 interviewees came from the five recycling plants used for site visits. Specifically, in Shanghai, Chongqing and Changsha, one project manager and one one-site worker from each site accepted to be interviewed. In Suzhou and Shenzhen, one project manager and two one-site workers participated in the interviews. The average duration of an interview is approximately one to two hours. Interview questions were sent to interviewees before the site visit. Interview questions were focused on four aspects: (1) characteristics of recycling plants, (2) government incentives, (3) local regulations, and (4) supervision and penalties. This research employs the qualitative research method, which consists of three processes. Firstly, All the responses were recorded as written format. Secondly, each interview was documented based on the recorded notes. Thirdly, content analysis was conducted which categorized the data into specific topics or themes. The original data collected at the five recycling plants were further organised and managed into Tables 13.1, 13.2, 13.3, and 13.4, based on their categories. According to these tables, a second round of analysis is conducted to explain and assess challenges hindering the current C&D waste management in the five selected cities.

## 13.3 Results and Discussions

### 13.3.1 Characteristics of the Five Recycling Plants

Table 13.1 describes characteristics of the five recycling plants. These five recycling plants are characterized by its area, ownership, business scope, source for recycling, recycling capability, plant type and major products. The details of these characteristics are discussed below.

#### 13.3.1.1 Types of Recycling Plants

According to Table 13.1, these five recycling plants are of different sizes. The largest recycling plant is located in Shanghai, covering an area of one million $m^2$, and has the capacity to recycle 4000 tonnes of C&D waste per day. The recycling plants in Suzhou and Chongqing are the second and third largest, occupying 78,000 and 65,333 $m^2$, respectively. The Suzhou recycling plant could recycle 4000 tonnes of C&D waste per day, while the plant in Chongqing could handle 2000 tonnes/day. The recycling plant in Shenzhen is 5000 $m^2$ in size, but it can recycle 7000 tonnes of C&D waste per day. The smallest recycling plant is located in Changsha, with a capacity to recycle only 420–480 tonnes of C&D waste per day. Fixed and mobile are two types of recycling plants, which adopt different levels of technology and produce recycled aggregates of different grades (Blengini & Garbarino, 2010). Fixed plants are permanently placed in a recycling centre and can produce high-quality recycled products (Ulubeyli et al., 2017), but they are forbidden to be built in urban areas in some Chinese cities, because of the emission of noise and dust (AECOM Asia Company Limited, 2018). The high initial cost and significantly increased cost of waste transportation from demolition site to a fixed recycling plant reduce the popularity of this type of recycling plants (Zhao et al., 2011). Mobile recycling plants are portable, and the process of recycling could be completed at demolition worksites (Ulubeyli et al., 2017). For instance, C&D waste can be directly separated and recycled at the mobile plants in Changsha. However, production in the Changsha industry is significantly influenced by the demolition speed. With the capability to deal with 150–200 tonnes of C&D waste per hour, the Changsha industry can only recycle 70–80 tonnes of the waste, because of demolition downtime. Despite mobile plants could reduce transportation costs by decreasing distances from recycling plants and construction sites (Zhao et al., 2011), complaints from surrounding residents regarding noise and dust limit the development of mobile plants in some cities, such as Chongqing. Along with decrease in the scale of urban renewal, surrounding residents might be affected by recycling activities. As large-scale construction activities are still undergoing in Changsha, surrounding residents have been relocated and are not affected by recycling activities. Interviewees from Changsha mentioned that without a city plan regarding land use for C&D

**Table 13.1** Basic information collected from recycling plants

| | Eastern China | | Southern-Central China | | Western China |
|---|---|---|---|---|---|
| | Suzhou | Shanghai | Shenzhen | Changsha | Chongqing |
| Area | 78,000 m$^2$ | 1,000,000 m$^2$ | 5000 m$^2$ | 2000–2667 m$^2$ | 65,333 m$^2$ |
| State-owned or private company | Private | Private | Private | State-owned | State-owned |
| Is demolition included in business | No | No | No | No | No |
| Source of C&D for recycling | Contract with contractors | Government contract; and/or purchase | Contract with contractors | Contract with contractors | Purchase |
| Quantity of waste recycled/day | 4000 tonnes/day | 4000 tonnes/day | 7000 tonnes/day | 420–480 tonnes/day | 2000 tonnes/day |
| Mobile or fixed | Fixed | Fixed | Fixed | Mobile | Fixed |
| Products | Recycled aggregate; bricks; and red brick powder | Recycled aggregate; and recycled aggregate concrete (C25 and below) | Recycled aggregate; light panel wall; and recycled aggregate concrete (C30 and below) | Recycled aggregate; and bricks | Recycled aggregate; and recycled aggregate concrete (C5-8) |

**Table 13.2** Information of incentives provided by local governments

| | Eastern China | | Southern-Central China | | Western China |
|---|---|---|---|---|---|
| | Suzhou | Shanghai | Shenzhen | Changsha | Chongqing |
| VAT deduction | 70% VAT deduction | 50% VAT deduction | 100% VAT exemption | 70% VAT deduction | 70% VAT deduction |
| Subsidy for waste recycling | Yes. Approximately 120 CNY/tonne (or US $17) | Yes | Yes | No | No |
| BOT right | Yes. Two BOT rights were obtained in 2019 | No | No | No | No |
| Cheaper price for electric and water use | No | No | No | No | No |
| Cheaper land use | Yes | No | Yes | No | No |

waste recycling, a mobile plant appears to be a better choice than a fixed plant.

#### 13.3.1.2 Source of C&D Waste for Recycling

Stability of the source of C&D waste for recycling is an important prerequisite to maintain long-term operation. The unstable source of C&D waste for recycling might result in problems in product supply and further cause absence of stability of revenues and market sustainability (AECOM Asia Company Limited, 2018). According to interview responses, there are two primary ways for acquiring C&D waste. (1) Companies in Shanghai, Shenzhen, Changsha and Suzhou have formed a stable partnership with local contractors. Contractors would deliver C&D waste produced at their demolition sites to these recycling plants and pay the associated transportation and recycling fees. (2) The industry in

**Table 13.3** Local regulations in China

| | Eastern China | | Southern-Central China | | Western China |
|---|---|---|---|---|---|
| | Suzhou | Shanghai | Shenzhen | Changsha | Chongqing |
| Waste reduction at plan stage | No | No | No | No | No |
| Forbidden waste export | No, but governmental approval is required | Yes | No, but governmental approval is required | No | No |
| Regulations on on-site sorting | Yes | Yes | No | No | No |
| Separate collection | Yes | Yes | No | Yes | No |
| Strict landfill ban | Yes | No | Yes | No | No |

**Table 13.4** Supervision and penalty

| | Eastern China | | Southern-Central China | Western China | |
|---|---|---|---|---|---|
| | Suzhou | Shanghai | Shenzhen | Changsha | Chongqing |
| Penalty on illegal dumping | A 10,000 CNY (or US $1430) fine; or/and sentence | Fines from 5000–50,000 CNY (or US $716–7166) | Fines from 5000–50,000 CNY (or US $716–7166); or/and sentence | Fines from 5000–50,000 CNY (or US $716–7166) | Fines from 5000–50,000 CNY (or US $716–7166) |
| Penalty on stakeholders' violation | Yes | Yes | Yes | Yes | Yes |
| Planned route for waste transportation | Yes | Yes | Yes | Yes | Yes |
| Method to calculate waste quantity | It is based on the disposal plan | It is based on the disposal plan | It is based on the disposal plan | It is based on the disposal plan | It is based on the disposal plan |
| Waste tracing system | Figure 13.3 | Figure 13.1 | Figure 13.2 | Figure 13.3 | Figure 13.3 |
| Report from public | Yes, by phone | Yes, by phone, e-mail, or mail | Yes, by phone | Yes, by phone | Yes, by phone |
| Consistent cooperation among government departments | | | | No | No |

Chongqing has to purchase C&D waste from contractors to maintain operation. The remote location of this industry has significantly increased its transportation costs. To reduce the disposal cost, contractors prefer to transport the waste to landfill sites. Interviewees from Changsha also showed concerns about the source of C&D waste. As described before, production of Changsha industry is dependent on demolition speed and the source of waste for recycling cannot be ensured.

#### 13.3.1.3 Products

Converting C&D waste into recycled aggregate for non-structural applications is the most prevalent method. As indicated by Table 13.1, the primary product from C&D waste is recycled aggregate. Recycled aggregate from C&D waste is expected to replace part or all of the natural aggregates in concrete (Guo et al., 2018). With appropriate mix design and construction, recycled-aggregate concrete can be safely used as a structural material in civil engineering

(Dimitriou et al., 2018). In Shanghai and Shenzhen, recycled aggregate can partly replace the natural aggregate in low-strength concrete for non-structural applications. Additionally, bricks including water permeable bricks, grass bricks, and bricks for tactile paving, are market-leading products.

### 13.3.2 Government Incentives

Introduction of preference policies, including financial subsidies, tax reduction, and low land rents, could help recycling companies to achieve economic feasibility and business profitability, and therefore fostering the recycling industry (Cheng et al., 2022; Jin et al., 2017). However, economic support is not attractive enough to recycling companies, because market for products recycled from C&D waste is underdeveloped (Huang et al., 2018). Promotion of the market is in urgent need (Huang et al., 2018). Table 13.2 presents government incentives in the five selected cities, with a focus on Value-added tax (VAT) deduction, subsidies, BOT contract, electricity use, water use and land use.

#### 13.3.2.1 Tax Deduction and Subsidies

VAT, as a consumption tax levied on a product, accounts for approximately 36% of the total tax revenues (Liu & Lu, 2015). In the beginning of 2009, a VAT reform was implemented at national wide, which allows tax deduction on the purchase of fixed assets and provides tax incentives to investors (Zhang et al., 2017). State Taxation Administration (2015) has applied VAT deduction to encourage recycling activities. C&D recycling companies, whose recycled aggregate is comprised of more than 90% C&D waste and satisfies technical specifications, are exempt from 50% VAT (State Taxation Administration, 2015). Application of VAT deduction is slightly different in practice. Interviewees from Shanghai stated that 50% VAT could be rebated, while interviewees in other cities could claim more than 50% VAT reductions. It is difficult for recycling companies to maintain long-term operation without governmental subsidies (Bao & Lu, 2020). Suzhou local government subsidises C&D waste recycling companies by 120 CNY/tonne (or US $16.9) recycled. Although Changsha municipal government has introduced a 3 CNY (or US $0.43) subsidy for every cubic meter of C&D waste recycled, interviewees from Changsha claimed that this policy was not executed in practice.

#### 13.3.2.2 BOT Contract

Private sector who obtains a BOT right is given responsibility to finance, build, operate infrastructure projects until the projects are transferred back to public sector (Song et al., 2017; Lu et al., 2022). BOT contracts can efficiently link the benefits or welfare of the private sector with the local government, ensure sufficient fund, combine strength of different sectors and optimize risk allocation (Wang et al., 2018). C&D waste recycling in Suzhou is developed on a BOT basis. Specifically, the Suzhou company obtained two BOT contracts in 2019 with duration of 20 years. One contract is related to C&D waste transportation, and the other is related to C&D waste recycling. Although a BOT contract provides recycling companies with a monopoly status and excludes them from competition for a set period (Qiu & Wang, 2011), it creates substantial entry barriers for new recyclers. In Shenzhen, the BOT was terminated to avoid a monopoly and to enhance the overall recycling capability.

#### 13.3.2.3 Land Use

C&D waste recycling is a complex system involving waste generation, transportation, recycling, disposal, and land use (Yuan, 2017). Land use is linked to adequate capacity for waste storage, production, and product preservation, while appropriate location is closely related to transportation cost. However, land use for C&D waste recycling is not guaranteed, because it is not included in urban construction plan in many cities, which hinders the development of recycling (AECOM Asia Company Limited, 2018). Land acquisition imposes an economic burden on investors for fixed plants. Many investors rent a piece of land to operate recycling plants. Suzhou local governments have allocated land to recycling companies and reduced the land rental costs. Specifically, interviewees from Shenzhen recycling plant stated that Shenzhen government charged only 1 CNY/$m^2$ (or US $0.14) per year for land use, to encourage the participation of new recyclers.

### 13.3.3 Local Regulations

It is important to investigate regulatory environment and current treatment of C&D waste. For instance, because landfilling is not a desirable approach to treat C&D waste, recycling strategies were formulated to reduce landfilling in some cities. As production of C&D waste is inevitable, the volume of the waste might exceed the city's recycling capability. Exporting waste to surrounding cities might be allowed. In addition, waste minimisation at the design stage is considered as one of the most effective method to control the waste generation and is encouraged by local governments (Yuan et al., 2022), but the execution is difficult. Table 13.3 describes information related to local regulations, with a focus on waste exportation, on-site sorting, and landfill policy.

#### 13.3.3.1 Waste Exportation

There is constant pressure from rapid urbanization and correspondingly increased C&D waste quantity. Exporting C&D waste to surroundings could relieve the pressure, but this action could cause waste management issues in other cities. In Shanghai, waste export is strictly prohibited. C&D waste produced in Shanghai should be disposed of within the same city. Waste exportation to surrounding cities or provinces is permitted in Shenzhen and Suzhou, but it requires approval from the local government. There is no restriction on waste exportation in Chongqing, but the transportation costs to other destinations are high. Therefore, waste export might not be a desirable option for waste holders. Specifically, in Changsha, a small amount of C&D waste would be transported to Jiangxi Province, because the landfill capacity of this city is nearly full.

#### 13.3.3.2 On-Site Sorting

Construction, renovation and demolition waste are identified as the three typical categories of C&D waste, based on its generation phase (Mahpour, 2018). Inert materials (including concrete, blocks and bricks) and some hazardous components (such as asbestos) are major components of C&D waste (Wu et al., 2019). However, because of the absence of regulations on on-site sorting, C&D waste mixed with unwanted fractions (such as domestic garbage) would be directly transported to recycling plants. It might potentially decrease quality of recycled aggregate and increase expenses of subsequent treatments, because recycling companies need to sort the waste (Wang et al., 2010). Interviewees from Suzhou and Shanghai industries mentioned that their local governments provided guidance on on-site sorting, including a rough classification of C&D waste and different treatments for each class. Although the operation conditions vary among these companies, a treatment principle has been developed to manage C&D waste. C&D waste can be roughly classified into three categories: (1) recyclables, (2) waste that can be burnt, and (3) waste that requires additional research. Steel, timber, and concrete waste belong to the first category. Steel should be separated from concrete waste at the demolition site and sold back to the market. Approximately 90% of steel and 50% of wood could be collected or sold on-site (Zhao et al., 2010). Although C&D waste cannot legally be mixed with municipal waste before landfilling or recycling, it is difficult to separate clean concrete waste from other classes of waste. For instance, concrete waste from road renovation is mixed with asphalt which would reduce the strength of recycled aggregate. The second category comprises municipal waste, which will end up in incineration plants. Proper methods to handle waste that is unrecyclable or incombustible (such as aerated blocks) need to be further investigated and developed. However, interviewees from other cities stated that there was an absence of relevant guidance.

#### 13.3.3.3 Landfill

Landfill is an undesirable option to treat C&D waste, because C&D waste can consume large spaces, and detrimental components can lead to harmful chemical leachate and other contaminants (Ulubeyli et al., 2017). Approximately 80% of C&D waste has the recycling potential (Wu et al., 2019; Ding & Xiao, 2014). Currently, C&D waste generated in Chongqing, Shanghai and Changsha could be lawfully destined in landfills. The C&D waste is poorly handled in Chongqing and Changsha, because landfill cost is rather low and landfill activities are not strictly regulated. Disposing C&D waste in landfills is strictly forbidden in Shenzhen and Suzhou. Waste landfilling should be approved by local government. Since local government prioritises waste recycling, only non-recyclable components could be landfilled in Shenzhen, but special treatment is required before landfilling.

### 13.3.4 Supervision and Penalty

Illegal dumping is rampant as this action is regularly reported (Duan et al., 2015). In addition, disobedience of related stakeholders increases difficulty in C&D waste management. Imposed penalties and enhanced supervision system can affect the behaviours of related stakeholders (Li et al., 2017). Table 13.4 presents current supervision system and penalties imposed on improper actions, including penalty policies, estimation of waste quantity and waste tracing system.

#### 13.3.4.1 Illegal Dumping

Illegally disposing C&D waste in unlawfully accepted places could cause detrimental damage to environment and human health (Du et al., 2020). Penalties can regulate illegal waste dumping behaviours. In Suzhou and Shenzhen, the severity of penalties depends on the degree of environmental damage caused by illegal dumping. The Municipal Comprehensive Law-enforcing Bureau of City Administration has the authority to determine penalties. The severity of penalty is determined by the location of the illegal dumping, the potential risks, and the waste volume. In Shenzhen and Suzhou, violators can be sentenced when the environment damage is severe. The other three cities follow their local rules or the national guidance (Ministry of Housing and Urban-Rural Development of the People's Republic of China, 2005).

#### 13.3.4.2 Waste Tracing and Quantity Estimation

Figures 13.1, 13.2, and 13.3 describe the waste tracing systems in five selected cities. All the vehicles used for transporting C&D waste should be equipped with a global positioning system (GPS), follow an approved route, and be under the supervision of local government. The estimation of the annual generation of C&D waste is based on demolition amount noted in the disposal plans which are submitted by contractors. This number could be further verified by recycling companies or landfill sites, according to the number of trucks and the truck volume to their sites (or by using a weighing scale). However, it is difficult to estimate the amount of C&D waste produced in Changsha, as numerous small-scale projects are not required to estimate waste quantity and submit disposal plans to the local government. The amount of C&D waste in Changsha is significantly underestimated.

Normally, before waste removal, contractors should submit a C&D waste disposal plan to local government and obtain permits for waste disposal. In Shanghai (Fig. 13.1), contractors should make a prepayment to an authorised account for disposal and transportation fees, according to the approved demolition amount in the disposal plan. The final payment can be made only after completion of transportation and waste disposal. The transportation and disposal fees are transferred directly from the governmental account after official verification. In Shenzhen, an electronic bill system is used to trace C&D waste (Fig. 13.2).

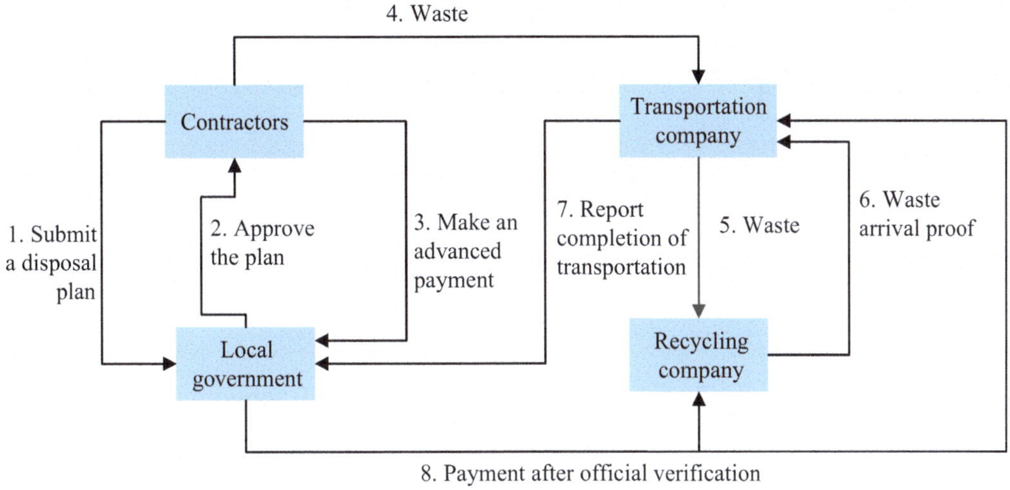


**Fig. 13.1** Waste tracing system in Shanghai

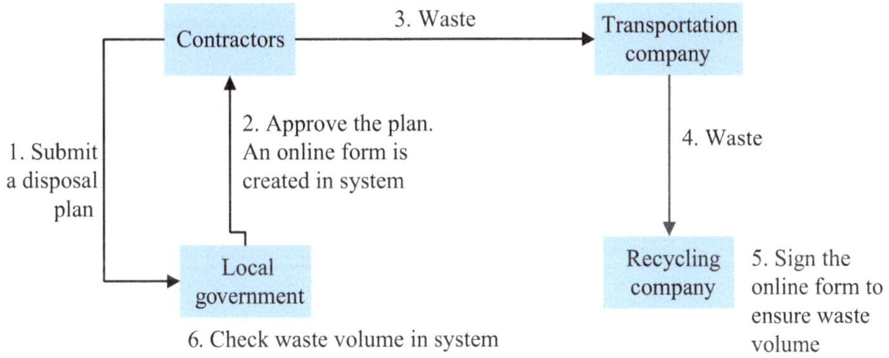


**Fig. 13.2** Waste tracing system in Shenzhen

Figure 13.3a could represent the C&D waste tracing system used in most Chinese cities. Specifically, Suzhou differs from those cities in that the definition of C&D waste in Suzhou is different from that in other cities. In Suzhou, C&D waste refers to clean concrete waste. In other cities, C&D waste consists of concrete, bricks, timber, glass, and furniture, which is defined as renovation waste in Suzhou. The recycling subsidy is only provided to companies who recycle renovation waste. The local government would block the demolition site and decide whether the waste produced is categorised as construction or renovation waste (Fig. 13.3b). After the assessment, the recycling industry can transport the waste to the recycling plant. For consistency, the term "C&D waste" is used to replace "renovation waste" in Suzhou in this study.

### 13.3.5 Challenges and Recommendations

It can be observed that some efforts have been made by local governments to promote C&D waste management in the past couple of years. However, the performance varies across different regions. Different cities are confronted with different challenges. C&D waste management needs to be further improved. Challenges are summerized in Table 13.5. Corresponding recommendations are discussed below.

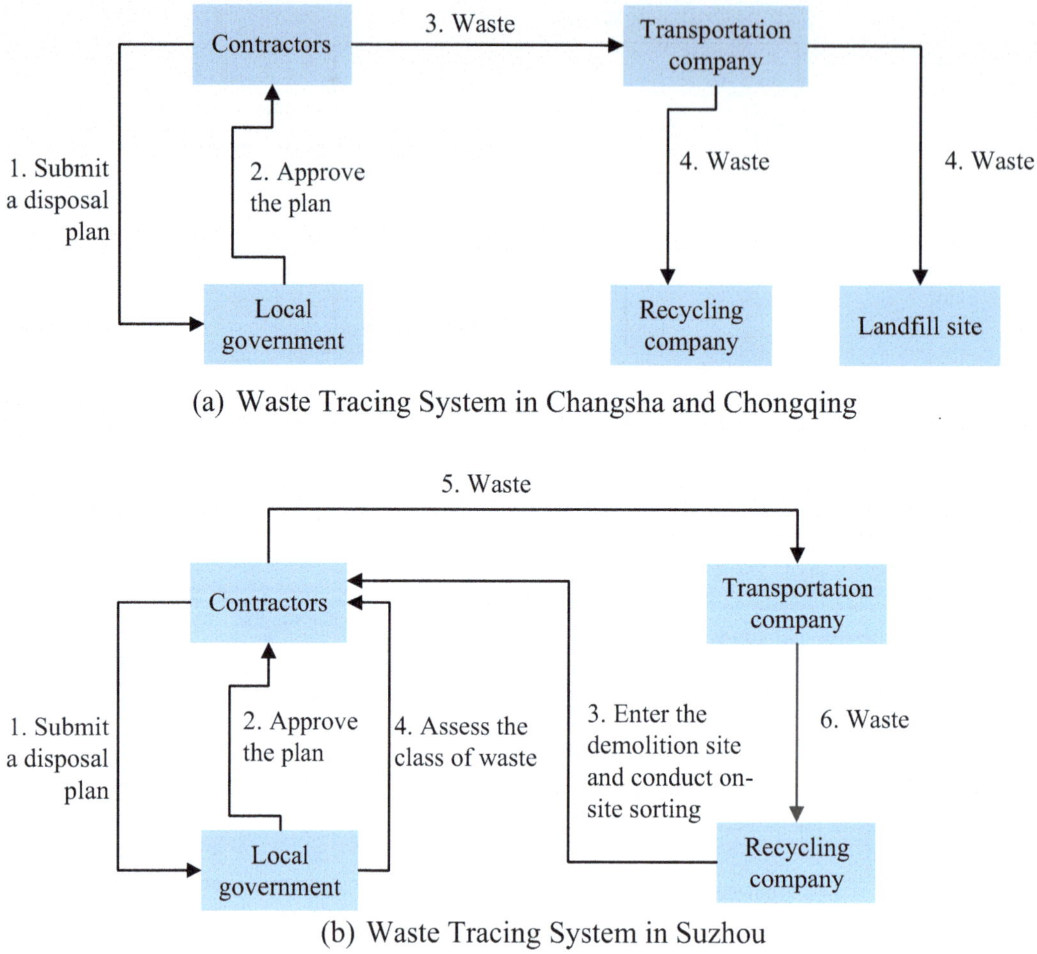


**Fig. 13.3** Waste tracing system in Suzhou, Changsha and Chongqing. (**a**) Waste tracing system in Changsha and Chongqing. (**b**) Waste tracing system in Suzhou

**Table 13.5** Challenges in C&D waste management in the five selected cities

| | Eastern China | | South-Central China | | Western China |
|---|---|---|---|---|---|
| Challenges | Suzhou | Shanghai | Shenzhen | Changsha | Chongqing |
| Unstable source of C&D waste for recycling | | | | × | × |
| Absence of subsidies for recycling activities | | | | × | × |
| High cost for land use | | × | | × | × |
| Absence of regulations on on-site sorting | | | × | × | × |
| Unregulated landfill activities | | × | | × | × |
| A lack of accurate estimation of waste quantity and distribution | × | × | | × | × |
| A lack of an effective waste tracing system | | | | × | × |

#### 13.3.5.1 Challenge One: Unstable Source of C&D Waste for Recycling

How to guarantee the stability of the waste sources for recycling activities is one challenge faced by recycling companies in Changsha and Chongqing. Short of C&D waste for recycling might be caused by the absence of mandatory recycling regulations or high transportation costs (AECOM Asia Company Limited, 2018). Local governments in Chongqing could learn experiences from other selected cities. Measures, including release of governmental contracts or franchise rights to recycling companies, could be taken. Mandatory policies on recycling are fundamental to achieve recycling target (Li et al., 2020a). Local government could provide some guidance on recycling process and enforce waste holders to deliver C&D waste to recycling plants. The transportation cost is directly influenced by the distance from demolition worksite and recycling plant. Because fixed recycling plants are not allowed to be established in urban areas and the number of recycling plants is significantly scare compared to landfill sites, waste holders usually need to pay higher transportation costs for delivery to recycling plants (Pan et al., 2020). The high transportation cost caused by the remote location of Chongqing recycling plant results in preferences for landfilling. A governmental subsidy could be applied to C&D waste transportation companies to reduce transportation costs. The government paying part of the transportation cost can help to shift waste holders' preferences from landfilling to recycling.

#### 13.3.5.2 Challenge Two: Absence of Subsidies for Recycling Activities and High Cost for Land Use

It is hard for C&D recycling companies to make profit without provision of economic preferences from government (AECOM Asia Company Limited, 2018). VAT deduction has been applied in the five selected cities. Some local governments, such as Changsha and Chongqing, should consider provision of subsidies for recycling activities to improve the economic feasibility of recycling companies. Land acquisition is one challenging issue for its high costs. In addition, the location of recycling plants would directly influence transportation cost. However, the economic condition of a municipality could determine the preference policies (Liu et al., 2020). For a few cities (such as Chongqing and Changsha), it is difficult to reduce rental prices for land use to the Shenzhen level (1 CNY/$m^2$/year or US $0.14/$m^2$/year), because of financial constraints. Franchise management, such as BOT, is applied in Suzhou, which could shelter recycling enterprises from competition and ensure the stability of C&D waste for recycling. Nevertheless, a franchised monopoly might emerge and prevent entry of small or medium-scale recycling companies. Therefore, local governments should consider the possible effects of franchise management.

#### 13.3.5.3 Challenge Three: Absence of Regulations on On-Site Sorting

On-site sorting is preferred, because it requires less labours and costs compared with sorting at recycling centres (Li et al., 2020a). On-site sorting could increase reusability and recyclability of C&D waste and therefore contribute to production of high-quality products (Bao & Lu, 2020). However, it seems very slow for the construction industry in China to embrace on-site sorting. Only governments in Suzhou and Shanghai provided a simple guidance on on-site sorting. There is an absence of detailed regulations on on-site sorting in the five selected cities. In most occasions, metals, such as steels, are picked out and directly sold on-site (Yuan, 2017). A mixture of non-inert and inert waste materials, hazardous components and some municipal waste is directly delivered to recycling companies. As some untreated waste in the mixture would impact the quality of recycled aggregate, recycling companies need to spend efforts to separate clean concrete from other kinds of waste. Relevant regulations should be executed to guide efficient on-site sorting. For instance, some recyclable materials, including concrete and bricks, should be carried to recycling plants, while municipal waste should be separately collected and delivered to incineration plants.

#### 13.3.5.4 Challenge Four: Unregulated Landfill Activities

Diverting C&D waste from landfill to recycling is urgent, because of the depletion of land resources (Li et al., 2020b). Landfilling of C&D waste generated in Chongqing, Shanghai and Changsha is lawfully accepted. Specifically, landfill activities are not strictly regulated in Chongqing and Changsha, which result in the heavy dependence on landfilling. A landfill ban could be employed to increase recycling rate (Jin et al., 2017). This ban can be applied to all the reusable or recyclable components in C&D waste. Landfill is only allowed to treat non-hazardous and unrecyclable components and should not be permitted without local government approval. However, fully implementation of the landfill ban is not advisable in a short-term, if there is no effective recycling system (Zhao et al., 2010). Economic penalties, such as a high landfill-tipping fee, could be another ideal approach to change industry's behaviours from landfill to recycling and therefore stimulate recycling activities (Li et al., 2020c). It could potentially increase the profits earned by recycling companies in the competition with high landfill charge fees. Specifically, mixing one class of waste with other substances or classes of waste should be prohibited to drive on-site sorting. A landfill charge scheme of different price levels for different class of waste could be considered. In United Kingdom and Germany, a higher landfill charge fee is applied to unsorted C&D waste than the sorted waste (Li et al., 2020c).

#### 13.3.5.5 Challenge Five: A Lack of Accurate Estimation of Waste Quantity and Distribution

The large generation of C&D waste in China is a big challenge in the development of an effective waste management system (Ding & Xiao, 2014). Estimation of C&D waste quantity is a prerequisite to understand the generation trend in a city and formulate reasonable recycling policies (Yu et al., 2018). In the five selected cities, estimation of C&D waste quantity is based on the waste quantity notified in waste disposal plans, which is submitted before demolition process. However, the actual amount is not accurately qualified, which might further cause inappropriate resource allocation and effectiveness of management (Menegaki & Damigos, 2018). Specifically, multiple factors, including project type, function of projects and composition of waste stream, could influence the accuracy of estimation (Yuan, 2017). The influences from these factors could be minimised through extensive researches in a large amount of sample projects (Yuan, 2017). Local governments can cooperate with universities or research institutes to establish a systematic data collection method (Akhtar & Sarmah, 2018). In addition, the regulation on small-scale construction projects should be tightened. In Changsha, small-scale construction projects are not required to submit waste disposal plan. The quantity of C&D waste from these projects remains unknown and this number is not included in the calculation of municipal C&D waste amount. Under this condition, the actual amount of C&D waste in Changsha is significantly underestimated. However, regulation on small-scale projects would incur additional labours and regulatory costs. Compliance costs brought by this regulation, including inspection charges and time spent on records, could become a burden on these small-scale projects in the industry (Parliament of Australia, 2003).

#### 13.3.5.6 Challenge Six: A Lack of An Effective Waste Tracing System

Some cities have implemented recycling policies, such as landfill ban and compulsory recycling requirements. Behaviours of related stakeholders in the construction industry can directly influence the effectiveness of these policies (Li et al., 2017). The obedience of related stakeholders is largely dependent on an effective supervision system (AECOM Asia Company Limited, 2018). A waste tracing system enables government to trace waste from its origin to its final disposal location. In addition, it could indirectly verify the waste quantity. Specifically, the electronic bill system in Shenzhen (Fig. 13.2) is a good example to follow. In the waste tracing system, responsibilities of related stakeholders in the recycling chain are emphasized. Bills are delivered to contractors from recycling companies or local governments. Contractors should fill these bills with relevant information such as the waste class, appearance, amount, and treatment. When receiving these bills, transportation companies should verify the waste class, waste quantity, and drop-off location. Managers from recycling companies and landfill sites should visually inspect the waste, determine whether it can be lawfully accepted, verify the volume, and sign the bills. To improve the management efficiency, an electronic bill system for data sharing could be developed. Contractors, transportation companies, recycling companies, and operators of landfill sites can log in to the system and record the waste details. In Shanghai's waste tracing systems (Fig. 13.1), waste holders have to make a payment in advance to transportation and recycling companies, to ensure the proper treatment of waste and prevent illegal dumping. Local governments can assess advantages and disadvantages of these systems and formulate a suitable C&D waste tracing system.

## 13.4 Conclusion

Recycling as an alternative approach to deal C&D waste could reduce dependence on landfills. However, the relatively low recovery rate of C&D waste indicates that certain challenges might prevent the successful implementation of recycling policies in practice in China. Therefore, identification of challenges is essential in the development of C&D management. Although C&D waste management in China has received increasing attention from researchers, the focus of past studies was on one specific city. Overall review of potential challenges in C&D waste management, particularly in promotion of recycling in China, is still inadequate.

This chapter adopted site visits to five different recycling plants and interviews with 12 industry practitioners to examine potential challenges in the current C&D waste management in China. Six challenges were identified: (1) unstable source of C&D waste for recycling, (2) absence of subsidies for recycling activities and high cost for land use, (3) absence of regulations on on-site sorting, (4) unregulated landfill activities, (5) a lack of accurate estimation of waste quantity and distribution, and (6) a lack of an effective waste tracing system. Recommendations to address these challenges were presented.

However, two challenges, i.e., a lack of higher value uses of recycled aggregate and the high cost for recycling, which were frequently discussed in past literatures, were not emphasized in this study. Investigating a higher value use of recycled aggregate could widen application of recycled aggregate and foster market, but most interviewees stated that market of recycled products was not their primary concern, because the demand for recycled aggregate or bricks has already surplus the supply. In addition, recycling is prohibitively expensive, because of rigorous environmental assessment approval, land acquisition and equipment purchase. Measures to alleviate their economic burden, including provision of subsidies, and lowering rent price, have been discussed in previous chapters.

This research has some limitations. Firstly, there might exist sampling biases. This study conducted site visits in five different Chinese cities which might be the most prosperous cities across different regions of China. However, this study has a relatively small sample size, which may not be representative of the large population. Secondly, interview was the main approach to obtain information, which might be subjective. The participants might provide responses based on what they think the interviewer wants to hear or the company wants them to talk about. Therefore, participants might give biased comments on certain questions.

The results of this study present current C&D waste management in China and provide a useful reference for researchers who are interested in C&D waste recycling industry. Some challenges, including 'unstable source of C&D waste' and 'a lack of an effective waste tracing system' which were not emphasized in previous studies, should receive additional attentions. In addition, recommendations raised in this study could help policy makers in China to improve the performance of C&D waste recycling.

**Acknowledgments** The authors wish to acknowledge the financial support from the Australian Research Council (ARC), Australian Government (No: DP190100559, DP200100057 and IH1501000006).

This article is from a reprint article in Waste Management, Vol 118, Page 610-625, 2020.

## References

Aecom Asia Company Limited. (2018). *People's Republic of China: Construction and demolition waste management and recycling [Online]*. The PRC Ministry of Housing and Urban-Rural Development and the Asian Development Bank. Available: https://www.adb.org/sites/default/files/project-documents/48105/48105-001-tacr-en.pdf

Akhtar, A., & Sarmah, A. K. (2018). Construction and demolition waste generation and properties of recycled aggregate concrete: A global perspective. *Journal of Cleaner Production, 186*, 262–281. https://doi.org/10.1016/j.jclepro.2018.03.085

Al-Otaibi, A., Bowan, P. A., Abdel Daiem, M. M., Said, N., Ebohon, J. O., Alabdullatief, A., Al-Enazi, E., & Watts, G. (2022). Identifying the barriers to sustainable management of construction and demolition waste in developed and developing countries. *Sustainability, 14*, 7532. https://doi.org/10.3390/su14137532

Au, L. S., Ahn, S., & Kim, T. W. (2018). System dynamic analysis of impacts of government charges on disposal of construction and demolition waste: A Hong Kong case study. *Sustainability, 10*, 1077. https://doi.org/10.3390/su10041077

Bao, Z., & Lu, W. (2020). Developing efficient circularity for construction and demolition waste management in fast emerging economies: Lessons learned from Shenzhen, China. *Science of the Total Environment, 724*, 138264. https://doi.org/10.1016/j.scitotenv.2020.138264

Bao, Z., Lu, W., Chi, B., Yuan, H., & Hao, J. (2019). Procurement innovation for a circular economy of construction and demolition waste: Lessons learnt from Suzhou, China. *Waste Management, 99*, 12–21. https://doi.org/10.1016/j.wasman.2019.08.031

Bao, Z., Lee, W. M. W., & Lu, W. (2020). Implementing on-site construction waste recycling in Hong Kong: Barriers and facilitators. *The Science of the Total Environment, 747*, 141091–141091. https://doi.org/10.1016/j.scitotenv.2020.141091

Benachio, G. L. F., Freitas, M. D. C. D., & Tavares, S. F. (2020). Circular economy in the construction industry: A systematic literature review. *Journal of Cleaner Production, 260*, 121046. https://doi.org/10.1016/j.jclepro.2020.121046

Blengini, G. A., & Garbarino, E. (2010). Resources and waste management in Turin (Italy): The role of recycled aggregates in the sustainable supply mix. *Journal of Cleaner Production, 18*, 1021–1030. https://doi.org/10.1016/j.jclepro.2010.01.027

Bureau of Housing and Urban-Rural Development of Changsha. (2019). Construction and demolition waste management in Changsha. In Bureau of Housing and Urban-Rural Development of Changsha (Ed.). Changsha, Hunan Province.

Campitelli, A., Kannengießer, J., & Schebek, L. (2022). Approach to assess the performance of waste management systems towards a circular economy: Waste management system development stage concept (WMS-DSC). *MethodsX, 9*, 101634–101634. https://doi.org/10.1016/j.mex.2022.101634

Changsha Municipal Government. (2018). Management on construction and demolition waste in Changsha. In Changsha Municipal Government (Ed.). Changsha, China.

Cheng, B., Huang, J., Li, J., Chen, S., & Chen, H. (2022). Improving contractors' participation of resource utilization in construction and demolition waste through government incentives and punishments. *Environmental Management, 70*, 666–680. https://doi.org/10.1007/s00267-022-01617-8

Deloitte. (2015). *Construction and demolition waste management in Germany*. European Commission.

Dimitriou, G., Savva, P., & Petrou, M. F. (2018). Enhancing mechanical and durability properties of recycled aggregate concrete. *Construction and Building Materials, 158*, 228–235. https://doi.org/10.1016/j.conbuildmat.2017.09.137

Ding, T., & Xiao, J. (2014). Estimation of building-related construction and demolition waste in Shanghai. *Waste Management, 34*, 2327–2334. https://doi.org/10.1016/j.wasman.2014.07.029

Ding, Z., Cao, X., Wang, Y., Wu, H., Zuo, J., & Zillante, G. (2022). Cost-benefit analysis of demolition waste management via agent-based modelling: A case study in Shenzhen. *Waste Management (Elmsford), 137*, 169–178. https://doi.org/10.1016/j.wasman.2021.10.036

Du, L., Feng, Y., Lu, W., Kong, L., & Yang, Z. (2020). Evolutionary game analysis of stakeholders' decision-making behaviours in construction and demolition waste management. *Environmental Impact Assessment Review, 84*, 106408. https://doi.org/10.1016/j.eiar.2020.106408

Duan, H., Wang, J., & Huang, Q. (2015). Encouraging the environmentally sound management of C&D waste in China: An integrative review and research agenda. *Renewable and Sustainable Energy Reviews, 43*, 611–620. https://doi.org/10.1016/j.rser.2014.11.069

Duan, H., Miller, T. R., Liu, G., & Tam, V. W. Y. (2019). Construction debris becomes growing concern of growing cities. *Waste Management, 83*, 1–5. https://doi.org/10.1016/j.wasman.2018.10.044

Ghisellini, P., Ji, X., Liu, G., & Ulgiati, S. (2018). Evaluating the transition towards cleaner production in the construction and demolition sector of China: A review. *Journal of Cleaner Production, 195*, 418–434. https://doi.org/10.1016/j.jclepro.2018.05.084

Guo, H., Shi, C., Guan, X., Zhu, J., Ding, Y., Ling, T.-C., Zhang, H., & Wang, Y. (2018). Durability of recycled aggregate concrete – a review. *Cement and Concrete Composites, 89*, 251–259. https://doi.org/10.1016/j.cemconcomp.2018.03.008

Guo, F., Wang, J., & Song, Y. (2022). How to promote sustainable development of construction and demolition waste recycling systems: Production subsidies or consumption subsidies? *Sustainable Production and Consumption, 32*, 407–423. https://doi.org/10.1016/j.spc.2022.05.002

Hossain, M. U., Ng, S. T., Antwi-Afari, P., & Amor, B. (2020). Circular economy and the construction industry: Existing trends, challenges and prospective framework for sustainable construction. *Renewable & Sustainable Energy Reviews, 130*, 109948. https://doi.org/10.1016/j.rser.2020.109948

Huang, B., Wang, X., Kua, H., Geng, Y., Bleischwitz, R., & Ren, J. (2018). Construction and demolition waste management in China through the 3R principle. *Resources, Conservation and Recycling, 129*, 36–44. https://doi.org/10.1016/j.resconrec.2017.09.029

Islam, R., Nazifa, T. H., Yuniarto, A., Shanawaz Uddin, A. S. M., Salmiati, S., & Shahid, S. (2019). An empirical study of construction and demolition waste generation and implication of recycling. *Waste Management, 95*, 10–21. https://doi.org/10.1016/j.wasman.2019.05.049

Jin, R., Li, B., Zhou, T., Wanatowski, D., & Piroozfar, P. (2017). An empirical study of perceptions towards construction and demolition waste recycling and reuse in China. *Resources, Conservation and Recycling, 126*, 86–98. https://doi.org/10.1016/j.resconrec.2017.07.034

Li, Y., Zhou, H., Su, L., Hou, H., & Dang, L. (2017). Investigation into the application of construction and demolition waste in urban roads. *Advances in Materials Science and Engineering, 2017*. https://doi.org/10.1155/2017/9510212

Li, J., Yao, Y., Zuo, J., & Li, J. (2020a). Key policies to the development of construction and demolition waste recycling industry in China. *Waste Management, 108*, 137–143. https://doi.org/10.1016/j.wasman.2020.04.016

Li, J., Zuo, J., Jiang, W., Zhong, X., Li, J., & Pan, Y. (2020b). Policy instrument choice for construction and demolition waste management: The case study of Shenzhen, China. *Engineering Construction and Architectural Management, 27*, 1283–1297. https://doi.org/10.1108/ECAM-11-2019-0632

Li, J., Zuo, J., Wang, G., He, G., & Tam, V. W. Y. (2020c). Stakeholders' willingness to pay for the new construction and demolition waste landfill charge scheme in Shenzhen: A contingent valuation approach. *Sustainable Cities and Society, 52*, 101663. https://doi.org/10.1016/j.scs.2019.101663

Li, Y., Li, M., & Sang, P. (2022). A bibliometric review of studies on construction and demolition waste management by using CiteSpace. *Energy and Buildings, 258*, 111822. https://doi.org/10.1016/j.enbuild.2021.111822

Lin, Z., Xie, Q., Feng, Y., Zhang, P., & Yao, P. (2020). Towards a robust facility location model for construction and demolition waste transfer stations under uncertain environment: The case of Chongqing. *Waste Management, 105*, 73–83. https://doi.org/10.1016/j.wasman.2020.01.037

Liu, Y., & Lu, Y. (2015). The economic impact of different carbon tax revenue recycling schemes in China: A model-based scenario analysis. *Applied Energy, 141*, 96–105. https://doi.org/10.1016/j.apenergy.2014.12.032

Liu, J., Nie, J., & Yuan, H. (2020). Interactive decisions of the waste producer and the recycler in construction waste recycling. *Journal of Cleaner Production, 256*, 120403. https://doi.org/10.1016/j.jclepro.2020.120403

Lockrey, S., Nguyen, H., Crossin, E., & Verghese, K. (2016). Recycling the construction and demolition waste in Vietnam: Opportunities and challenges in practice. *Journal of Cleaner Production, 133*, 757–766. https://doi.org/10.1016/j.jclepro.2016.05.175

Lu, W., Webster, C., Peng, Y., Chen, X., & Zhang, X. (2017). Estimating and calibrating the amount of building-related construction and demolition waste in urban China. *International Journal of Construction Management, 17*, 13–24. https://doi.org/10.1080/15623599.2016.1166548

Lu, W., Du, L., & Feng, Y. (2022). Decision making behaviours and management mechanisms for construction and demolition waste recycling based on public–private partnership. *Environmental Science and Pollution Research International, 29*, 82078–82097. https://doi.org/10.1007/s11356-022-21221-x

Mahpour, A. (2018). Prioritizing barriers to adopt circular economy in construction and demolition waste management. *Resources, Conservation and Recycling, 134*, 216–227. https://doi.org/10.1016/j.resconrec.2018.01.026

Menegaki, M., & Damigos, D. (2018). A review on current situation and challenges of construction and demolition waste management. *Current Opinion in Green and Sustainable Chemistry, 13*, 8–15. https://doi.org/10.1016/j.cogsc.2018.02.010

Ministry of Housing and Urban-Rural Development of the People's Republic of China. (2005). Regulation on construction and demolition waste management. In Minister of Housing and Urban-Rural Development of the People's Republic of China (Ed.). China.

Ministry of Housing and Urban-Rural Development of the People's Republic of China. (2018). Experimental

programs of construction waste management in selected provinces. In Ministry of Housing and Urban-Rural Development of the People's Republic of China (Ed.).
Ministry of Land Infrastructure Transport and Tourism (JAPAN). (2018). Survey on construction and demolition waste in 2018. In Ministry of Land Infrastructure Transport and Tourism (Japan) (Ed.). Japan.
Nanchang Municipal Ecology and Environment Bureau. (2020). *It is imperative to promptly initiate the regulations for managing C&D waste*. Nanchang Municipal Ecology and Environment Bureau.
Pan, X., Xie, Q., & Feng, Y. (2020). Designing recycling networks for construction and demolition waste based on reserve logistics research field. *Journal of Cleaner Production, 260*, 120841. https://doi.org/10.1016/j.jclepro.2020.120841
Parliament of Australia. (2003). Reducing the burden of regulation. In Employment, S. S. C. O. E. A. (Ed.).
Pires, A., & Martinho, G. (2019). Waste hierarchy index for circular economy in waste management. *Waste Management, 95*, 298–305. https://doi.org/10.1016/j.wasman.2019.06.014
Qiu, L. D., & Wang, S. (2011). BOT projects: Incentives and efficiency. *Journal of Development Economics, 94*, 127–138. https://doi.org/10.1016/j.jdeveco.2009.11.008
Shanghai Municipal People's Government. (2017). Management on construction and demolition waste in Shanghai. In Shanghai Municipal People's Government (Ed.). Shanghai, China.
Shi, Q., Ren, H., Ma, X., & Xiao, Y. (2019). Site selection of construction waste recycling plant. *Journal of Cleaner Production, 227*, 532. https://doi.org/10.1016/j.jclepro.2019.04.252
Shooshtarian, S., Caldera, S., Maqsood, T., Ryley, T., Wong, P. S. P., & Zaman, A. (2022). Analysis of factors influencing the creation and stimulation of the Australian market for recycled construction and demolition waste products. *Sustainable Production and Consumption, 34*, 163–176. https://doi.org/10.1016/j.spc.2022.09.005
Song, J., Jin, L., Zhao, Y., & Hu, W. (2017). Using bargaining-game model to negotiate compensation for the early termination of BOT highway projects. *Transportation Research Part A, 105*, 197–209. https://doi.org/10.1016/j.tra.2017.06.017
State Taxation Administration. (2015). Catalogue of value-added tax deduction for comprehensive utilization of resources. In Administration, S. T. (Ed.). China.
Tam, V. W. Y., Kotrayothar, D., & Loo, Y.-C. (2009). On the prevailing construction waste pecycling practices: A south east Queensland study. *Waste Management & Research, 27*, 167–174. https://doi.org/10.1177/0734242X08091864
Ulubeyli, S., Kazaz, A., & Arslan, V. (2017). Construction and demolition waste recycling plants revisited: Management issues. *Procedia Engineering, 172*, 1190–1197. https://doi.org/10.1016/j.proeng.2017.02.139
Wafula, F., Abuya, T., Amin, A., & Goodman, C. (2014). The policy-practice gap: describing discordances between regulation on paper and real-life practices among specialized drug shops in Kenya. *BMC Health Services Research, 14*, 394. https://doi.org/10.1186/1472-6963-14-394
Wang, J., Yuan, H., Kang, X., & Lu, W. (2010). Critical success factors for on-site sorting of construction waste: A china study. *Resources, Conservation and Recycling, 54*, 931–936. https://doi.org/10.1016/j.resconrec.2010.01.012
Wang, D., Qi, D., Mao, H., Xu, X., Xue, C., & Hu, M. (2013). Comparison of construction waste management in Chongqing and Shenzhen. *Environmental Sanitation Engineering, 21*, 59–62. http://kns.cnki.net//KXReader/Detail?TIMESTAMP=637101133056976250&DBCODE=CJFQ&TABLEName=CJFD2013&FileName=HJWS201303024&RESULT=1&SIGN=RO6zCKgGcoKUXoOVs3kPiotgZSM%3d
Wang, F., Xiong, M., Niu, B., & Zhuo, X. (2018). Impact of government subsidy on BOT contract design: Price, demand, and concession period. *Transportation Research Part B: Methodological, 110*, 137–159. https://doi.org/10.1016/j.trb.2018.02.006
Wang, Z., Li, H., & Yang, X. (2020). Vision-based robotic system for on-site construction and demolition waste sorting and recycling. *Journal of Building Engineering, 32*, 101769. https://doi.org/10.1016/j.jobe.2020.101769
Wu, H., & Yuan, H. (2016). An investigation of demolition waste management: Case of Shenzhen in China. In Y. Wu, S. Zheng, J. Luo, W. Wang, Z. Mo, & L. Shan (Eds.), *Proceedings of the 20th international symposium on advancement of construction management and real estate*.
Wu, H., Zuo, J., Zillante, G., Wang, J., & Yuan, H. (2019). Status quo and future directions of construction and demolition waste research: A critical review. *Journal of Cleaner Production, 240*, 118163. https://doi.org/10.1016/j.jclepro.2019.118163
Wu, Z., Yu, A. T. W., & Poon, C. S. (2020). Promoting effective construction and demolition waste management towards sustainable development: A case study of Hong Kong. *Sustainable Development, 28*, 1713–1724. https://doi.org/10.1002/sd.2119
Xu, B., Chen, S., Tai, J., & Tao, Q. (2012). Management policies of urban construction and demolition waste. *Environmental Sanitation Engineering, 3*, 10.14181/j.cnki.1002-851x.2011.04.025.
Yong, J., Li, M., Gu, B., & Wan, Y. (2016). Chongqing municipal building utilization of landfill management. *Environment and Sustainable Development, 41*, 202–203. https://doi.org/10.19758/j.cnki.issn1673-288x.2016.04.055

Yu, D., Duan, H., Song, Q., Li, X., Zhang, H., Zhang, H., Liu, Y., Shen, W., & Wang, J. (2018). Characterizing the environmental impact of metals in construction and demolition waste. *Environmental Science and Pollution Research, 25*, 13823–13832. https://doi.org/10.1007/s11356-018-1632-z

Yu, S., Awasthi, A. K., Ma, W., Wen, M., Di Sarno, L., Wen, C., & Hao, J. L. (2022). In support of circular economy to evaluate the effects of policies of construction and demolition waste management in three key cities in Yangtze River Delta. *Sustainable Chemistry and Pharmacy, 26*, 100625. https://doi.org/10.1016/j.scp.2022.100625

Yuan, H. (2017). Barriers and countermeasures for managing construction and demolition waste: A case of Shenzhen in China. *Journal of Cleaner Production, 157*, 84–93. https://doi.org/10.1016/j.jclepro.2017.04.137

Yuan, H., Shen, L., & Wang, J. (2011). Major obstacles to improving the performance of waste management in China's construction industry. *Facilities, 29*, 224–242. https://doi.org/10.1108/02632771111120538

Yuan, H., Wang, Z., Shi, Y., & Hao, J. (2022). A dissipative structure theory-based investigation of a construction and demolition waste minimization system in China. *Journal of Environmental Planning and Management, 65*, 514–535. https://doi.org/10.1080/09640568.2021.1889484

Zhang, Y. (2009). China's economic zones: A regional round-up. *The Economic Observer*.

Zhang, L., Chen, Y., & He, Z. (2017). The effect of investment tax incentives: Evidence from China's value-added tax reform. *International Tax and Public Finance, 25*, 913–945. https://doi.org/10.1007/s10797-017-9475-y

Zhao, W., Leeftink, R. B., & Rotter, V. S. (2010). Evaluation of the economic feasibility for the recycling of construction and demolition waste in China—The case of Chongqing. *Resources, Conservation and Recycling, 54*, 377–389. https://doi.org/10.1016/j.resconrec.2009.09.003

Zhao, W., Ren, H., & Rotter, V. S. (2011). A system dynamics model for evaluating the alternative of type in construction and demolition waste recycling center – The case of Chongqing, China. *Resources, Conservation and Recycling, 55*, 933–944. https://doi.org/10.1016/j.resconrec.2011.04.011

# 14 Architects' Behaviour Towards Implementing Construction and Demolition Waste Minimisation Strategies: The Case of Australia

Kamyar Kabirifar, Mohammad Mojtahedi, and Cynthia Changxin Wang

## 14.1 Introduction

The construction industry generates a plethora of employment, shapes the built environment, and contributes remarkably to economic growth; however, it is inherently environmentally unfriendly, particularly considering its direct consequence, construction and demolition waste (CDW) (Bao & Lu, 2020). Around ten billion tonnes of CDW are generated annually (Wu et al., 2019; Yazdani et al., 2021), throughout the construction, renovation, and demolition activities (Kabirifar et al., 2020a; Menegaki & Damigos, 2018). Similarly, over 27 million tonnes of CDW was produced in Australia in 2018–2019 of which one-third was disposed of (Kabirifar et al., 2021b), which is a significant source of environmental pollution, economic loss, and excessive raw material extraction (Goel et al., 2021; Kong & Ma, 2020). For instance, each ton of landfilled CDW produces almost 90 Kilograms of potential global warming emissions (Ding et al., 2016; Kabirifar et al., 2020b). Furthermore, global resource extraction has risen from 7 gigatonnes in 1900 to 89 gigatonnes in 2015 (Zhang et al., 2022). Therefore, careful attention should be paid to construction and demolition waste management (CDWM) (Aslam et al., 2020).

Various factors, such as waste policies and regulations (e.g., landfill levy) (Wu et al., 2020), economic factors (e.g., reusing, recycling, recovering, and disposal costs) (Ghafourian et al., 2021; Marrero et al., 2017; Shivaraj et al., 2018), social factors (e.g., public health) (Blaisi, 2019), technical and technological factors (e.g., advanced recycling technologies) (Huang et al., 2018), etc. have been identified pivotal in managing CDW. However, until recent years, the critical role of human resources, mostly referred to their attitudes and behaviour towards waste minimisation, has received less level of attention (Kabirifar et al., 2021a; Li et al., 2018; Udawatta et al., 2015). Governments, for instance, are more concerned with the enforcement of policies and regulations (Du et al., 2020; Su, 2020), whereas the clients pay significant attention to economic matters (Ghaffar et al., 2020), and the consultants are more neutral or passive about waste management as they believe waste management is the contractors' duty during construction and/or demolition stages (Kabirifar et al., 2021b).

Unlike other contributing factors to CDWM, the role of stakeholders' attitudes and behaviour in CDW minimisation are complex in nature; thus, they should be explained with the support

K. Kabirifar (✉) · M. Mojtahedi · C. C. Wang
School of Built Environment, University of New South Wales, Sydney, Australia
e-mail: kamyar.kabirifar@unsw.edu.au; m.mojtahedi@unsw.edu.au; cynthia.wang@unsw.edu.au



J. Zuo et al. (eds.), *Circular Economy for Buildings and Infrastructure*, Sustainable Development Goals Series, https://doi.org/10.1007/978-3-031-56241-9_14

of the robust and relevant theoretical grounding (Salvia et al., 2021). In this regard, a well-recognised social psychological theory, the theory of planned behaviour (TPB), was initially employed by Teo and Loosemore (2001) to explore the influence of contractors' attitudes on reducing CDW. The TPB was later applied to investigate individuals' behaviour in reducing, reusing, and recycling CDW (Begum et al., 2009), to assess the recycling behaviour of CDW stakeholders (Jain et al., 2020; Mak et al., 2019), and to study the waste reduction behaviour of contractors, project managers and construction workers (Li et al., 2018; Liu et al., 2017; Wu et al., 2017; Yang et al., 2020; Yuan et al., 2018). However, it should be noted that these studies have mainly focused on the attitudes and behaviour of construction site workforce during construction and/or demolition stages, while less attention has been given to the role of architects and designers in minimising CDW.

It is crucial to consider that almost one-third of CDW can be avoided during the planning and design phase if waste minimisation measures, such as standard dimensions and units, detailed design, dimensional coordination, low-waste technologies' application, etc. are adopted by the architects (Gupta et al., 2020; Li et al., 2018; Osmani et al., 2006). Thus, architects play an essential role in CDW minimisation in the construction industry (Kim Soo et al., 2020). Nevertheless, the number of studies focusing on the architects' perception and their practical waste minimisation strategies during the planning and design phase is limited (e.g., Li et al., 2015). Also, the existing studies lack in presenting a sound theoretical grounding based on the TPB, which can assess the attitudes and behaviour of architects thoroughly, as the TPB has been criticised for failing to sufficiently explain certain environmental behaviours, and other variables (e.g., moral norms) should be included in the model (Li et al., 2018). Thus, this research aims at establishing a conceptual model that empirically investigates the architects' predictors who apply waste minimisation measures through the planning and design phase of construction projects and their impacts on waste minimisation behaviour.

## 14.2 Literature Review

### 14.2.1 The Role of Architects in Construction and Demolition Waste Management

Architects have different attitudes towards implementing waste minimisation strategies in the planning and design phase. For instance, while some studies have addressed the positive attitudes of architects towards implementing waste minimisation strategies (Wang et al., 2019), others have proven that architects generally believe that waste is produced during construction activities and its contractors' duty to manage it (Udawatta et al., 2015). Moreover, the architects' attitudes towards implementing waste minimisation strategies might be affected by other stakeholders, i.e., clients, who are reluctant to undertake waste minimisation strategies by themselves or ask architects to consider these strategies in their design (Kim Soo et al., 2020). In addition, a lack of formal training, knowledge, and awareness, along with the absence of frameworks, standards and systems have contributed to the inefficiency of architects in implementing waste minimisation strategies (Li et al., 2015). Thus, in order to analyse the architects' real behaviour (RB) in implementing waste minimisation strategies, the architects' impact on CDWM should primarily be identified. Then, a robust theoretical model is required to predict and explain the waste minimisation behaviour of the architects.

Related literature indicates that in the planning and design phase, a significant volume of CDW can be avoided by considering various measures from architects, such as i) precise implementation of waste management regulations (e.g., CDW sorting and collection techniques) (Nußholz et al., 2019; Parisi Kern et al., 2015), and plans (e.g., the waste management report in the preconstruction or the construction phases) (Jiménez-Rivero et al., 2017), ii) design for waste prevention (e.g., size, type, and quantity of materials can accurately be estimated by the application of BIM technology) (Akanbi et al., 2018; Esa et al., 2017; Li et al., 2018; López Ruiz et al., 2020), iii) building standards (e.g., sustainable green building standards, i.e., LEED) (Yeheyis et al., 2013), and iv) design for disassembly/deconstruction (e.g., by the application of prefabricated components) (Hossain et al., 2020; Jaillon & Poon, 2014; López Ruiz et al., 2020). Table 14.1 represents the main measures and detailed descriptions of architects' impact on CDWM.

### 14.2.2 Theoretical Grounding

Human is considered the most important resource in construction projects considering their capacity to link all other resources in order to achieve the final project objectives (Zeng et al., 2015). In this regard, various responsibilities, roles, concerns, and goals associated with human resources result in different attitudes and behaviours toward CDWM (Teo et al., 2000). For instance, Udawatta et al. (2015) investigated the attitudinal

**Table 14.1** Main measures and detailed descriptions of architects' impact on CDWM

| Code | Measures | description | References |
|---|---|---|---|
| RB 1 | Waste regulations and plans implemented in the planning and design phase | 1. Standard waste collection and sorting techniques (e.g., rubbish chute for construction waste in high-rise buildings)<br>2. Standard waste bins (e.g., size and availability)<br>3. Improvement in collaboration and communication among stakeholders (e.g., regular meetings and training)<br>4. Proper handling and storage of materials | Jiménez-Rivero et al. (2017), Nußholz et al. (2019), Ordoñez and Rahe (2013), Parisi Kern et al. (2015), and Singh and Ordoñez (2016) |
| RB 2 | Design for waste prevention | 1. Optimum size design of buildings with the help of prefabrication technology<br>2. Accurate estimation of the volume and type of construction materials with the help of BIM technology<br>3. Proper material selection, by focusing on materials that could extend raw material conservation and reduce $CO_2$ emissions<br>4. Having a thorough knowledge of the material market concerning their reusability, recyclability, and recoverability<br>5. Having a thorough knowledge about the availability of recycling facilities and the materials within which can be recycled or recovered<br>6. Set a target for allowable wastage | Akanbi et al. (2018), De Silva and Vithana (2008), Dumlao-Tan and Halog (2017), Esa et al. (2017), Huang et al. (2013), Lachimpadi et al. (2012), Li et al. (2018), López Ruiz et al. (2020), Low et al. (2014), Poon et al. (2004), Urio and Brent (2006), Yean Yng Ling and Song Anh Nguyen (2013), and Zhang et al. (2012) |
| RB 3 | Building standards | Sustainable green buildings standards, e.g., LEED | Yeheyis et al. (2013) |
| RB 4 | Design for disassembly or deconstruction | 1. Prefabricated components and modular design allow easy collection, sorting and thereafter, reuse, recycling or recovering CDW<br>2. Deconstruction or sequential demolition technology | Gangolells et al. (2014), Hossain et al. (2020), Jaillon and Poon (2014), Kerdlap et al. (2019), and López Ruiz et al. (2020) |

and behavioural impacts of different stakeholders towards CDWM and came up with the contractors who are more interested in financial incentives compared to the developers who are more price driven. However, as stated, the attitudes and behaviour of stakeholders in CDWM are complex and need to be explained with the help of potent theoretical grounding. In this regard, there are more than eighty various theories of behaviour and behavioural change in the fields of sociology, psychology, economics, and anthropology (Darnton, 2008; Davis et al., 2015). Nevertheless, when it comes to environmental behaviours, the application of theory-based behavioural intervention is limited (Parajuly et al., 2020). The theory of planned behaviour (TPB), derived from the theory of reasoned action (TRA) (Ajzen, 1985), is one of the most extensively used psychological theories in environmental behaviour research to predict the stakeholders' attitudes and behaviour regarding CDWM (e.g., Jain et al., 2020).

Regarding an individual's "intention" to conduct a specific behaviour in the centre, the TPB proposes that such intentions can be anticipated from "attitudes" towards the "behaviour" as well as "subjective norms" and "perceived behavioural control," and that actual behaviour is the consequence of these intentions coupled with the individual's perception of behavioural control (Ajzen, 1991). Despite widespread acceptance of the TPB, some researchers have argued that more variables should be incorporated into the model to improve its explanatory power (Jain et al., 2020; Li et al., 2018). For instance, Turaga et al. (2010) stated that environmental behaviours are not limited to rational as morality also plays an important role in predicting stakeholders' environmental behaviour. In this regard, personal norms refer to an individual's specific environmental behaviours, which are formed based on their moral interests, responsibilities, and obligations (Sia & Jose, 2019; Wang et al., 2020). Furthermore, in the research conducted by Bortoleto et al. (2012), the personal norm was shown to be one of the key predictors of waste avoidance behaviour. In this regard, the personal norm is considered the primary element of the value-belief-norm (VBN) theory by which environmental behaviours are stimulated (Sia & Jose, 2019). Stern (2000) developed the value-belief-norms (VBN) theory, which was originally developed from Schwartzs' norm activation model (NAM) (Schwartz & Ben David, 1976), to explain environmental behaviour from the viewpoints of self-interest, responsibility, and obligation. It is perhaps the most prevalent theory in studying environmental behaviour (Bronfman et al., 2015; Janmaimool & Denpaiboon, 2016).

It has also been proven that an individual's intention is influenced by the views of others, particularly in the context of an organisation, where the philosophy and aims of the organisation have a substantial impact on employees' intentions (Li et al., 2018). Regarding the organisational influence on stakeholders' behaviour (Herold, 2018), different theories, such as legitimacy theory, stakeholder theory, and institutional theory, have been applied to investigate and predict the relationship between the organisation and its shareholders (Fernando & Lawrence, 2014). For instance, stakeholder theory has been applied to explain why companies intend to practice sustainability (Meixell & Luoma, 2015; Yuen et al., 2017). Also, legitimacy theory has been used to predict the social-environmental performance of organisations (Crossley et al., 2021). However, since the institutional theory tends to take a wide macro perspective to explain why an organisation chooses a certain pro-environmental action, it is favoured above other theories (Fernando & Lawrence, 2014). In the context of CDWM, the institutional theory may aid in understanding and explaining the many forms of pressures organisations exert on their employees (e.g., social, political, etc.). Based on an institution's regulations, rules, and cultural cognition, institutional pressure can be divided into three aspects of coercive pressure, normative pressure, and mimetic pressure (Jain et al., 2020; Martínez-Ferrero & García-Sánchez, 2017; Zeng et al., 2017).

## 14.3 Conceptual Framework and Hypotheses Development

Figure 14.1 depicts the suggested model for assessing architects' real behaviour while applying waste minimisation measures based on the literature review. In addition to the TPB's broad framework, the suggested model takes into consideration VBN and Institutional theories. The hypothesised relationships among the variables in the model are discussed below.

$H_1$: Architects' attitude toward CDW minimisation has a significant effect on their behavioural intention to implement CDW minimisation strategies

According to the TPB, an individual with a favourable attitude towards a certain behaviour is more likely to conduct that behaviour (Ajzen, 1991). The positive attitude of individuals has been proven effective in their behavioural intention towards implementing CDW minimisation strategies i.e., contractors' attitudes towards reducing CDW in the construction phase (Begum et al., 2009; Ding et al., 2016), or architects' attitudes in implementing waste minimisation strategies in the planning and design phase (Li et al., 2015). On the other hand, some studies have also assessed the negative attitude of contractors towards CDWM which has led to the overproduction of CDW (Li et al., 2018). Thus, consistent with the literature $H_1$ is proposed to investigate whether architects' attitude has a significant effect on their behavioural intention towards implementing waste minimisation strategies.

$H_2$: Architects' subjective norm has a significant effect on their behavioural intention towards implementing waste minimisation strategies

According to the TPB, the subjective norm is a social factor that refers to an individual's perceived social pressure to conduct or refrain from performing a certain behaviour, which performs as a predictor of specific behaviour via behavioural intention (Ajzen, 1991). In this regard, Li et al. (2015) asserted that the degree of approval/disapproval from key stakeholders in the organisation i.e., the clients, managers and colleagues could motivate/demotivate architects to adopt waste minimisation strategies. In this regard, the power and interest of key stakeholders have a decisive role in forming architects' subjective norms on their behavioural intention to perform waste minimisation strategies (Li et al.,

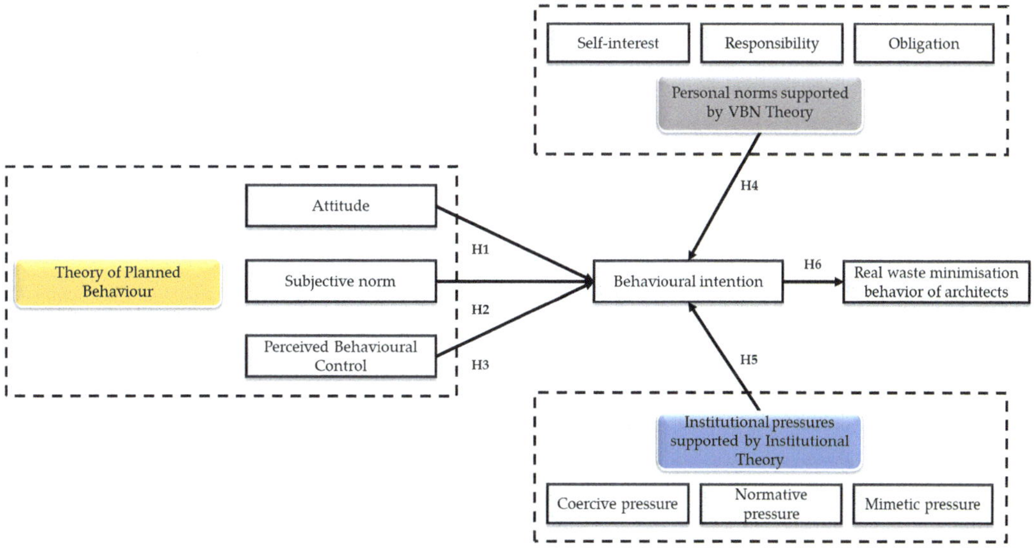


**Fig. 14.1** The conceptual framework for architects' real behaviour in implementing waste minimisation strategies

2021). Thus, consistent with the literature $H_2$ is proposed to investigate whether subjective norm has a significant effect on architects' behavioural intention towards implementing waste minimisation strategies.

$H_3$: Architects' perceived behavioural control has a significant effect on their behavioural intention towards implementing waste minimisation strategies

Perceived behavioural control refers to the perceived ease or difficulty (ability) to perform a certain behaviour (Ajzen, 1991). Perceived behavioural control accounts for individuals' constraints (e.g., control, capability, experience, and opportunity) that limit people's ability to convert their intention into action (Rauch & Hulsink, 2015). In the context of CDWM, Begum et al. (2009) emphasised lack of experience as the critical factor in the contractors' behaviour in waste management. Similarly, Li et al. (2015) pointed out capability constraints as determinants of designers' perceived behavioural in implementing waste minimisation behaviour. Thus, consistent with the literature $H_3$ is proposed to investigate whether architects' perceived behavioural control has a significant effect on their behavioural intention towards implementing waste minimisation strategies.

$H_4$: Architects' personal norms have a significant effect on their behavioural intention towards implementing waste minimisation strategies

As discussed earlier, the TPB was chastised for ignoring moral issues, particularly in morally relevant domains (Turaga et al., 2010). Personal norm, described as a firmly internalised moral attitude, is an essential driver of environmental behaviour. It differs from the subjective norm in that the consequences of violating or supporting it are linked to one's self-concept (Li et al., 2018). The application of personal norm in CDWM is supported by the value-belief-norm (VBN) theory, which explains behaviour as a predictor of environmental protection from the perspectives of self-interest, responsibility, and obligation, which are reflected in the form of commitment to perform a certain behaviour (Janmaimool & Denpaiboon, 2016). Commitment has been proven effective in explaining behavioural intention towards CDWM in the work of some scholars (Bakshan et al., 2017; Ramayah et al., 2012). Thus, consistent with the literature $H_4$ is proposed to investigate whether architects' personal norms have a significant effect on their behavioural intention towards implementing waste minimisation strategies.

$H_5$: Institutional pressures have a significant effect on architects' behavioural intention towards implementing waste minimisation strategies

The institutional theory in the domain of CDWM focuses on the active role of individuals within organisations to explain how the pressure on individuals lead them to undertake certain behaviour (Andersson, 2021; Beunen et al., 2017). These pressures are mainly attributed to the rules and regulations that might impact an organisation's adoption of sustainable practices (Jain et al., 2020). In this regard, coercive pressures include government-enforced standards and frameworks, normative pressures are established by various stakeholders, and mimetic pressures are created by competing enterprises (Alshbili & Elamer, 2020; Chithambo et al., 2020).

$H_6$: Behavioural intention has a significant effect on the real waste minimisation behaviour of architects

According to the TPB, behavioural intention, which reflects how motivated an individual is to perform a certain behaviour, is the immediate antecedent of the behaviour in question (Ajzen, 1991). This theoretical assumption has been supported by findings considering the relationship between behavioural intention and actual behaviour in CDWM (Begum et al., 2009). Since it is almost impossible to measure behavioural intention and the actual behaviour of architects towards implementing waste minimisation strategies, this study has considered four categories of CDWM regulations and plans, design for waste prevention, building standards, and design for deconstruction/disassembly to measure the degree of architects' waste minimisation strategies implementation, which have been proven effective in minimising CDW (Table 14.1). Thus, consistent with the literature $H_6$ is proposed to investigate whether architects' behavioural intention has a significant effect on their real behaviour towards implementing waste minimisation strategies.

## 14.4 Research Methods

### 14.4.1 Questionnaire Preparation and Distribution

In order to assess the architects' attitudes and behaviours towards implementing CDW minimisation strategies in Australia, a questionnaire was designed as it is a proper method of data collection from the target population promptly, providing less biased results when compared to other methods (Babbie, 2015; Creswell & Creswell, 2017; Phillips & Stawarski, 2008). Questionnaire surveys help researchers collect responses from industry experts and generalise the results to the population of interest (Babbie, 2015; Sekaran & Bougie, 2016). In order to avoid systematic bias in the data collection process, which is a major source of concern in research pertinent to human factors, the questionnaire was developed in accordance with the procedural remedies given by Podsakoff et al. (2003).

The questionnaire (Appendix 14.1) contained two main sections: section one included 6 questions on the respondents' demographic characteristics, and section two included 27 close-ended questions in seven sub-sections to seek respondents' opinions about the degree of acceptance of different factors of the architects' attitude and behaviour towards implementing waste minimisation strategies. A five-point Likert Scale was used for questions of part 2, sections 1 to 6, from 1 = strongly disagree to 5 = strongly agree. For part 2, section 7, which aimed at measuring the real behaviour of architects in implementing waste minimisation strategies, a five-point Likert Scale from 1 = never, to 5 = always was used. A five-point Likert scale was utilised in this study since it is a standard and easy measure that motivates respondents in responding to questions (Joshi et al., 2015). Once the questionnaire design was finalised, it was considered by the UNSW HREAP Executive and granted ethical approval No HC220017 on 25 February 2022.

Upon the completion of the questionnaire design, the pilot testing (Babbie, 2015) was performed with the help of three design managers with over 10 years of experience and extensive knowledge of waste management, who were responsible for managing architects, civil, mechanical, and electrical engineers etc. within organisations, as they were the most involved individuals in the design process of construction projects. In the pilot testing process, the structure, clarity of questions, language, and measurement capability within an appropriate time were tested and improvements were made. The questionnaire was then amended and readied for the main distribution. The target population with specific criteria, such as availability, accessibility, and willingness to participate in the study, were secured by adopting the convenience sampling approach. The convenience sampling method ensured that relevant firms, meaningful data, and a similar emphasis on the study purpose were fulfilled (Etikan et al., 2016).

To formulate a convenience sampling method in this study, architects from top architecture firms in Australia were initially identified and approached through their LinkedIn profiles. An official invitation letter and the link to the online questionnaire were sent to 86 architects. Finally, 30 usable and valid responses were received. The data collection process commenced in February 2022 and finished in May 2022. The 35% response rate was considered satisfactory concerning the construction management research (Fellows & Liu, 2008; Teck Heng Lim et al., 2010).

### 14.4.2 Statistical Tool

In this study, Structural Equation Modeling (SEM) was applied because it provides ample information on any modeling error, constructs unobservable latent variables, tests priori theoretical assumptions against the empirical data, and provides models among the multiple predictors and variables (Hair et al., 2013; Hair Joseph et al., 2019; Hair Jr et al., 2017). Partial Least Squares (PLS-SEM) and Covariance-Based (CB-SEM) are the two main types of SEM. This study utilised PLS-SEM as it deals well with small sample sizes, works effectively with complex models, dominates missing data, has strong statistical power, and last but foremost works well with non-normally distributed data through bootstrapping, a nonparametric procedure that allows testing the statistical significance of various PLS-SEM results (Hair Joseph et al., 2019; Hair Jr et al., 2017; Jörg et al., 2012). Furthermore, when theory is less established or there is a concern about validating a conceptual framework from a prediction standpoint, PLS-SEM is more appropriate (Hair Joseph et al., 2019). Therefore, SmartPLS software was used in this study for data analysis.

### 14.4.3 Post-data Collection Analysis

The collected data were checked against four criteria: (1) missing data, (2) suspicious response pattern, (3) outliers, and (4) common method bias (CMB) (Hair Jr et al., 2017; Jordan & Troth, 2019). Since the respondents were prohibited from moving forward to the next question without answering the former question through a forced-answer strategy in the online questionnaire design, no missing data was ensured (Hair Jr et al., 2017). Also, after transferring data from the Qualtrics platform to Microsoft Excel, data were checked against suspicious response patterns (straight or diagonal lining), and no suspicious response pattern was detected (Hair Jr et al., 2017). Besides, stem-and-leaf and box plots analysis using IBM SPSS 25 software were performed to ensure there were no significant outliers (Hair Joseph et al., 2019; Sarstedt & Mooi, 2019). In the last step, Harmon's single factor test using IBM SPSS 25 software was performed to detect Common Method Bias (CMB) as it is a cause of concern in research involving the measurement of behavioural sciences (Podsakoff et al., 2011). In this test, all the latent variables were loaded into one factor and if the total variance of that factor was less than 50% it could be concluded that CMB has not affected the results (Podsakoff et al., 2003, 2011). This was verified as no rotation resulted in other factors rather than the single factor, which had a low variance. Figure 14.2 illustrates the research model in SmartPLS. The blue circles refer to the latent variables, which form the structural model, while the yellow rectangles refer to the indicators and their relationship with their respective latent variables form the measurement model.

### 14.4.4 Post-data Collection Analysis

Both aspects of the research model, the measurement model and the structural model, were assessed in this step (Hair Joseph et al., 2019). The reliability and validity of the measures were assessed in the measurement model assessment. In this respect, indicator reliability, internal consistency, convergent validity (AVE), and discriminant validity were assessed (Hair Joseph et al., 2019; Henseler et al., 2016). Once these four criteria were assessed, the structural model assessment criteria including collinearity, coefficient of determination ($R^2$), effect size ($f^2$), structural model path (t-value), and blindfolding and predictive relevance ($Q^2$) were evaluated (Hair Jr et al., 2017, 2019).

## 14.5 Results

### 14.5.1 Respondents' Background

Over two third of the respondents were male, while 46.67% were between 34 and 41 years old. Also, more than half of the respondents had over 15 years of work experience. Besides, all of the respondents had at least a bachelor's degree. Almost 73% of the respondents were managers and consultants and the majority of them were

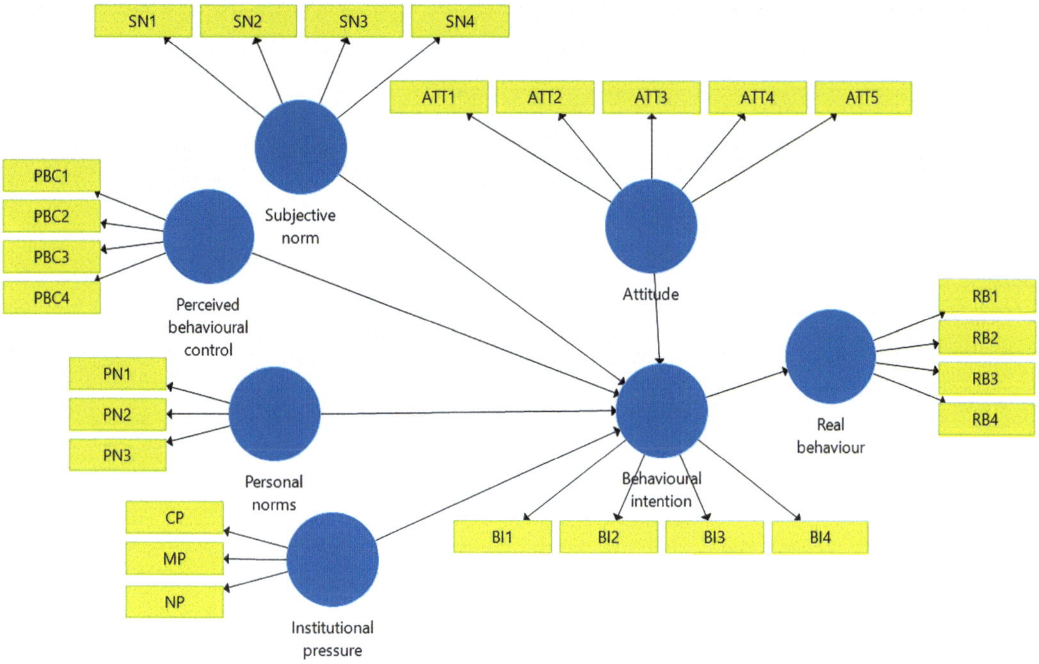


**Fig. 14.2** Research model in SmartPLS

involved in multi-unit residential housing. The respondents' profile is illustrated in Table 14.2.

### 14.5.2 Measurement Model Assessment

Initially, the PLS-SEM algorithm was run in several rounds to reach the accepted threshold of 0.5 and above for indicator reliability (equivalent to 0.708 for outer loadings). It is recommended to remove values below 0.4, whereas values between 0.4 and 0.708 can only be removed if their removal improves the values of internal consistency reliability (Cronbach's alpha and composite reliability), convergent validity (AVE) and discriminant validity (HTMT) (Hair Joseph et al., 2019). In this step, RB1 and RB2 were removed from further analysis as their values were below 0.5 and internal consistency reliability of "Real behaviour" could not be fulfilled without their removal. Table 14.3 illustrates results for outer loadings, Cronbach's alpha, composite reliability, and AVE.

To assess the discriminant validity, Heterotrait–monotrait ratio (HTMT) was obtained from complete bootstrapping with 5000 iterations (Hair Jr et al., 2019; Henseler et al., 2015). It is recommended that values below 0.85 for latent variables are accepted for HTMT assessment (Hair Joseph et al., 2019). Table 14.4 illustrates the results for HTMT.

### 14.5.3 Structural Model Assessment

In the first step of the structural model assessment, VIF values were calculated to assess the collinearity. Since all values were below 3, no collinearity was observed among latent variables (Hair Joseph et al., 2019). Then, the exploratory/predictability power of the research model was assessed by calculating the coefficient of determination ($R^2$). Behavioural intention (BI) was considered substantial as its $R^2$ and $R^2$ adjusted values were 0.727 and 0.670, respectively. However, Real behaviour's (RB) $R^2$ and $R^2$ adjusted with values of 0.209 and 0.181, respectively were considered weak (Hair Joseph et al., 2019). In the

**Table 14.2** Profile of respondents

| | | | | | |
|---|---|---|---|---|---|
| 1. Gender | Male | Female | Non-binary/third gender | Prefer not to say | |
| Percentage | 70% | 30% | 0 | 0 | |
| 2. Age | 18–25 | 26–33 | 34–41 | 42–49 | 50 years and above |
| Percentage | 10% | 26.67% | 46.67% | 16.67% | 0 |
| 3. Work experience (years) | 0–5 | 6–10 | 11–15 | 16–20 | 20 years and above |
| Percentage | 3.33% | 13.33% | 26.67% | 33.33% | 23.33% |
| 4. Highest academic qualification | Certificate | Diploma | Bachelor's degree | Master's degree | PhD |
| Percentage | – | – | 26.67% | 60% | 13.33% |
| 4. Working category | Consultant (Inc. government) | Design manager | Freelance architect | Other | |
| Percentage | 40% | 33.33% | 26.67% | – | |
| 6. Company field | Multi-unit residential | Commercial | Industrial | Infrastructure | Other (Inc. research) |
| Percentage | 46.67% | 26.67% | 13.33% | 10% | 3.33% |

next step, effect size ($f^2$) including direct and indirect effects, to assess whether removing a construct has an effective impact on the endogenous construct was evaluated (Hair Jr et al., 2017). Results indicated that "Attitude" has large impacts on both "Behavioural intention" (directly- $f^2$: 0.807) and "Real behaviour" (indirectly- $f^2$: 0.318), whereas the impacts of "Personal norm" on "Behavioural intention" (directly-$f^2$:0.132) and "Real behaviour" (indirectly-$f^2$:0.052) were medium and small, respectively. Also, the impact of "Behavioural intention" on "Real behaviour" (0.393) was considered large (Cohen, 1992; Hair Joseph et al., 2019). All other effect sizes, direct or indirect, were revealed as negative (no effect size). In the fourth step, the predictive relevance ($Q^2$) of the endogenous constructs, which refers to the predictive suitability of the model, was calculated (Hair Joseph et al., 2019; Hair Jr et al., 2017). Regarding the Omission Distance of D = 7, since both endogenous constructs of BI (0.269) and RB (0.092) were values other than zero, they were considered acceptable (Hair Joseph et al., 2019). In the last step and in order to accept or reject the developed hypotheses, *t*-values were calculated. To do this, two-tailed bootstrapping with a significance level of 0.1 and 5000 iterations were performed through which the *t*-values of 1.65 and above were considered as a threshold for hypothesis acceptance (Hair Joseph et al., 2019; Kabirifar et al., 2021a). Although $H_1$ (*t*:4.315) and $H_6$ (*t*: 2.158) were accepted, other hypotheses were not accepted as their *t*-values were below the accepted threshold (*t*-value: 1.65). Table 14.5 illustrates the results of the structural model assessment.

## 14.6 Discussion

The results revealed that the first hypothesis, architects' attitude toward CDW minimisation has a direct effect on their behavioural intention to implement CDW minimisation strategies, is accepted. Apart from the acceptable *t*-value (Table 14.5), attitudes' large impact on the behavioural intention is another indication that acknowledges this acceptance. It is also worth mentioning that none of the attitudes' indicators was removed in the measurement model assessment, which is another indication of the significance of this factor in explaining architects' real behaviour. This finding is consistent with studies performed globally (Li et al., 2015) or in the Australian context (Zhao et al., 2021) in which the role of architects' and/or designers' attitudes have been proven effective in their behavioural

**Table 14.3** Outer loadings, Cronbach's alpha, composite reliability and, AVE of the research model

| NO | Construct | Indicator | Outer loadings* | CA[a] | CR[b] | AVE[c] |
|---|---|---|---|---|---|---|
| 1 | Attitude | ATT 1 | 0.882 | 0.871 | 0.908 | 0.667 |
| 2 | | ATT 2 | 0.918 | | | |
| 3 | | ATT 3 | 0.653 | | | |
| 4 | | ATT 4 | 0.726 | | | |
| 5 | | ATT 5 | 0.873 | | | |
| 6 | Subjective norm | SN 1 | 0.818 | 0.853 | 0.902 | 0.698 |
| 7 | | SN 2 | 0.736 | | | |
| 8 | | SN 3 | 0.822 | | | |
| 9 | | SN 4 | 0.952 | | | |
| 10 | Perceived behavioural control | PBC 1 | 0.741 | 0.795 | 0.839 | 0.573 |
| 11 | | PBC 2 | 0.895 | | | |
| 12 | | PBC 3 | 0.539 | | | |
| 13 | | PBC 4 | 0.806 | | | |
| 14 | Personal norms | PN 1 | 0.859 | 0.742 | 0.847 | 0.651 |
| 15 | | PN 2 | 0.851 | | | |
| 16 | | PN 3 | 0.701 | | | |
| 17 | Institutional pressure | CP | 0.992 | 0.858 | 0.859 | 0.677 |
| 18 | | NP | 0.783 | | | |
| 19 | | MP | 0.659 | | | |
| 20 | Behavioural intention | BI 1 | 0.864 | 0.778 | 0.859 | 0.605 |
| 21 | | BI 2 | 0.651 | | | |
| 22 | | BI 3 | 0.768 | | | |
| 23 | | BI 4 | 0.813 | | | |
| 24 | Real behaviour | **RB 1** | **0.410** | 0.748 | 0.887 | 0.797 |
| 25 | | **RB 2** | **0.448** | | | |
| 26 | | RB 3 | 0.870 | | | |
| 27 | | RB 4 | 0.915 | | | |

*Acceptance range: X > 0.708
[a]CA: Cronbach's alpha – X > 0.7
[b]CR: Composite reliability – X > 0.6
[c]AVE: Average Variance Extracted – X > 0.5

**Table 14.4** Discriminant validity

| | HTMT (bootstrapping) | | | | | |
|---|---|---|---|---|---|---|
| Constructs | Attitude (AT) | Behavioural intention (BI) | Institutional pressure (IP) | Perceived behavioural control (PBC) | Personal norms (PN) | Real behaviour (RB) |
| Behavioural intention | 0.883 | | | | | |
| Institutional pressure (IP) | 0.396 | 0.351 | | | | |
| Perceived behavioural control (PBC) | 0.631 | 0.463 | 0.353 | | | |
| Personal norms (PN) | 0.604 | 0.594 | 0.349 | 0.427 | | |
| Real behaviour (RB) | 0.375 | 0.575 | 0.333 | 0.458 | 0.565 | |
| Subjective norm (SN) | 0.605 | 0.358 | 0.546 | 0.612 | 0.376 | 0.263 |

**Table 14.5** Results of the structural model assessment

| Hypotheses | Structural path | VIF | $R^2$ and $R^2$ adjusted | Total effect size ($f^2$) | | $Q^2$ | $t$-value | Decision |
|---|---|---|---|---|---|---|---|---|
| $H_1$ | AT -> BI | 2.090 | BI ($R^2$): 0.727 | AT -> BI | 0.807 | BI:0.269 | 4.315 | Accepted |
| | | | BI ($R^2$ adjusted): 0.670 | AT -> RB | 0.318 | RB:0.092 | | |
| $H_2$ | SN -> BI | 1.634 | RB ($R^2$): 0.209 | SN -> BI | -0.042 | | 0.235 | Not Accepted |
| | | | RB ($R^2$ adjusted): 0.181 | SN -> RB | -0.016 | | | |
| $H_3$ | PBC -> BI | 1.658 | | PBC -> BI | -0.078 | | 0.458 | Not Accepted |
| | | | | PBC -> RB | -0.031 | | | |
| $H_4$ | PN -> BI | 1.393 | | PN -> BI | 0.132 | | 0.835 | Not Accepted |
| | | | | PN -> RB | 0.052 | | | |
| $H_5$ | IP -> BI | 1.227 | | IP -> BI | -0.318 | | 1.443 | Not Accepted |
| | | | | IP -> RB | -0.125 | | | |
| $H_6$ | BI -> RB | 1.00 | | BI -> RB | 0.393 | | 2.158 | Accepted |

intention in minimising CDW. The second hypothesis, architects' subjective norm has a significant effect on their behavioural intention towards implementing waste minimisation strategies, was not accepted. In other words, the degree of approval/disapproval from key stakeholders, such as local governments, clients, managers, and colleagues could not fully motivate architects to adopt behavioural intention towards implementing waste minimisation strategies in the Australian context, which is contrary to the findings of Li et al. (2021). This inconsistency might be attributed to inadequate executive orders issued by the government to implement CDWM in the design phase, unclear or unintegrated organisational goals concerning CDWM in the design phase, various perceptions of clients towards CDWM in the design phase, and architects' dependency on choosing/declining waste minimisation strategies in the design phase in the Australian construction industry, which have been addressed to some extent in some studies (Kabirifar et al., 2021b; Udawatta et al., 2015). The third hypothesis, architects perceived behavioural control has a significant effect on their behavioural intention towards implementing waste minimisation strategies, was not accepted. As discussed in the literature, individual restrictions, such as control, capability, experience, and opportunity that limit an individual's ability to translate their intention into action are accounted for by perceived behavioural control. Although several studies have come up with a significant impact of perceived behavioural control on the behavioural intention in different phases of CDWM (e.g., Mhatre et al., 2021), this concept has received less attention in the Australian context. One possible reason for this could refer to the lack of experience and capability of architects to implement waste minimisation strategies in the design phase, which was observed in the findings of (Begum et al., 2009; Li et al., 2015). The fourth hypothesis, architects' personal norms have a significant effect on their behavioural intention towards implementing waste minimisation strategies, was not accepted. Related literature indicated that the application of personal norm on CDWM, which is supported by the value-belief-norm (VBN) theory is a critical element in driving environmental behaviour from self-interest, responsibility, and obligation perspectives in order to shape a certain behaviour. Inconsistent with the global acceptability of this impact in similar studies (e.g., Parajuly et al., 2020), this concept was not proven in the Australian context. One possible reason lies in the fact that although self-interest might be a decisive role in adopting waste minimisation strategies by architects, responsibility and obligation toward behavioural intention in adopting waste minimisation strategies depend highly on the organisational/governmental waste management policies, regulations, and standards, which might affect personal norms to some extent, which has been addressed in similar studies (e.g., Yuan et al., 2018). The fifth hypothesis, institutional pressures have a significant effect on architects' behavioural intention towards implementing waste minimisation strategies, was not accepted. The presence of institutional pressures and their impacts on the behavioural intention in the context of CDWM is justified by the institutional theory, which emphasises the role of pressure in forming individuals' certain behaviour. Although part of this finding is consistent with the work of other researchers, for instance, normative and mimetic pressures have been proven ineffective on behavioural intention in CDWM in a study performed by Jain et al. (2020), the critical role of coercive pressure on behavioural intention to adopt waste minimisation strategies cannot be ignored (Liu et al., 2022). The sixth hypothesis, the behavioural intention has a significant impact on the real waste minimisation behaviour of architects was accepted. Apart from an acceptable *t*-value of 2.158, which was above the minimum threshold of 1.65, the large effect size of behavioural intention on the real behaviour was another reason denoting that behavioural intention has a significant impact on the real behaviour. Thus, it is confirmed that motivation has an essential role in shaping an individual's certain waste minimisation behaviour, which is consistent with the findings of several studies in this area (e.g., Wang & Mangmeechai, 2021).

## 14.7 Conclusion

Construction waste has substantial environmental, social, and economic consequences. Implementing an appropriate waste management mechanism in the design stage of construction projects contributes significantly to minimising construction waste at the source. Architects play a critical role in CDWM by designing buildings that are environmentally sustainable and minimising waste generation. By incorporating sustainable design principles and circular economy concepts into their plans, architects can help reduce the amount of waste generated during the construction and demolition process and throughout the building's lifecycle. However, it should be noted that the critical role of architects' attitudes and behaviour in waste minimisation has been overlooked.

Thus, this study developed a model to empirically explore the architects' real behaviour in implementing waste minimisation strategies based on the correlation of the theory of planned behaviour (attitude, subjective norm, and perceived behavioural control), value-belief-norms theory (self-interest, responsibility, and obligation), and institutional theory (coercive, normative, and mimetic pressure). Results indicated that attitude has a great impact on the behavioural intention toward implementing waste minimisation strategies ($H_1$). Also, the behavioural intention was proven influential in explaining architects' real behaviour in implementing waste minimisation strategies ($H_6$). Conversely, subjective norms ($H_2$), perceived behavioural control ($H_3$), personal norm ($H_4$), and institutional pressures ($H_5$) were revealed ineffective on behavioural intention.

Research findings yield worthwhile insights for the governments and organisations involved in CDWM from design perspective. For instance, by legislating mandatory BIM implementation in the design phase, in which interdisciplinary design clashes are detectable and an accurate estimation of construction materials is accessible, a huge volume of waste, which will be produced in the following phases of construction projects, can be kept away from landfills. Although architects should maintain their independency during planning and design phase of the construction projects, relevant policies, regulations, and standards to minimise CDW generation should be enforced to support architects and to strengthen their awareness and capability to implement waste minimisation strategies. Apart from policies, regulations, and standards, encouraging architects to adopt waste minimisation strategies, implementing effective communication among all stakeholders through regular meetings during the design phase, training courses, and visual demonstration of the positive/negative impacts of proper/improper waste management mechanisms in the design phase can stimulate architects to implement waste minimisation strategies by enhancing their personal norms. Besides, the atmosphere of organisations involved in the design of construction projects should propel more collaboration and participation among architects and other stakeholders including managers and colleagues to improve waste minimisation strategies' implementation. Evidently, implementing waste minimisation strategies in the design phase, i.e., by designing prefabricated building components, specifying sustainable materials, designing for disassembly, etc. contributes significantly to reducing waste which is the principal component of circular economy. Future studies can also benefit from this study by expanding the applied theories and developed framework to identify potential attitudinal/behavioural factors that might be effective in waste minimisation from human resource perspective. By adopting this approach, the construction industry can contribute to a more sustainable future and reduce its environmental impact that can eventually contribute to net zero carbon emission by 2050.

This study has also some limitations. The main limitation of this study can be attributed to the small sample size due to the difficulties associated with data collection in the Australian construction industry. A larger sample size could present more reliable and generalisable results. Also, although the developed model to explore architects' real behaviour to implement waste minimisation strategies included the most applicable factors supported by the most prevalent theories in the CDW research domain, correlation of other theories or factors, such as norm activation theory, costs associated with CDWM in the design phase, etc. can be incorporated into the model in future studies.

## Appendix 14.1: Questionnaire Survey

### HC no: HC220017, Approved on 25 February 2022

### Section 1: Personal Information

| |
|---|
| 1.1 Respondent's gender: |
| [] Male<br>[] Female<br>[] Non-binary/third gender<br>[] Prefer not to say |
| 1.2 Respondent's age: |
| [] 18 to 25 |
| [] 26 to 33 |
| [] 34 to 41 |
| [] 42 to 49 |
| [] 50 years and above |
| 1.3 Work experience in CDWM (Please check only one) |
| [] 0 to 5 years |
| [] 6 to 10 years |
| [] 11 to 15 years |
| [] 16 to 20 years |
| [] 21 year and above |
| 1.4 Highest Academic Qualification (Please check only one) |
| [] Certificate |
| [] Diploma |
| [] Bachelor's degree |
| [] Master's degree |
| [] PhD |
| 1.5 Working category<br>[] Consultancy<br>[] Head of design management |
| [] Freelance architect<br>[] Other |
| 1.6 Project type |
| [] Multi-unit residential |
| [] Commercial |
| [] Industrial |
| [] Infrastructure |
| [] Other |

## Section 2: Indicators' Rating

(1) = strongly disagree, (2) = disagree, (3) = neutral, (4) = agree, and (5) = strongly agree

| No | 1 | Attitude | 1 | 2 | 3 | 4 | 5 |
|---|---|---|---|---|---|---|---|
| 1 | ATT1 | CDW minimisation via design is beneficial to environmental conservation. | | | | | |
| 2 | ATT2 | CDW minimisation via design enhances the sustainability development of the society. | | | | | |
| 3 | ATT3 | CDW minimisation via design helps to develop an environmentally friendly corporate image. | | | | | |
| 4 | ATT4 | CDW minimisation via design improves the social image of the project. | | | | | |
| 5 | ATT5 | CDW minimisation is worthy to be implemented. | | | | | |
| | **2** | **Subjective Norm** | | | | | |
| 6 | SN1 | The local governments approve/support me to implement CDW minimisation strategies. | | | | | |
| 7 | SN2 | The project clients approve/support me to implement CDW minimisation strategies. | | | | | |
| 8 | SN3 | My direct manager (s) approve/support me to implement CDW minimisation strategies. | | | | | |
| 9 | SN4 | My colleagues or project team approve/ support me to implement CDW minimisation strategies. | | | | | |

| No | 1 | Attitude | 1 | 2 | 3 | 4 | 5 |
|---|---|---|---|---|---|---|---|
| | **3** | **Perceived Behavioural Control** | | | | | |
| 10 | PBC1 | I have adequate control over employing CDW minimisation strategies. | | | | | |
| 11 | PBC2 | I have adequate capability to employ CDW minimisation strategies. | | | | | |
| 12 | PBC3 | I have adequate experience to employ CDW minimisation strategies. | | | | | |
| 13 | PBC4 | I have adequate opportunity to employ CDW minimisation strategies. | | | | | |
| | **4** | **Personal Norms** | | | | | |
| 14 | PN1 | I feel I have a self-interest to implement CDW minimisation strategies. | | | | | |
| 15 | PN2 | I feel I am morally responsible to implement CDW minimisation strategies. | | | | | |
| 16 | PN3 | I believe I have a moral obligation to implement CDW minimisation strategies. | | | | | |
| | **5** | **Coercive, Normative and Mimetic Pressure** | | | | | |
| 17 | CP | I feel I am under pressure from government organisations to adopt CDW minimisation strategies. | | | | | |
| 18 | NP | I feel I am under pressure from my organisation to adopt CDW minimisation strategies. | | | | | |

(continued)

| No | 1 | Attitude | 1 | 2 | 3 | 4 | 5 |
|---|---|---|---|---|---|---|---|
| 19 | MP | I feel I am under competitive pressure from my organisations' competitors to adopt CDW minimisation strategies. | | | | | |
| | **6** | **Behavioural Intention** | | | | | |
| 20 | BI1 | I intend to adopt CDWM regulations and plans in the future. | | | | | |
| 21 | BI2 | I intend to adopt design for waste prevention in the future. | | | | | |
| 22 | BI3 | I intend to adopt building codes and standards in the future. | | | | | |
| 23 | BI4 | I intend to adopt design for deconstruction/ disassembly in the future. | | | | | |
| | **7** | **Waste minimisation real behaviour** | | | | | |
| 24 | RB1 | I have implemented CDWM regulations and plans. | | | | | |
| 25 | RB2 | I have implemented design for waste prevention. | | | | | |
| 26 | RB3 | I have implemented building codes and standards. | | | | | |
| 27 | RB4 | I have implemented design for deconstruction/ disassembly. | | | | | |

(1) = never, (2) = seldom, (3) = sometimes, (4) = frequently, and (5) = always

## References


Ajzen, I. (1985). *From intentions to actions: A theory of planned behavior, action control* (pp. 11–39). Springer.

Ajzen, I. (1991). The theory of planned behavior. *Organizational Behavior and Human Decision Processes, 50*(2), 179–211.

Akanbi, L. A., Oyedele, L. O., Akinade, O. O., Ajayi, A. O., Davila Delgado, M., Bilal, M., & Bello, S. A. (2018). Salvaging building materials in a circular economy: A BIM-based whole-life performance estimator. *Resources, Conservation and Recycling, 129*, 175–186.

Alshbili, I., & Elamer, A. A. (2020). The influence of institutional context on corporate social responsibility disclosure: A case of a developing country. *Journal of Sustainable Finance & Investment, 10*(3), 269–293.

Andersson, R. (2021). *Institutional changes in construction waste management*. Chalmers University of Technology, Department of Architecture and Civil . . . .

Aslam, M. S., Huang, B., & Cui, L. (2020). Review of construction and demolition waste management in China and USA. *Journal of Environmental Management, 264*, 110445.

Babbie, E. R. (2015). *The basics of social research*. Nelson Education.

Bakshan, A., Srour, I., Chehab, G., El-Fadel, M., & Karaziwan, J. (2017). Behavioral determinants towards enhancing construction waste management: A Bayesian Network analysis. *Resources, Conservation and Recycling, 117*, 274–284.

Bao, Z., & Lu, W. (2020). Developing efficient circularity for construction and demolition waste management in fast emerging economies: Lessons learned from Shenzhen, China. *Science of the Total Environment, 724*, 138264.

Begum, R. A., Siwar, C., Pereira, J. J., & Jaafar, A. H. (2009). Attitude and behavioral factors in waste management in the construction industry of Malaysia. *Resources, Conservation and Recycling, 53*(6), 321–328.

Beunen, R., Patterson, J., & Van Assche, K. (2017). Governing for resilience: The role of institutional work. *Current Opinion in Environmental Sustainability, 28*, 10–16.

Blaisi, N. I. (2019). Construction and demolition waste management in Saudi Arabia: Current practice and roadmap for sustainable management. *Journal of Cleaner Production, 221*, 167–175.

Bortoleto, A. P., Kurisu, K. H., & Hanaki, K. (2012). Model development for household waste prevention behaviour. *Waste Management, 32*(12), 2195–2207.

Bronfman, N. C., Cisternas, P. C., López-Vázquez, E., Maza, C. D. I., & Oyanedel, J. C. (2015). Understanding attitudes and pro-environmental behaviors in a Chilean Community. *Sustainability, 7*(10), 14133–14152.

Chithambo, L., Tingbani, I., Agyapong, G. A., Gyapong, E., & Damoah, I. S. (2020). Corporate voluntary greenhouse gas reporting: Stakeholder pressure and the mediating role of the chief executive officer. *Business Strategy and the Environment, 29*(4), 1666–1683.

Cohen, J. (1992). A power primer. *Psychological Bulletin, 112*(1), 155–159.

Creswell, J. W., & Creswell, J. D. (2017). *Research design: Qualitative, quantitative, and mixed methods approaches*. Sage Publications.

Crossley, R. M., Elmagrhi, M. H., & Ntim, C. G. (2021). Sustainability and legitimacy theory: The case of sustainable social and environmental practices of small and medium-sized enterprises. *Business Strategy and the Environment, 30*, 3740–3762.

Darnton, A. (2008). *Reference report: An overview of behaviour change models and their uses*. Government Social Research Behaviour Change Knowledge Review.

Davis, R., Campbell, R., Hildon, Z., Hobbs, L., & Michie, S. (2015). Theories of behaviour and behaviour change across the social and behavioural sciences: A scoping review. *Health Psychology Review, 9*(3), 323–344.

De Silva, N., & Vithana, S. B. K. H. (2008). Use of PC elements for waste minimization in the Sri Lankan construction industry. *Structural Survey, 26*(3), 188–198.

Ding, Z., Yi, G., Tam, V. W. Y., & Huang, T. (2016). A system dynamics-based environmental performance simulation of construction waste reduction management in China. *Waste Management, 51*, 130–141.

Du, L., Feng, Y., Lu, W., Kong, L., & Yang, Z. (2020). Evolutionary game analysis of stakeholders' decision-making behaviours in construction and demolition waste management. *Environmental Impact Assessment Review, 84*, 106408.

Dumlao-Tan, M. I., & Halog, A. (2017). Moving towards a circular economy in solid waste management: Concepts and practices. In S. Goel (Ed.), *Advances in solid and hazardous waste management* (pp. 29–48). Springer International Publishing.

Esa, M. R., Halog, A., & Rigamonti, L. (2017). Strategies for minimizing construction and demolition wastes in Malaysia. *Resources, Conservation and Recycling, 120*, 219–229.

Etikan, I., Musa, S. A., & Alkassim, R. S. (2016). Comparison of convenience sampling and purposive sampling. *American Journal of Theoretical and Applied Statistics, 5*(1), 1–4.

Fellows, R., & Liu, A. (2008). Impact of participants' values on construction sustainability. *Proceedings of the Institution of Civil Engineers – Engineering Sustainability, 161*(4), 219–227.

Fernando, S., & Lawrence, S. (2014). A theoretical framework for CSR practices: Integrating legitimacy theory, stakeholder theory and institutional theory. *Journal of Theoretical Accounting Research, 10*(1), 149–178.

Gangolells, M., Casals, M., Forcada, N., & Macarulla, M. (2014). Analysis of the implementation of effective waste management practices in construction projects and sites. *Resources, Conservation and Recycling, 93*, 99–111.

Ghaffar, S. H., Burman, M., & Braimah, N. (2020). Pathways to circular construction: An integrated management of construction and demolition waste for resource recovery. *Journal of Cleaner Production, 244*, 118710.

Ghafourian, K., Kabirifar, K., Mahdiyar, A., Yazdani, M., Ismail, S., & Tam, V. W. Y. (2021). A synthesis of Express Analytic Hierarchy Process (EAHP) and Partial Least Squares-Structural Equations Modeling (PLS-SEM) for sustainable construction and demolition waste management assessment: The case of Malaysia. *Recycling, 6*(4), 73.

Goel, A., Ganesh, L. S., & Kaur, A. (2021). Sustainability assessment of construction practices in India using inductive content analysis of research literature. *International Journal of Construction Management, 21*(8), 802–817.

Gupta, S., Jha, K. N., & Vyas, G. (2020). Proposing building information modeling-based theoretical framework for construction and demolition waste management: Strategies and tools. *International Journal of Construction Management, 22*, 1–11.

Hair Joseph, F., Risher Jeffrey, J., Sarstedt, M., & Ringle Christian, M. (2019). When to use and how to report the results of PLS-SEM. *European Business Review, 31*(1), 2–24.

Hair, J. F., Jr., Hult, G. T. M., Ringle, C., & Sarstedt, M. (2017). *A primer on partial least squares structural equation modeling (PLS-SEM)*. Sage Publications.

Hair, J. F., Jr., Page, M., & Brunsveld, N. (2019). *Essentials of business research methods*. Routledge.

Hair, J. F., Ringle, C. M., & Sarstedt, M. (2013). Partial least squares structural equation modeling: Rigorous applications, better results and higher acceptance. *Long Range Planning, 46*(1–2), 1–12.

Henseler, J., Ringle, C. M., & Sarstedt, M. (2015). A new criterion for assessing discriminant validity in variance-based structural equation modeling. *Journal of the Academy of Marketing Science, 43*(1), 115–135.

Henseler, J., Hubona, G., & Ray Pauline, A. (2016). Using PLS path modeling in new technology research: Updated guidelines. *Industrial Management & Data Systems, 116*(1), 2–20.

Herold, D. M. (2018). Demystifying the link between institutional theory and stakeholder theory in sustainability reporting. *Economics, Management and Sustainability, 3*(2), 6–19.

Hossain, M. U., Ng, S. T., Antwi-Afari, P., & Amor, B. (2020). Circular economy and the construction industry: Existing trends, challenges and prospective framework for sustainable construction. *Renewable and Sustainable Energy Reviews, 130*, 109948.

Huang, T., Shi, F., Tanikawa, H., Fei, J., & Han, J. (2013). Materials demand and environmental impact of buildings construction and demolition in China based on dynamic material flow analysis. *Resources, Conservation and Recycling, 72*, 91–101.

Huang, B., Wang, X., Kua, H., Geng, Y., Bleischwitz, R., & Ren, J. (2018). Construction and demolition waste management in China through the 3R principle. *Resources, Conservation and Recycling, 129*, 36–44.

Jaillon, L., & Poon, C. S. (2014). Life cycle design and prefabrication in buildings: A review and case studies in Hong Kong. *Automation in Construction, 39*, 195–202.

Jain, S., Singhal, S., Jain, N. K., & Bhaskar, K. (2020). Construction and demolition waste recycling: Investigating the role of theory of planned behavior, institutional pressures and environmental consciousness. *Journal of Cleaner Production, 263*, 121405.

Janmaimool, P., & Denpaiboon, C. (2016). Evaluating determinants of rural Villagers' engagement in conservation and waste management behaviors based on integrated conceptual framework of Pro-environmental behavior. *Life Sciences, Society and Policy, 12*(1), 12.

Jiménez-Rivero, A., De Guzmán-Báez, A., & García-Navarro, J. (2017). Enhanced on-site waste Management of Plasterboard in construction works: A case study in Spain. *Sustainability, 9*(3), 450.

Jordan, P. J., & Troth, A. C. (2019). Common method bias in applied settings: The dilemma of researching in organizations. *Australian Journal of Management, 45*(1), 3–14.

Jörg, H., Christian, M. R., & Marko, S. (2012). Using partial least squares path modeling in advertising research: Basic concepts and recent issues. In *Handbook of research on international advertising*. Edward Elgar Publishing.

Joshi, A., Kale, S., Chandel, S., & Pal, D. K. (2015). Likert scale: Explored and explained. *Current Journal of Applied Science and Technology, 7*, 396–403.

Kabirifar, K., Mojtahedi, M., Changxin Wang, C., & Vivian, W. Y. T. (2020a). A conceptual foundation for effective construction and demolition waste management. *Cleaner Engineering and Technology, 1*, 100019.

Kabirifar, K., Mojtahedi, M., Wang, C., & Tam, V. W. Y. (2020b). Construction and demolition waste management contributing factors coupled with reduce, reuse, and recycle strategies for effective waste management: A review. *Journal of Cleaner Production, 263*, 121265.

Kabirifar, K., Mojtahedi, M., Changxin Wang, C., & Tam, V. W. Y. (2021a). Effective construction and demolition waste management assessment through waste management hierarchy; a case of Australian large construction companies. *Journal of Cleaner Production, 312*, 127790.

Kabirifar, K., Mojtahedi, M., & Wang, C. C. (2021b). A systematic review of construction and demolition waste Management in Australia: Current practices and challenges. *Recycling, 6*(2), 34.

Kerdlap, P., Low, J. S. C., & Ramakrishna, S. (2019). Zero waste manufacturing: A framework and review of technology, research, and implementation barriers for enabling a circular economy transition in Singapore. *Resources, Conservation and Recycling, 151*, 104438.

Kim Soo, Y., Nguyen Minh, V., & Luu Van, T. (2020). A performance evaluation framework for construction and demolition waste management: Stakeholder perspectives. *Engineering, Construction and Architectural Management, 27*(10), 3189–3213.

Kong, L., & Ma, B. (2020). Evaluation of environmental impact of construction waste disposal based on fuzzy set analysis. *Environmental Technology & Innovation, 19*, 100877.

Lachimpadi, S. K., Pereira, J. J., Taha, M. R., & Mokhtar, M. (2012). Construction waste minimisation comparing conventional and precast construction (Mixed System and IBS) methods in high-rise buildings: A Malaysia case study. *Resources, Conservation and Recycling, 68*, 96–103.

Li, J., Tam, V. W. Y., Zuo, J., & Zhu, J. (2015). Designers' attitude and behaviour towards construction waste minimization by design: A study in Shenzhen, China. *Resources, Conservation and Recycling, 105*, 29–35.

Li, J., Zuo, J., Cai, H., & Zillante, G. (2018). Construction waste reduction behavior of contractor employees: An extended theory of planned behavior model approach. *Journal of Cleaner Production, 172*, 1399–1408.

Li, G., Sun, F., Fu, B., Deng, L., Yang, K., Liu, Y., & Che, Y. (2021). How to promote the public participation in eco-compensation in transboundary river basins: A case from Planned Behavior perspective. *Journal of Cleaner Production, 313*, 127911.

Liu, J., Li, G., Li, H., & Wang, Z. (2017). Planned behavior theory-based study on the influencing factors in construction waste reducing willingness—with construction workers as an example. *Ekoloji, 26*(102), 13–28.

Liu, J., Chen, Y., & Wang, X. (2022). Factors driving waste sorting in construction projects in China. *Journal of Cleaner Production, 336*, 130397.

López Ruiz, L. A., Roca Ramón, X., & Gassó Domingo, S. (2020). The circular economy in the construction and demolition waste sector – A review and an integrative model approach. *Journal of Cleaner Production, 248*, 119238.

Low, S. P., Gao, S., & See, Y. L. (2014). Strategies and measures for implementing eco-labelling schemes in Singapore's construction industry. *Resources, Conservation and Recycling, 89*, 31–40.

Mak, T. M. W., Yu, I. K. M., Wang, L., Hsu, S.-C., Tsang, D. C. W., Li, C. N., Yeung, T. L. Y., Zhang, R., & Poon, C. S. (2019). Extended theory of planned behaviour for promoting construction waste recycling in Hong Kong. *Waste Management, 83*, 161–170.

Marrero, M., Puerto, M., Rivero-Camacho, C., Freire-Guerrero, A., & Solís-Guzmán, J. (2017). Assessing the economic impact and ecological footprint of construction and demolition waste during the urbanization of rural land. *Resources, Conservation and Recycling, 117*, 160–174.

Martínez-Ferrero, J., & García-Sánchez, I.-M. (2017). Coercive, normative and mimetic isomorphism as determinants of the voluntary assurance of sustainability reports. *International Business Review, 26*(1), 102–118.

Meixell, M. J., & Luoma, P. (2015). Stakeholder pressure in sustainable supply chain management. *International*

*Journal of Physical Distribution & Logistics Management, 45*(1/2), 69–89.
Menegaki, M., & Damigos, D. (2018). A review on current situation and challenges of construction and demolition waste management. *Current Opinion in Green and Sustainable Chemistry, 13*, 8–15.
Mhatre, P., Gedam, V., Unnikrishnan, S., & Verma, S. (2021). Circular economy in built environment – Literature review and theory development. *Journal of Building Engineering, 35*, 101995.
Nußholz, J. L. K., Nygaard Rasmussen, F., & Milios, L. (2019). Circular building materials: Carbon saving potential and the role of business model innovation and public policy. *Resources, Conservation and Recycling, 141*, 308–316.
Ordoñez, I., & Rahe, U. (2013). Collaboration between design and waste management: Can it help close the material loop? *Resources, Conservation and Recycling, 72*, 108–117.
Osmani, M., Glass, J., & Price, A. (2006). Architect and contractor attitudes to waste minimisation. *Proceedings of the Institution of Civil Engineers – Waste and Resource Management, 159*(2), 65–72.
Parajuly, K., Fitzpatrick, C., Muldoon, O., & Kuehr, R. (2020). Behavioral change for the circular economy: A review with focus on electronic waste management in the EU. *Resources, Conservation & Recycling: X, 6*, 100035.
Parisi Kern, A., Ferreira Dias, M., Piva Kulakowski, M., & Paulo Gomes, L. (2015). Waste generated in high-rise buildings construction: A quantification model based on statistical multiple regression. *Waste Management, 39*, 35–44.
Phillips, P. P., & Stawarski, C. A. (2008). *Data collection: Planning for and collecting all types of data*. Wiley.
Podsakoff, P. M., MacKenzie, S. B., Lee, J.-Y., & Podsakoff, N. P. (2003). Common method biases in behavioral research: A critical review of the literature and recommended remedies. *Journal of Applied Psychology, 88*(5), 879–903.
Podsakoff, P. M., MacKenzie, S. B., & Podsakoff, N. P. (2011). Sources of method bias in social science research and recommendations on how to control it. *Annual Review of Psychology, 63*(1), 539–569.
Poon, C. S., Yu, A. T. W., & Jaillon, L. (2004). Reducing building waste at construction sites in Hong Kong. *Construction Management and Economics, 22*(5), 461–470.
Ramayah, T., Lee, J. W. C., & Lim, S. (2012). Sustaining the environment through recycling: An empirical study. *Journal of Environmental Management, 102*, 141–147.
Rauch, A., & Hulsink, W. (2015). Putting entrepreneurship education where the intention to act lies: An investigation into the impact of entrepreneurship education on entrepreneurial behavior. *Academy of Management Learning & Education, 14*(2), 187–204.
Salvia, G., Zimmermann, N., Willan, C., Hale, J., Gitau, H., Muindi, K., Gichana, E., & Davies, M. (2021). The wicked problem of waste management: An attention-based analysis of stakeholder behaviours. *Journal of Cleaner Production, 326*, 129200.
Sarstedt, M., & Mooi, E. (2019). *A concise guide to market research* (3rd ed.). Springer.
Schwartz, S., & Ben David, A. (1976). Responsibility and helping in an emergency: Effects of blame. *Ability and Denial of Responsibility. Sociometry, 39*(4), 406–415.
Sekaran, U., & Bougie, R. (2016). *Research methods for business: A skill building approach*. Wiley.
Shivaraj, S., Abd Halid, A., Sasitharan, N., Kumanan, K., Haritharan, M., & Zawawi, D. (2018). Investigate how construction waste generation rate is different for every types of project in peninsular Malaysia using site visit method. *International Journal of Integrated Engineering, 10*(1), 150–156.
Sia, S. K., & Jose, A. (2019). Attitude and subjective norm as personal moral obligation mediated predictors of intention to build eco-friendly house. *Management of Environmental Quality: An International Journal, 30*(4), 678–694.
Singh, J., & Ordoñez, I. (2016). Resource recovery from post-consumer waste: Important lessons for the upcoming circular economy. *Journal of Cleaner Production, 134*, 342–353.
Stern, P. (2000). Toward a coherent theory of environmentally significant behavior. *Journal of Social Issues, 56*(3), 407–424.
Su, Y. (2020). Multi-agent evolutionary game in the recycling utilization of construction waste. *Science of the Total Environment, 738*, 139826.
Teck Heng Lim, B., Lan Oo, B., & Ling, F. (2010). The survival strategies of Singapore contractors in prolonged recession. *Engineering, Construction and Architectural Management, 17*(4), 387–403.
Teo, M. M. M., & Loosemore, M. (2001). A theory of waste behaviour in the construction industry. *Construction Management and Economics, 19*(7), 741–751.
Teo, M. M., Loosemore, M., Masosszeky, M., & Karim, K. (2000). Operatives attitudes towards waste on a construction project. In *Annual conference-ARCOM* (pp. 509–517).
Turaga, R. M. R., Howarth, R. B., & Borsuk, M. E. (2010). Pro-environmental behavior: Rational choice meets moral motivation. *Annals of the New York Academy of Sciences, 1185*(1), 211–224.
Udawatta, N., Zuo, J., Chiveralls, K., & Zillante, G. (2015). Attitudinal and behavioural approaches to improving waste management on construction projects in Australia: Benefits and limitations. *International Journal of Construction Management, 15*(2), 137–147.
Urio, A. F., & Brent, A. C. (2006). Solid waste management strategy in Botswana: The reduction of construction waste: Technical note. *Journal of the South African Institution of Civil Engineering = Joernaal van die Suid-Afrikaanse Instituut van Siviele Ingenieurswese, 48*(2), 18–22.

Wang, H., & Mangmeechai, A. (2021). Understanding the gap between environmental intention and pro-environmental behavior towards the waste sorting and management policy of China. *International Journal of Environmental Research and Public Health, 18*(2), 757.

Wang, J., Yu, B., Tam, V. W. Y., Li, J., & Xu, X. (2019). Critical factors affecting willingness of design units towards construction waste minimization: An empirical study in Shenzhen, China. *Journal of Cleaner Production, 221*, 526–535.

Wang, S., Wang, J., Yang, S., Li, J., & Zhou, K. (2020). From intention to behavior: Comprehending residents' waste sorting intention and behavior formation process. *Waste Management, 113*, 41–50.

Wu, Z., Yu, A. T. W., & Shen, L. (2017). Investigating the determinants of contractor's construction and demolition waste management behavior in Mainland China. *Waste Management, 60*, 290–300.

Wu, H., Zuo, J., Zillante, G., Wang, J., & Yuan, H. (2019). Status quo and future directions of construction and demolition waste research: A critical review. *Journal of Cleaner Production, 240*, 118163.

Wu, Z., Yu, A. T. W., & Poon, C. S. (2020). Promoting effective construction and demolition waste management towards sustainable development: A case study of Hong Kong. *Sustainable Development, 28*(6), 1713–1724.

Yang, B., Song, X., Yuan, H., & Zuo, J. (2020). A model for investigating construction workers' waste reduction behaviors. *Journal of Cleaner Production, 265*, 121841.

Yazdani, M., Kabirifar, K., Frimpong, B. E., Shariati, M., Mirmozaffari, M., & Boskabadi, A. (2021). Improving construction and demolition waste collection service in an urban area using a simheuristic approach: A case study in Sydney, Australia. *Journal of Cleaner Production, 280*, 124138.

Yean Yng Ling, F., & Song Anh Nguyen, D. (2013). Strategies for construction waste management in Ho Chi Minh City, Vietnam. *Built Environment Project and Asset Management, 3*(1), 141–156.

Yeheyis, M., Hewage, K., Alam, M. S., Eskicioglu, C., & Sadiq, R. (2013). An overview of construction and demolition waste management in Canada: A lifecycle analysis approach to sustainability. *Clean Technologies and Environmental Policy, 15*(1), 81–91.

Yuan, H., Wu, H., & Zuo, J. (2018). Understanding factors influencing project managers' behavioral intentions to reduce waste in construction projects. *Journal of Management in Engineering, 34*(6), 04018031.

Yuen, K. F., Wang, X., Wong, Y. D., & Zhou, Q. (2017). Antecedents and outcomes of sustainable shipping practices: The integration of stakeholder and behavioural theories. *Transportation Research Part E: Logistics and Transportation Review, 108*, 18–35.

Zeng, S. X., Ma, H. Y., Lin, H., Zeng, R. C., & Tam, V. W. Y. (2015). Social responsibility of major infrastructure projects in China. *International Journal of Project Management, 33*(3), 537–548.

Zeng, H., Chen, X., Xiao, X., & Zhou, Z. (2017). Institutional pressures, sustainable supply chain management, and circular economy capability: Empirical evidence from Chinese eco-industrial park firms. *Journal of Cleaner Production, 155*, 54–65.

Zhang, X., Wu, Y., & Shen, L. (2012). Application of low waste technologies for design and construction: A case study in Hong Kong. *Renewable and Sustainable Energy Reviews, 16*(5), 2973–2979.

Zhang, C., Hu, M., Di Maio, F., Sprecher, B., Yang, X., & Tukker, A. (2022). An overview of the waste hierarchy framework for analyzing the circularity in construction and demolition waste management in Europe. *Science of the Total Environment, 803*, 149892.

Zhao, X., Webber, R., Kalutara, P., Browne, W., & Pienaar, J. (2021). Construction and demolition waste management in Australia: A mini-review. *Waste Management & Research, 40*(1), 34–46.

# Building Occupancy and Smart Metering

# 15

Wei Wang

## 15.1 Introduction

Building sector has received increasing attentions in recent years for its energy-saving potential. Energy conservation in buildings require effective facility management and a good understanding of occupants'energy demand and buildings' capacity. Therefore, occupant, as building energy end-user, has been widely recognized as the key energy-efficient determinant. With the rapid development of the Internet of Things (IoT) technology armed in buildings, it becomes more feasible to sense the characteristic of the building occupancy, especially through the pre-installed signal networks in the building. The research presented in this chapter can be useful for optimizing the use of signals from low-cost networks in buildings, and also how this can be used to optimize building energy control systems. This chapter would like to answer several concerns to energy-savings and occupancy, (i) by discussing the types of occupancy resolution and their corresponding application in energy-saving projects; (ii) introducing experiment and model construction of Wi-Fi signal networks to sense building occupancy; (iii) investigating the linkage of Wi-Fi enabled occupancy sensing to building energy simulation and system control. (iv) implication of the proposed system in future building engineering.

## 15.2 Occupancy Resolutions for Building Energy-Savings

Occupancy information refers to the number- and time-based schedules of building occupants, which were categorized on different scales to quantify occupancy resolution. One of the most popular definitions of occupancy resolutions was provided in (Melfi et al., 2011) and can be seen in Fig. 15.1: (1) space, including building, floor, and room levels; (2) occupants, including occupancy, count, identity, and activity level; and (3) time span, including days, hours, minutes, and seconds. In (Christensen et al., 2014), another popular definition includes presence, which is used to determine whether an occupant is present in a space; count, or the number of occupants present in a given space; location, meaning the occupant's exact physical position; track, which is the movement of occupants; and identity, which indicates who is in the room. In large spaces or spaces with collaborative working air-conditioners, the "count" level in the room's spatial resolution is usually not accurate enough for HVAC control optimization. In the study (Du et al., 2016), Du et al. proposed one method to optimize collaborative air-conditioners by finding the relationship between energy

W. Wang (✉)
Southeast University, Nanjing, Jiangsu Province, China
e-mail: weiwang@seu.edu.cn



J. Zuo et al. (eds.), *Circular Economy for Buildings and Infrastructure*, Sustainable Development Goals Series, https://doi.org/10.1007/978-3-031-56241-9_15

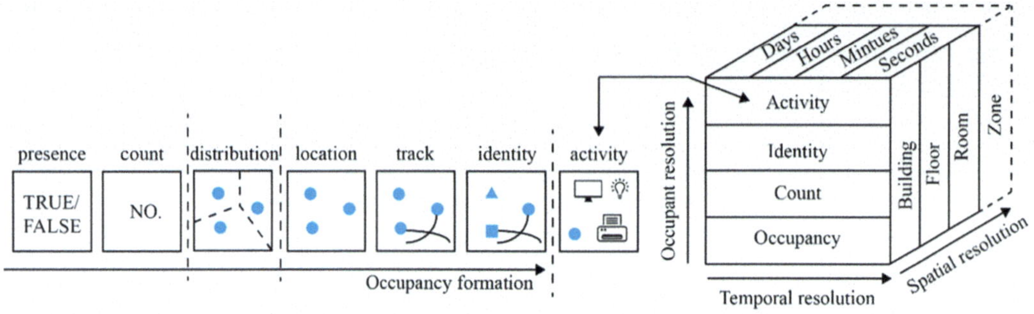


**Fig. 15.1** Occupancy resolution, including occupancy distribution levels

efficiency, temperature, and user location so that real-time "location" levels of occupant information can be used to optimize HVAC systems. However, technology to accurately locate occupants is currently undeveloped, and HVAC systems cannot respond to such accurate temperature control in one exact location. To address this limitation, this book proposes adding zone occupancy distribution to occupancy resolution, which is illustrated in Fig. 15.1. The zone occupancy level can be defined with the "count" level and "location" level of occupants, in which the system must find the number of occupants distributed in each thermal zone but not their exact location, thus avoiding complex positioning algorithms. In spatial resolution, a zone level would be added to describe occupancy distribution information.

This form of occupancy resolution has major benefits: (1) high occupancy resolution with low detection cost, (2) efficient control of multiple HVAC system terminals, and (3) energy saving and space management based on occupancy distribution preferences.

## 15.3 Occupancy & Circular Economy at Building Sector

Buildings and their internal and external built environment have put major pressure on the natural environment because of their high energy consumption and carbon emission, therefore, the concept of "circular buildings" becomes important and fundamental for its role in transitioning to a circular economy (CE) (Dabaieh, 2022; Pomponi & Moncaster, 2017). However, "Buildings don't use energy, but people do", and in this context, occupants can influence a building's energy consumption in three ways (Liang et al., 2016): (1) the participating heat balance of the building through occupants' body heat release; (2) occupancy-based demand including thermal comfort and indoor air quality (Wang et al., 2017a, b; Goyal et al., 2013; Korkas et al., 2016); (3) occupant interactions with building systems and building controls (Hong et al., 2016; Yan et al., 2015; Wang et al., 2017c). The needs of the occupants require the building to provide a more green, comfortable, efficient and healthy environment, which also requires a more sustainable life cycle assessment of the building (González et al., 2021). The whole life cycle of a green building includes design and construction (pre-occupancy assessment), operation and maintenance (occupant-centered), and demolition and renewal (post-occupancy assessment). Figure 15.2 shows the relationship between occupancy, green building, circular buildings, circular economy. Gonzalez has proposed the methodology to calculate and assess the circularity in different building sectors (González et al., 2021), that counts the circularity index of building energy, materials, water, social activity, and economic. Moreover, "circular buildings" can extend beyond the project itself to the surrounding area, contributing to economic and social development (Foster, 2020). Therefore, occupancy information becomes an important determinant in circular building and circular economy.

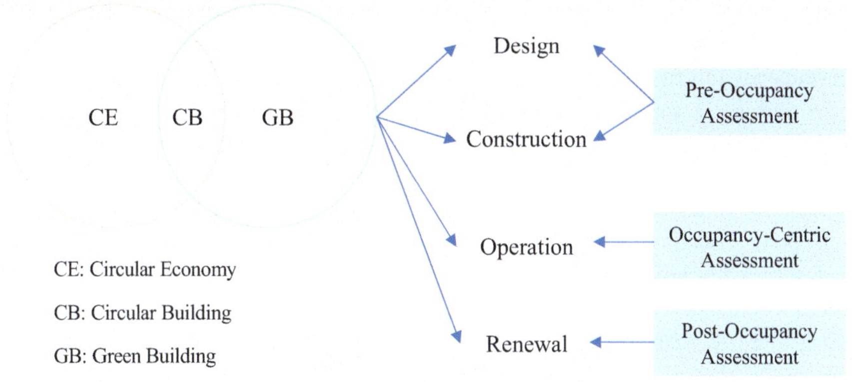


**Fig. 15.2** The relationship between occupancy, green building, circular buildings, circular economy

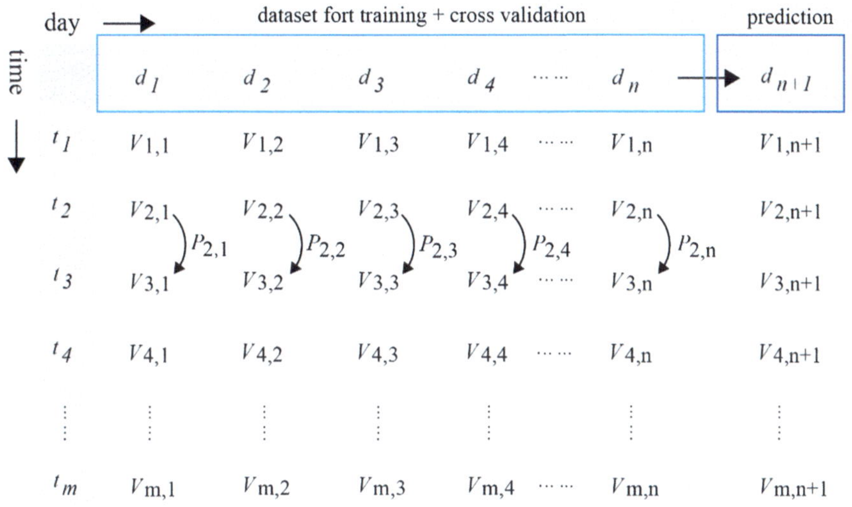


**Fig. 15.3** A typical probability correlation matrix of the proposed occupancy model

## 15.4 Occupancy Characteristics

The occupant behavior in a building is closely related to its actual energy consumption during its operation stage (Liang et al., 2016; Wang et al., 2016a; Wang & Shao, 2017). As a presentation of the utilization of building functions, occupancy contains unique features. Researchers have discussed and summarized these features into five characteristics (Zhao et al., 2015; Dodier et al., 2006; Chen et al., 2015; Jiang et al., 2016). (1) Time series characteristic. Occupancy is associated with time of a day and shows a periodic pattern. (2) Statistic characteristic. Historical occupancy data can infer the current occupancy pattern. (3) Stochastic characteristic. Current occupancy status is determined probabilistically according to the previous status. (4) Limited occupancy statuses. The occupancy status is simple and commonly contains two types, occupied or unoccupied (in or out). (5) Chronological distribution. The probability of occupant presence varies at different times of day. In addition, when occupants have interrelationships, such as being co-workers, the probability of their presence is a joint probability or conditional probability. Figure 15.3 illustrates the typical data format of the occupancy transfer probability matrix for the proposed model. In Fig. 15.3, $t_m$ and $d_n$ stand for the $m$ th and $n$ th day in the dataset. The $V_{m,\ n}$ means the occupancy information at time interval $m$ and day $n$ and $P_{2,\ n}$ stands for the transfer probability from time interval $t2$ to $t3$ in $n$ th day. The matrix shows the three dimensions of occupancy, i.e., its time, date, and probability among each other, treating a building space usage in a day as a Markov chain process, and the prediction for the next day $d_{n\ +\ 1}$ can be divided into a Markov chain prediction process for each time of this day.

## 15.5 Current Occupancy Smart Sensing

In the main topics of International Energy Agency (IEA) Energy in Buildings and Community (EBC) Annex 66, occupancy Behavior (OB) has been regarded as one of most significant impacts on building energy consumptions and energy-saving technology (Yan et al., 2017), and it summarized highlighted detailed occupancy model issues, including occupancy sensing and model evaluation. Occupancy information is crucial for energy-efficient building design, and American Society of Heating, Refrigerating and Air-Conditioning Engineers (ASHRAE) recommends fixed occupancy diversity factors in standard 90.1-2007 (ASHRAE, n.d.) when actual occupancy information is unavailable. However, significant discrepancies were found given uncertainties in occupants' behaviors and occupancy patterns, which can result in an estimation error of up to 40% (Gunay et al., 2013). Therefore, researchers proposed various occupancy measurement and modeling approaches.

Following Fig. 15.1, the occupancy information can be divided into different resolutions, and to acquire occupancy information at different resolutions, various sensor technologies and occupancy algorithms have been utilized to detect occupancy profiles and improve data resolution.

### 15.5.1 Methods for Sensing Presence, Count, and Identity of Occupancy Studies

To identify the presence level, passive infrared (PIR) sensors, motion sensors, and lighting sensors can respond to the presence/absence of occupants within their field-of-view (Dodier et al., 2006), but these sensors cannot detect stationary occupants (Benezeth et al., 2011). Studies also employ building security cameras to receive information regarding building or room presence and compile occupancy information (Davis & Nutter, 2010; Li & Dong, 2018; Zou et al., 2017). Regarding the count level, the ASHRAE standard 90.1-2007 (ASHRAE, n.d.) recommends designing diverse occupancy factors for different building types. This designed occupancy schedule is useful for size operation when actual data are unknown but can cause 40% variation compared to the actual occupancy information (Gunay et al., 2013). Some researchers built an agent-based stochastic occupancy simulator and provided a web application for the simulation and visualization of occupancy information (Chen et al., 2017a, b; Luo et al., 2017). Another one of the most widely applied approaches is using carbon dioxide (CO2) concentration. Wang et al. developed several dynamic CO2-based models that estimate and predict occupancy in commercial buildings using CO2 concentration (Shan et al., 2012; Wang et al., 2016b; Wang & Jin, 1998). Jiang et al. proposed a feature-scaled extreme learning machine approach based on CO2 concentration to predict occupancy and reported 50% accuracy (Jiang et al., 2016). To improve accuracy in occupancy estimation, some researchers integrate CO2 sensor data with other environmental sensors, such as temperature, humidity, lighting, and sound sensors. (Ekwevugbe et al., 2013; Ekwevigbe et al., 2013). Diaz and Jimenez conducted an experiment based on the power consumption of computers under occupancy variation estimated by CO2, suggesting that CO2 concentration is a good indicator of occupancy (Díaz & Jiménez,

2017). These approaches are not ideal due to constraints of time delay, high cost, and the inaccuracy of indirect detection (Garg & Bansal, 2000). However, CO2-based approaches have several constraints, such as low sensitivity to occupant mobility and slow response to drastic occupancy changes (Yang et al., 2016). Some studies also utilized existing Wi-Fi signal networks to profile the number of occupants inside buildings with Markov chain and artificial neural network algorithms, resulting in about 80% accuracy (Wang et al., 2017b; Martani et al., 2012).

The "identity" level is also important when building energy-saving studies. Since occupants are the end-users of building energy, providing occupants with energy use information is an efficient feedback control method. For example, one residential eco-feedback study in (Ueno et al., 2006) shows 10% energy savings when users were provided historical consumption data. Normative comparison features have been recently added to eco-feedback systems to provide users and peers with socially contextualized feedback (Chen et al., 2013; Peschiera et al., 2010; Jain et al., 2012). Cameras can sense identification of occupant via image analysis, which is a traditional and effective tool. In recent years, the development of smart devices enables identification using a unique service set identifier (SSID) or the Mac address of smart devices (Wang et al., 2017b). However, as human involvement is necessary, such detection correlates with ethical considerations and privacy concerns.

### 15.5.2 Methods for Sensing Location, Tracking, and Activity of Occupancy Studies

RFID sensors can usually report location and tracking levels using the spatial coordinates of occupants in an RFID network deployed space. When using active RFID systems, Li et al. (2012) found an average positioning accuracy of 88% for stationary occupants and 62% for mobile occupants, although the active network infrastructure is expensive (Yang et al., 2016). However, RFID systems require occupants carry a dedicated receiver tag and require additional cost investments in implementing supporting facilities and infrastructures. Some researchers also investigated the application of motion sensors to track occupancy movement and behaviors, which achieves an average accuracy of 96% and 95% in offline and online modes, respectively (Soltanaghaei & Whitehouse, 2017). At the Massachusetts Institute of Technology, Fadel Adib et al. have researched how to track an occupant through walls via Wi-Fi signals and body radio reflections using Radio Frequency (RF)-capture and WiTrack methods (Adib et al., 2015). However, this method faces some constraints when cooperating with an HVAC system, including high cost when detecting multiple occupants in a large space where Wi-Fi signal reflection will be distributed by other bodies, and real-time tracking detection is not very applicable in HVAC controls.

The International Energy Agency's Energy in Buildings and Communities Programme Annex 66 project highlighted significant roles of occupant behavior in a building performance study (Yan et al., 2017). The American Time Use Survey (ATUS), conducted by the U.S. Bureau of Labor Statistics, provides access to records of respondents' activities and locations during a regular day (U.S. Bureau of Labor Statistics, 2017). Diao et al. (2017) proposed identifying and classifying occupants' activities with direct energy consumption outcomes and energy time use data through k-modes clustering, probability neural networks, and an inhomogeneous Markov chain model based on ATUS data. Researchers concluded that building energy use patterns could be observed and investigated with occupancy behavior information. Yu et al. (2011) employed k-Means clustering analysis to examine the effects of different behavior patterns on energy consumption, revealing four identified behavior clusters can be significantly attributed to some end-use loads, such as the HVAC load. Sun and Hong (2017) defined five occupant activity measures in office buildings—lighting, plug load, thermal comfort criteria, HVAC control, and window control—and simulated and

**Table 15.1** Overview of typical occupancy sensing technologies and example studies

| Occupancy Resolution | Sensing technologies | Applications |
|---|---|---|
| Presence | PIR, Video/Camera, Switch (e.g. Door), RFID, Wi-Fi/Bluetooth | ON/OFF control of lighting, HVAC, and other systems. |
| Count | PIR, Video/Camera, Switch (e.g. Door), RFID, CO2, Consumption sensing (e.g. Energy), Wi-Fi/Bluetooth | Demand-based controls over HVAC systems |
| Identity | Video/Camera, RFID, Wi-Fi/Bluetooth | Individual feedback |
| Location | Video/Camera, RFID, Wi-Fi/Bluetooth | Review on various indoor positioning systems |
| Tracking | Video/Camera, RFID, Wi-Fi/Bluetooth | Indoor positioning systems |
| Activity | Video/Camera, Switch (e.g. Door), Consumption sensing (e.g. Energy) | Indirect activity identification |

analyzed the individual and integrated impact of these measures on building energy use. Table 15.1 gives an overview of related studies and typical sensing technologies for different occupancy resolutions.

## 15.6 Design of Building occupancy Smart Sensing with Wi-Fi Network

### 15.6.1 Experiment Design of Low-Invasive Occupancy Sensing

To obtain occupancy profiles, the Wi-Fi probe enabled occupancy sensing experiment is conducted in two open office rooms at the City University of Hong Kong campus and both are in use for Ph.D. research students in the department of architecture and civil engineering. Three types of days (weekday, Saturday, and holiday) were tested. During the scan of Wi-Fi probe devices, the operation frequency of those devices is set as 1 minute, which means it can observe Wi-Fi requests and responses from the occupancy every minute, and in this case, the MAC address is regarded as the identification of occupants in the area. Figure 15.4 shows the schematic diagrams of the two types of rooms. During the test period, there are about 6–7 available assess points (APs) covering the places serving with some AP SSIDs on this campus, such as University WiFi, CityUGuest, and so on. Those APs have already created the overlapping and free signal network in those areas. Meanwhile, computers in those zones would work with Wi-Fi signal aside from the occupants' mobile devices, and it is inaccessible to figure out which AP the indoor users would connect to. Open office room A is a large-scale room with a gross area of about 200 square meters and a design maximum of 68 seats, but only about 25 users available during the experiment period. Room B is one small-scale room with only about 25 square meters and a design maximum of 10 seats, but only 3 users available during the experiment period.

To acquire the actual occupancy data, this research installed cameras around the doors of test office rooms. Figure 15.5 shows the installation of the Wi-Fi probe and cameras.

### 15.6.2 The Theory of Low-Invasive Occupancy Sensing

In modern buildings, there are many installed Wi-Fi signals, providing users with various points of access to the internet. Often, users also bring at least one mobile device to connect with. The coupled systems prototype in this experiment between users, access points, and the Wi-Fi probe is drawn in Fig. 15.6. Within general types of connection scanning from devices, there are two common types, active scanning (including direct scanning and broadcast scanning) and passive scanning (Sidiropoulos et al., 2012). In

**Fig. 15.4** Schematic diagrams of open office room A (left) and room B (right)

**Fig. 15.5** Schematic diagrams of Wi-Fi (left) and camera (right)

direct scanning, the client sends a probe request to the named and specific SSID AP that user would like to connect to, and only this requested SSID AP can reply with a probe response. In the broadcast scanning, the clients would broadcast a probe request with a null SSID that all APs in the zone would receive and respond with a probe response. The client could choose one preferred AP to connect to. In passive scanning, available APs would act as a beacon to broadcast around, and the client decides if one connection is requested, instead of actively sending a probe request. The request and response would be captured by the Wi-Fi probe with a timestamp and the MAC address of each client, and the information would be remotely uploaded to the server.

When the Wi-Fi logs indicate a client's device sends a connection request to one AP or could receive the response, we infer the client is within the detection area of the Wi-Fi probe. The following pseudocode shows the detection processing. In this case, each MAC address would represent the one-to-one client. The time t0 shows detection start time as the sign that the first device communicates an authentication request to one AP through active scanning or passive scanning, and the MAC address would be recorded. In the interval time counted from t0, keep note down the number, MAC address of all the clients. At the end, the detection procedure would finish with a note of the MAC address of the last client.

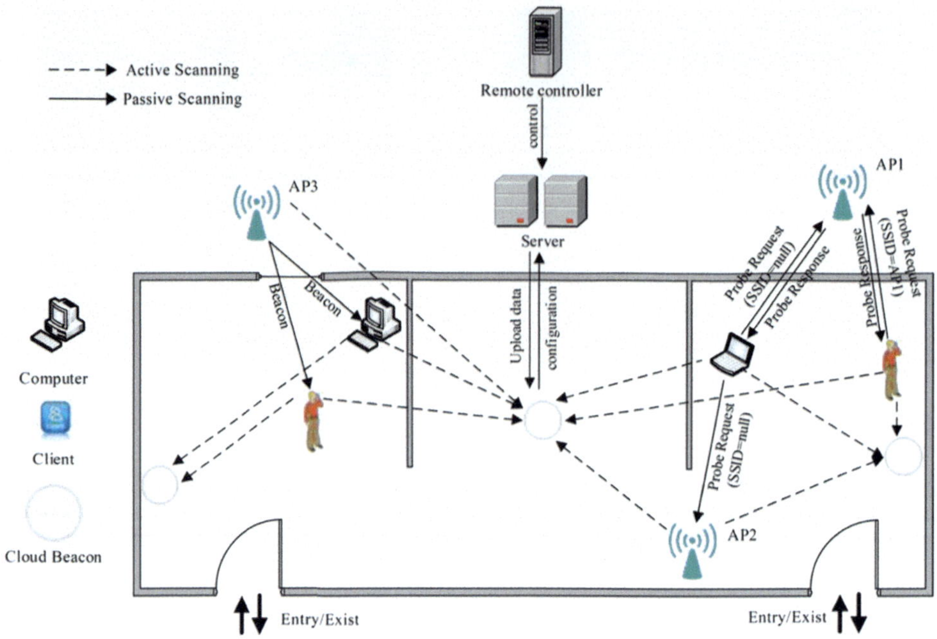


**Fig. 15.6** Profile of request and response between occupants and APs

Pseudocode:

```
Set N(t) = 0, Tbegin = t0, Tend = t1, T =
test time series(0,t1-t0), Δt = test time
interval
Detect:
For t =t0:
  If SSID1 (probe request at time t0 ):
  N(t) =1, t(SSID)= Tbegin; MAC (client1)
Else for t in time series T:
  for i in time series T/Δt
  if SSID (probe request at time t0+ i*Δt):
 N(t) = N(SSID), t (SSID) = Tbegin + i*Δt,
Mac(client)
 return N(t), t(SSID), Mac(client)
 end
Else for t = t1:
 if SSID2 (probe request at time t1):
 N(t) = 1, t(SSID) = Tend; Mac(client2)
Return N(t), t(SSID), Mac(client)
End
```

### 15.6.3 Raw Data Preprocessing and Duration Time Functions

In this case, the Wi-Fi probe detection system would automatically record MAC addresses during the request and response from clients at the fixed resolution time interval. To filter computers and temporary occupants, we choose a 20 minute time window, given the occupancy data has a resolution of 1 minute, and the 20 minute time window could include a maximum of 20 occupied times for each possible client. As shown in Fig. 15.7, we could formulate the number of times that one client (c1) is detected at every record for the detection of the client in one window at the time interval:

$$l_{\Delta t} = l_{(n+1)}\big|_{t=tn+\Delta t} - l_{(n)}\big|_{t=tn} \tag{15.1}$$

In one window, we could format

$$l_{\Delta t} \xrightarrow{count} n_{\Delta t} = 1 \tag{15.2}$$

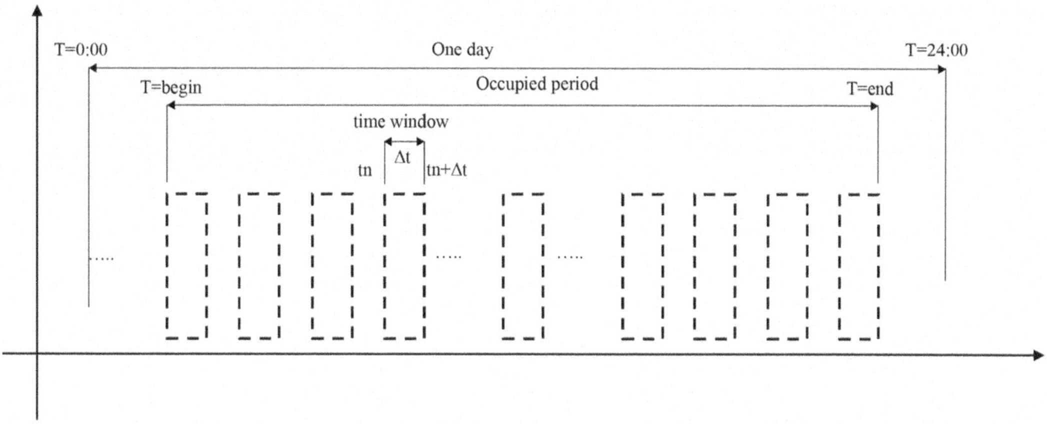


**Fig. 15.7** Illustrated time window approach during occupied period

$$N_{\Delta t} = \sum n_{\Delta t} \tag{15.3}$$

where $l_n$ represents the number of times the client is detected from T = 0:00 to tn, and $l_{\Delta t}$ is the number of times one client (c1) is detected in the time window. That $n_{\Delta t} = 1$ means the MAC address of $l_{\Delta t}$ for client c1 counts the number of occupant as one in this window. $N_{\Delta t}$ counts the total number of occupants in this time window.

The occupancy schedule in the time window can be profiled with one two-dimension coordinate of (Δt, $N_{\Delta t}$). For the whole day, the occupancy schedule can be profiled at each time window, and such occupancy would be used in the next occupancy profile models.

For this client c1, the duration time Δt in this time window can be defined as

$$\Delta t = (l_{\Delta t} - 1) * t_R \tag{15.4}$$

while $t_R$ is the resolution in this case.

Define the duration time function for client c1 as $f(t)$

$$f(t)|_{client} = \sum_{i=1}^{k=\frac{24*60}{\Delta T}} \Delta t = \sum_{i=1}^{k} \times \left(l_{n+1}|_{t=tn+\Delta t} - l_n|_{t=tn} - 1\right) * t_R \tag{15.5}$$

In this case, the occupied patterns from computers, temporary occupants, and permanent occupants are different in the duration time, dramatically affecting the occupancy schedule profile. Depending on user habits, some occupants might keep computers in working status over 24 hours in case of personal need, while others might turn off computers when they leave and re-open the computers in the morning. However, computers, especially desktops, would not move in or out of the room more frequently than occupants, meaning computers would be under the detection of the Wi-Fi probe for a long time during the occupied time, while the occupants could be out of the room for the toilet, rest, lunch, or a visit to another room. Therefore, in this case, the time slot can be divided into two parts, occupied time (from T = begin to T = end) and unoccupied time (from T = 0:00 to T = begin and T = end to T = 24:00). The duration time function for one client can be described as one kind of piecewise function. It would be obvious when the duration time function $f(t)$ is continuous in the unoccupied time; the MAC address of the client should be the computer's or the non-occupant-related clients in this case. While in the occupied time, it would be confused about these clients. So in this book, it is defined that the temporary occupants are those which the duration time in one day is less than 20 minutes. The occupancy schedule would take no account of the corresponding MAC addresses of those clients. However, if the duration time of clients takes up the whole occupancy time, the corresponding MAC addresses of those clients would be ignored since those devices were

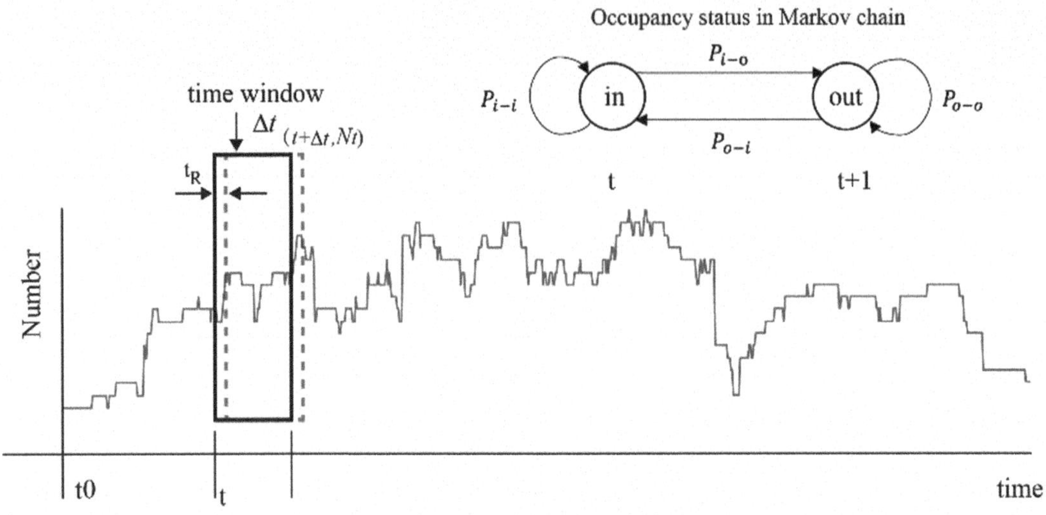


**Fig. 15.8** Illustration of the time-window based Markov chain model

observed during the detection periods as having an extremely long connection period, which is approximate to 14 hours in this book (the whole open schedule from 7:00 to 23:00) because clients in the office might go out for dinners or other private absences. Therefore, as this book sought to observe occupancy profiled by permanent occupants, such MAC addresses in the detection would be excluded.

The Markov chain is a classical method to simulate stochastic processes through analyzing the transitions from one state to another (Dodier et al., 2006; Chen et al., 2015; Yang & Becerik-Gerber, 2014). In the Markov chain model, there are two important input parameters considered in the process: the initial status and transfer probability. This book considered two occupancy statuses as "in" and "out." If the MAC address of one client was detected by the Wi-Fi probe detection, we describe this client as "in," otherwise, we describe this client as "out." The conventional stochastic model would underperform if occupancy status only considers the current status of MAC addresses. Some MAC addresses, which might be "out of room" at the current time, would still be "in the room" with much probability at the next time. To overcome this problem, the Markov chain proposed in this book would like to apply the moving time window approach, and all possible MAC addresses in the time window would be considered. It assumed that if the MAC address of one client has been detected in the current time window, it would still possibly be detected in the next time window. Figure 15.8 shows the illustration of the time-window based Markov chain model.

Since there are two initial occupancy statuses, in or out, the transfer probability indicates the probability of status changes from time t to time t + 1, denoted as

$$TPM|_t = \begin{bmatrix} P_t^{i-o} & P_t^{i-i} \\ P_t^{o-o} & P_t^{o-i} \end{bmatrix} \ (t > 1) \tag{15.6}$$

where $TPM|_t$ means the transition probability matrix at the current time $t$ of one occupant. In the transfer matrix, $P_t^{i-i}$ denotes the observed probability that one occupant whose status is "in" at the current time $t$ would be still "in" at the next time, and $P_t^{o-i}$ denotes the observed probability that one occupant whose status is "out" at the current time $t$ would be "in" at the next time. The probability could be calculated by the observable conditional probability. For example, it could calculate $P^{o-i}$ as

$$P_t^{o-i} = P(observed\ state = \mathrm{i} | observed\ state = o)\ (t > 1) \tag{15.7}$$

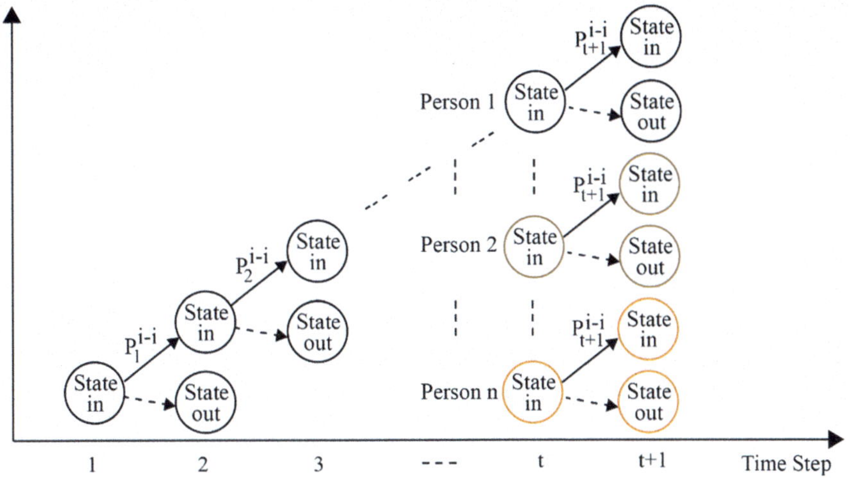


**Fig. 15.9** A sample process of the Markov chain

Figure 15.9 shows a sample process of the Markov chain. Supposing there are totally n1, n2 detected MAC addresses in the current time window and next time window, respectively, and occupancy resolution is $t_R$. All possible MAC addresses could be included in the MAC address Hamming matrix, and the occupancy status of MAC addresses at one time in the matrix should equal two values, in or out.

Moving Window

$t \rightarrow t + t_R$

$$
\begin{matrix}
M_{1,t} & M_{1,t+1} & \cdots & M_{1,t+j} & \cdots & M_{1,t+\Delta t} & & M_{1,t+t_R} & M_{1,t+1+t_R} & \cdots & M_{1,t+j+t_R} & \cdots & M_{1,t+\Delta t+t_R} \\
M_{2,t} & M_{2,t+1} & \cdots & M_{2,t+j} & \cdots & M_{2,t+\Delta t} & & M_{2,t+t_R} & M_{2,t+1+t_R} & \cdots & M_{2,t+j+t_R} & \cdots & M_{2,t+\Delta t+t_R} \\
\vdots & \vdots & & \vdots & & \vdots & & \vdots & \vdots & & \vdots & & \vdots \\
M_{i,t} & M_{i,t+1} & \cdots & M_{i,t+j} & \cdots & M_{i,t+\Delta t} & & M_{i,t+t_R} & M_{i,t+1+t_R} & \cdots & M_{i,t+j+t_R} & \cdots & M_{i,t+\Delta t+t_R} \\
\vdots & \vdots & & \vdots & & \vdots & & \vdots & \vdots & & \vdots & & \vdots \\
M_{n_1,t} & M_{n_1,t+1} & \cdots & M_{n_1,t+j} & \cdots & M_{n_1,t+\Delta t} & & M_{n_2,t+t_R} & M_{n_2,t+1+t_R} & \cdots & M_{n_2,t+j+t_R} & \cdots & M_{n_2,t+\Delta t+t_R}
\end{matrix}
$$

We simplify it that "in" equals "1" and "out" equals "0," so we can change the status of the MAC address in the transfer matrix as 1—1,1—0,0—0,0—1. Based on Eq.15.7, the transfer probability can be calculated:

$$
\begin{aligned}
P_{i-i} &= P_{1-1} \\
&= \frac{\sum N_{1-1}}{\sum N_{1-1} + \sum N_{1-0}} \qquad P_{o-o} \\
&= P_{0-0} = \frac{\sum N_{0-0}}{\sum N_{0-0} + \sum N_{0-1}}
\end{aligned}
\tag{15.8}
$$

However, for the whole day, transition probability would be different. For example, if one occupant occupies during working time, he or she would still occupy with higher probability. If one occupant occupies during overtime, he or she would still occupy with lower probability. In order to eliminate the impact of occupancy behaviors, this book adopted the piecewise calculation of transfer probability in the Markov chain model. In the study of Simona (D'Oca & Hong, 2015), occupancy schedules for a typical office

**Table 15.2** The divided occupancy pattern in this book

| Pattern | Going to work | Working | Lunch | Going off work |
|---|---|---|---|---|
| Schedule | 8–10:00 | 10–12:00, 13:30–18:00 | 12–13:30 | 18–23:00 |

could be clustered into the main typical working activities for four mined patterns: (1) going to work (P1) increases in the global occupancy curve; (2) working (P2) stablizes the global occupancy curve; (3) lunch/breakfast (P3) one valley decreases in global occupancy curve; (4) going off work (P4) decreases in the global occupancy curve. Based on those, we divide our office schedule as follows (Table 15.2):

Therefore, transition probability would be calculated in a sectional pattern for the whole day. Finally, in the chosen time window, we get the total number of MAC address detected in this time window ($\Delta t$, $N_{\Delta t}$) and the occupancy status at the time ($t$, $N_t$). In this case, we can get

$$N_{t+1} = \sum_{n=1}^{N(t)} 1 * P_{1-1}^{n} + \sum_{n=1}^{N_{\Delta t} - N(t)} 1 * P_{0-1}^{n} \quad t \in [P1, P2, P3, P4] \tag{15.9}$$

Those occupancy models are completely based on the inputs of observed occupancy results from Wi-Fi probe devices and applied in profiling occupancy schedules of two rooms and each type of day.

### 15.6.4 Predicted Occupancy for HVAC Operation

Although the presence of occupants may change temporarily, it also could be observed that habitual occupancy profiles would follow a similar pattern over a long period of time, and it is possible to develop specific occupancy profiles for different types of days. In the occupancy results from the Wi-Fi probe devices, the occupancy pattern is too undulatory to input in the occupancy-related HVAC studies. Therefore, in this book, we proposed a dynamic Markov time-window prediction model considering five mentioned occupancy characteristics. The other two modeling techniques, pattern regression and time-series model, are chosen for the comparison of proposed occupancy models. The profile process could be concluded as follows. First, occupancy profile is the presence accumulation of occupants in a time series, while unoccupied status or predicted unoccupied status would be excluded. Second, since the occupancy input would influence the HVAC adjustment frequency and frequent adjustment should be avoided, a 5 minute resolution of occupancy is selected for profile design. Thirdly, the 5 minute size of the moving time window would be used to profile the possible occupancy. Finally, four mined patterns (P1-P4) are chosen in sectionalized calculations of the transfer matrix of each detected and valid MAC address. In this case, we only modeled the occupancy profiles of large-scale room A from 08:00 am to 23:00 pm and three types of experiment days (weekday, Saturday and holiday) were chosen. Figures 15.10, 15.11, and 15.12 show the occupancy profile modeling results of a weekday, holiday, and Saturday.

## 15.7 Energy-Efficient Applications of Building occupancy Sensing

### 15.7.1 Demand-Based Control for Outdoor Air Circulation

This strategy, proposed in this book, is based on the predictive occupancy information through the received Wi-Fi signals. Since the selection of the critical zone could dramatically affect the system settings and energy consumption, the proposed strategy utilizes the detected and predicted occupancy level rather than CO2 concentration as the reference. Given the slow dilution process of CO2, the change of CO2 concentration level often significantly lags behind human breathing and mechanical ventilation. Therefore, this book utilized a more responsible Wi-Fi detected occupancy profiles. Figure 15.13 illustrates a scheme

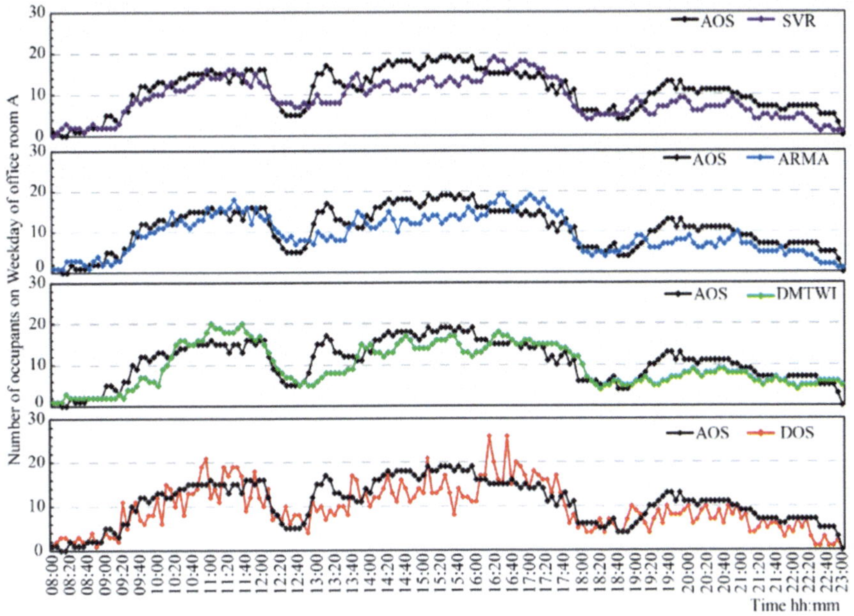


**Fig. 15.10** Occupancy results under different models on a weekday

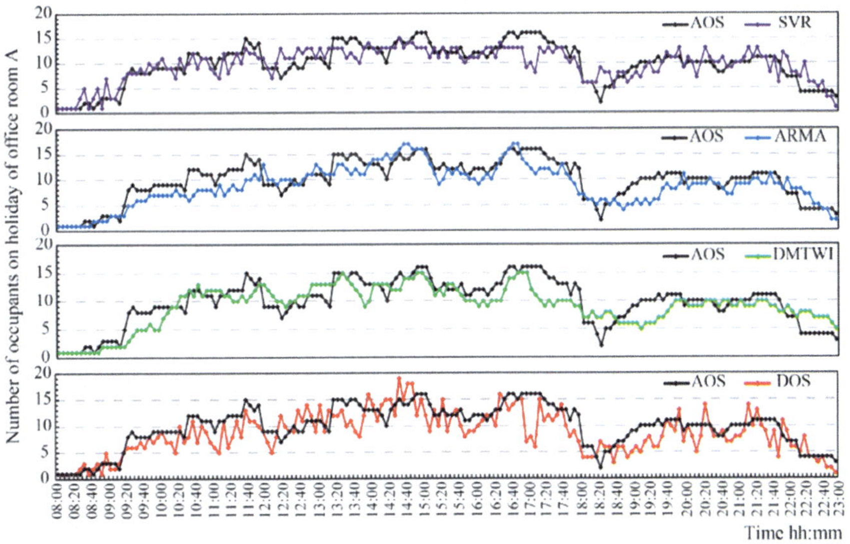


**Fig. 15.11** Occupancy results under different models on a holiday

of occupancy detection systems based on Wi-Fi probe and environmental sensors. In the proposed method, the Wi-Fi probe actively attempts to record the signal request and response from users, even if no connection between the Access Points (AP) and user tags. Also, the unique MAC addresses of users' device is logged as the occupants' identity. With a Markov time window, the signal processing system can estimate occupants' locations in a given space or zone (Wang et al., 2017b, d).

In conventional single-zone Markov chain model, either an "in" tag or an "out" tag is assigned to the temporal occupancy statuses. A

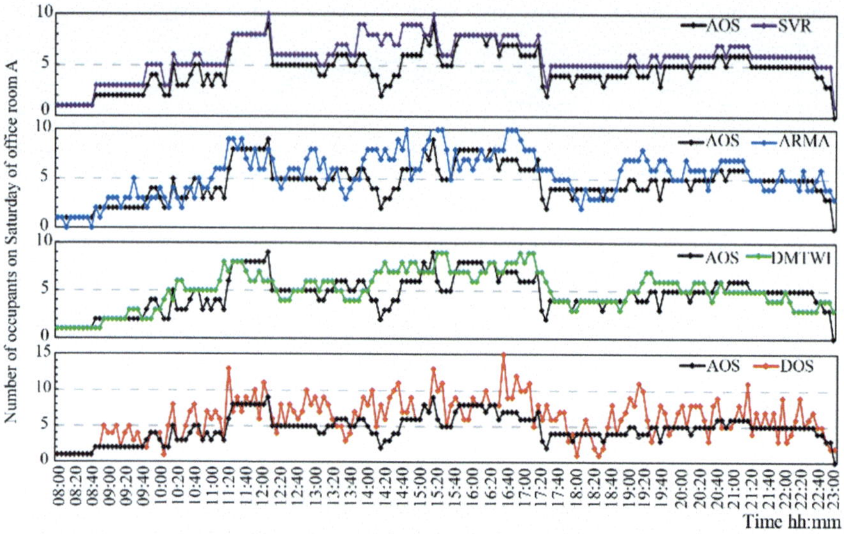


**Fig. 15.12** Occupancy results under different models on a Saturday

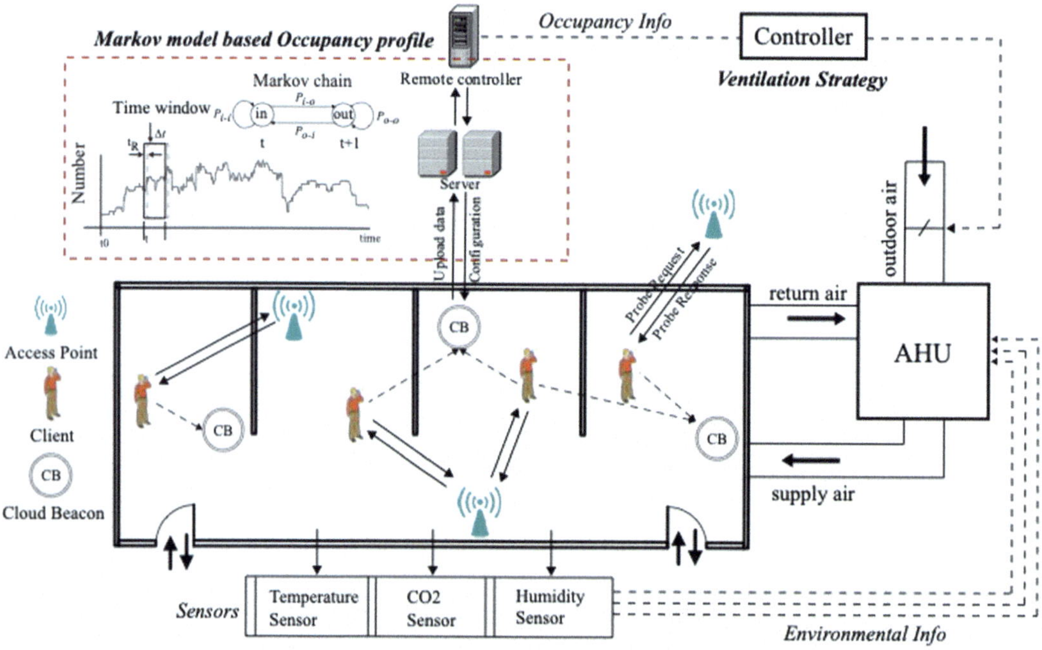


**Fig. 15.13** The illustration of occupancy sensing for demand-based outdoor air control

transfer probability matrix is calculated with historical records to represent the probability of status changes from time $t$ to time $t + 1$.

$$TPM|_t = \begin{bmatrix} P_t^{in-out} & P_t^{in-in} \\ P_t^{out-out} & P_t^{out-in} \end{bmatrix} \ (t > 1) \qquad (15.10)$$

where $TPM|_t$ is the transfer matrix at $t$. $P_t^{in-out}$ and $P_t^{in-in}$ denote the observed probability that one occupant whose status is "in" will change to "out" or remain "in" at the next time period. $P_t^{out-in}$ and $P_t^{out-out}$ denote the observed probability that one occupant whose status is "out" will change to "in" or remain "out" at the next time period. These probabilities can be calculated with observable conditional probabilities. For example,

$$P_t^{out-in} = P(observed\ state = in | observed\ state = out)\ (t > 1) \quad (15.11)$$

Then, with Bayes' theorem,

$$p^{in-in} = \frac{\sum N_{in-in}}{\sum N_{in-in} + \sum N_{in-out}} \quad p^{out-in} = \frac{\sum N_{out-in}}{\sum N_{out-out} + \sum N_{out-in}} \quad (15.12)$$

To extend the current approach for a multi-zone space, for zone $i$, the total number of occupants in one time window is $P_{z,i}^{\Delta t}$. Then, the probability of $(P_{z,i}^{t+1})$ at time $t + 1$ is

$$P_{z,i}^{t+1} = \sum_{n=1}^{P_{z,i}^t} 1 * p^{in-in} + \sum_{n=1}^{P_{z,i}^{\Delta t} - P_{z,i}^t} 1 * p^{out-in} \quad (15.13)$$

Outdoor air flow rate requirement in zone $i$ can be formulated as:

$$X_{z,i} = \frac{C_{z,i} - C_o}{P_{z,i} * S} * V_{sa,z,i} \quad (15.14)$$

With Eq. 15.14, it is possible to find out the necessary ventilation requirements for each zone. If all zones served by the same AHU, then the total required outdoor air amount is

$$V_{oa,tot} = \sum_i V_{oa,z,i} \quad (15.15)$$

Therefore, it could find outdoor air requirement proportion for zone $i$ as

$$f_{z,i} = \frac{V_{oa,z,i}}{V_{oa,tot}} \quad (15.16)$$

The aggregated uncorrected outdoor air fraction for the AHU is

$$X_{uncorrect} = \sum_i f_{z,i} * X_{z,i} \quad (15.17)$$

The predicted outdoor air requirements for zone $i$ at next time period $(t + 1)$ are

$$V_{oa,z,i}^{t+1} = R_p * P_{z,i}^{t+1} + R_a * A_{z,i} \quad (15.18)$$

$$X_{z,i}^{t+1} = \frac{V_{oa,z,i}^{t+1}}{V_{sa,z,i}^{t+1}} \quad (15.19)$$

For the critical zone selection, this book proposed to select two critical zones, who have the highest and second highest demand based on predicted occupancy level. The fraction of at maximum load for first critical zone is

$$X_{critical1} = max\{X_{z,i}^{t+1}\} = max\left\{\frac{V_{o,z,i}^{t+1}}{V_{s,z,i}^{t+1}}\right\} \quad (15.20)$$

For second critical zone is

$$X_{critical2} = max\{X_{z,i}^{t+1} - \{X_{critical1}\}\} \quad (15.21)$$

Then, the averaged fraction for both critical zone is

$$Z = \frac{1}{2}(X_{critical1} + X_{critical2}) \quad (15.22)$$

The selection of second critical zone intends to avoid the exaggerated and biased demand when the demand the first critical zone is much higher than all other zones. Then the corrected fraction can be calculated through

$$X_{OA} = \frac{X_{uncorrect}}{1 + X_{uncorrect} - Z} = \frac{\sum_i f_{z,i} * X_{z,i}}{1 + \sum_i f_{z,i} * X_{z,i} - \frac{1}{2}(X_{critical1} + X_{critical2})} \quad (15.23)$$

### 15.7.2 Linking Building Energy-Cyber-Physical Systems with Building Occupancy

A fundamental e-CPSs framework includes at least a physical building system and cyber model for energy management and optimization. The physical building model reflects the actual conditions and performance of a building while the cyber model is a digital twin that can be used for various computational processes. The physical buildings usually have sensors and sensor network installed which allows acquiring the various types of environmental information, such as temperatures, CO2 concentration, and relative humidity (RH), and system operation information, such as supply/outdoor air flow rate and temperature, pump efficiency, and instantaneous energy load. The building information model is the key to associate both components and to create a dependable digital twin for the actual building. The building information model contains static features and dynamic operation settings. The static features include building materials, geometry, location, system type, and etc., while the dynamic operation settings include the operation schedule, efficiency, and settings of HVAC, lighting, and security systems.

To extend conventional e-CPSs, this book proposes to integrate dynamic occupancy information to enable data exchange between the physical building and cyber model. As the physical infrastructure of the building system, Wi-Fi networks were utilized to obtain the signal strength of occupants' device/tag. The obtained occupancy information serves as the inputs for a cyber model for data analysis and system optimization. To connect both components of e-CPSs, this book also developed an occupancy interpreter based on ensemble algorithms to convert Wi-Fi signal strengths to occupant number and schedule. Once detailed occupancy information is captured, the cyber model can conduct energy simulation with the building information model and suggest proper operational settings for the facility/mechanical systems. The Fig. 15.14 shows the structure of the proposed occupancy linked e-CPSs.

Figure 15.15 shows the energy cyber model applied in this book. The model was developed

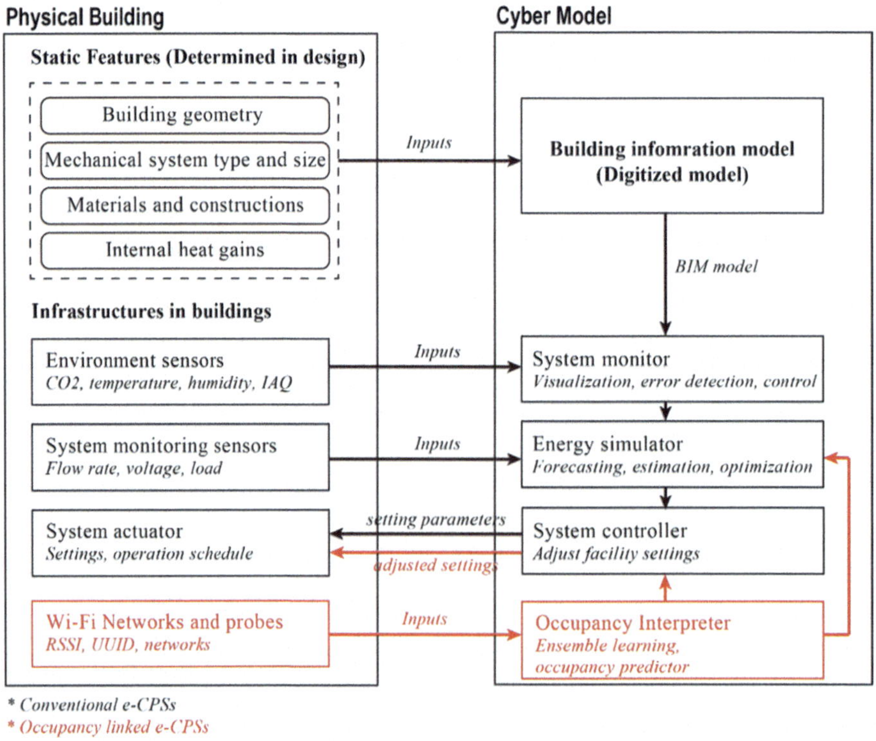


**Fig. 15.14** The scheme of the occupancy linked e-CPSs

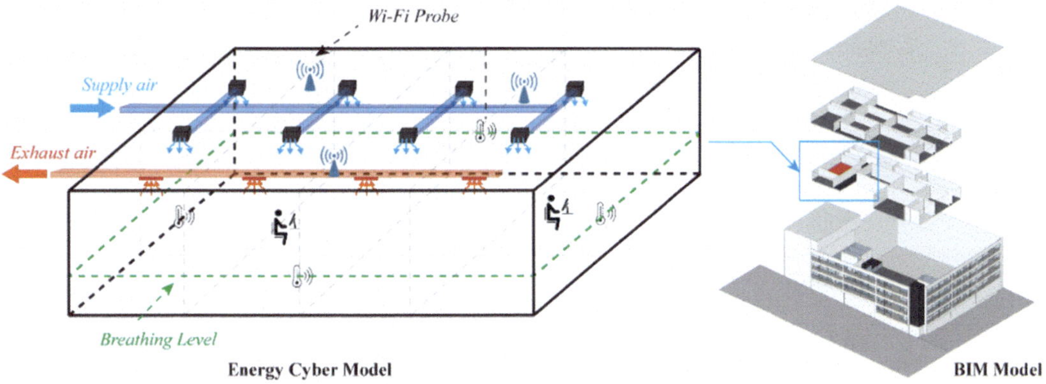


**Fig. 15.15** The energy cyber model for the experiment testbed

with EnergyPlus and DOE2 to optimize facility operation. Based on BIM models, the energy cyber model is able to incorporate construction materials, building geometries, and schedule of operation to estimate the energy consumption of the building. With co-simulation with other programming languages, such as Matlab or Python, the model is capable of tuning system settings to minimize energy consumption. This book employed Eppy, a Python package that can manipulate EnergyPlus IDF files (Eppy Tutorial — eppy 0.5.44 documentation, n.d.), to searching for the optimal system settings. It takes full advantage of the rich data structure and idioms that are available in Python and provide availability of designing expected energy model and algorithm to integrate physical and cyber models. Eppy can help programmatically navigate, search, and modify EnergyPlus IDF files. Users can use Eppy to create one or multi new IDF files, or makes changes with original IDF files, e.g. to change construction for all walls, occupancy information in all the interior zones, and read data from the output files after EnergyPlus simulation run. Related to occupancy linked e-CPSs, Eppy provides a direct way to link occupancy results from ensemble models as the input of cyber energy model through Python.

The cyber model matches the physical room with a size of 20 m (length) x 10 m (width) x 3 m (height) and 20 occupants. Internal heat sources were set as 75 W for per person, 150 W for per computer, and 35 W for per lamp. The light schedule followed the on/off schedule and the schedule for computers was assumed to same as the occupancy schedule. Hong Kong has a subtropical climate and is unique for its high-density high-rise urban form. According to statistics (Ang et al., 2017), the typical mean, minimum, maximum values of monthly average temperature are around 23.4 °C, 13.3 °C, and 29.8 °C, respectively. Also, relative humidity (RH) of Hong Kong is high and minimum, maximum values of monthly average RH are 78.2%, 60%, and 90%, respectively. The weather boundary used typical Hong Kong weather condition and the heat transfers from wall, floor, and ceiling were ignored since the experiment was one inner zone adjacent to conditioned zones. The cooling temperature setting point is 24 °C and heating condition is ignored.

### 15.7.3 Energy-Efficient HVAC Control Through Dynamic Spatial Occupancy Distribution

Conventional occupancy data usually quantifies the number of occupants in a given space as a single integer. However, occupants in a large space are not necessarily evenly distributed. In addition, they could frequently relocate themselves inside the space without the total number of occupants ever changing. It is essential to interpret occupancy profile with a spatial dimension that considers the synergetic operation of

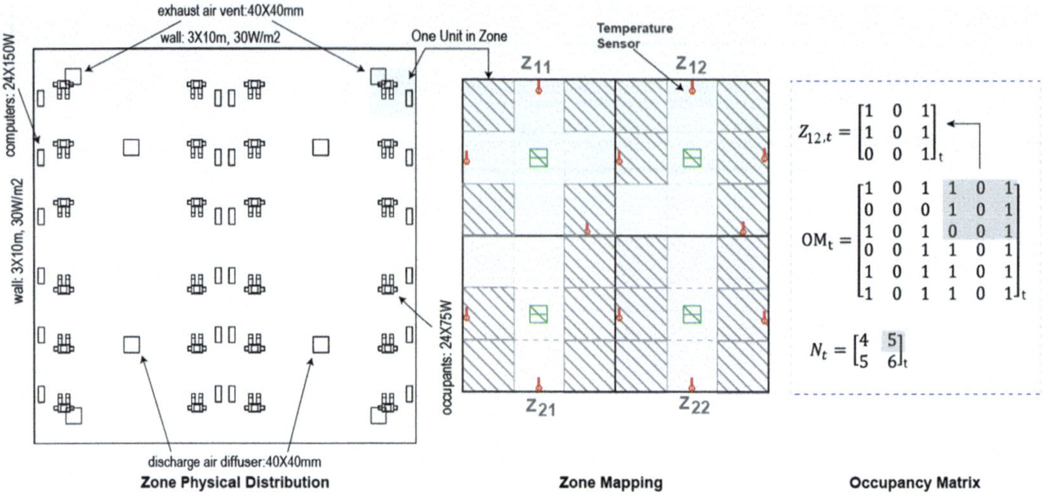


**Fig. 15.16** An illustrated example of the occupancy matrix in one zone

multi-zone VAV boxes in order to avoid overcooling or insufficient cooling. Thus, this book quantifies occupancy distribution as a temporally blocked sparse matrix (occupancy matrix). To retrieve the occupancy matrix, a 2-D space is divided into two scales: the zone scale and the patch scale. First, the 2-D space is divided into multiple zones based on the coverage of VAV terminals or thermal zones as in Fig. 15.16. All elements in a non-occupancy zone (zones with no residents, such as furniture spaces, structural columns, and inaccessible spaces) are defined as a zero matrix. Second, the occupancy zones (zones that may have occupants) are further divided into smaller patches. When there is an occupant in a patch, the element represents the patch as a one; when no occupant is in the patch, the element is a zero. Finally, the occupancy matrix is composed of multiple blocks (zones) with many zero elements (patches). The occupancy matrix is represented as a sparse matrix with non-zero blocks.

$$\mathrm{OM_t} = \begin{bmatrix} Z_{11} & \dots & Z_{1C} \\ \dots & Z_{cd} & \dots \\ Z_{C1} & \dots & Z_{CD} \end{bmatrix}, Z_{cd} = \begin{bmatrix} h_{11} & \dots & h_{1J} \\ \dots & h_{ij} & \dots \\ h_{I1} & \dots & h_{IJ} \end{bmatrix} \tag{15.24}$$

$$\mathrm{OM_t}((c-1)*I+i,\ (d-1)*J+j) = 1, \mathrm{N}_{Z_{cd}} = \sum_{1}^{i}\sum_{1}^{j}\delta(h_{ij}) \tag{15.25}$$

Where for $\forall Z_{cd} \neq 0$, $h_{ij} \neq 0$, $\mathrm{OM_t}((c-1)*I+i,$ $(d-1)*J+j)$ represents occupied patch of i-th row and j-th column of $Z_{cd}$ and $\delta(h_{ij}) = 1$ when $h_{ij} \neq 0$, otherwise, $\delta(h_{ij}) = 0$.

Therefore, no matter the shape of a large space, it can always be presented as a square matrix by adding zero blocks without affecting the computational complexity of the sparse matrix. The size of patches is determined by the positioning precision of the IPS. For example, in the BLE network, the patch is 1 meter by 1 meter square. Since the proposed occupancy matrix is temporal, it can be updated with dynamic occupancy.

The schematic confirmation of the distributed air handling unit (AHU) system of the test bed is shown in Fig. 15.17. As a typical multi-zone office room, each VAV box serves a signal zone. The system mixes the return air with the outdoor air at the fraction of $\gamma$. Before being delivered to the room, the supply air is conditioned to an operative supply air temperature $T^{SA}$ and an operative supply air humidity $H^{SA}$.

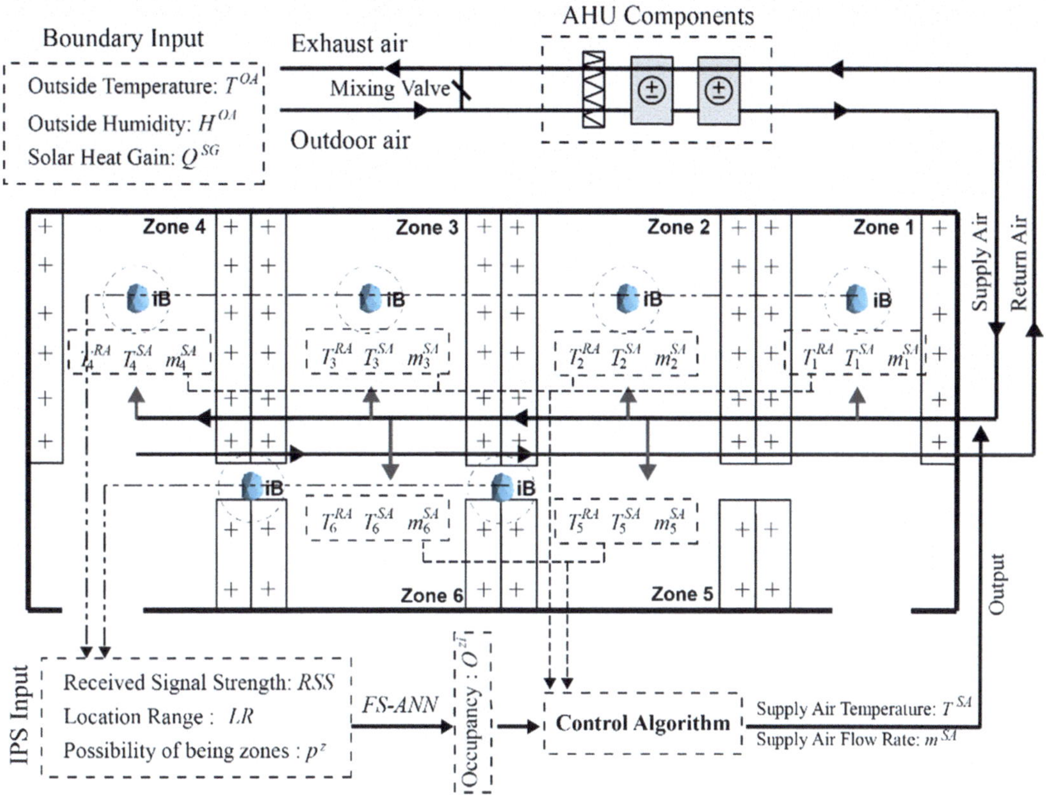


**Fig. 15.17** The control parameter configuration for each zone

$T^{OA}$ and $H^{OA}$ are used to represent the outdoor air temperature and humidity. The energy consumption of the system is calculated based on the supply airflow rate $m^{SA}$ during a given time.

## 15.7.4 CFD Simulation for Building Energy Assessment

In commercial buildings, the separation of thermal zones and corresponding subsystems allows facility managers to provide a flexible thermal comfort level with regard to space occupancy status. To examine the validity of the proposed control algorithm, Fluent Airpak was employed to simulate the temperature distribution of airflow patterns under different control strategies in a typical cooling season of Hong Kong. Fluent Airpak is one of the most popular commercial software programs used in simulating airflow, air quality, and contamination. It can construct realistic boundary conditions and predict the air spread and penetration in a confined room. The office room model is 20 m long, 10 m wide, and 3 m high. Occupancy information was collected from the IPS. To simplify the model, this book concludes the three assumptions: (1) heat transfer from adjacent walls was ignored since we assume that the model room was surrounded by conditioned areas. This assumption can help focus on that the temperature distribution is only influenced by the occupancy distribution and air terminals. (2) the VAV box in each zone is independent that they can be controlly individually. (3) for temperature distribution, the steady state simulation would be conducted for each case and the heat transfer mechanism between zones would not be included. Inner heat gain was generated by occupants (75 W each) and computers (150 W each). Supply air in each zone was delivered by side diffusers, and supply air temperature was controlled and kept at 19 °C.

For each controller, the temperature parameter requirement was diverse, and the temperature distribution field was uneven. However, the temperature distribution field was under the same occupancy distribution field, since the operation of air-conditioning or facilities inside the areas determines the temperature field. The specifications and physical conditions of the model room are illustrated in Fig. 15.18 and Table 15.3.

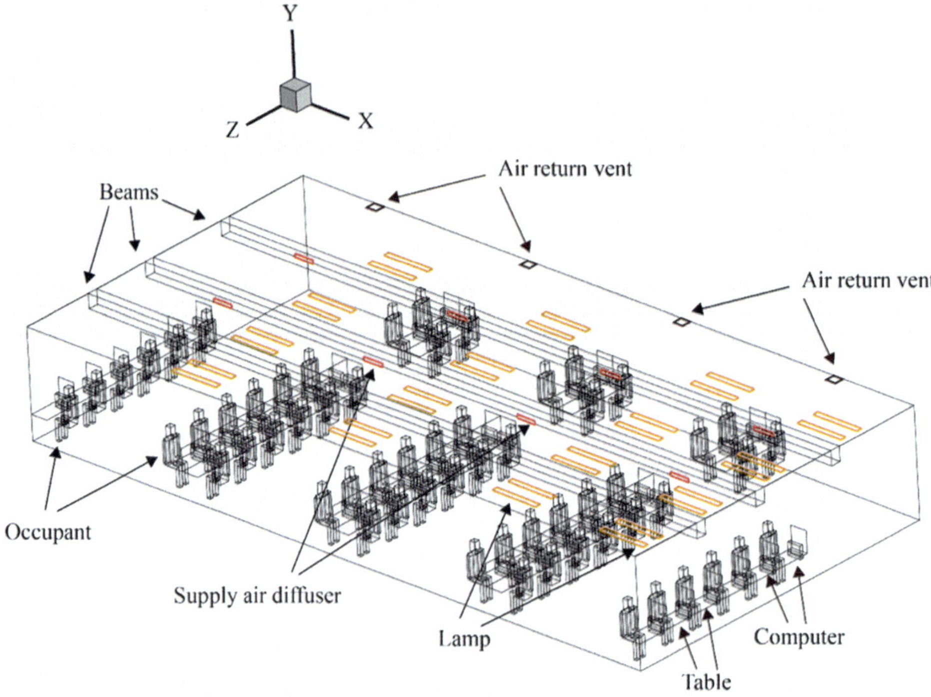


**Fig. 15.18** The CFD model configuration of the office room

**Table 15.3** Boundary condition setup

| | |
|---|---|
| Ventilation type | Mixing ventilation |
| Geometry | 20 m (length) x 10 m (width) x 3 m (height) |
| Diffuser type,<br>Dimensions and quantity | Square ceiling diffuser,<br>0.6 m x 0.1 m, 6 |
| Mesh type | Structured, one million |
| Discretization scheme | Second-order, SIMPLE, implicit |
| Air exhaust vent and quantity | 0.2 m x 0.2 m, 4, Normal pressure out |
| Heat sources | Occupant, 75 W/person<br>Computer, 150 W/each<br>Lamp, 35 W/per |
| Wall, floor and ceiling | Heat adiabat |
| Radiation | Discrete Ordinates (DO) |
| Air supply temperature | 19 °C |
| Air supply flow rate | 1950 $m^3$/h |
| Ambient temperature | 30 °C |
| Room temperature set | 24 ° C± 0.5 °C |
| Turbulence model | k-ε standard equations and standard wall equation |
| Residuals of convergence | Continuity, momentum, turbulent kinetic 10−4<br>Energy turbulent dissipation rate 10−6 |

### 15.7.5 Indoor Air Temperature Distribution

The aim is to match the conditioned temperature distribution with the occupancy distribution, thus, we proposed two standard scenarios as a demonstration for the comparison of all the controllers. Three controller were included, the proposed DTC, which is DSOD-Based Temperature Controller, conventional scenario-RATC, which is return air temperature based controller, and idle scenario-ATOC, which is actual occupancy based temperature controller.

The same parameters from the CFD model were used to examine the performance of the AOTC and DTC models. Figure 15.19 illustrates the simulated temperature distribution under the RATC, AOTC, and DTC at the height of 1.1 m (the breathing level). The figure reveals that the temperature distribution of the RATC is lower and more even than that of the other two controllers. The temperature is also maintained at a low level in unoccupied areas. In Case 1, all zones are occupied, and the temperature differences between the RATC and the AOTC and DTC were kept between 0.3 °C and 0.4 °C.

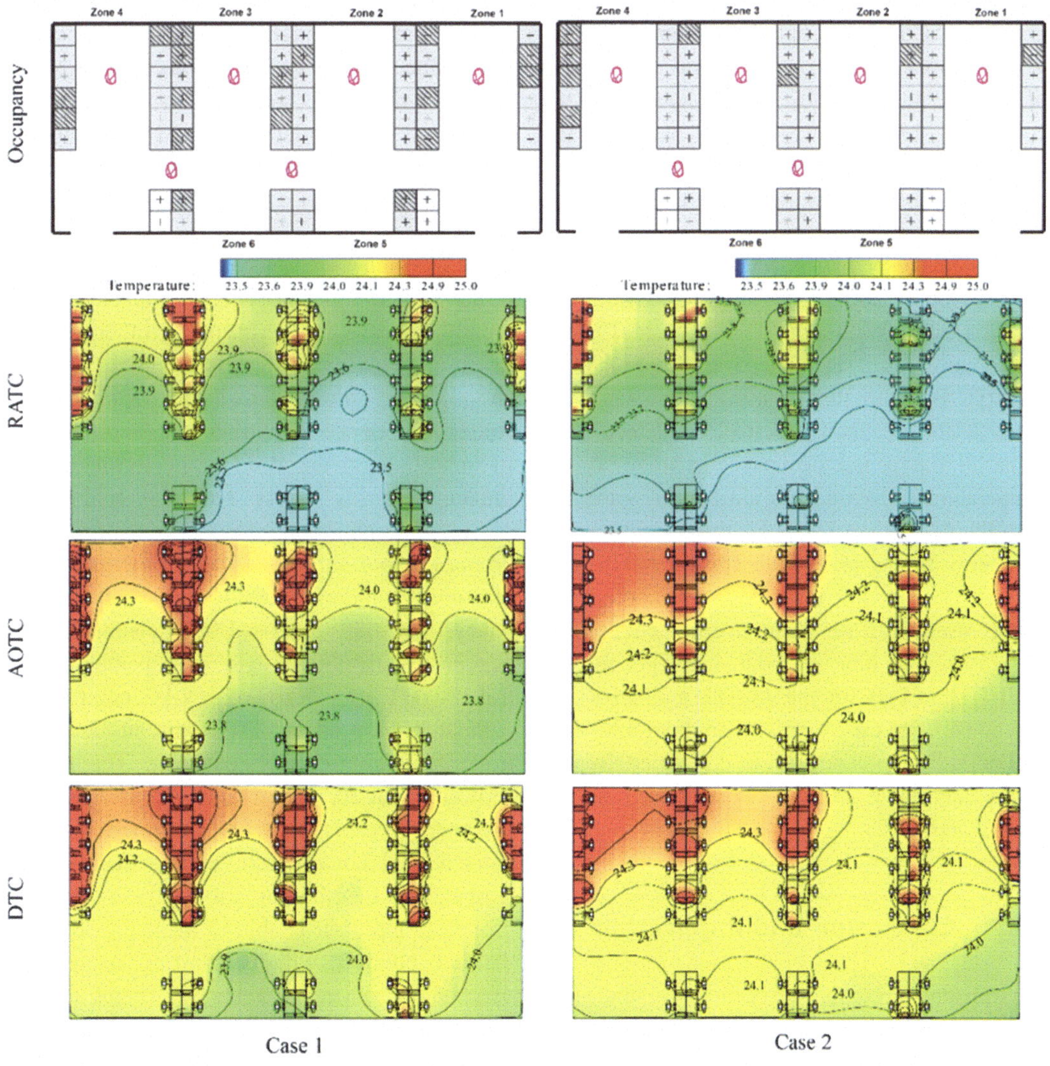


**Fig. 15.19** Temperature distribution simulation results for the three controllers

**Table 15.4** The representative temperature in occupied and unoccupied zones under two cases (°C)

| | | Case 1 | | | Case 2 | | |
|---|---|---|---|---|---|---|---|
| | | RATC | AOTC | DTC | RATC | AOTC | DTC |
| Zone 1 | $T^{occu}$ | 23.8 | 24.1 | 24.2 | 23.5 | 24.1 | 24.1 |
| Zone 2 | $T^{occu}$ | 23.8 | 24.1 | 24.2 | 23.7 | 24.1 | 24.1 |
| Zone 3 | $T^{occu}$ | 23.9 | 24.3 | 24.3 | 23.8 | 24.2 | 24.3 |
| Zone 4 | $T^{occu}$ | 23.9 | 24.3 | 24.3 | 23.8 | 24.3 | 24.3 |
| Zone 5 | $T^{occu}$ | 23.5 | 23.8 | 24 | | | |
| | $T^{unocc}$ | – | – | – | 23.5 | 24 | 24 |
| Zone 6 | $T^{occu}$ | 23.5 | 23.9 | 23.9 | – | – | – |
| | $T^{unocc}$ | – | – | – | 23.2 | 24 | 24 |

In Case 2, only Zones 1 through 4 were occupied, and temperature differences were kept between 0.4 °C and 0.6 °C for Zone 5 and between 0.5 °C and 0.8 °C for Zone 6, the unoccupied zones.

Table 15.4 provides a numerical comparison of average temperatures for all zones and all controllers. In addition, it shows based on the results of occupancy detection by the FS-ANN model that Zone 5 is an unoccupied zone in Case 1. Therefore, the detected occupancy matrix and actual occupancy matrix are different, and the cooling supply airflow rate in Zone 5 from the DTC is zero. With regard to Fig. 15.19, the temperature result in Zone 5 of Case 1 from the DTC is higher than it is from the AOTC. In Case 2, the occupancy matrix of detection is equal to the actual occupancy matrix, so the temperature distribution is almost the same for both the AOTC and DTC. Also, the result shows an energy saving potential from the AOTC and DTC by conserving the amount of cooling air supply used. In Case 1, the average cooling air amount in unit time is about 4.78 kg/s for the RATC, 3.56 kg/s for the AOTC, and 3.16 kg/s for the DTC. In Case 2, a total of about 4.63 kg/s of cooling supply airflow is consumed by the RATC and about 2.68 kg/s is used for both the AOTC and DTC.

## 15.8 Conclusion and Implication

This chapter presents the occupancy studies in aspects of WiFi experiment, data collection and pre-processing, model and prediction, application. With several experiments and simulations, this chapter can conclude the following conclusions. This research utilized the time-series and stochastic characteristics of detected WiFi request to develop a reliable and automatic occupancy detection and prediction mechanism. MAC addresses of devices were used as the identification tags to differentiate occupants and facilities. Conventional SVR ARMA is also used to benchmark the performance of the proposed Markov chain model. There are tolerances of occupancy about 2, 3, and 4 when accuracy is over 80%. The results suggest the proposed occupancy model greatly improve the performance of occupancy detection. Following up, this research proposed an occupancy-profile based multi-zone control strategy to operate a building ventilation system. In this strategy, the WiFi probe technology is implemented to detect real-time and high-resolution occupancy information to determine proper outdoor air portions for all thermal zones of a given building space. Under detected occupancy, this study proposed one dynamic spatial occupancy distribution-based temperature controller (DTC) for improve HVAC control and compared the proposed DTC with an ideal controller and an existing controller. The results showed about 20.14% of energy conservation in the energy simulation when compared with the conventional controller in 2 days. The spatial temperature distribution also suggests that the proposed approach is able to maintain occupant thermal comfort without wasting energy by overcooling unoccupied zones. Therefore, the proposed IPS-based demand-driven HVAC control mechanism has been proven to be valid and effective.

As far as contributions of this research, there are still some unsolved limitations and questions, which significantly constitutes the future works and implications in terms of expanding occupancy sensing experiment, comparisons, application. This section discusses some potential and significant future works. Currently, the occupancy detection experiment in this research is only conducted in one large-scale room. However, the multi-room experiments for occupancy sensing are quite important to (1) validate the occupancy detection and prediction methods in different types and capacities of rooms and (2) study occupancy movement information between different rooms and levels inside buildings. One important future work is to expand the WiFi-based experiment. Therefore, the first future work should investigate those questions by conducting more experiments in multi-rooms and levels insides buildings. Secondly, occupancy acquires many attentions in many years and has become one integrated research topic, which has involved various sensing technologies and model techniques. WiFi technology enabled occupancy sensing and model is one relatively novel research and should be compared with conventional methods comprehensively. On the other hand, how to get more accurate occupancy information through WiFi is also one essential work by combining WiFi technology with conventional ones. Those issues are not discussed as limitations in the current study, however, they inspire interesting works in the future to combine WiFi technology with indoor air parameters, such as CO2 concentration, temperature, and so on, to create more accurate and flexible occupancy detection method.

Moreover, since WiFi is one efficiently and widely used in modern buildings, it provides the convenient ways to interact with the online services. On the other hand, to provide a graphical user interface (GUI) of the building- or urban-related information is proved as a significant and effective way for building and urban energy management. Therefore, one important future work is to build and provide the WiFi enabled occupancy information interface to visualize the building occupancy information. The contribution of occupancy interface can facilitate the building facility manager to acquire occupancy status of building spaces in real time and accurately, in response to (1) reassign the building space for a better use and energy savings, (2) close or minimize the building facility in unoccupied or low occupied building space, (3) adjust the operations of building facility based on occupancy demand to avoid the extra energy supply. Another potential future work continuing the second one is to visualize and analyze the occupancy information and building energy pattern. One of the most important contributions of occupancy information is to benefit the building energy management, making it valuable to achieve the pattern recognition of occupancy information and building energy information. Data mining and pattern recognition are the key ways to learn occupancy related energy information.

## References

Adib, F., Hsu, C-Y., Mao, H., Katabi, D., & Durand, F. (2015). *RF-Capture: Capturing the human figure through a wall.* http://rfcapture.csail.mit.edu/index.html

Ang, B. W., Wang, H., & Ma, X. (2017). Climatic influence on electricity consumption: The case of Singapore and Hong Kong. *Energy, 127*, 534–543. https://doi.org/10.1016/j.energy.2017.04.005

ASHRAE. (n.d.). *ASHRAE Standard 90.1-2007: Energy standard for buildings except low-rise residential buildings.*

Benezeth, Y., Laurent, H., Emile, B., & Rosenberger, C. (2011). Towards a sensor for detecting human presence and characterizing activity. *Energy and Buildings, 43*, 305–314. https://doi.org/10.1016/j.enbuild.2010.09.014

Chen, J., Jain, R. K., & Taylor, J. E. (2013). Block Configuration Modeling: A novel simulation model to emulate building occupant peer networks and their impact on building energy consumption. *Applied Energy, 105*, 358–368. https://doi.org/10.1016/j.apenergy.2012.12.036

Chen, Z., Xu, J., & Soh, Y. C. (2015). Modeling regular occupancy in commercial buildings using stochastic models. *Energy and Buildings, 103*, 216–223. https://doi.org/10.1016/j.enbuild.2015.06.009

Chen, Y., Hong, T., & Luo, X. (2017a). An agent-based stochastic Occupancy Simulator. *Building Simulation.* https://doi.org/10.1007/s12273-017-0379-7

Chen, Y., Liang, X., Hong, T., & Luo, X. (2017b). Simulation and visualization of energy-related occupant behavior in office buildings. *Building Simulation, 10*, 785–798. https://doi.org/10.1007/s12273-017-0355-2

Christensen, K., Melfi, R., Nordman, B., Rosenblum, B., & Viera, R. (2014). Using existing network infrastructure to estimate building occupancy and control plugged-in devices in user workspaces. *International Journal of Communication Networks and Distributed Systems, 12*, 4. https://doi.org/10.1504/IJCNDS.2014.057985

D'Oca, S., & Hong, T. (2015). Occupancy schedules learning process through a data mining framework. *Energy and Buildings, 88*, 395–408. https://doi.org/10.1016/j.enbuild.2014.11.065

Dabaieh, M. (2022). Circular architecture and building design. In *Circular design for zero emission architecture and building practice*. Woodhead Publishing.

Davis, J. A., & Nutter, D. W. (2010). Occupancy diversity factors for common university building types. *Energy and Buildings, 42*, 1543–1551.

Diao, L., Sun, Y., Chen, Z., & Chen, J. (2017). Modeling energy consumption in residential buildings: A bottom-up analysis based on occupant behavior pattern clustering and stochastic simulation. *Energy and Buildings, 147*, 47–66. https://doi.org/10.1016/J.ENBUILD.2017.04.072

Díaz, J. A., & Jiménez, M. J. (2017). Experimental assessment of room occupancy patterns in an office building. Comparison of different approaches based on CO 2 concentrations and computer power consumption. *Applied Energy, 199*, 121–141. https://doi.org/10.1016/j.apenergy.2017.04.082

Dodier, R. H., Henze, G. P., Tiller, D. K., & Guo, X. (2006). Building occupancy detection through sensor belief networks. *Energy and Buildings, 38*, 1033–1043. https://doi.org/10.1016/j.enbuild.2005.12.001

Du, X., Song, M., & Li, L. (2016). The energy efficiency assessment and optimization of collaborative working air-conditions. *Energy and Buildings, 133*, 88–95. https://doi.org/10.1016/j.enbuild.2016.09.015

Ekwevigbe, T., Brown, N., Pakka, V. H., & Fan, D. (2013). Real-time building occupancy sensing for supporting demand driven HVAC operations. *International Conference for Enhanced Building Operations*.

Ekwevugbe, T., Brown, N., Pakka, V., & Fan, D. (2013). Real-time building occupancy sensing using neural-network based sensor network. In *2013 The 7th IEEE Conference on Information Science and Technology* (pp. 114–119). IEEE. https://doi.org/10.1109/DEST.2013.6611339

Eppy Tutorial — eppy 0.5.44 documentation. (n.d.). https://pythonhosted.org/eppy/Main_Tutorial.html

Foster, G. (2020). Circular economy strategies for adaptive reuse of cultural heritage buildings to reduce environmental impacts. *Resources, Conservation and Recycling, 152*, 104507. https://doi.org/10.1016/j.resconrec.2019.104507

Garg, V., & Bansal, N. K. (2000). Smart occupancy sensors to reduce energy consumption. *Energy and Buildings, 32*, 81–87. https://doi.org/10.1016/S0378-7788(99)00040-7

González, A., Sendra, C., Herena, A., Rosquillas, M., & Vaz, D. (2021). Methodology to assess the circularity in building construction and refurbishment activities. *Resources, Conservation & Recycling Advances, 12*. https://doi.org/10.1016/j.rcradv.2021.200051

Goyal, S., Ingley, H. A., & Barooah, P. (2013). Occupancy-based zone-climate control for energy-efficient buildings: Complexity vs. performance. *Applied Energy, 106*, 209–221. https://doi.org/10.1016/j.apenergy.2013.01.039

Gunay, H. B., O'Brien, W., & Beausoleil-Morrison, I. (2013). A critical review of observation studies, modeling, and simulation of adaptive occupant behaviors in offices. *Building and Environment, 70*, 31–47. https://doi.org/10.1016/j.buildenv.2013.07.020

Hong, T., Taylor-Lange, S. C., D'oca, S., Yan, D., Corgnati, S. P., D'Oca, S., et al. (2016). Advances in research and applications of energy-related occupant behavior in buildings. *Energy and Buildings, 116*, 694–702. https://doi.org/10.1016/j.enbuild.2015.11.052

Jain, R. K., Taylor, J. E., & Peschiera, G. (2012). Assessing eco-feedback interface usage and design to drive energy efficiency in buildings. *Energy and Buildings, 48*, 8–17. https://doi.org/10.1016/J.ENBUILD.2011.12.033

Jiang, C., Masood, M. K., Soh, Y. C., & Li, H. (2016). Indoor occupancy estimation from carbon dioxide concentration. *Energy and Buildings, 131*, 132–141. https://doi.org/10.1016/j.enbuild.2016.09.002

Korkas, C. D., Baldi, S., Michailidis, I., & Kosmatopoulos, E. B. (2016). Occupancy-based demand response and thermal comfort optimization in microgrids with renewable energy sources and energy storage. *Applied Energy, 163*, 93–104.

Li, Z., & Dong, B. (2018). Short term predictions of occupancy in commercial buildings—Performance analysis for stochastic models and machine learning approaches. *Energy and Buildings, 158*, 268–281. https://doi.org/10.1016/J.ENBUILD.2017.09.052

Li, N., Calis, G., & Becerik-Gerber, B. (2012). Measuring and monitoring occupancy with an RFID based system for demand-driven HVAC operations. *Automation in Construction, 24*, 89–99. https://doi.org/10.1016/j.autcon.2012.02.013

Liang, X., Hong, T., & Shen, G. Q. (2016). Improving the accuracy of energy baseline models for commercial buildings with occupancy data. *Applied Energy, 179*, 247–260. https://doi.org/10.1016/j.apenergy.2016.06.141

Luo, X., Lam, K. P., Chen, Y., & Hong, T. (2017). Performance evaluation of an agent-based occupancy simulation model. *Building and Environment, 115*, 42–53. https://doi.org/10.1016/J.BUILDENV.2017.01.015

Martani, C., Lee, D., Robinson, P., Britter, R., & Ratti, C. (2012). ENERNET: Studying the dynamic relationship between building occupancy and energy consumption. *Energy and Buildings, 47*, 584–591. https://doi.org/10.1016/j.enbuild.2011.12.037

Melfi, R., Rosenblum, B., Nordman, B., & Christensen, K. (2011). Measuring building occupancy using existing network infrastructure. In *2011 International Green Computing Conference and Workshops, IEEE* (pp. 1–8). https://doi.org/10.1109/IGCC.2011.6008560

Peschiera, G., Taylor, J. E., & Siegel, J. A. (2010). Response–relapse patterns of building occupant electricity consumption following exposure to personal, contextualized and occupant peer network utilization data. *Energy and Buildings, 42*, 1329–1336.

Pomponi, F., & Moncaster, A. (2017). Circular economy for the built environment: A research framework. *Journal of Cleaner Production, 143*, 710–718. https://doi.org/10.1016/j.jclepro.2016.12.055

Shan, K., Sun, Y., Wang, S., & Yan, C. (2012). Development and In-situ validation of a multi-zone demand-controlled ventilation strategy using a limited number of sensors. *Building and Environment, 57*, 28–37. https://doi.org/10.1016/j.buildenv.2012.03.015

Sidiropoulos, N., Mioduszewski, M., & Oljasz, P., & Schaap Edwin Schaap, E. (2012). *Open Wifi SSID Broadcast vulnerability SSN Project Assessment 2012.*

Soltanaghaei, E., & Whitehouse, K. (2017). Practical occupancy detection for programmable and smart thermostats. *Applied Energy*. https://doi.org/10.1016/J.APENERGY.2017.11.024

Sun, K., & Hong, T. (2017). A simulation approach to estimate energy savings potential of occupant behavior measures. *Energy and Buildings, 136*, 43–62. https://doi.org/10.1016/J.ENBUILD.2016.12.010

U.S. Bureau of Labor Statistics. *American Time Use Survey 2009*. https://www.bls.gov/tus/#data (accessed January 10, 2017).

Ueno, T., Sano, F., Saeki, O., & Tsuji, K. (2006). Effectiveness of an energy-consumption information system on energy savings in residential houses based on monitored data. *Applied Energy, 83*, 166–183.

Wang, S., & Jin, X. (1998). CO2-based occupancy detection for on-line outdoor air flow control. *Indoor and Built Environment, 7*, 165–181. https://doi.org/10.1159/000024577

Wang, Y., & Shao, L. (2017). Understanding occupancy pattern and improving building energy efficiency through Wi-Fi based indoor positioning. *Building and Environment, 114*, 106–117. https://doi.org/10.1016/j.buildenv.2016.12.015

Wang, Q., Augenbroe, G., Kim, J.-H., & Gu, L. (2016a). Meta-modeling of occupancy variables and analysis of their impact on energy outcomes of office buildings. *Applied Energy, 174*, 166–180. https://doi.org/10.1016/j.apenergy.2016.04.062

Wang, S., Burnett, J., & Chong, H. (2016b). *Experimental validation of CO2-based occupancy detection for demand-controlled ventilation*. https://doi.org/10.1177/1420326X9900800605

Wang, W., Chen, J., Lu, Y., & Wei, H.-H. (2017a). Energy conservation through flexible HVAC management in large spaces: An IPS-based demand-driven control (IDC) system. *Automation in Construction, 83*, 91–107. https://doi.org/10.1016/J.AUTCON.2017.08.021

Wang, W., Chen, J., & Song, X. (2017b). Modeling and predicting occupancy profile in office space with a Wi-Fi probe-based Dynamic Markov Time-Window Inference approach. *Building and Environment, 124*, 130–142. https://doi.org/10.1016/J.BUILDENV.2017.08.003

Wang, W., Chen, J., Huang, G., & Lu, Y. (2017c). Energy efficient HVAC control for an IPS-enabled large space in commercial buildings through dynamic spatial occupancy distribution. *Applied Energy*. https://doi.org/10.1016/J.APENERGY.2017.06.060

Wang, W., Lin, Z., & Chen, J. (2017d). Promoting energy efficiency of HVAC operation in large office spaces with a Wi-Fi probe enabled Markov time window occupancy detection approach. In *World Engineers Summit – Applied Energy Symposium Forum Low Carbon Cities Urban Energy Joint Conference WES-CUE 2017*. Singapore.

Yan, D., O'Brien, W., Hong, T., Feng, X., Burak Gunay, H., Tahmasebi, F., et al. (2015). Occupant behavior modeling for building performance simulation: Current state and future challenges. *Energy and Buildings, 107*, 264–278. https://doi.org/10.1016/j.enbuild.2015.08.032

Yan, D., Hong, T., Dong, B., Mahdavi, A., D'Oca, S., Gaetani, I., et al. (2017). IEA EBC Annex 66: Definition and simulation of occupant behavior in buildings. *Energy and Buildings, 156*, 258–270. https://doi.org/10.1016/J.ENBUILD.2017.09.084

Yang, Z., & Becerik-Gerber, B. (2014). Modeling personalized occupancy profiles for representing long term patterns by using ambient context. *Building and Environment, 78*, 23–35. https://doi.org/10.1016/j.buildenv.2014.04.003

Yang, J., Santamouris, M., & Lee, S. E. (2016). Review of occupancy sensing systems and occupancy modeling methodologies for the application in institutional buildings. *Energy and Buildings, 121*, 344–349. https://doi.org/10.1016/j.enbuild.2015.12.019

Yu, Z., Fung, B. C. M., Haghighat, F., Yoshino, H., & Morofsky, E. (2011). A systematic procedure to study the influence of occupant behavior on building energy consumption. *Energy and Buildings, 43*, 1409–1417. https://doi.org/10.1016/J.ENBUILD.2011.02.002

Zhao, Y., Zeiler, W., Boxem, G., & Labeodan, T. (2015). Virtual occupancy sensors for real-time occupancy information in buildings. *Building and Environment, 93*, 9–20. https://doi.org/10.1016/j.buildenv.2015.06.019

Zou, J., Zhao, Q., Yang, W., & Wang, F. (2017). Occupancy detection in the office by analyzing surveillance videos and its application to building energy conservation. *Energy and Buildings, 152*, 385–398. https://doi.org/10.1016/J.ENBUILD.2017.07.064

# 16 Building Eco-feedback System and Occupants' Energy Using Behavior

Jiayu Chen

## 16.1 Introduction

The circular economy aims to address the challenges of resource scarcity and environmental degradation to promote economic growth. Energy as the major resource required by the whole society is a paramount consideration to reform the current economic cycle to the circular economy. In recent years, carbon neutrality and rising energy prices become one of the most significant topics of the circular economy. Over 40% of energy was consumed by building sectors in most developed countries and such consumption is steadily increasing over the past decades (U.S. Energy Information Administration, 2022). An increasing number of researchers have agreed that promoting energy efficiency is the key to reducing energy consumption in the building sector. Different from centrally controlled commercial buildings, residential buildings have a higher potential for energy efficiency improvement, as occupants have full control over their energy use. Due to different lifestyles and living habits, energy consumption can be drastically different in buildings with similar scales and settings. Moreover, there is a famous observed "take back" effect, which suggests the energy saving achieved through physical improvements and retrofitting can be severely diminished by the inefficient behaviors of the consumers (Haas et al., 1998). When people realized their buildings become more energy-efficient, they tend to use more energy. Therefore, sustained energy-saving relies on both retrofit and behavioral interventions (Jain et al., 2013). Many research has suggested that behavior intervention and behavior policy are effective strategies for non-price-based energy efficiency promotion (Allcott & Mullainathan, 2010). A comprehensive national research has shown that the energy-saving and carbon emission potential with behavioral intervention is around 7.4% (Dietz et al., 2009). One of the most effective behavior intervention strategies is the proper delivery of information through accessible channels. Eco-feedback system is such a novel platform to implement behavior interventions for residential building occupants. Combining circular economy principles with energy eco-feedback systems, individuals and organizations have the potential to improve their energy efficiency and reduce their environmental impact at a lower cost. At the same time, by reducing energy consumption and minimizing waste, the whole society will benefit from a more sustainable and circular economy.

In summary, the major research gap in existing studies of eco-feedback systems is identifying

J. Chen (✉)
School of Civil Engineering, Tsinghua University, Beijing, China
e-mail: jiayuchen@tsinghua.edu.cn



J. Zuo et al. (eds.), *Circular Economy for Buildings and Infrastructure*, Sustainable Development Goals Series, https://doi.org/10.1007/978-3-031-56241-9_16

structuralized and effective channels of information delivery and developing proper feedback components to maximize the impact of eco-feedback systems. This chapter is organized as follows, Sect. 16.2 introduces the concept of the eco-feedback system and its theoretical foundations; Sect. 16.3 introduces the fundamentals of developing a usable eco-feedback system; Sect. 16.4 discusses the major design principles and components of the system; Sect. 16.5 illustrates the eco-feedback system with two case studies; and Sect. 16.6 concludes and discusses the application of the eco-feedback system.

## 16.2 Eco-feedback System and Its Theoretical Foundation

Advancements in electronics and sensing technologies enable access to a multitude of information during building operations. Occupancy sensors and smart meters can aggregate fundamental building energy consumption and residency information for comprehensive analysis and usage. Eco-feedback is an intervention platform that utilizes such aggregated data and provides tailored information to change the energy-using behavior of building residents. Most of the eco-feedback system design relies on the premise that occupants are often unaware of their energy using a profile with high time resolution (Attari et al., 2010). Like the speedometer to drivers, once the actual consumption is accessible with high time resolution, such as day-to-day basis or hourly basis, they can take actions to reduce their energy consumption (Abrahamse et al., 2007; Wever et al., 2008; Wilson & Dowlatabadi, 2007). Therefore, the theoretical foundation of eco-feedback systems includes occupants'involvement in energy consumption, the occupant's psychological recognition of information, and physical behavior patterns.

### 16.2.1 The Role of Occupants and Occupants'Behaviors in Building Energy Use

Occupants' behaviors are the results of complex social, biological, and psychological contexts (Yoshino et al., 2017). International Energy Agency's Annex 66 (Yan et al., 2017) reports occupants' profile directly determines their standard of comfort and their choice of how to use energy. Even the most simple behavior, their presence, can passively affect the energy balance in buildings (Harputlugil et al., 2019). Moreover, other ubiquitous behaviors, such as changing the temperature setting of the thermostat, turning lights or other appliances on and off, adjusting the fan ventilation, etc. can finally result in comprehensive behavior-related energy consumption. These characteristics of the energy consumers make it extremely difficult to directly measure or track event or appliance-associated energy consumption or provide effective intervention. Given the rapid development modern internet of things technologies, even if the occupants are not present in buildings also can affect the building energy consumption. For example, many smart buildings allows the building maintain residents' personal comfort by pre-condition approach to main the residents' well-being and productivity (Page et al., 2008). Extending this concept, many buildings' operation take such uncertain occupancy into decision-making process for building facility control (Feriadi & Wong, 2004).

With such complications in the role of building occupancy, behavior change or intervention is a formidable task with difficulties and uncertainties. Each individual has his/her unique personal profile, attitude, and sociocultural attributes, which significantly affect their lifestyle, preference, beliefs, and attitudes to energy conservation. An understanding and implementation of applying anthropological, sociological, and geographical methods can obtain exceptional outcomes (Lévy & Belaïd, 2018). A recent report from European Environment affirms the behavior change associated with energy-saving can reach as high as 20% (European Environment Agency, 2013).

### 16.2.2 The Development of Eco-feedback Systems

The earliest attempt of using an eco-feedback system can be traced back to the 1970s. Without the help of modern smart metering and internet technologies, early energy consumption feedbacks are manually prepared with limited input data. In the earlier studies, Brown and Macey, and Becker organized the studies by exposing residents to various types of energy feedback and setting goals based on psychological and behavioral theories (Becker, 1978; Brown & Macey, 1983). The experiment results suggest that 2–20% of the goals are achieved. After the 2000s, sustainability and reduction of carbon emissions becomes a key topic of environmental and political discussion and inspires researchers to find new solutions for energy resource-saving. The installation of centrally monitored distribution systems and a web-enabled platform allows more sophisticated eco-feedback system design. For example, Ueno et al. developed a web-based eco-feedback system that can visualize appliance-level of energy reports on a day-by-day basis (Ueno et al., 2006). The connected appliances have a reported 12% energy consumption drop and also cause energy conservation in other appliances. One significant study was conducted by Petersen et al., who provide a comparison of different feedback frequencies with web-based eco-feedback platforms (Petersen et al., 2007). Based on their experiments on residential dormitories, the occupants who received real-time and internet-based feedback can save up to 55% of energy during the experiment period. The encouraging results were followed by many residential and commercial endeavors in the following years (ComEd, 2022; EnergyHub, 2022; OPower, 2022; PowerMeter, 2022). Also, an increasing number of research teams have conducted studies on the design and implementation of eco-feedback systems.

The development of the platform mainly has two directions, technocentric systems, and socio-centric systems. The technocentric systems mainly rely on improving the energy dynamic monitoring and friendly visualization interface. It targets developing individual professional energy management systems and provides opportunities for occupants to improve their own energy-efficient profile. On the other hand, the socio-centric systems are investigated and proposed by the academic community to focus on providing feedback in groups or local communities. It usually includes social components and peer comparison (Grevet et al., 2010; Gulbinas et al., 2014; Schultz et al., 2007). In recent years, with the improvement of cost-effective monitoring and internet of things technologies, combining both the technocentric and socio-centric systems become a new trend. For example, Watt's Watts, a closed-loop feedback platform, allows tailored information visualization through a user dashboard (Gulbinas et al., 2014; Jain et al., 2012). The administrator can decide which modules should be displayed to maximize the users' energy-saving potential. It also includes the social components that administer allowed to make social influence over users' social networks and change users' behaviors. Among the latest studies and applications of eco-feedback systems, the robust data collection and delivery ability is the starting point and the effective and strategic design of feedback components with economical and sociological incentives are the core of future system evolution.

## 16.3 The Minimum System of Eco-feedback Approach

### 16.3.1 Architecture of a Minimum System

A typical eco-feedback system follows a closed feedback loop, which continuously posts all types of information to influence occupants' behaviors. Due to the uncertainty of human preference and decision-making, continuously updated information allows adaptive and immediate feedback to have a sustained effect. Based on the finding in the previous studies (Peschiera & Taylor, 2012; Petkov et al., 2011), in general, higher feedback frequency and a social component can effectively improve the outcomes of intervention.

Figure 16.1 shows the minimum architecture of the eco-feedback system. The system requires data acquisition from the internet of things (IoT) system to acquire physical energy usage data. The streamed data then need to be structuralized and combined with users' behavior information from a user interface and combined with social network information. Then, based on the networked relationships among users, selected information can be delivered to users with different incentives (such incentives are not necessarily associated with monetary value). Finally, the changed behavior will be captured by the IOT sensing system and server as new inputs for data acquisition. The modern eco-feedback system needs physical equipment to capture the energy consumption information either from smart meters or internet attached transduces that connect to a power distribution board. The data logger should be able to transmit or upload acquired data to a structuralized database for the ease of data visualization and post-processing. The captured apparent power of the entire house, apartment, or single facility should be recorded with high-resolution reading devices. For the resident context, the major energy consumers should be included, such as air-conditioners, refrigerators, computers, and lights. The populated feedback information after aggregation and translation should be visualized by the users with a user interface or dashboard. The modularized and customizable platform is recommended to allow tailored feedback based on the users' needs or profiles. Often, historical comparison, baseline comparison, and potential benefit of retrofitting are the most fundamental information that should be included. Usually, the tricky part of the design for the system is whether the feedback can effectively change the occupants' behaviors. In most scenarios, a newly established eco-feedback system may not work well due to the unique background of each user. The adaptive and closed-loop system can improve itself by testing various feedback components over time. The most effective and simple approach is to redesign the interface to leverage social norms or comparative incentives. Therefore, the eco-feedback system always requires high system intelligence to test hypotheses while updating system settings.

### 16.3.2 Baseline Assessment

The premise of designing proper eco-feedback is securing sufficient qualitative and quantitative data. Qualitative data includes building occupants' opinions, attitudes, perceptions, and social contexts. Quantitative data can be used to examine and validate different choices in design and acquire a better understanding of platform incentives. With the help of both information, the baseline of energy consumption and energy-saving potential can be roughly estimated. In addition, by knowing occupants' ability to control building facilities, a fundamental profile of occupants can be drawn. Such ability can be defined into four categories (Hong et al., 2016), including (1) occupants' sensation and control based on their thermal comfort, (2) occupants' presence and location, (3) occupants' operation over building components, such as windows, shadings, doors, and blinds, and (4) occupants' control over building facilities, such as air conditioners, fans, and lights. To quire these data for baseline assessment, previous studies have recommended the following approaches.

- Survey/Questionnaires. This approach is the most popular and direct method to gather specific data that is needed. Survey/Questionnaires can be sent to the focus group or amalgamated with people to acquire both qualitative and quantitative data. Demographics are also important to construct relevant datasets. The results can be utilized to quantify the occupants' baseline and test hypotheses to gain knowledge of the relationships of variables. However, survey/questionnaires are usually subjective options or feelings of survey takers.
- Direct Monitoring. The monitored outcomes are the most reliable and objective performance of occupants' energy usage. The

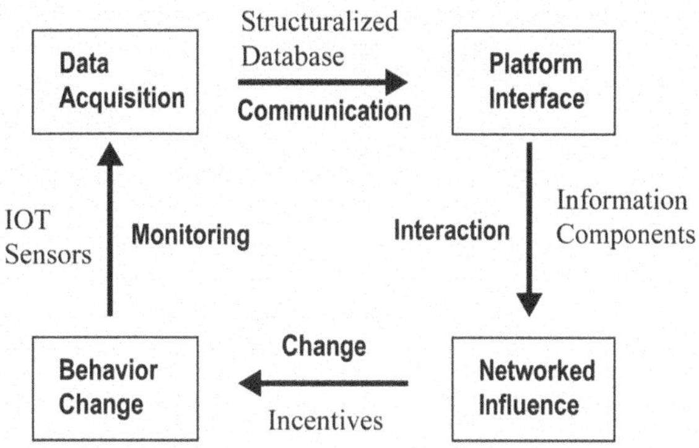


**Fig. 16.1** Configuration of a minimum closed-loop eco-feedback system

monitoring data can be as simple as monthly electricity bills and as complicated as spatial movements that are tracked by indoor positioning systems (Wang et al. 2017a, b). It does not constrain energy consumption, and environmental or physical records, such as $CO_2$ concentration, temperature, humanity, sound intensity, and air pressures, from multi-sensor networks, should also be included. Time resolution is a critical concern of monitoring, due to the configuration of monitoring sensors and sampling rate, the time interval of the morning can vary from 1 min to daily or monthly.

- Numerical Simulations. Simulation utilized a modeling approach to estimate energy load and occupants' behaviors to guide the design of eco-feedback systems. Before a large-scale of implementation, simulation can test the possible scenarios and predict the responses of occupants. Different temporal and spatial analyses on occupancy can identify a proper combination of feedback elements, data resolution, target group, and spatial distribution. With the data collected from direct monitoring, many sophisticated numerical methods can be used to roughly estimate the energy-saving potential of a feedback platform. In addition, with a modern simulation approach, such as agent-based simulations, the occupants' behavior rationale, and decision processes can be emulated (Chen et al., 2012, 2013).
- Post Occupancy Evaluation. This is a more objective evaluation that alternative to the survey/questionnaire approach. It evaluates the building's energy performance in accordance with occupants' behaviors and occupancy records. It is a powerful tool that can properly access domestic energy use and the lifestyles of residents (Vale & Vale, 2010).

### 16.3.3 Key Factors and References for Development

To develop proper eco-feedback platforms, previous studies have investigated various design factors. Chatzigeorgiou and Andreous' recent study has an excellent summary of these studies (Chatzigeorgiou & Andreou, 2021). Extending their review, the following table reports the possible key factors investigated by previous studies. With the findings from these studies, a comprehensive eco-feedback system design can be established (Table 16.1).

## 16.4 Design of the Eco-feedback Components

### 16.4.1 Components of Feedback

How and what information will be delivered to energy users is the key factor to spurring and sustaining behavior change to save energy use (Jacucci et al., 2009). Based on Chatzigeorgiou and Andreous' summary, there are seven effective components (Chatzigeorgiou & Andreou, 2021). They can work independently or combined

**Table 16.1** Some studies of the feedback key factors

| Building type | Scope of design | No. of studies | Findings | References |
|---|---|---|---|---|
| Residential | Electricity feedback | 156 | Monetary incentives have limited impact or are even counterproductive and face-to-face feedback is preferable | Delmas et al. (2013) |
| Residential | Energy feedback | 189 | Communities goal setting and community-level feedback are more powerful | Šćepanović et al. (2017) |
| Residential | Feedback component | 38 | Higher frequency feedback is preferable; personalized feedback is more effective than social comparison | Abrahamse et al. (2005) |
| All | Feedback component | 42 | Goal setting and combined strategy are more effective | Karlin et al. (2015) |
| All | Feedback component | 72 | Compared with positive feedback, the impact of neutral or negative feedback is unclear | Hermsen et al. (2016) |
| Residential | Feedback component | 29 | Community building and Social comparison strategies are more effective than non-social ones | Abrahamse and Steg (2013) |
| All | Feedback component | 87 | A combined intervention strategy is more effective, including (1) rewards and goal setting; (2) instructions and goal setting; (3) Commitment and goal setting; (4) Prompts and making it easy; (5) Prompts and justifications; (6) Cognitive dissonance and justifications | Osbaldiston and Schott (2012) |
| All | Feedback component | 12 | Aggregated feedback is more effective than disaggregated feedback; Disaggregated electricity feedback can be helpful for sub-group of the population | Petersen et al. (2007) |
| Residential | Feedback component | 83 | Social network feedback can effectively change occupants' behavior under peer pressure | Peschiera et al. (2010) |
| All | Eco-feedback design | 10 | Gamification can engage consumers in pro-environmental behaviors; social network feedback can increase awareness of energy conservation | Morganti et al. (2017) |
| All | Eco-feedback design | 25 | Gamification leads to more self-reported and actual energy conservation behavior but if the behaviors are sustainable is unclear | Johnson et al. (2017) |
| Residential | Eco-feedback design | 25 | Portable and multiple platforms are effective; understandable visualization is more helpful | Murugesan et al. (2015) |
| Residential | Eco-feedback design | 45 | Feedback outcomes depend on the cultural bias; the impacts of feedback are subject to rebound effects (conservation activities are effective last couple of weeks) | Liebe et al. (2018) |
| Residential | Eco-feedback design | 44 | Social comparison can significantly reduce energy consumption | Andor and Fels (2018) |

(proper design is required to avoid counter-effective combination).

- High-resolution data. Frequent feedback provides users with accurate and reliable information. However, the data resolution is not linearly associated with saving amounts.
- Historical comparison. Historical feedback serves as a comparison with the users themselves. It provides better reference, given all the personal profiles and building physical settings would be similar. Due to seasonal weather changes, a comparison with past years would be helpful. However, some studies suggest this approach is less effective than setting a certain goal.
- Monetary translation. Most studies in the past suggested the monetary incentive is not effective and sometimes counterproductive. As a possible option in the feedback toolbox, this strategy could be helpful to certain groups with proper design.

- Peer pressure and social norms. One of the most well-studied strategies in recent years is social network influence. Most relevant studies suggest this approach is astonishingly effective and sustainable. The design of this strategy depends on comparison among target groups and the weight of members' interrelationships.
- Disaggregated items. This strategy focuses on providing itemized information components based on categorization. The energy consumption can be broken into individual appliance consumption or zone-based spatial consumption. It is a popular strategy that is usually combined with other strategies that allow flexible customization and tailoring.
- Gamification. Some previous studies utilized this strategy and validated its feasibility. Gamification usually provides awarding feedback to provide a direct and visible impact on users. Under a properly designed contest environment among users, the game can provide a valuable non-monetary price for users. With a more friendly interface and interactive mechanism, this strategy poses higher requirements on eco-feedback system design.
- Goal settings. Most studies that investigate the impact of goal settings suggested this strategy has an unclear efficacy, although it may lead to a relatively high energy-saving potential. One of the major concerns of this strategy is the relapse effect after the satisfaction of goals.

### 16.4.2 Metric of Studying the Eco-feedback System Design

Proper eco-feedback system design has a substantial impact on the final performance of behavior intervention. Khosrowpour et al. have a comprehensive review of the most recent occupant energy feedback research (Khosrowpour et al., 2018). In addition to conventional methodologies, they proposed a meta-analysis and highly recommend methodology fusion. Their studies suggested five inter-sectional approaches as follows:

- Analytics-surveys. One major lack of knowledge in design is the validation of occupants' actual actions. Such validation is crucial to understanding the perception and actual energy usage. Often, the occupants are willing to make behavior changes, but their actual outcomes are inconsistent with their intentions (Chen et al., 2014). The major benefit of integrated survey and analytics is bridging such a knowledge gap. It also can enhance the overall performance of measurement algorithms.
- Experiments-analytics. Occupant behaviors are well studied during the experiment period, but lack of close watch on the variation of patterns during or after experiments. With help of high-resolution monitoring data (which is not necessarily displayed to users), embedded data analysis during the experiments provides a more systematic and deeper understanding.
- Experiment-analytics-survey. Combing energy analysis and surveys can formulate more practical feature sets and classify occupants more accurately (Viegas et al., 2016). Fusing insights that concluded from analytics (such as patterns of total load, disaggregated load, and occupancy) with surveys allows for valid experiment design and testing.
- Simulations-experiments. Simulation serves as an alternative and convenient validation when the real experiment is not possible. It is also regarded as predictive validation when robust behavioral and structural models were built. A solo simulation model is not powerful enough to accurately estimate behavioral response and energy saving. A deeper link between simulation and experiments can improve both prediction power and accuracy. It is also beneficial to create a reliable model to emulate occupants' behavior and response with cost-effective solutions.
- Analytics-simulation. The current simulation approach (such as agent-based simulations)

of energy feedback mainly incorporates social network principles. However, these models haven't been properly fine-tuned as it is extremely difficult to emulate human behaviors and opinions with mathematical models. Therefore, coordinating analytics provides a better opportunity for designing appropriate underlying mathematical models.

### 16.4.3 Principles of Eco-feedback System Design

Wield with different components and strategies, different possible designs can be validated with the validation metrics. These tested hypotheses and conclusions can be summarized as design principles or specific platform interfaces and functionalities. Mäkivierikko et al. extracted 24 design principles from the literature and identified the most important principles through experiments (Mäkivierikko et al., 2019). The top ten principles are listed as follows:

- Device-specific feedback: Provide per-appliance feedback. It is difficult to have per-appliance consumption, this value can be estimated with statistical analysis or simulation. For shared devices, non-personalized consumption could lower motivation.
- Collective goals—good: Give the user a feeling of being helpful and contributing. The collective goals can increase the users' positive attitude toward contribution.
- Use different types of feedback: Combining different forms of feedback. The feedback can be in form of graphic, numerical, textural, or game avatars.
- Use positive feedback: Provide positive affections, such as a feeling of competence, a sense of autonomy, and responsibility.
- Use comparative feedback: The comparative feedbacks exploit two types of social norm, descriptive norm, and injunctive norm. Knowing what other's performances and being known by others for their performance.
- Fair goals: The content and standards of feedback should be fair to enable the intervention to be trustworthy and meaningful.
- Avoid information overload: Showing too much information will increase the data interpreting difficulty. Abstract and direct information delivery is preferable.
- Actionable feedback: The feedback should assure the action is feasible and the control is perceivable. For example, tips for energy-saving actions should be provided.
- Design for the whole family: The feedback should be accessible and appealing to every family member, including children. Gamification components can be added to attract children.
- Personalized feedback: Tailored feedback increases effectiveness and accommodates personal preference. It can associate personal behavior and its consequences.
- Frequent or real-time feedback: High-resolution data can help the user recognize how their behaviors are associated with a certain amount of energy consumption. This enables the users with a clear expectation of their control of appliances.
- Show the actions that can make a difference: Users can visualize explanatory screens about energy saving/load shifting. It is helpful when collective goals on the household/neighborhood level are shown to everyone.
- Relevance: Give references to make sure the feedback is understandable and relatable. Avoid using technical terms and use comparative explanations.

## 16.5 Case Study

### 16.5.1 Case 1: Apartment-Level Eco-feedback System for a Residential Building in New York City

In 2017, a research team composed of researchers from Columbia University and the University of Notre Dame conducted a comprehensive study on designing eco-feedback for large-scale building implementation (Meinrenken et al., 2021). The research team recruited experts from data science, mechanical engineering, civil engineering, and computer science and developed a sophisticated based on the findings of previous studies. To validate the designed platform, an experiment was deployed for 125 apartments. The platform considered both the exogenous effects (weather, daytime, etc.) and idiosyncratic effects (lifestyles, habits, regular behaviors, etc.). This case study intends to demonstrate a general workflow for eco-feedback system deployment. The following paragraphs include a general process to implement the designed eco-feedback platform.

#### 16.5.1.1 Setup Infrastructures

Each apartment installed a high-resolution (15-min reading rate) multi-channel electric power meter with scan transponders (Quadlogic Controls Corporation). A remote Microsoft Azure server based on HTTP/UDP was configured to receive data readings. The system has an automated diagnosis system to ensure reliability and accuracy during data transmission. A pre-processing middleware is installed to determine the peak usage at a specific time window.

#### 16.5.1.2 Design Feedback Schemes

The eco-feedback was designed to be accessible at any time and responsible to attract users' attention by sending reminder messages. To avoid overloading the users, the message was randomized and automatically generated approximately twice a week. The feedback includes contextualized information, a positive sentiment of the text, and peer usage comparison. The content and features of feedback changed over the period of the experiment. In other words, the same residents received different formats of feedback over time and the feedback type was improved gradually. Eight of ten features were binary and adjustable to configure if a certain piece of information would be present or absent, including graph display, peer average usage, historical usage, consumption projection, cost, projected cost, relevant article/news, and peak usage time. Another two features were sentiments of communication (positive, negative, neutral) and form of quantity display (kWh, equivalent tree cut down, $CO_2$, the equivalent mile of driven, greenhouse gas pollution).

#### 16.5.1.3 Generate and Deliver Feedback

A flexible, template-based natural language generation model is necessary to create textural feedback. The template allows the combination of features based on the quantitative response of users. Given flexible elements, the message developed by this platform can have a total of 55,560 possible different message types. The content also can be delivered in different sentiments and referral URLs. An SMTP server was able to handle the automated message delivery or allow subscription/unsubscription of the message with different frequencies.

#### 16.5.1.4 Remove/Isolate Outliers

The error in monitoring devices or the absence of occupants will result in outliers among users. They will cause inaccurate and inefficient aggregated comparative information. Therefore, unrealistic consumption or abnormal data points should be detected by the checking mechanism to identify and remove outliers.

#### 16.5.1.5 Select Reduction Metrics

A metric to measure the efficacy of feedback and prompt reductions is critical to formulating closed-loop feedback. This platform includes four metrics, including average treatment vs. control on a specific date (any date in baseline or feedback period), baseline adjustment for a certain resident in the target group, baseline adjustment for the entire target group, average treatment vs. control on a specific date (during feedback period only). Following Fig. 16.2 shows the overview of the end-to-end eco-feedback system.

### 16.5.2 Case 2: Watt's Watt

Watt's Watts is one of the earliest social network-enabled eco-feedback platforms designed for research purposes by researchers from Virginia

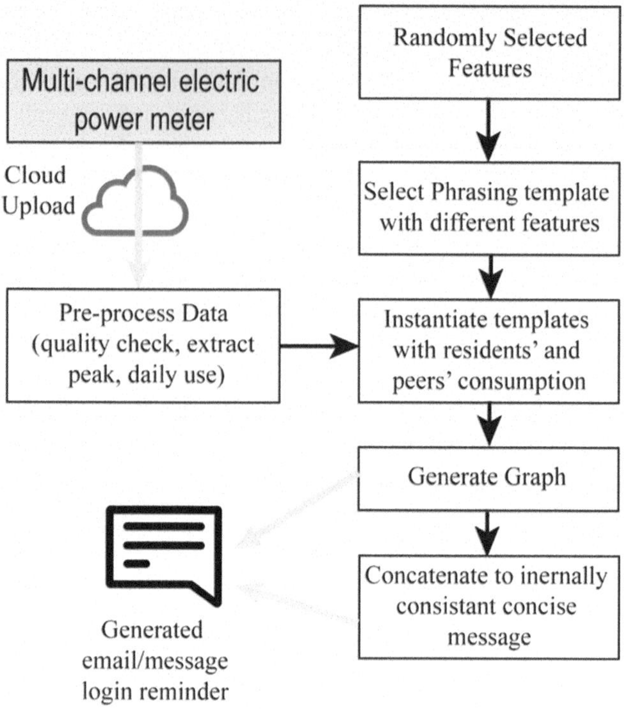


**Fig. 16.2** Feedback workflow of Case 1

Polytechnic Institute and State University, Columbia University, New York University, and the University of Illinois Urbana-Champaign (Gulbinas et al., 2014). The platform has industry-level energy monitoring capabilities and a data interface for occupants' social networks. It was designed to be modularized to validate the correlation between feedback components and energy-saving potential. More practical usage of this platform is to examine the user interface design to maximize energy-saving potential. The platform administrators can determine the components and recognized social networks that users can access. Figure 16.3 shows a sample interface of Watt's Watt. This case study intends to demonstrate how to set up the user interface to incorporate eco-feedback strategies. The interface can be divided into three functional units:

#### 16.5.2.1 Visualization

Social norms had been reported to strongly affect human behavior and normative comparisons involving people with closer relationships are preferred over anonymous users. A stronger sense of group identity improves the impact of normative feedback. The design principles of Watt's Watts follow the above conclusions by adopting real-time, individual-level, and normative compressions. Well-designed function units are located at different segments on the entire interface canvas. Contents that bear higher influence are displayed in brighter colors. A peer comparison checkbox control panel allows users to overlay the consumption history of each peer for direct and convenient comparison.

#### 16.5.2.2 Social Component

Normative comparison under a social network context potentially has a negative impact due to assimilation (using more energy than other comparison peers). Effective social network feedback components should assimilate users who use relatively low energy together or contrast with users with high energy consumption. One of the constraints of experiments is experiment targets usually have similar physical building configurations (size, facility system, geolocation).

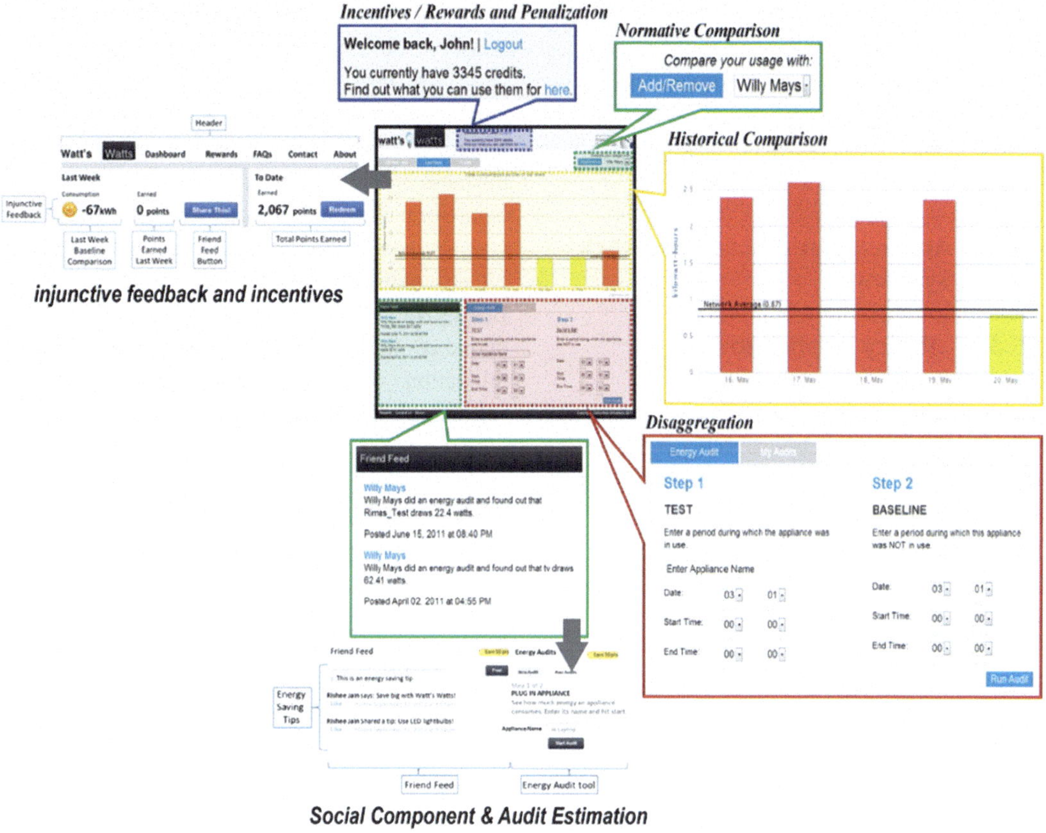


**Fig. 16.3** The user interface of Watt's Watt

This is a barrier that prohibits efficacy in large-scale applications. Therefore, such assimilating is crucial for effective feedback. Watt's Watts was designed to encourage assimilations in upward and downward comparisons with low-consumption users. The chance of sustained energy-saving behavior depends on long-term behavior/lifestyle change. Intimate relationship has larger power in shifting people's attitude and permanently changing their behaviors. Therefore, it is always necessary to develop a widget that allows users to share and generate energy-saving tips. When users are reluctant to share feeds, administrators can employ an anonymous avatar. Often, instead of saving tips, shared appliance-level energy consumption is helpful for users to diagnose the source of high consumption and identify the major energy consumers at home.

#### 16.5.2.3 Incentives

Although monetary incentives had been proven yielding limited impact, incentives are still introduced in Watt's Watt for data collection and statistical validation purposes. Watt's Watts has a unique incentive structure with rewarding points. The points can be used to redeem gift certificates from local businesses. Users can accrue points by modifying behavior and achieving certain energy-saving goals. Points also can be earned by completing required actions with the interface and sharing feeds with other users. The point rewarding system has a similar concept to gamification that quickly and directly shows the results when the user takes action. The number of points that award to users can be set by administrators and the points can be taken away from users who abuse them.

Although many eco-feedback systems have different interface designs, their theoretical foundations are the same. By combining proper strategies and clean interfaces, a functional eco-feedback system can be established.

## 16.6 Application of Eco-feedback Systems

### 16.6.1 Policy Implications

The first policy implication for the government is providing funds or technical support to startups with a similar platform. Given the influence of the social network on the internet, eco-feedbacks suggest a novel energy conservation potential that does not introduce a negative impact. The cost of deployment of eco-feedback is exceptionally low compared to regular retrofitting or equipment upgrading solutions. Its impact can be easily scaled up to millions of users. Encouraging environmentally friendly and responsible lifestyles and behaviors can be the most sustainable and clean solution for the global carbon-neutral target. In addition, eco-feedback systems require support from remote and smart metering and intelligent reporting infrastructures. The development of all relevant industries can benefit from the wide application of eco-feedback systems.

The second policy consideration is information disclosure and privacy. One of the most powerful feedback strategies is social networked comparison. Previous studies suggested that closer relationships usually have a higher impact. To exploit such an effect, the release of the name and consumption information is necessary. Therefore, how such information should be used and secured is a crucial factor that needs to be carefully considered, even if the users themselves agreed to publish their information. How to avoid companies misleading users and acquiring irrelevant and sensitive personal data should be carefully reviewed.

The third policy implication is encouraging and extending scientific research and development to industrial applications and scaling up their impacts. Studies on behavior changes to conserve building energy consumption is not a new idea, however, until recent years, practical applications are emerging in the markets. An integrated effort of scientists, policy-makers, and business operations can translate these theoretical behavioral insights into quantifiable energy and cost savings. Therefore, encouraging such joints efforts among industry pioneers, such as ComEd, EnergyHub, Oracle OPower, etc., and the academic community is the key to rapid growth and utilization of these new energy conservation solutions.

### 16.6.2 e-CPSs: New Trends in Eco-feedback Systems

Eco-feedback systems are designed for integrated information delivery to interfere with occupants'behaviors. It highly relies on information acquired from various sources and sensing systems. In recent years, cutting-edge digitization technologies have been proposed to fuse different computational and sensing processing with integrated control and communication system (Shi et al., 2021). Such a system is called a cyber-physical system (CPSs), which can model, visualize, and operate complex building systems with various computing tools. When CPS's function mainly targeting on optimize building energy systems, the system is called an energy-cyber-physical system (e-CPS) (Gupta et al., 2011). e-CPSs optimize building facility operation and reduce total energy consumption. However, the major challenge for e-CPSs implementation is occupancy, which is the primary determinant of energy load and system demand. An emerging trend is using occupancy to connect the physical loads and cyber digital model (Wang et al., 2019). Eco-feedback systems are one inevitable subsystem that can enrich e-CPSs. On one hand, the e-CPSs require reliable and sophisticated monitoring and actuation components and digital building information models. It serves as the essential infrastructure for eco-feedback data acquisition. On the other hand, an eco-feedback system can observe and interfere with occupants'behaviors, which finally results in a variation in energy demand. The comprehensive functionality and its lack of embedment of occupants' information and behavior patterns reveal significant research potential and inspiring trends in building energy conservation.

## 16.7 Summary and Conclusions

In summary, existing studies suggest that eco-feedback systems can effectively promote sustainable behavior and reduce energy consumption. However, their effectiveness depends on various factors, including the type and context of feedback provided, and an individual's motivation to change their behavior. Tailored personal feedback is more effective than generic feedback since it is more relevant and meaningful, and more likely to motivate behavior change. Moreover, social comparison with peers and neighbors can be a powerful motivator. Relevant studies indicate that eco-feedback systems can be used in various settings, such as residential, commercial, and public, with a well-designed feedback interface and content.

However, existing studies have apparent limitations. First, there is a lack of long-term impact. While eco-feedback systems help raise awareness about energy usage, their impacts have shown to be uncertain and unsustainable over time. Some studies show that their impacts diminish over time. Second, there is a lack of standardization. People may be more motivated by different types of feedback or only responsive to certain types, making it challenging to develop a standardized feedback system suitable for everyone. Third, the scope is limited. Most eco-feedback systems focus on energy usage, but it is worth investigating other possibilities, such as water usage or waste reduction.

For future studies, some research directions are highly recommended. First, integration with sensing systems. Eco-feedback systems rely on high-resolution home smart metering, but integration with sensing technologies that can automatically stream energy usage data and patterns can help provide more relevant and tailored information. Second, combining gamification and social influence. Both types of feedback have been proven effective in previous studies, and combining them with other potential feedback components may create an extraordinary impact on residents' behavior decisions. Third, data privacy and security. Eco-feedback systems may collect sensitive user data, so investigating and developing secure and privacy-preserving mechanisms is paramount to protect user data from unauthorized access and misuse.

## References


Abrahamse, W., & Steg, L. (2013). Social influence approaches to encourage resource conservation: A meta-analysis. *Global Environmental Change, 23*(6), 1773–1785. https://doi.org/10.1016/j.gloenvcha.2013.07.029

Abrahamse, W., Steg, L., Vlek, C., & Rothengatter, T. (2005). A review of intervention studies aimed at household energy conservation. *Journal of Environmental Psychology, 25*(3), 273–291. https://doi.org/10.1016/j.jenvp.2005.08.002

Abrahamse, W., Steg, L., Vlek, C., & Rothengatter, T. (2007). The effect of tailored information, goal setting, and tailored feedback on household energy use, energy-related behaviors, and behavioral antecedents. *Journal of Environmental Psychology, 27*(4), 265–276. https://doi.org/10.1016/j.jenvp.2007.08.002

Allcott, H., & Mullainathan, S. (2010). Behavior and energy policy. *Science, 327*(5970), 1204–1205. https://doi.org/10.1126/science.1180775

Andor, M. A., & Fels, K. M. (2018). Behavioral economics and energy conservation – A systematic review of non-Price interventions and their causal effects. *Ecological Economics, 148*(June), 178–210. https://doi.org/10.1016/j.ecolecon.2018.01.018

Attari, S. Z., DeKay, M. L., Davidson, C. I., Bruine, W., & de Bruin. (2010). Public perceptions of energy consumption and savings. *Proceedings of the National Academy of Sciences, 107*(37), 16054–16059. https://doi.org/10.1073/pnas.1001509107

Becker, L. J. (1978). Joint effect of feedback and goal setting on performance: A field study of residential energy conservation. *Journal of Applied Psychology, 63*(4), 428–433. https://doi.org/10.1037/0021-9010.63.4.428

Brown, M. A., & Macey, S. M. (1983). Understanding residential energy conservation through attitudes and beliefs. *Environment and Planning A: Economy and Space, 15*(3), 405–416. https://doi.org/10.1068/a150405

Chatzigeorgiou, I. M., & Andreou, G. T. (2021). A systematic review on feedback research for residential energy behavior change through Mobile and web interfaces. *Renewable and Sustainable Energy Reviews, 135*(January), 110187. https://doi.org/10.1016/j.rser.2020.110187

Chen, J., Taylor, J. E., & Wei, H.-H. (2012). Modeling building occupant network energy consumption decision-making: The interplay between network structure and conservation. *Energy and Buildings, 47*(April), 515–524. https://doi.org/10.1016/j.enbuild.2011.12.026

Chen, J., Jain, R. K., & Taylor, J. E. (2013). Block configuration modeling: A novel simulation model to emulate building occupant peer networks and their impact on building energy consumption. *Applied Energy, 105*(May), 358–368. https://doi.org/10.1016/j.apenergy.2012.12.036

Chen, V. L., Delmas, M. A., & Kaiser, W. J. (2014). Real-time, appliance-level electricity use feedback system: How to engage users? *Energy and Buildings, 70*(February), 455–462. https://doi.org/10.1016/j.enbuild.2013.11.069

ComEd. (2022). *Comed an Exelon company*. https://www.comed.com/Pages/default.aspx

Delmas, M. A., Fischlein, M., & Asensio, O. I. (2013). Information strategies and energy conservation behavior: A meta-analysis of experimental studies from 1975 to 2012. *Energy Policy, 61*(October), 729–739. https://doi.org/10.1016/j.enpol.2013.05.109

Dietz, T., Gardner, G. T., Gilligan, J., Stern, P. C., & Vandenbergh, M. P. (2009). Household actions can provide a behavioral wedge to rapidly reduce US carbon emissions. *Proceedings of the National Academy of Sciences, 106*(44), 18452–18456. https://doi.org/10.1073/pnas.0908738106

EnergyHub. (2022). *EnergyHub*. https://www.energyhub.com/

European Environment Agency. (2013). *Achieving energy efficiency through behaviour change: What does it take?* Copenhagen. https://doi.org/10.2800/49941

Feriadi, H., & Wong, N. H. (2004). Thermal comfort for naturally ventilated houses in Indonesia. *Energy and Buildings, 36*(7), 614–626. https://doi.org/10.1016/j.enbuild.2004.01.011

Grevet, C., Mankoff, J., & Anderson, S. D. (2010). Design and evaluation of a social visualization aimed at encouraging sustainable behavior. In *2010 43rd Hawaii international conference on system sciences* (pp. 1–8). IEEE. https://doi.org/10.1109/HICSS.2010.135

Gulbinas, R., Jain, R. K., Taylor, J. E., Peschiera, G., & Golparvar-Fard, M. (2014). Network Ecoinformatics: Development of a social Ecofeedback system to drive energy efficiency in residential buildings. *Journal of Computing in Civil Engineering, 28*(1), 89–98. https://doi.org/10.1061/(ASCE)CP.1943-5487.0000319

Gupta, S. K. S., Mukherjee, T., Varsamopoulos, G., & Banerjee, A. (2011). Research directions in energy-sustainable Cyberphysical systems. *Sustainable Computing: Informatics and Systems, 1*(1), 57–74. https://doi.org/10.1016/j.suscom.2010.10.003

Haas, R., Auer, H., & Biermayr, P. (1998). The impact of consumer behavior on residential energy demand for space heating. *Energy and Buildings, 27*(2), 195–205. https://doi.org/10.1016/S0378-7788(97)00034-0

Harputlugil, G. U., Harputlugil, T., Pedergnana, M., & Sarıoğlu, E. (2019). A novel approach for renovation of current social housing stock based on energy consumption in Turkey: Significance of occupant behaviour. *Architectural Science Review, 62*(4), 323–337. https://doi.org/10.1080/00038628.2019.1615862

Hermsen, S., Frost, J., Renes, R. J., & Kerkhof, P. (2016). Using feedback through digital technology to disrupt and change habitual behavior: A critical review of current literature. *Computers in Human Behavior, 57*(April), 61–74. https://doi.org/10.1016/j.chb.2015.12.023

Hong, T., Taylor-Lange, S. C., D'Oca, S., Yan, D., & Corgnati, S. P. (2016). Advances in research and applications of energy-related occupant behavior in buildings. *Energy and Buildings, 116*(March), 694–702. https://doi.org/10.1016/j.enbuild.2015.11.052

Jacucci, G., Spagnolli, A., Gamberini, L., Chalambalakis, A., Björksog, C., Bertoncini, M., Torstensson, C., & Monti, P. (2009). Designing effective feedback of electricity consumption for mobile user interfaces. *PsychNology Journal, 7*(3), 265–289.

Jain, R. K., Taylor, J. E., & Peschiera, G. (2012). Assessing eco-feedback Interface usage and design to drive energy efficiency in buildings. *Energy and Buildings, 48*(Supplement C), 8–17. https://doi.org/10.1016/j.enbuild.2011.12.033

Jain, R. K., Gulbinas, R., Taylor, J. E., & Culligan, P. J. (2013). Can social influence drive energy savings? Detecting the impact of social influence on the energy consumption behavior of networked users exposed to normative eco-feedback. *Energy and Buildings, 66*(Supplement C), 119–127. https://doi.org/10.1016/j.enbuild.2013.06.029

Johnson, D., Horton, E., Mulcahy, R., & Foth, M. (2017). Gamification and serious games within the domain of domestic energy consumption: A systematic review. *Renewable and Sustainable Energy Reviews, 73*(June), 249–264. https://doi.org/10.1016/j.rser.2017.01.134

Karlin, B., Zinger, J. F., & Ford, R. (2015). The effects of feedback on energy conservation: A meta-analysis. *Psychological Bulletin, 141*(6), 1205–1227. https://doi.org/10.1037/a0039650

Khosrowpour, A., Jain, R. K., Taylor, J. E., Peschiera, G., Chen, J., & Gulbinas, R. (2018). A review of occupant energy feedback research: Opportunities for methodological fusion at the intersection of experimentation, analytics, surveys and simulation. *Applied Energy,*

*218*(May), 304–316. https://doi.org/10.1016/j.apenergy.2018.02.148

Lévy, J.-P., & Belaïd, F. (2018). The determinants of domestic energy consumption in France: Energy modes, habitat, households and life cycles. *Renewable and Sustainable Energy Reviews, 81*(January), 2104–2114. https://doi.org/10.1016/j.rser.2017.06.022

Liebe, U., Gewinner, J., & Diekmann, A. (2018). What is missing in research on non-monetary incentives in the household energy sector? *Energy Policy, 123*-(December), 180–183. https://doi.org/10.1016/j.enpol.2018.08.036

Mäkivierikko, A., Shahrokni, H., & Kordas, O. (2019). Designing energy feedback for a local social network. *Energy and Buildings, 199*(September), 88–101. https://doi.org/10.1016/j.enbuild.2019.06.038

Meinrenken, C. J., Abrol, S., Gite, G. B., Hidey, C., McKeown, K., Mehmani, A., Modi, V., Turcan, E. C., Xie, W., & Culligan, P. J. (2021). Residential electricity conservation in response to auto-generated, multi-featured, personalized eco-feedback designed for large scale applications with utilities. *Energy and Buildings, 232*(February), 110652. https://doi.org/10.1016/j.enbuild.2020.110652

Morganti, L., Pallavicini, F., Cadel, E., Candelieri, A., Archetti, F., & Mantovani, F. (2017). Gaming for earth: Serious games and gamification to engage consumers in pro-environmental Behaviours for energy efficiency. *Energy Research & Social Science, 29*(July), 95–102. https://doi.org/10.1016/j.erss.2017.05.001

Murugesan, L. K., Hoda, R., & Salcic, Z. (2015). Design criteria for visualization of energy consumption: A systematic literature review. *Sustainable Cities and Society, 18*(November), 1–12. https://doi.org/10.1016/j.scs.2015.04.009

OPower. (2022). *Oracle Opower*. https://www.oracle.com/hk/industries/utilities/opower-energy-efficiency/what-is-opower/

Osbaldiston, R., & Schott, J. P. (2012). Environmental sustainability and behavioral science. *Environment and Behavior, 44*(2), 257–299. https://doi.org/10.1177/0013916511402673

Page, J., Robinson, D., Morel, N., & Scartezzini, J.-L. (2008). A generalised stochastic model for the simulation of occupant presence. *Energy and Buildings, 40*(2), 83–98. https://doi.org/10.1016/j.enbuild.2007.01.018

Peschiera, G., & Taylor, J. E. (2012). The impact of peer network position on electricity consumption in building occupant networks utilizing energy feedback systems. *Energy and Buildings, 49*(June), 584–590. https://doi.org/10.1016/j.enbuild.2012.03.011

Peschiera, G., Taylor, J. E., & Siegel, J. A. (2010). Response–relapse patterns of building occupant electricity consumption following exposure to personal, contextualized and occupant peer network utilization data. *Energy and Buildings, 42*(8), 1329–1336. https://doi.org/10.1016/j.enbuild.2010.03.001

Petersen, J. E., Shunturov, V., Janda, K., Platt, G., & Weinberger, K. (2007). Dormitory residents reduce electricity consumption when exposed to real-time visual feedback and incentives. *International Journal of Sustainability in Higher Education, 8*(1), 16–33. https://doi.org/10.1108/14676370710717562

Petkov, P., Köbler, F., Foth, M., & Krcmar, H. (2011). Motivating domestic energy conservation through comparative, community-based feedback in Mobile and social media. In *Proceedings of the 5th international conference on communities and technologies – C&T '11* (p. 21). ACM Press. https://doi.org/10.1145/2103354.2103358

PowerMeter, Google. (2022). *Google PowerMeter*. https://sites.google.com/site/powermeterpartners/google-meter-api

Šćepanović, S., Warnier, M., & Nurminen, J. K. (2017). The role of context in residential energy interventions: A meta review. *Renewable and Sustainable Energy Reviews, 77*(September), 1146–1168. https://doi.org/10.1016/j.rser.2016.11.044

Schultz, P. W., Nolan, J. M., Cialdini, R. B., Goldstein, N. J., & Griskevicius, V. (2007). The constructive, destructive, and reconstructive power of social norms. *Psychological Science, 18*(5), 429–434. https://doi.org/10.1111/j.1467-9280.2007.01917.x

Shi, M., He, H., Li, J., Han, M., & Jia, C. (2021). Multi-objective tradeoff optimization of predictive adaptive cruising control for autonomous electric buses: A cyber-physical-energy system approach. *Applied Energy, 300*(October), 117385. https://doi.org/10.1016/j.apenergy.2021.117385

U.S.Energy Information Administration. (2022). *Annual energy outlook 2022*. https://www.eia.gov/outlooks/aeo/

Ueno, T., Sano, F., Saeki, O., & Tsuji, K. (2006). Effectiveness of an energy-consumption information system on energy savings in residential houses based on monitored data. *Applied Energy, 83*(2), 166–183. https://doi.org/10.1016/j.apenergy.2005.02.002

Vale, B., & Vale, R. (2010). Domestic energy use, lifestyles and POE: Past lessons for current problems. *Building Research & Information, 38*(5), 578–588. https://doi.org/10.1080/09613218.2010.481438

Viegas, J. L., Vieira, S. M., Melício, R., Mendes, V. M. F., & Sousa, J. M. C. (2016). Classification of new electricity customers based on surveys and smart metering data. *Energy, 107*(July), 804–817. https://doi.org/10.1016/j.energy.2016.04.065

Wang, W., Chen, J., Huang, G., & Yujie, L. (2017a). Energy efficient HVAC control for an IPS-enabled large space in commercial buildings through dynamic spatial occupancy distribution. *Applied Energy, 207*-(December), 305–323. https://doi.org/10.1016/j.apenergy.2017.06.060

Wang, W., Chen, J., & Song, X. (2017b). Modeling and predicting occupancy profile in office space with a Wi-fi probe-based dynamic Markov time-window inference approach. *Building and Environment,*

*124*(November), 130–142. https://doi.org/10.1016/j.buildenv.2017.08.003

Wang, W., Hong, T., Li, N., Wang, R. Q., & Chen, J. (2019). Linking energy-cyber-physical systems with occupancy prediction and interpretation through WiFi probe-based ensemble classification. *Applied Energy, 236*(February), 55–69. https://doi.org/10.1016/j.apenergy.2018.11.079

Wever, R., van Kuijk, J., & Boks, C. (2008). User-centred Design for Sustainable Behaviour. *International Journal of Sustainable Engineering, 1*(1), 9–20. https://doi.org/10.1080/19397030802166205

Wilson, C., & Dowlatabadi, H. (2007). Models of decision making and residential energy use. *Annual Review of Environment and Resources, 32*(1), 169–203. https://doi.org/10.1146/annurev.energy.32.053006.141137

Yan, D., Hong, T., Dong, B., Mahdavi, A., D'Oca, S., Gaetani, I., & Feng, X. (2017). IEA EBC annex 66: Definition and simulation of occupant behavior in buildings. *Energy and Buildings, 156*(December), 258–270. https://doi.org/10.1016/j.enbuild.2017.09.084

Yoshino, H., Hong, T., & Nord, N. (2017). IEA EBC annex 53: Total energy use in buildings—Analysis and evaluation methods. *Energy and Buildings, 152*(October), 124–136. https://doi.org/10.1016/J.ENBUILD.2017.07.038

# 17 Towards the Circular Economy: Improving Construction and Demolition Waste Management Through Stakeholder Engagement

Yuchen She, Nilupa Udawatta, Olubukola Tokede, and Chunlu Liu

## 17.1 Introduction

The construction industry is criticised due to its poor resource management (Shooshtarian et al., 2022). It consumes large amounts of natural resources (World Economic Forum, 2016) and creates a significant amount of construction and demolition waste (CDW) (Islam et al., 2019; Wu et al., 2019a, b). Even though most CDW can be recycled (Hyvärinen et al., 2020) and some countries have a high CDW recycling rate, most CDW worldwide continues to be sent to landfill sites (Menegaki & Damigos, 2018). The results include severe environmental issues (Ding et al., 2016; Hou et al., 2018); damage to human health (Mah et al., 2018); and potential economic loss (Ding et al., 2016; Osobajo et al., 2020).

Construction and demolition waste (CDW) is generated throughout the entire life of building projects, even during the planning and design stages, due to the lack of a well-thought-out consideration for reducing and managing waste during the early stages of building projects (Esa et al., 2017). The initiation of new building construction involves obtaining raw materials from nature, which are subsequently converted into construction materials and assembled at the construction site (Mangialardo & Micelli, 2018). In this way, constructed buildings are not able to be deconstructed and become obsolete at the end of their life cycle. The materials, as well as all waste generated throughout the process, are ultimately disposed of in landfills or incinerated (Mangialardo & Micelli, 2018). Without the potential for reuse, the end of the life cycle becomes the phase with the most impact, being responsible for half of the total amount of CDW generated (Kibert, 2008). Managing CDW is not really about the physical and linear practices that begin after waste generation. The linear "take-make-use-dispose" process is based on high consumption of raw materials and low recovery of resources (Anastasiades et al., 2020). Instead, CDW management practices need to be treated as a priority in construction, by applying circular economy (CE) principles starting from the planning stages to predict and avoid possible future waste in the entire building life cycle.

However, the linear "take-make-use-dispose" process (i.e., the linear economy) continues to dominate in the construction industry (Mangialardo & Micelli, 2018). In contrast to the linear economy, the CE has emerged as an alternative way to reach zero waste generation in the construction industry (Eberhardt et al., 2022;

Y. She · N. Udawatta (✉) · O. Tokede · C. Liu
School of Architecture and Built Environment, Deakin University, Geelong, Australia
e-mail: shesh@deakin.edu.au; nilupa.udawatta@deakin.edu.au; olubukola.tokede@deakin.edu.au; chunlu.liu@deakin.edu.au



J. Zuo et al. (eds.), *Circular Economy for Buildings and Infrastructure*, Sustainable Development Goals Series, https://doi.org/10.1007/978-3-031-56241-9_17

Hossain et al., 2020; Munaro et al., 2020). Thus, to achieve a sustainable construction industry, transition from the linear economy to a sustainable system by applying CE thinking is urgent (Munaro et al., 2020). The CE emphasises system innovations to design out waste, optimise resource utilisation, minimise negative impacts and achieve a balance among the environment, society and the economy (Geissdoerfer et al., 2017; Kristensen & Mosgaard, 2020). Notably, the CE, an evolutionary concept from industries and societies, is accepted as a closed-loop economy for pursuing complete sustainability through a no-waste culture (De los Rios & Charnley, 2017) and by holding the value of materials and components for the longest time possible in the life cycle of products (Kirchherr et al., 2017). Furthermore, the CE is an ideal system-oriented model for addressing the overuse of raw materials and the overproduction of waste. The construction industry largely appears to be a feasible field for CE application (Brambilla et al., 2019; López Ruiz et al., 2020), especially for building system-oriented CDW management. The CE, thus, has great potential for developing truly sustainable CDW management that minimises the impacts of the business-as-usual scenario. It could thoroughly alter the way that a building is designed, constructed and used, as well as the way building components and used materials are brought back into circulation.

Effective CDW management inevitably involves various stakeholders (Udawatta et al., 2015a) who can influence the CE application in construction projects (Bajaj, 2018). All stakeholders must reconsider and rethink their practices to improve the current CDW management in construction projects (Aslam et al., 2020). The term "stakeholders" refers to organisations or individuals influencing, or being influenced by, CDW management throughout the building life cycle. This includes internal stakeholders (e.g., clients, designers, builders and suppliers) and external stakeholders (e.g., the general public, governments, etc.) (Park & Tucker, 2017). Interactions between these different types of stakeholders will vary based on the changes in their roles during the entire life cycle of building projects (Zhao, 2021). This can lead to changes in behaviour, attitude, awareness and interest (Kim et al., 2020), thereby presenting challenges to the implementation of CDW management (Lingard et al., 2000). In this context, harnessing people's power through the pervasive connectivity of stakeholder engagement (SE) is a promising development for CE implementation in CDW management. In addition, with individual stakeholders, SE processes can promote reform and accelerate transformation (Salvioni & Almici, 2020a). Moreover, the use of SE processes by organisations can involve stakeholders in achieving explicit purposes to fulfil agreed results (Franklin, 2020; Leopizzi, 2020). Acting as a fundamental mechanism of responsibilities, SE allows organisations to have their own stakeholders to identify, understand and reply to sustainability problems, and to then report, define and respond to stakeholders on decisions, actions and performance. This supports the engagement of both organisations and stakeholders in learning from each other through this mutual contact (Leopizzi, 2020). By this means, SE can assist organisations to enable cultural reforms, practise sustainability principles and reach expectations in relation to the economy, society and the environment (Salvioni & Almici, 2020b). With projects treated as temporary organisations (Turner & Müller, 2003), SE can also be effectively applied in construction projects. Thus, SE can unlock the application of CE-based CDW management practices to achieve a waste-free culture in the construction industry.

Despite SE playing an essential role in forming and firming CE implementation in CDW management, a research gap is evident in the understanding of how SE can be effectively utilised to facilitate the transition from the linear economy to the CE in the construction industry. Thus, this chapter aims to address this research gap by conducting a comprehensive narrative literature review on the application of SE in implementing CE principles to improve CDW management. The findings of this literature review contribute to developing a waste-free culture in the construction industry by providing insights into how various stakeholders can effectively collaborate in the transition towards CE thinking.

## 17.2 Research Method

The primary research approach adopted by this chapter is the narrative literature review method which is used to explore circular economy (CE) practices in construction and demolition waste (CDW) management and stakeholder engagement (SE) in CE-based CDW management. A literature review is an essential research method that allows researchers to identify, evaluate and interpret existing research related to a particular topic or research question (Fink, 2020). This method is particularly suitable in this chapter as it enables the gathering and synthesis of existing knowledge from various sources, identifying gaps in the literature and providing recommendations for future research and practice.

The data collection for this literature review involved a systematic search of electronic databases and academic search engines, comprising Web of Science, Scopus, Google Scholar and ScienceDirect. The search strategy was adapted for each database, with results limited to peer-reviewed journal articles, books, book chapters, conference papers and reports published in English. Relevant studies were identified using specific keywords and phrases, such as "circular economy", "construction and demolition waste management" and "stakeholder engagement". Boolean operators (i.e., AND, OR) and truncation (*) were used to enhance the search results and capture relevant literature. In addition to the electronic database search, a manual search of reference lists of the chosen studies was performed to identify other relevant publications. This snowballing technique is an essential approach to ensure the comprehensiveness of the literature review (Greenhalgh & Peacock, 2005). The search process focused on literature published from 2000 onwards to ensure the inclusion of the most recent and relevant studies on the topic.

The inclusion and exclusion criteria for the literature review were based on the relevance of the studies to the research question and their methodological quality. To select the most relevant studies for this chapter, explicit inclusion and exclusion criteria were applied. The inclusion criteria were: (1) studies focusing on CDW management; (2) studies addressing CE practices in waste management or the construction industry; (3) studies discussing SE in waste management or construction; and (4) studies published in peer-reviewed journals, conference proceedings, books or reports. Studies were excluded if they: (1) did not focus on CDW management or the construction industry; (2) did not address CE practices or SE; or (3) were published in non-peer-reviewed sources (e.g., opinion articles, blog posts, newspaper articles, etc.); and (4) studies of poor methodological quality or with insufficient information. Studies that were not available in full-text format were also excluded.

The selected studies were analysed using a thematic analysis approach. A wide range of qualitative research methods are used to identify, analyse and present/report patterns within data (Braun & Clarke, 2006). This method involves a six-step process: acquainting oneself with the data, developing initial codes, exploring themes, appraising themes, establishing and naming themes and presenting the report (Braun & Clarke, 2006). The analysis started with a detailed examination of the selected studies to identify key concepts, theories and methodologies related to SE and CE-based CDW management. The studies were coded using an open coding approach which allowed emerging themes and categories to be identified. The codes were then organised into themes which were reviewed and refined to ensure their coherence and relevance to the research question. The final step involved synthesising the findings and discussing the implications of the literature review for gaining an understanding of SE in CE-based CDW management.

## 17.3 Construction and Demolition Waste (CDW) Management

The term "construction and demolition waste (CDW)" refers to the solid waste produced in construction and demolition processes (Bilal et al., 2020; Kibert, 2008). This waste is contingent on numerous determinants, including the project type (e.g., buildings, bridges, road/rail); the project size (e.g., low rise, high rise); and the project operation approach (e.g., construction, refurbishment, demolition), as well as the project operation site (Diyamandoglu & Fortuna, 2015; Shen et al., 2004; Silva et al., 2017; Zhao et al., 2010). Menegaki and Damigos (2018) classified CDW as from human-made sources (e.g., building construction works, and building renovation and demolition works) and from nature-made sources (e.g., earthquakes, floods and tsunamis). Demolition waste is a compound of different building materials and resources, such as aggregates, concrete, timber, paper products, metals, insulating materials and glass which are generally contaminated by paint, fasteners, cement, soil and wall coverings (El-Haggar, 2007). This type of waste is created from entirely or selectively demolishing existing structures through human activities or natural disasters (Al-Ansary et al., 2004) and partly from the waste derived from renovation works (El-Haggar, 2007). Construction waste is treated as relatively less contaminated, comprising mixed building materials from numerous construction works (Pacheco-Torgal et al., 2013) and mainly comprises glass, plastics, timber, steel, cement, mortar, concrete, bricks, organic waste and soil from excavations (Noor et al., 2020).

In the prior literature, a conventional CDW management hierarchy was established in the pursuit of solutions to decrease waste generation (Yuan & Shen, 2011). This hierarchy comprises four key strategies (i.e., reduction, reusing, recycling and disposal) or seven strategies (i.e., avoidance, reduction, reusing, recycling, recovery, treatment and disposal). In this traditional hierarchy, the environmental consequences of implementing these strategies are arranged in an ascending sequence, from the least to the most significant impact (Yuan & Shen, 2011), with preference given to using the strategies in descending order from the most favourable one to the least favourable one (Kabirifar et al., 2020a; Yuan & Shen, 2011). Of these CDW management strategies, reduce, reuse and recycle are well known as the 3Rs and are treated as fundamental principles in waste management (Huang et al., 2018). The waste treatment or recovery strategy is regarded as a general category of waste management, while disposal is the last option is an inefficient waste management strategy (Kabirifar et al., 2020a). Notably, despite applying the recycling option, which is the preference over the disposal option in the hierarchy, recycling is not always suitable for all types of CDW (Minunno et al., 2018), so the waste created is sent to landfill sites.

Many sustainable environmental management practices for CDW are being initiated to avoid waste disposal, such as designing out waste, efficiency in the use of materials, building deconstruction, and site waste management and prevention (Gálvez-Martos et al., 2018). However, waste and other materials that do not appear suitable for utilisation, such as construction stage excess materials and demolition stage residual materials, are still discarded even with sustainable CDW management (Sapuay, 2016). This phenomenon exposes poor waste management behaviours and inappropriate stakeholders' attitudes when dealing with the high volume of CDW produced in the construction industry (Kabirifar et al., 2020a, b). For example, clients may demonstrate inappropriate attitudes towards CDW management by prioritising profit ahead of the use of sustainable options to deal with waste (Lu et al., 2015). Inadequate training and poor quality CDW management result in CDW generation (Udawatta et al., 2015a). However, project stakeholders can achieve high waste diversion rates and waste elimination in construction projects by introducing appropriate sustainable CDW management viewpoints, for example, focusing on restorative and regenerative construction practices (Udawatta et al., 2021) and shifting the CDW management hierarchy from linear to circular (Kabirifar et al., 2020b). In this sense, bringing restorative, regenerative and circular thought to waste management practices is vital to the current construction industry. Thus, the application of CE principles to improve CDW management is explored in the next section.

## 17.4 Application of Circular Economy (CE)-Based Principles in Construction and Demolition Waste (CDW) Management

Circular economy (CE) principles and effective waste management are two major components of sustainable resource management that are critical for reducing environmental impacts and promoting economic growth through construction activities. The three design principles of CE, eradicating waste and pollution, maintaining the optimal value of products and resources in circulation and fostering the restoration of natural ecosystems, are extensively promoted by organisations and governments worldwide (European Commission, 2019; Organisation for Economic Co-operation and Development, 2020). The three Rs (i.e., reduce, reuse and recycle) are also the CE's foundational principles (Ghisellini et al., 2016) and are prioritised in an effective waste management hierarchy (Kabirifar et al., 2020a). In Australia, the effective waste management hierarchy has been extended to include "recover" as a fourth R, making it a 4Rs framework (Kirchherr et al., 2017). Researchers have expanded the R-based framework to increase circularity beyond the 4Rs to include 5Rs, 6Rs and even 9Rs. According to Kristensen and Mosgaard (2020), the 12 main types of CE principles that commonly emerge are "refuse", "rethink", "reduce", "reuse", "repair", "refurbish", "remanufacture", "repurpose", "recycle", "recover", "regenerate" and "remine". These principles are often combined with different orders and hierarchies by various categories, with some having been implemented for many years (Reike et al., 2018).

The CE principles have a close relationship with the effective CDW management hierarchy, as shown in Table 17.1.

The CE principles emphasise the importance of eliminating waste and pollution, consistent with "waste prevention", the first principle of the waste management hierarchy. Waste prevention is crucial for reducing CDW generated in construction and demolition process. In addition, circulating products and materials at their highest value is essential, as this aligns with "reusing", the second principle of the waste management hierarchy. By reusing materials and products, the amount of CDW that ends up in landfill sites can be minimised. The CE principle of "regenerate" is similar to "recovery", the fourth principle of the waste management hierarchy, both of which emphasise the need to regenerate the natural environment by promoting the recovery of materials and resources. By recovering materials through closed-loop systems, waste generation in the construction and demolition process can be minimised.

Thus, it can be argued that embracing the principles of circularity in the CDW management hierarchy can lead to a shift towards sustainable waste management practices. For example, designing for disassembly, reusing materials and products, and recycling materials can effectively reduce waste generation during the construction and demolition process (Eberhardt et al., 2022). By following the CE principles and integrating them into the CDW management hierarchy, sustainable waste management practices can be achieved. This shift towards CE thinking promotes the efficient use of resources, reduces waste generation and ensures sustainable waste management practices.

## 17.5 Stakeholder Engagement (SE) in Circular economy (CE)-Based Construction and Demolition Waste (CDW) Management

The construction industry involves diverse stakeholders, including clients, architects, engineers, contractors, suppliers, regulators and local communities, each of whom has different interests, perspectives and levels of influence over waste management practices (Gluch et al., 2014). These stakeholders, with their various purposes, occupy different positions in CDW management (Udawatta et al., 2015a), leading to different behaviours in CDW management in practice (Lu et al., 2015). Their reluctance to support

**Table 17.1** CE R-based principles and CDW management hierarchy

| CE R-based principles | Description | CDW management hierarchy |
|---|---|---|
| Refuse | Avoiding the use of materials or products that generate waste | Waste prevention |
| Rethink | Re-evaluating the necessity of certain materials, products or processes | Waste prevention |
| Reduce | Minimising the amount of materials used, thereby reducing waste generation | Waste prevention |
| Reuse | Using products and materials again in their original form or for a similar purpose | Reusing |
| Repair | Fixing damaged products or materials to extend their life | Reusing |
| Refurbish | Restoring materials or products to a like-new condition | Reusing |
| Remanufacture | Rebuilding materials or products to extend their life and improve performance | Reusing |
| Repurpose | Adapting materials or products for a new purpose | Reusing |
| Recycle | Converting waste materials into new materials or products | Recycling |
| Recover | Extracting valuable materials and resources from waste | Recovery |
| Regenerate | Restoring or improving the natural environment | Recovery |
| Remine | Extracting valuable materials from waste deposited in landfill or other waste disposal sites | Recovery |

certain project targets can result in failure to achieve these goals (Bal et al., 2013). Achieving sustainable waste management targets needs the participation and involvement of these construction project stakeholders (Joseph, 2006). Construction and demolition waste (CDW) management currently needs a method to involve and an agreement to bind project stakeholders in closed-loop construction (Ghaffar et al., 2020), thus ensuring a more sustainable approach to resource use and waste minimisation. The absence of or deficiency in effective SE during the project life cycle can negatively impact on accomplishment of the expected project targets, especially in the planning and implementation phases (Bahadorestani et al., 2020). Thus, it is widely accepted that successful sustainable CDW management requires effective engagement and collaboration with key stakeholders, even though they have varied responsibilities and accountability in construction projects (Alamgir et al., 2012; Li et al., 2015; Udawatta et al., 2015b).

Engaging with key construction project stakeholders is essential for developing and implementing effective CE-based CDW management. This engagement can help to build a shared understanding of the challenges, opportunities and potential solutions associated with waste reduction and resource efficiency (Shooshtarian et al., 2022). Stakeholder engagement (SE) in CE-based CDW management enables the identification and prioritisation of stakeholder needs and concerns, which can inform the development of more targeted and context-specific waste management policies, regulations and practices (Ajayi & Oyedele, 2018). By involving stakeholders in decision-making processes and promoting transparency and accountability, SE can also help to build trust and commitment among stakeholders, facilitating the implementation of sustainable waste management practices that are aligned with CE principles (Bocken et al., 2016; Gluch et al., 2014). Furthermore, SE can foster innovation in waste management practices by facilitating the exchange of knowledge, ideas and best practices among stakeholders (Bocken et al., 2016). This can contribute to the identification and implementation of novel waste management technologies and approaches that are aligned with CE principles. Finally, SE can contribute to developing a supportive environment for adopting CE principles in CDW management (Ghisellini et al., 2016). By engaging with stakeholders and raising awareness of the benefits of circular approaches, a shared understanding of the importance of waste reduction and resource efficiency can be achieved (Ajayi & Oyedele,

2018). This can lead to a shift in attitudes and behaviours within the construction industry, encouraging the adoption of sustainable practices, such as designing for disassembly, reusing materials and products, and recycling materials.

Although SE plays a crucial role in implementing CE principles in CDW management, several challenges need to be addressed to fully realise its potential. One of the main challenges is the lack of awareness and understanding among stakeholders of CE principles and their relevance to CDW management (Oluleye et al., 2022). This can hinder the development of a shared vision and commitment to waste reduction and resource efficiency, limiting the effectiveness of SE efforts (Bocken et al., 2016). Another challenge is the potential for conflicts of interest between different stakeholder groups which can result in resistance to adopting CE principles in CDW management (Charef, 2022). For example, some stakeholders may prioritise short-term economic gains over long-term sustainability goals, leading to reluctance to implement more sustainable waste management practices (Lu et al., 2015). Addressing these conflicts requires effective communication, negotiation and collaboration among stakeholders to balance competing interests and achieve mutually beneficial outcomes (Gluch et al., 2014). The fragmentation of the construction industry and the complexity of its supply chains also present challenges for SE in CE-based CDW management (Ghisellini et al., 2016). The involvement of numerous stakeholders, with varying degrees of influence and decision-making authority, can make it challenging to coordinate and align waste management efforts across the entire construction process (Oluleye et al., 2022). This challenge can be addressed by adopting a more systematic and integrated approach to SE which involves establishing clear roles, responsibilities and communication channels between stakeholders (Ajayi & Oyedele, 2018).

Despite these challenges, several opportunities exist that can enhance SE in implementing CE principles in CDW management. Firstly, the growing recognition of the environmental, social and economic benefits of CE-based waste management practices provides a strong incentive for stakeholders to collaborate and adopt more sustainable approaches (Sparrevik et al., 2021). This can help to create a sense of urgency and shared commitment among stakeholders, facilitating the transition towards a circular construction industry. Secondly, the increasing availability of digital technologies and data-driven tools can support more effective collaboration and coordination among diverse stakeholders in CE-based CDW management (Yu et al., 2022). For example, Building Information Modelling (BIM) can facilitate communication, collaboration and decision making among stakeholders throughout the construction process, enabling the identification and implementation of waste reduction and resource efficiency measures (Razali et al., 2019). Similarly, digital platforms and social media can be leveraged to facilitate knowledge sharing, networking and innovation among stakeholders, promoting the adoption of CE principles in CDW management (Bocken et al., 2016). Lastly, significant potential exists for policy interventions and regulatory frameworks to support the various stakeholders involved in CE-based CDW management (Giorgi et al., 2022). Governments and regulatory bodies can play a crucial role in establishing clear waste management targets, incentives and penalties and by providing guidance, resources and support for stakeholders to engage in more sustainable construction practices (Benachio et al., 2020; Kirchherr et al., 2017). By fostering a supportive policy environment, SE efforts can be more effectively channelled towards implementing CE principles in CDW management. Therefore, conducting CE-based CDW management relies on stakeholders and SE.

## 17.6 Conclusion

This chapter aimed to investigate the application of SE in applying CE principles to improve CDW management in construction projects. Circular economy (CE) principles offer a promising strategy to reduce waste generation, increase resource efficiency and mitigate environmental impacts in construction and demolition. However, implementing CE practices in CDW management requires collaboration between diverse stakeholders, including designers, builders, waste management companies, regulators and users, with their varied interests, perspectives and influence levels to ensure effective waste reduction and resource efficiency in the construction industry. Stakeholder engagement (SE) can foster innovation, build trust and develop a supportive environment for adopting CE approaches in CDW management. However, challenges, such as lack of awareness, conflicts of interest and the construction industry's fragmented nature, need to be addressed to fully harness the potential of SE in implementing CE principles. Despite these challenges, opportunities, such as the growing recognition of the environmental, social and economic benefits of circular approaches, the increasing availability of digital technologies and supportive policy interventions can enhance SE in CDW management. By fostering collaboration and shared commitments between stakeholders, the construction industry can transition towards more sustainable waste management practices that are aligned with CE principles.

A limitation of the current study is that the narrative literature review method may not capture the most recent advances or emerging trends in SE and CE practices. In these fields, research is constantly evolving, so keeping up to date with the latest developments and findings is crucial for a better understanding of the field. In light of this limitation, several recommendations can be made for future research and practice. Future research could explore the challenges of and barriers to adopting digital technologies, CE practices and SE in CDW management. This exploration would provide a more balanced view of the applicability and effectiveness of these approaches and contribute to the development of tailored strategies and solutions for specific contexts and circumstances. Lastly, it is recommended that practitioners and policy makers consider the findings and insights from this book chapter when developing and implementing CDW management strategies. By fostering a culture of collaboration among stakeholders, embracing innovative technological solutions and integrating CE principles into waste management practices, they can better address the pressing issue of CDW, ultimately contributing to more sustainable and environmentally responsible practices in the construction industry.

## References


Ajayi, S. O., & Oyedele, L. O. (2018). Critical design factors for minimising waste in construction projects: A structural equation modelling approach. *Resources, Conservation & Recycling, 137*, 302–313.

Al-Ansary, M. S., El-Haggar, S. M., & Taha, M. A. (2004). *Sustainable guidelines for managing demolition waste in Egypt*. Paper presented at the International RILEM [International Union of Laboratories and Experts in Construction Materials, Systems and Structures] Conference, Barcelona, 9–11 November 2004.

Alamgir, M., Bidlingmaier, W., & Cossu, R. (2012). Successful waste management strategies in developing countries require meaningful involvement of the concerned stakeholders. *Waste Management, 32*(11), 2007–2008.

Anastasiades, K., Blom, J., Buyle, M., & Audenaert, A. (2020). Translating the circular economy to bridge construction: Lessons learnt from a critical literature review. *Renewable and Sustainable Energy Reviews, 117*, 109522.

Aslam, M. S., Huang, B., & Cui, L. (2020). Review of construction and demolition waste management in China and USA. *Journal of Environmental Management, 264*, 110445.

Bahadorestani, A., Naderpajouh, N., & Sadiq, R. (2020). Planning for sustainable stakeholder engagement based on the assessment of conflicting interests in projects. *Journal of Cleaner Production, 242*, 118402.

Bajaj, S. (2018). In support of the circular economy: An exercise in stakeholder mapping. *Matériaux & Techniques, 105*(5–6), 512.

Bal, M., Bryde, D., Fearon, D., & Ochieng, E. (2013). Stakeholder engagement: Achieving sustainability in the construction sector. *Sustainability, 5*(2), 695–710.

Benachio, G. L. F., Freitas, M. C. D., & Tavares, S. F. (2020). Circular economy in the construction industry: A systematic literature review. *Journal of Cleaner Production, 260*, 121046.

Bilal, M., Khan, K. I. A., Thaheem, M. J., & Nasir, A. R. (2020). Current state and barriers to the circular economy in the building sector: Towards a mitigation framework. *Journal of Cleaner Production, 276*, 123250.

Bocken, N. M. P., de Pauw, I., Bakker, C., & van der Grinten, B. (2016). Product design and business model strategies for a circular economy. *Journal of Industrial and Production Engineering, 33*(5), 308–320.

Brambilla, G., Lavagna, M., Vasdravellis, G., & Castiglioni, C. A. (2019). Environmental benefits arising from demountable steel-concrete composite floor systems in buildings. *Resources, Conservation and Recycling, 141*, 133–142.

Braun, V., & Clarke, V. (2006). Using thematic analysis in psychology. *Qualitative Research in Psychology, 3*(2), 77–101.

Charef, R. (2022). Supporting construction stakeholders with the circular economy: A trans-scaler framework to understand the holistic approach. *Cleaner Engineering and Technology, 8*, 100454.

De los Rios, I. C., & Charnley, F. J. S. (2017). Skills and capabilities for a sustainable and circular economy: The changing role of design. *Journal of Cleaner Production, 160*, 109–122.

Ding, Z., Wang, Y., & Zou, P. X. W. (2016). An agent based environmental impact assessment of building demolition waste management: Conventional versus green management. *Journal of Cleaner Production, 133*, 1136–1153.

Diyamandoglu, V., & Fortuna, L. M. (2015). Deconstruction of wood-framed houses: Material recovery and environmental impact. *Resources, Conservation and Recycling, 100*, 21–30.

Eberhardt, L. C. M., Birkved, M., & Birgisdottir, H. (2022). Building design and construction strategies for a circular economy. *Architectural Engineering and Design Management, 18*(2), 93–113.

El-Haggar, S. M. (2007). Sustainability of construction and demolition waste management. In *Sustainable industrial design and waste management* (pp. 261–292). Elsevier/Academic.

Esa, M. R., Halog, A., & Rigamonti, L. (2017). Developing strategies for managing construction and demolition wastes in Malaysia based on the concept of circular economy. *Journal of Material Cycles and Waste Management, 19*(3), 1144–1154.

European Commission. (2019). *Sustainable products in a circular economy – Towards an EU product policy framework contributing to the circular economy*. European Commission. Available via European Commission. https://ec.europa.eu/environment/circular-economy/pdf/sustainable_products_circular_economy.pdf. Accessed 5 Apr 2022.

Fink, A. (2020). *Conducting research literature reviews: From the internet to paper* (5th ed.). Sage.

Franklin, A. L. (2020). *Stakeholder engagement*. Springer International Publishing AG.

Gálvez-Martos, J.-L., Styles, D., Schoenberger, H., & Zeschmar-Lahl, B. (2018). Construction and demolition waste best management practice in Europe. *Resources, Conservation and Recycling, 136*, 166–178.

Geissdoerfer, M., Savaget, P., Bocken, N. M. P., & Hultink, E. J. (2017). The circular economy – A new sustainability paradigm? *Journal of Cleaner Production, 143*, 757–768.

Ghaffar, S. H., Burman, M., & Braimah, N. (2020). Pathways to circular construction: An integrated management of construction and demolition waste for resource recovery. *Journal of Cleaner Production, 244*, 118710.

Ghisellini, P., Cialani, C., & Ulgiati, S. (2016). A review on circular economy: The expected transition to a balanced interplay of environmental and economic systems. *Journal of Cleaner Production, 114*, 11–32.

Giorgi, S., Lavagna, M., Wang, K., Osmani, M., Liu, G., & Campioli, A. (2022). Drivers and barriers towards circular economy in the building sector: Stakeholder interviews and analysis of five European countries' policies and practices. *Journal of Cleaner Production, 336*, 130395.

Gluch, P., Gustafsson, M., Thuvander, L., & Baumann, H. (2014). Charting corporate greening: Environmental management trends in Sweden. *Building Research & Information, 42*(3), 318–329.

Greenhalgh, T., & Peacock, R. (2005). Effectiveness and efficiency of search methods in systematic reviews of complex evidence: Audit of primary sources. *BMJ, 331*(7524), 1064–1065.

Hossain, M. U., Ng, S. T., Antwi-Afari, P., & Amor, B. (2020). Circular economy and the construction industry: Existing trends, challenges and prospective framework for sustainable construction. *Renewable and Sustainable Energy Reviews, 130*, 109948.

Hou, Z., Liu, Y., & Wei, C. (2018). Influence of construction and demolition waste on fitness and community structure of cicada nymphs: New bioindicators of soil pollution. *PLoS One, 13*(9), e0203744.

Huang, B., Wang, X., Kua, H., Geng, Y., Bleischwitz, R., & Ren, J. (2018). Construction and demolition waste management in China through the 3R principle. *Resources, Conservation & Recycling, 129*, 36–44.

Hyvärinen, M., Ronkanen, M., & Kärki, T. (2020). Sorting efficiency in mechanical sorting of construction and demolition waste. *Waste Management & Research, 38*(7), 812–816.

Islam, R., Nazifa, T. H., Yuniarto, A., Shanawaz Uddin, A. S. M., Salmiati, S., & Shahid, S. (2019). An empirical study of construction and demolition waste generation and implication of recycling. *Waste Management, 95*, 10–21.

Joseph, K. (2006). Stakeholder participation for sustainable waste management. *Habitat International, 30*(4), 863–871.

Kabirifar, K., Mojtahedi, M., Wang, C. C., & Tam, V. W. Y. (2020a). Construction and demolition waste management contributing factors coupled with reduce, reuse, and recycle strategies for effective waste management: A review. *Journal of Cleaner Production, 263*, 121265.

Kabirifar, K., Mojtahedi, M., Wang, C. C., & Tam, V. W. Y. (2020b). A conceptual foundation for effective construction and demolition waste management. *Cleaner Engineering and Technology, 1*, 100019.

Kibert, C. J. (2008). *Sustainable construction: Green building design and delivery* (2nd ed.). Wiley.

Kim, S. Y., Nguyen, M. V., & Luu, V. T. (2020). A performance evaluation framework for construction and demolition waste management: Stakeholder perspectives. *Engineering, Construction and Architectural Management, 27*(10), 3189–3213.

Kirchherr, J., Reike, D., & Hekkert, M. (2017). Conceptualizing the circular economy: An analysis of 114 definitions. *Resources, Conservation and Recycling, 127*, 221–232.

Kristensen, H. S., & Mosgaard, M. A. (2020). A review of micro level indicators for a circular economy – Moving away from the three dimensions of sustainability? *Journal of Cleaner Production, 243*, 118531.

Leopizzi, R. (2020). Stakeholder engagement. In S. Idowu, R. Schmidpeter, N. Capaldi, L. Zu, M. Del Baldo, & R. Abreu (Eds.), *Encyclopedia of sustainable nanagement* (pp. 1–5). Springer International Publishing.

Li, J., Tam, V. W. Y., Zuo, J., & Zhu, J. (2015). Designers' attitude and behaviour towards construction waste minimization by design: A study in Shenzhen, China. *Resources, Conservation and Recycling, 105*, 29–35.

Lingard, H., Graham, P., & Smithers, G. (2000). Employee perceptions of the solid waste management system operating in a large Australian contracting organization: Implications for company policy implementation. *Construction Management and Economics, 18*(4), 383–393.

López Ruiz, L. A., Roca Ramón, X., & Gassó Domingo, S. (2020). The circular economy in the construction and demolition waste sector – A review and an integrative model approach. *Journal of Cleaner Production, 248*, 119238.

Lu, W., Peng, Y., Webster, C., & Zuo, J. (2015). Stakeholders' willingness to pay for enhanced construction waste management: A Hong Kong study. *Renewable and Sustainable Energy Reviews, 47*, 233–240.

Mah, C. M., Fujiwara, T., & Ho, C. S. (2018). Environmental impacts of construction and demolition waste management alternatives. *Chemical Engineering Transactions, 63*, 343–348.

Mangialardo, A., & Micelli, E. (2018). Rethinking the construction industry under the circular economy: Principles and case studies. In A. Bisello, D. Vettorato, P. Laconte, & S. Costa (Eds.), *Smart and sustainable planning for cities and regions, march 2017* (Green Energy and Technology) (pp. 333–334). Springer.

Menegaki, M., & Damigos, D. (2018). A review on current situation and challenges of construction and demolition waste management. *Current Opinion in Green and Sustainable Chemistry, 13*, 8–15.

Minunno, R., O'Grady, T., Morrison, G., Gruner, R., & Colling, M. (2018). Strategies for applying the circular economy to prefabricated buildings. *Buildings, 8*, 125.

Munaro, M. R., Tavares, S. F., & Bragança, L. (2020). Towards circular and more sustainable buildings: A systematic literature review on the circular economy in the built environment. *Journal of Cleaner Production, 260*, 121134.

Noor, T., Javid, A., Hussain, A., Bukhari, S. M., Ali, W., Akmal, M., & Hussain, S. M. (2020). Types, sources and management of urban wastes. In P. Verma, P. Singh, R. Singh, & A. S. Raghubanshi (Eds.), *Urban ecology: Emerging patterns and social-ecological systems* (pp. 239–263). Elsevier.

Oluleye, B. I., Chan, D. W. M., & Olawumi TO. (2022). Barriers to circular economy adoption and concomitant implementation strategies in building construction and demolition waste management: A PRISMA and interpretive structural modeling approach. *Habitat International, 126*, 102615.

Organisation for Economic Co-operation and Development (OECD). (2020). *Improving resource efficiency and the circularity of economies for a greener world.* OECD. Available via OECDiLibrary. https://www.oecd-ilibrary.org/deliver/1b38a38f-en.pdf?itemId=%2Fcontent%2Fpaper%2F1b38a38f-en&mimeType=pdf. Accessed 5 Apr 2022.

Osobajo, O. A., Oke, A., Omotayo, T., & Obi, L. I. (2020). A systematic review of circular economy research in the construction industry. *Smart and Sustainable Built Environment, 11*(1), 39–64.

Pacheco-Torgal, F., Tam, V. W. Y., Labrincha, J., Ding, Y., & de Brito, J. (Eds.). (2013). *Handbook of recycled concrete and demolition waste*. Woodhead Publishing.

Park, J., & Tucker, R. (2017). Overcoming barriers to the reuse of construction waste material in Australia: A review of the literature. *International Journal of Construction Management, 17*(3), 228–237.

Razali, M., Haron, N., Salihudin, H., Alias, A., Harun, A., & Abubakar, A. (2019). A review: Application of building information modelling (BIM) over building life cycles. In F. Hejazi, R. S. Muhammad Rashid, & M. S. Jaafar (Eds.), *Sustainable civil and construction engineering conference, University Putra Malaysia, Kuala Lumpur, 25–27 August 2018* (IOP (Institute of Physics) conference series: Earth and environmental science, vol 357).

Reike, D., Vermeulen, W. J. V., & Witjes, S. (2018). The circular economy: New or refurbished as CE 3.0? – Exploring controversies in the conceptualization of the circular economy through a focus on history and resource value retention options. *Resources, Conservation and Recycling, 135*, 246–264.

Salvioni, D., & Almici, A. (2020a). Circular economy and stakeholder engagement strategy. *Symphonya Emerging Issues in Management, 1*, 26–44.

Salvioni, D., & Almici, A. (2020b). Transitioning toward a circular economy: The impact of stakeholder engagement on sustainability culture. *Sustainability, 12*(20), 8641.

Sapuay, S. E. (2016). Construction waste – Potentials and constraints. *Procedia Environmental Sciences, 35*, 714–722.

Shen, L. Y., Tam, V. W. Y., Tam, C. M., & Drew, D. (2004). Mapping approach for examining waste management on construction sites. *Journal of*

*Construction Engineering & Management, 130*(4), 472–481.

Shooshtarian, S., Maqsood, T., Caldera, S., & Ryley, T. (2022). Transformation towards a circular economy in the Australian construction and demolition waste management system. *Sustainable Production and Consumption, 30*, 89–106.

Silva, R. V., de Brito, J., & Dhir, R. K. (2017). Availability and processing of recycled aggregates within the construction and demolition supply chain: A review. *Journal of Cleaner Production, 143*, 598–614.

Sparrevik, M., de Boer, L., Michelsen, O., Skaar, C., Knudson, H., & Fet, A. M. (2021). Circular economy in the construction sector: Advancing environmental performance through systemic and holistic thinking. *Environment Systems and Decisions, 41*(3), 392–400.

Turner, J. R., & Müller, R. (2003). On the nature of the project as a temporary organization. *International Journal of Project Management, 21*(1), 1–8.

Udawatta, N., Zuo, J., Chiveralls, K., & Zillante, G. (2015a). Attitudinal and behavioural approaches to improving waste management on construction projects in Australia: Benefits and limitations. *International Journal of Construction Management, 15*(2), 137–147.

Udawatta, N., Zuo, J., Chiveralls, K., & Zillante, G. (2015b). Improving waste management in construction projects: An Australian study. *Resources, Conservation and Recycling, 101*, 73–83.

Udawatta, N., Zuo, J., Chiveralls, K., & Zillante, G. (2021). From green buildings to living buildings? Rating schemes and waste management practices in Australian educational buildings. *Engineering, Construction and Architectural Management, 28*(4), 1278–1294.

World Economic Forum. (2016). *Shaping the future of construction: A breakthrough in mindset and technology*. World Economic Forum. Available via World Economic Forum. http://www3.weforum.org/docs/WEF_Shaping_the_Future_of_Construction_full_report__pdf. Accessed 5 Apr 2022.

Wu, H., Zuo, J., Yuan, H., Zillante, G., & Wang, J. (2019a). A review of performance assessment methods for construction and demolition waste management. *Resources, Conservation and Recycling, 150*, 104407.

Wu, H., Zuo, J., Zillante, G., Wang, J., & Yuan, H. (2019b). Status quo and future directions of construction and demolition waste research: A critical review. *Journal of Cleaner Production, 240*, 118163.

Yu, Y., Yazan, D. M., Junjan, V., & Iacob, M.-E. (2022). Circular economy in the construction industry: A review of decision support tools based on information & communication technologies. *Journal of Cleaner Production, 349*, 131335.

Yuan, H., & Shen, L. (2011). Trend of the research on construction and demolition waste management. *Waste Management, 31*(4), 670–679.

Zhao, W., Leeftink, R. B., & Rotter, V. S. (2010). Evaluation of the economic feasibility for the recycling of construction and demolition waste in China – The case of Chongqing. *Resources, Conservation and Recycling, 54*(6), 377–389.

Zhao, X. (2021). Stakeholder-associated factors influencing construction and demolition waste management: A systematic review. *Buildings, 11*(4), 149.

# 18 Circular Economy Outcomes in Civil Infrastructure Projects – A Case Study at the Regency to Pym Street Project

Anna Kordahi, Jian Zuo, Liancheng Li, and Cathy Chesson

## 18.1 Introduction

Civil construction infrastructure projects, such as road projects, have significant environmental impacts. Roads involve a large quantity of material use, which is associated with substantial amount of Greenhouse Gas (GHG) emissions. Considering the large quantity of material use, it is necessary to take the embodied emissions of materials into consideration. Indeed, sustainability has been taken into consideration in the infrastructure developments in South Australia. For instance, there is the Green Procurement Guideline released by Procurement Services SA that covers infrastructure projects. The South Australian Government has a climate change action plan and a Waste strategy that are relevant to carbon reduction. For instance, the South Australian Government has committed to net zero emissions and a more circular economy. Infrastructure construction and maintenance activities are a significant source of (scope 3) emissions for SA Government. Department for Infrastructure and Transport (The Department) is seeking to drive low emission design and construction and circular economy outcomes through contract requirements, specifications and procurement.

This chapter uses the Regency to Pym Street (R2P) project to showcase some of the practices adopted to achieve emission reduction and circular economy objectives. R2P project is recently completed with a budget of more than A$350 million. The outcomes and benefits of adopting circular economy principles and practices are provided. These include the reuse of some materials collected from other sites, the innovative use of waste materials on road construction, etc., which leads to the diversion of waste from landfill and the reduction of associated GHG emissions in many cases. Key lessons learnt were drawn. Reflections are also provided for future directions and considerations in terms of adopting circular economy principles in infrastructure projects.

## 18.2 Circular Economy in Australia/South Australia

### 18.2.1 Global Circular Economy

Circular economy (CE) is currently a topic of great interest to government agencies, industries, and individuals. Circular economy is defined by the United Nations Economic Commission for

A. Kordahi · C. Chesson
McConnell Dowell Diona Joint Venture, Adelaide, SA, USA

J. Zuo (✉) · L. Li
School of Architecture and Civil Engineering, The University of Adelaide, Adelaide, SA, Australia
e-mail: jian.zuo@adelaide.edu.au



J. Zuo et al. (eds.), *Circular Economy for Buildings and Infrastructure*, Sustainable Development Goals Series, https://doi.org/10.1007/978-3-031-56241-9_18

Europe (UNECE) as "a new and inclusive economic paradigm that aims to minimize pollution and waste, extend product lifecycles, and enable broad sharing of physical and natural assets" (CIRCULAR ECONOMY, 2022). This perception took decades to form and is still evolving (Nikolaou & Tsagarakis, 2021). The concept of CE arose from Kenneth E. Boulding's essay "The Economics of the Coming Spaceship Earth" (Blewitt, 2015). In his description, the earth is not an open system that could generate and provide unlimited resources for human needs and drain generated waste by human activities. Instead, it is a closed system where any economic growth and development are highly bonded with resource consumption and waste disposal, which build upon the scarification of its environment (Boulding, 1966).

Kirchherr et al. (2017) examined 114 definitions of CE and generated a one-sentence definition for circular economy:

> *"A circular economy describes an economic system that is based on business models which replace the 'end-of-life' concept with reducing, alternatively reusing, recycling and recovering materials in production/distribution and consumption processes, thus operating at the micro level (products, companies, consumers), meso level (eco-industrial parks) and macro level (city, region, nation and beyond), with the aim to accomplish sustainable development, which implies creating environmental quality, economic prosperity and social equity, to the benefit of current and future generations."*

CE is a purposed social operational mechanism, which attempts to balance human development and the natural environment. It is achieved through the efficient usage of natural resources and preventing resources ending up in the environment in the form of pollutants at their end-of-life. It is an indispensable component of sustainable development (OECD, 2020). Waste management and recycling attracted wide attention in the last decade. Examples include to excessive packaging, and how resources are managed in the construction- and automation industries (Morseletto, 2020). Notably, remanufacturing plays a vital role in CE as it is the critical segment to close the loop, through reusing materials and using recycled products where possible (Singhal et al., 2020; Yuan et al., 2020).

### 18.2.2 Differences in Circular economy Across Nations

From the implementation perspective, CE frameworks have been established on three levels: (1) company, (2) industry, and (3) region and nation (Nikolaou & Tsagarakis, 2021). To establish the macro-level CE, strategies and roadmaps have been developed to map out the corresponding actions. In general, the European Union is the leader of circular economy (Betancourt Morales & Zartha Sossa, 2020). However, the transition progress could vary significantly at a city or region level (OECD, 2020).

CE has also been practiced in developing countries. It has been suggested that necessary adjustments have to be made to the CE framework with a consideration of local conditions (Betancourt Morales & Zartha Sossa, 2020). CE had a long history in China aiming to maximise the use of resources, which could date back to the 1950s and has gained rapid development since 2000 (Zhu et al., 2018).

The adoption of CE also varies according to the scale of business. Compared to large companies, small and medium-sized enterprises have been paid more attention when considering the implementation of CE (Betancourt Morales & Zartha Sossa, 2020).

### 18.2.3 Circular Economy in Australia

The governance of CE on a national level can be categorised into three types: (1) dominated by the environment authority, (2) dominated by multiple authorities, and (3) guided by a central government. In general, the environment authority plays a critical role in the implementation of circular economy (e.g. Chile, Japan and New Zealand). Secondly, either an industry-focused authority or an economy-focused authority shares the responsibility with the environmental authority (e.g. Colombia, Denmark, Italy, Belgium). In Australia, principles of circular economy have been taken into consideration when developing and implementing policies by the Department of

Industry, Science, Energy and Resources (DISER) and the Department of Agriculture, Water and the Environment (DAWE), other than a dedicated national agency (OECD, 2020).

To facilitate the circular economy, Australia attempts to integrate waste management and recycling with manufacturing to enable so-called "waste to have its second life". Waste management has been placed on the national level in the form of the National Waste Policy since 1992, which focused on waste minimisation. The following versions issued in 2009 and 2018, have been titled "Less waste, more resources" (Commonwealth Environment Protection Agency, 1992; Environment Protection and Heritage CounciL, 2009). However, the latest issued version (2018) was the first time to adopt "Circular Economy" as the key principle (Commonwealth of Australia, 2018b). More recently, the National Waste Policy Action Plan was published in 2019 to coordinate various levels of governments (e.g. the Commonwealth, state and territory, local councils) on waste management and investment. Council of Australian Governments (COAG) has released Waste Export Ban on unsorted waste plastic, paper, glass and tyres.

Product stewardship (oil, television and computer, tyres, food waste, etc.) and the Australian Packaging Covenant Agreement are the other two critical components of waste management strategy in Australia. The Commonwealth Scientific and Industrial Research Organisation (CSIRO) developed a CE roadmap for plastics, glass, paper, and tyres, which aims to facilitate the transformation of waste management and recycling sector towards sustainable development (Schandl et al., 2020). The National Manufacturing Priority Road Map was issued in 2021 (Australian Government, 2021). Table 18.1 summarises waste management schemes and programmes that are being implemented in Australia.

In Australia, state and territory governments take more charge of environmental protection. Each state could make their own decisions on voluntary regimes, such as Container Deposit Scheme. South Australia is the first state that implements this scheme which dates to 1978 (Levitzke, 2020).

South Australia (SA) is the leading state in waste management and recycling activities (Levitzke, 2020). The latest report by Green Industries SA (2021) associated reuse with the circular economy for the first time. 83.3% of waste has been diverted from landfill, which consists of masonry, soil, organics, metals, cardboard & paper, glass, plastics, and other materials. The value of those recovered resources was estimated at $340 million, where 4800 employment opportunities are provided.

Meanwhile, construction and demolition waste and plastic waste have been recycled into construction products. The reprocessing data presented in *Analysis of Australia's municipal recycling infrastructure capacity* (Commonwealth of Australia, 2018a), reveals that SA has a robust waste plastic recovery in Australia and has become a recovery hub for waste plastics. In the financial year 2019–20, SA ranked the second in terms of recovery rate (20%) for plastics. It recovered not only waste plastics generated in SA, but also 17,900 tonnes of waste plastics imported from interstate and overseas (Green Industries, 2021). Meanwhile, these waste plastics have been used as wood plastic composite (WPC), bitumen additives and modifiers to manufacture outdoor fixtures (e.g. wheel stoppers and outdoor furniture) and various types of asphalt.

## 18.3 Overview of Resource Management at the Regency to Pym Street (R2P) Project

### 18.3.1 Background of R2P Project

In 2018, the Australian and South Australian Governments committed $354 million to deliver the Regency Road to Pym Street (R2P) Project. It connects the South Road Superway and Torrens Road to River Torrens Motorway, and provides a continuous 47-kilometre Motorway along the North-South Corridor. Located approximately

**Table 18.1** Being implemented waste management schemes and programmes

| Waste management schemes and programs | | Executive authority/Governmental departments |
|---|---|---|
| Product Stewardship | Oil | Department of Agriculture, Water and the Environment (DAWE) |
| | Tyres | Tyre Stewardship Australia |
| | Large plastic bags | Big bag recovery |
| | Batteries | Battery Stewardship Scheme |
| | Aluminium Cladding | Fairview aluminium cladding scheme |
| | Television and computer recycling scheme | Department of the Environment |
| | Plastic and packaging | Australian Packing Covenant (APC) |
| | polyvinyl chloride coated polyester textiles | Department of Agriculture, Water and the Environment (DAWE); Vinyl Council of Australia (VCA) |
| | MobileMuster | Australian Mobile Telecommunications Association (AMTA) |
| | PODcycle—Coffee capsule and pod recycling | Department of Agriculture, Water and the Environment (DAWE); Planet Ark Environmental Foundation |
| | Recycling commercial furniture | Department of Agriculture, Water and the Environment (DAWE) |
| | Furniture 360—Designing out waste for commercial furniture and furnishings | Australasian Furnishing Association (AFA) |
| | ResiLoop—Stewardship for resilient flooring | Australian Resilient Flooring Association (ARFA) |
| Waste Levy Schemes | | States' Environment Protection Authority |
| Waste Export Ban | | Council of Australian Governments |
| Stop Food Waste Australia | | Department of the Environment and Energy (DEE) |
| Voluntary Commitment Program (Food waste) | | Department of Agriculture, Water and the Environment (DAWE); Food Innovation Australia Limited (FIAL) |
| Australasian Recycling Label (ARL) | | Australian Packing Covenant Organisation (APCO) |
| Kerbside recycling system | | Local councils |
| Environment Protection beverage container Deposit and Recovery Scheme | | State governments |
| Recycling Modernisation Fund (RMF) | | Australian Government Department of Agriculture, Water and Environment (DAWE) in conjunction with State and Territory environment agencies |
| Modern Manufacturing Initiative (MMI) | | Australian Government Department of Industry, Science, Energy and Resources (DISER) |
| ReMade in Australia brand | | Department of Climate Change, Energy, the Environment and Water (DCCEEW) |
| Sustainable procurement | | Department of Climate Change, Energy, the Environment and Water (DCCEEW) |

five kilometres to the west of the Adelaide's central business district, this project included a new six-lane, 1.8-kilometre non-stop Motorway, along the South Road corridor, between Regency Road and Pym Street (Fig. 18.1). It forms part of the future 78-kilometre North-South Corridor.

The Department of Infrastructure and Transport (The Department) appointed the R2P Alliance to design and construct the following activities to deliver the project:

- A 1.8-kilometre section of non-stop roadway, providing three lanes in each direction,
- An overpass over Regency Road,
- Full access to and from the non-stop motorway at Regency Road,
- Intersection upgrades at Regency Road,

**Fig. 18.1** The constructed R2P Project Overpass which connects 47 km of motorway, with access at grade to local traffic

- Two lanes in each direction on the South Road arterial road to provide access to the surrounding community and local businesses,
- A grade separated pedestrian and cycle overpass over South Road in the vicinity of Pym Street,
- Left in and left out only access at Pym Street,
- Improved cycling and pedestrian facilities,
- Landscaping and noise barriers (where required),
- Traffic management, signage and Intelligent Transport Systems (ITS),
- Relocation of utility services away from non-stop carriageway and located in a position to minimise impact of maintenance activities on traffic flows, and
- A heavy vehicle bypass road connecting Exeter Crescent and Pedder Crescent.

Apart from the traditional construction objectives, there was a drive to obtain sustainable outcomes as part of this project, in line with The Department's commitment to "deliver effective planning policy, transport, infrastructure and programs in a manner that balances economic, environmental and social needs, and aligns with The Department's Climate Change Priorities". This approach further supports Australia's Zero Emissions Strategy that was targeted for 2050 at the time the R2P Project was planned, through seeking a reduction in carbon emissions through the implementation of more sustainable outcomes. In doing this, The Department was seeking to minimise the impacts of its activities on, and where practicable enhance, the environment, and deliver lasting benefits to the community. For this project, The Department required the R2P Alliance to achieve an Infrastructure Sustainability Rating (v2.0) for the design and construction phases, through the implementation of the IS tool (www.iscouncil.org).

The R2P Alliance's design methodology was undertaken to maximise the achievement of the following sustainability objectives; and to implement these objectives during construction:

- Minimise the generation of greenhouse gases,
- Minimise generation of waste through the application of the waste management hierarchy,
- Minimise use of mains water across the full asset lifecycle,
- Minimise materials' lifecycle impacts,
- Mitigate sustainability risks and drive improved sustainability performance in the project's supply chains,
- Protect and where possible enhance water quality,
- Avoid mobilisation of contaminants and where feasible, remediate contaminated land,
- Minimise the destruction or disturbance of native flora and amenity vegetation and where possible to improve biodiversity outcomes,
- Design assets having regard to future climate change impacts, and
- Contribute to improved community wellbeing.

For this case study, the focus will be on sustainable material use initiatives.

### 18.3.2 Resource Efficiency Strategy

In line with Resource Efficiency Strategy and Management (Rso-1) credit of the IS rating tool, a Resource Efficiency Strategy (RES) was developed during the start of the Design Phase to assist in highlighting potential resource management opportunities that could be achieved during the Design, Construction and Operational Phases. Input into this strategy was received from various disciplines and entities such as engineering, design, construction, operations, sustainability, environmental, The Department and key stakeholders. The RES addressed the following:

- Evaluation of project inputs and outputs to provide an overview of project resources.
- Identification of resource management opportunities through evaluation of Business as Usual (BAU) and market opportunities, such as reuse of materials, use of recycled materials, and recycling instead of disposal. The BAU on material use was determined through benchmarking of the sustainable outcomes of other major civil construction projects in South Australia, which were recently completed at the time of assessment.
- Identification of Specific, Measurable, Achievable, Relevant, and Time Bound (SMART) Targets, and how to raise the bar on BAU where possible.

Firstly, the key resource inputs (materials required to construct the project, and conduct maintenance during the operational phase) and outputs (typical waste streams that will originate from the project and the operational phase) were identified. These inputs and outputs were visually presented in a spider chart (Fig. 18.2), to make the

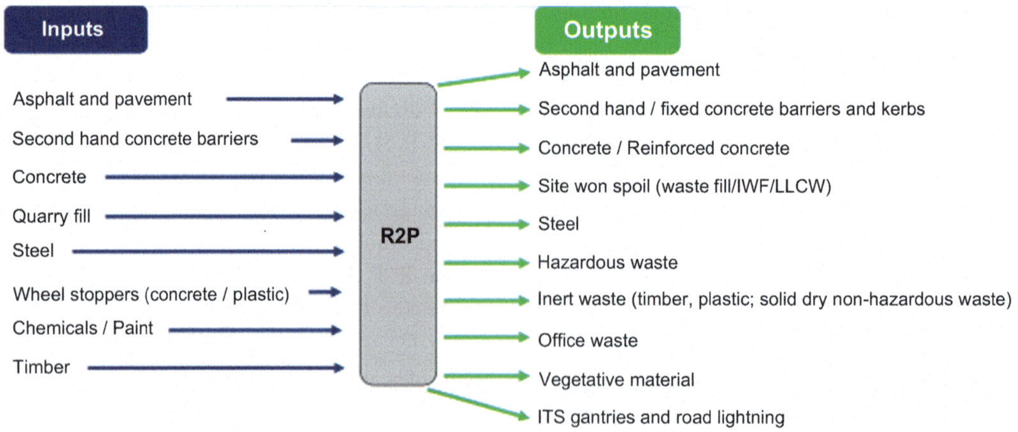


**Fig. 18.2** Overview of R2P project resources, based on inputs and outputs

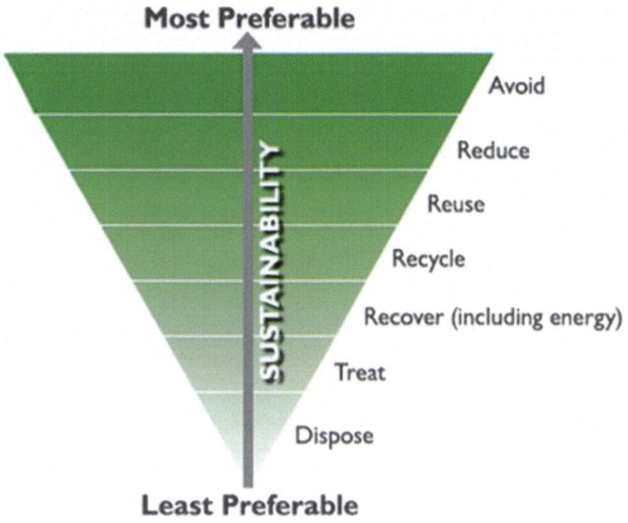


**Fig. 18.3** The Waste Management Hierarchy, indicating the most to least preferable way of managing materials. (www.epa.sa.gov.au/environmentl_info/waste_recycling)

evaluation process for potential resource management as per the waste management hierarchy, easier.

This was followed by conducting an in-depth assessment of resource management opportunities according to the waste management hierarchy that is prescribed by the Environmental Protection Authority (EPA) (Fig. 18.3).

Considerations assessed included:

- Evaluating the inputs/outputs chart to determine potential circular economy opportunities i.e. can any output materials from the project be reused on the project. Set targets regarding those opportunities such as reusing more than 90% of site won material on site or at a local spoil management facility.
- Assessing the BAU approach, for example, for supplementary cementitious materials (SCMs) used in concrete BAU for structural concrete SCMs were found to be 25% at the time of the assessment—the R2P Alliance set a target for achieving 30% SCMs in structural concrete.
- Drawing on knowledge from specialists where required (such as subject matter experts, project engineers and designers, local government agencies, the client).
- Obtaining input from suppliers on solutions proposed by the market (i.e. availability of recycled material for use on this project, such as recycled glass in PM2/20, soft plastics in asphalt, and 30% recycled aggregate in kerbs).
- Potential partnerships to meet the project's resource management targets (i.e. local Council to reuse all mulch at a location close by the R2P Project).
- Legal approvals and requirements for any of the resource management strategies (e.g. can the project legally reuse site won spoil on site).
- The volume input material required, and output material generated were assessed to identify the key focus areas where the biggest impact will be achieved. The team did not focus on every single resource required or provided by the project, but rather on the big-ticket items where more sustainable outcomes can be achieved—i.e. minimal time and financial resources required for maximum outcomes achieved.
- Evaluation of the project milestone dates, including potential project scheduling risks, including the financial and practical implications for double handling of materials where required (i.e. when site won spoil is excavated from site, the area there it is required for backfilling may not yet be available).

The resource management targets set in the RES (Table 18.2) were used to develop the Resource Management Action Plan, which provided a

**Table 18.2** Examples of SMART targets set for the R2P Project during the development of the RES

| Resource | SMART Target |
|---|---|
| Pavement and asphalt | 20% reduction in embodied Greenhouse Gas Emissions in pavement material use from base case to final design and as built (including asphalt and crush rock) |
| Asphalt | Use 90% recycled material in the car park at the Main Compound |
| Asphalt | Implement a trial on the use of soft plastics in asphalt; report on feasibility for future projects—The Department to adopt in future projects if feasible |
| Gantries | Reuse 100% of gantries within the footprint of the project that are structurally fit for purpose |
| Variable Message Sign (VMS) | Reuse 100% pre-project installed VMS on site that comply with standard requirements from The Department |
| Road lighting (light poles) | Reuse 100% of road light poles within the footprint of the project that are fit for purpose |
| Noise walls | Reuse the noise walls that is installed in the vicinity of Pym Street that is structurally suitable for reuse and relocation, within the R2P Project footprint |
| Intermediate waste fill (IWF) | >90% IWF diverted from landfill (reuse on site; or send for recycling)<br>> Signed off Multi Criteria Assessment (MCA) on intermediate waste reuse and recycling options; and 100% compliance with the target outcome of the MCA |
| Low level contaminated waste (LLCW) | >90 % LLCW reused on site during construction phase for backfilling in the ramps<br><10% LLCW disposed at landfill |
| Asbestos contaminated material | 100% asbestos contaminated material disturbed on site is removed and disposed at EPA approved landfill |
| Concrete supply | Use >50% supplementary cementitious materials (SCM) in all non-structural concrete <32 MPa<br>**Note:** non-structural concrete is mainly used in foot paths for this project. Non-structural concrete is a low strength concrete (< 32 MPa) |
| Concrete kerbs and permanent concrete barriers on site | 90% kerb and reinforced concrete that is generated on site and placed in bins or trucks to be taken off-site is recycled |
| Concrete barriers in Superway Casting Yard (Temporary construction barriers) | 100% reuse of legal complaint concrete barriers (136 units) from the Superway Casting Yard as temporary barriers |
| Materials (e.g. steel, concrete, fill) for new ramp alignment and bridge design | 5% reduction in embodied Greenhouse Gas Emissions in Regency Bridge material use from base case to final design and as built |
| Wheel stoppers | 51 concrete beams in Superway casting yard are re-used at the main compound car park |
| Trees removed | 100 % reuse/recycling of vegetation removal material |
| Clean/Inert material (e.g. bricks, concrete, paper, plastics, glass, metal (e.g. steel and copper), copper timber, asphalt, tyres, pallets, material from construction activities) | 90% clean/inert material diverted from landfill (may include on-site reuse) |
| Office resource outputs (e.g. paper, cardboard, plastic, food waste) | >95% office waste diverted from landfill |
| Spoil (uncontaminated waste fill—WF) | >90% WF diverted from landfill (reuse on site; or reuse externally) |

tracking tool for achieving these targets during the construction and operational phases of the project.

### 18.3.3 Resource Management Action Plan

A template was developed for the Resource Management Action Plan (REAP) to ensure that the required information was captured for all targets set in the RES. Key headings in the REAP were:

- Project input or output
- Description of the initiative
- Objective or goal as per the Waste Management Hierarchy
- Justification for the Waste Management Hierarchy level achieved
- Potential constrains, risks, opportunities, and costs
- Project lifecycle applicable (e.g. design, construction, operational)
- BAU approach
- Project target
- Action plan
- Responsible person
- Due date

An example of the REAP template is attached in Appendix 18.1.

The REAP was tracked monthly as part of project reporting, to ensure that progress against targets was occurring. An extract from a monthly report is provided (Table 18.3) to indicate how monthly tracking was reported. Where targets were lagging, the responsible person was consulted to determine potential constraints and opportunities to correct. Additional overall targets were set for meeting at least 100% of the design targets; and 80% of the construction targets.

Resources to support the implementation of the REAP, such as management support, were utilised where necessary to drive progress to meet these targets. For the operational targets, a copy of the REAP was provided to The Department during the project handover phase. This will allow The Department to implement the actions required and for tracking of the REAP after project handover.

### 18.3.4 Resource Management at the R2P Project

Example of sustainable resource management outcomes achieved on the R2P Project are outlined below.

#### 18.3.4.1 Recycled PM2/20 in Lower Layers of Road Base

Early during the design phase (at 30% design), the volume of imported fill including PM2/20 (road base material with particles ranging from 2 to 20 mm in size) was estimated to be approximately 32,600 tons. Recycled PM2/20 was identified as a substitute to virgin material and proposed to the client for approval. This was approved for use by The Department, as it met the road base requirements as outlined in the Master Specification for Supply of Pavement Materials RD-PV-S1 (Department for Infrastructure and Transport, 2020), and was implemented in construction. Note that the location of the lower layers of the road base is illustrated in Fig. 18.4, as the sub-base course.

#### 18.3.4.2 Use of 5% Glass in Recycled PM2/20 in Lower Layers of Road Base

A trial of 5% recycled glass in PM2/20 was implemented in the lower layers of the road base (this layer is referenced as the sub-base course in Fig. 18.4). This resulted in the use of 170 tons of the 5% recycled glass in PM2/20 (Fig. 18.5). It was the first time that recycled glass was used in the lower layers of road base in South Australia.

This resulted in circular economy outcomes of approximately 8.5 ton of recycled glass that was used on site, instead of potentially ending up in landfill for disposal. Another outcome of this initiative was the approval of using recycled glass in road base as a trial by a major road authority. This assist to provide evidence that this product is viable and can be used in road base in future projects.

**Table 18.3** An extract of the monthly REAP tracking to drive the successful implementation of project resource targets from one of the project's month end reports

| No | Project specific performance target | Target status | Action plan (link to any relevant documents/ plans) | Responsible person | Due date | Action plan status |
|---|---|---|---|---|---|---|
| 1 | 20% reduction is embodied Greenhouse Gas Emissions in pavement material use from base case to final design and as built (including asphalt and crush rock)—reference ISCA Materials Calculator | Materials used are tracked on a monthly basis. Ongoing | Design to maximize retention of existing asphalt. | Pavements Design Lead | R2P: 30 April 2020 Pedder Cre: 31 July 2020 | Completed (designs at IFC) |
| 2 | | | Approval from commissioner's rep | Pavements Design Lead | New due date: 30 April 2020 | Completed (designs at IFC) |
| | | | Include 20 year design life into future projects for increased used on existing pavements | The Department Operations | TBD | In progress |
| 3 | | | Include requirement of recycled aggregate in pavements in the Quarry Supply contract | Quarry procurement package owner/ Environmental manager | 28 Feb 2020 | Completed (costing was requested and included in Bid Evaluation—for Pk21a and 21b) |

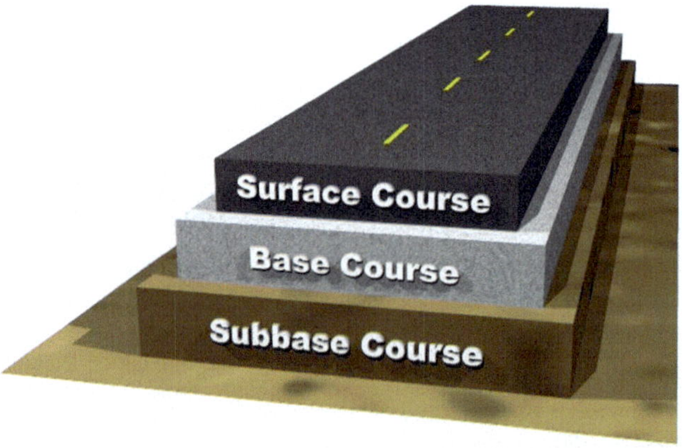


**Fig. 18.4** Basic diagram of the layers of an asphalt road. (Source: Components Of Road Pavement Structure—Engineering Discoveries)

#### 18.3.4.3 Recycled Materials in Asphalt (Soft Plastics, Printer Cartridges, RAP, Canola Oil)

Reconophalt™, an asphalt product that uses soft plastics as a partial binder replacement in asphalt, was used in a trial on the R2P Project. It was placed on the south bound arterial road section of the project (Fig. 18.6, right). This was the first use of the product on a Department of Infrastructure and Transport managed road and has led to changes in The Department's Masters Specification requirements for Sustainability in Design—

**Fig. 18.5** Stockpile of 5% recycled glass in PM2/20, about to be placed on site (the Environmental and Sustainability Manager, Anna Kordahi, is on the left; with the responsible project engineer, Joey Vu, in the middle, and a representative from RecoureCo, Andrew Silvestri, on the right)

**Fig. 18.6** Temporary carpark (left), and the installation of the soft plastics in asphalt trial on the arterial road (right) for which both a Reconophalt™ product was used

PC-ST1 (Department for Infrastructure and Transport, 2021a) and Sustainability in Construction—PC-ST2 (Department for Infrastructure and Transport, 2021b), regarding recycled material initiatives to be investigated and implemented where possible on subsequent major projects.

The site office car park also utilised Reconophalt™ as part of a 100% recycled asphalt mix (Fig. 18.6, left). Reconophalt™ was further used in shared user paths. For all these Reconophalt™ applications, the circular economy outcomes achieved on the project are presented in Table 18.4 below.

#### 18.3.4.4 Use of 30% Recycled aggregate in Concrete for Kerbs

Recycled content was considered for use where possible. The use of a 30% recycled aggregate content in kerbs were assessed through a

**Table 18.4** Circular economy outcomes provided using Reconophalt™ on the R2P project

| Application | Recycled asphalt plannings (RAP) (ton) | Recycled canola oil (kg) | Waste toner cartridges (number) | Plastic bags (number) |
|---|---|---|---|---|
| South Australian first application on an arterial road | 90 | 470 | 7500 | 372,105 |
| Temporary carpark | 205 | 324 | 2500 | 110,000 |
| Shared user paths | 114 | 0 | 9230 | 316,000 |
| Total | **409** | **794** | **19,230** | **798,105** |

multicriteria analysis process, to evaluate the environmental, social, environmental, technical and cost implications to determine the most sustainable outcome for the project. The outcome supported the use of 30% recycled aggregate in kerbs.

An average of 900 kg of aggregate was used per meter cubed of concrete mix, that was used for kerbs. By using 30% recycled aggregate, 300 kg of waste aggregate per cubic meter is diverted from landfill, or 300 kg of virgin material per cubic meter was saved.

The total of 442 cubic meters of concrete with 30% recycled aggregate was utilised for kerbs on the R2P Project. Hence 133 tons of natural aggregate was saved; or 133 tons of aggregate was diverted from landfill.

Note that the use of recycled content did not impact carbon emissions, as it merely replaced natural aggregate. The carting distances and the methodology used to obtain the recycled content was similar than the distance and methodology used to obtain the natural aggregate. With natural aggregate being a finite resource, future availability may be further away from urban projects, resulting in a potential future reduction in carbon emissions through transporting distances between the use of natural aggregate, or recycled content.

The use of recycled content met the design and technical specification requirements, and hence there is no basis to believe that it will impact the design life or kerb durability across the design life differently than what natural aggregate will.

#### 18.3.4.5 Reuse of Noise Walls

During the design phase, an approximate 80 m of noise wall panels were identified that had to be demolished to allow for the construction of the pedestrian bridge ramp. The waste management hierarchy, required urban design outcomes, cost, technical requirements and practicality were considered in in the decision-making process. The most viable option that was implemented was to reuse this 80 m portion of noise wall on the R2P Project on an adjacent location (Fig. 18.7), instead of off-site recycling or disposal.

#### 18.3.4.6 Reuse/Recycling of Spoil on Site

The earthworks volumes estimated indicated a deficit in fill on the project, meaning more material was required to be imported than disposed of. The site geotechnical and contamination assessment were evaluated to determine the possibility of reusing any material on site—both technical and legal requirements were evaluated. The decision-making process used to determine if Waste Fill and Intermediate Waste Fill could be reused on site, is presented in Fig. 18.8.

The Low-Level Contaminated Waste (LLCW) could be reused on site as it was placed under the new constructed roadway in the lower layers of the pavement and capped with asphalt. Note that the National Environmental Protection Assessment of Site Contamination (ASC NEMP) health investigation levels for commercial or industrial land use was not exceeded—hence the LLCW could legally be placed under the asphalt in the lower layers of the pavement. The asphalt is an impermeable layer, which assist to limit leaching from the LLCW. Note that the soil contamination report also confirmed that the LLWC has a low leachability. Hence, the risk of contaminating surrounding areas and groundwater was low.

The approved approach resulted in the reuse of a large portion of site won spoil on site. The spoil that could not be reused on site due to staging constraints, reluctance to double handling of material that would have resulted in excessive

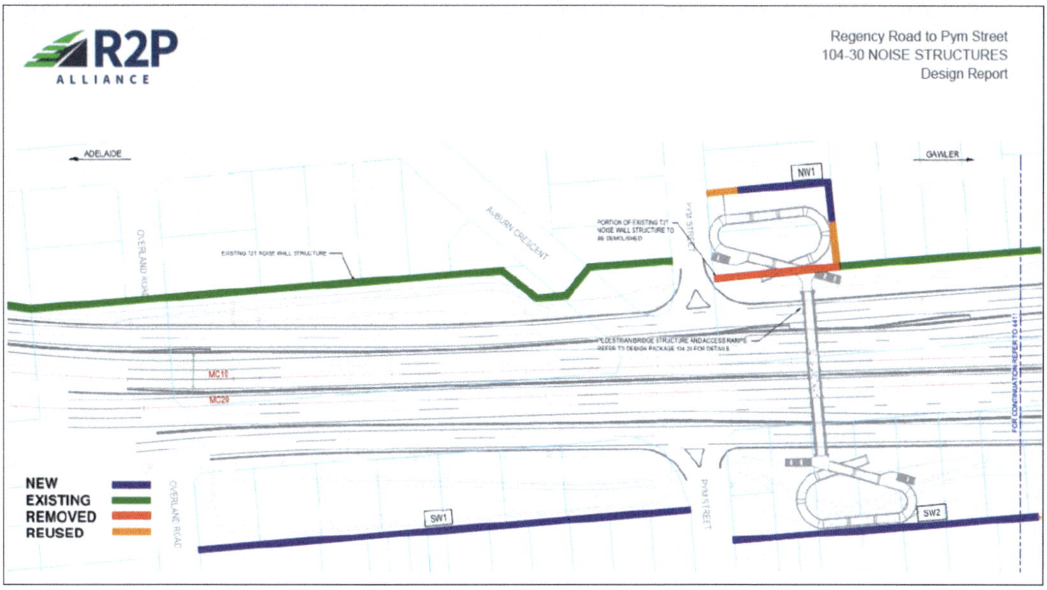


**Fig. 18.7** Reuse of existing noise wall instead of off-site recycling or disposal

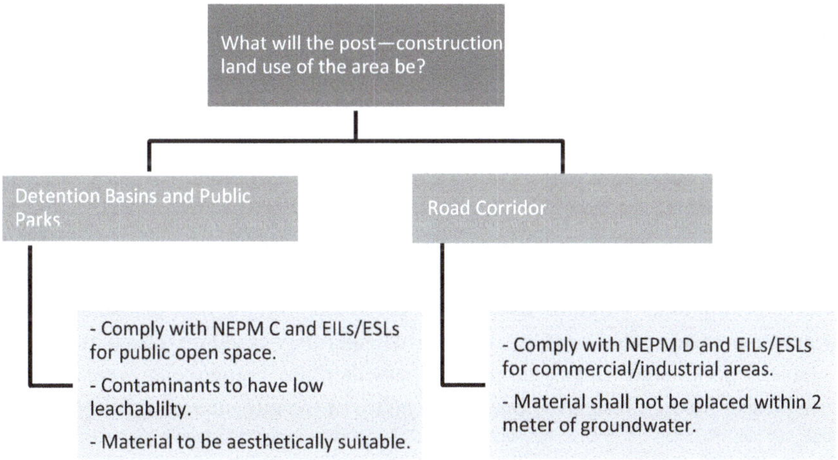


**Fig. 18.8** Spoil decision making process for WF and IWF

cost, and a local recycling facility who agreed to take any Waste Fill or Intermediate Waste Fill free of charge for reuse purposes, was diverted from landfill. The volumes and percentages reused spoil are presented in Table 18.5.

#### 18.3.4.7 Mulching and Reuse of Trees Impacted

Civil construction projects often require the clearing of vegetation to allow for the project to be constructed. On the R2P Project, the tree that required removal or pruning were mulched. The mulch was provided to the City of Port Adelaide Enfield (Fig. 18.9) for reuse at the local golf course. The material was directly placed in an agreed location within the golf course boundary, to minimise unnecessary rehandling and transport of the mulch by Council. The R2P Alliance provided approximately 260 tons of mulch to Council for use.

**Table 18.5** Site won spoil reused on the R2P Project

| Soil classification | Waste fill | Intermediate waste fill | Low level contaminated waste |
|---|---|---|---|
| Total mass reused (tons) | 778,170 | 41,410 | 20 |
| Percentage reused | 100% | 100% | 100% |

**Fig. 18.9** Mulch originating from the R2P Project tree removal, stockpiled in an area indicated by local Council

#### 18.3.4.8 Reuse of Tanks to Harvest Rainwater for Use

The construction phase for the R2P Project was anticipated to be at least 18 months. Given this duration, the opportunity to harvest rainwater for toilet flushing was evaluated to assist in reducing potable water consumption. Circular economy outcomes were also investigated i.e., the availability of viable second-hand tanks within the vicinity of the project for use, instead of purchasing new tanks.

Three 5000-liter tanks were sourced from a nearby project and connected on the R2P Project site, to the site compound roofs, to harvest rainwater run-off. This water was plumbed in for flushing of toilets. Water mains were connected as a back-up to ensure that the sewerage system remained functional during low rainfall periods.

The reuse of the second-hand water tanks was successful in achieving circular economy outcomes. Water harvesting and reuse was also successful in using 266,000 litres of harvested rainwater instead of potable water for toilet flushing.

#### 18.3.4.9 Reuse of ITS (Intelligent Transport Systems)

Various ITS components were identified during the initial output assessment (Fig. 18.2). The technical specifications of the existing ITS components were evaluated to determine if they complied with the new design specifications, and whether they could be reused as is, refurbished or whether new components were required. A summary of the outcome of existing ITS components is presented in Table 18.6.

Not all ITS outputs from the R2P Project were reused or refurbished for the R2P Project because it was old infrastructure that was no longer compliant to required technical specifications; or it did not comply to the project specific design requirements. Where the ITS outputs could not be used on the R2P Project due to design requirements, but were complaint with the technical specifications, it was provided back to The Department for refurbishment to allow for future reuse on other infrastructure projects. The ITS outputs that were no longer viable for future use, were recycled (e.g. the electrical components and steel).

**Table 18.6** ITS outputs reused or refurbished

| ITS component | Output number | Reused/Refurbished number |
|---|---|---|
| Variable Message Sign (VMS) | 3 | 1 (1 kept for potential future use by The Department) |
| Traffic Signals | 9 | 0 (9 were reused for temporary works) |
| Light Poles | 63 | 17 |
| ITS Gantries | 5 | 1 (4 kept for potential future use by The Department) |

#### 18.3.4.10 Increased SCMs

When the Resource Efficiency Strategy was developed, the BAU SCM percentage for recently completed civil construction project in South Australia at that stage, were evaluated. It was noted that the SCM percentage achieved for structural concrete was up to 20%; and for non-structural concrete usually up to 25%. The BAU SCM percentage was challenged during the early design phase to identify opportunities for increasing the SCM content, and hence reducing the Portland Cement addition to concrete. This process entailed various aspects such as:

- Determine the SCM in concrete availability in the South Australian market; and ability to deliver the required SCM percentage. This was driven through the procurement process.
- Obtaining market input on reasonable SCM percentages in various concrete mixes, and how increased SCM content can be achieved.
- Evaluate limitations placed on increased SCM content by technical requirements and specifications, and if the increased SCM percentage will still result in a compliant product.
- Conducting on-site trials to evaluate the workability and curing time of concrete with an increased SCM content.
- Interact in interstate knowledge share opportunities with projects that previously achieved higher SCM percentages in their concrete mixes.

Key lessons learnt regarding the use of increased SCMs in concrete include:

- The investigation of increased SCM content in concrete is crucial during the early design phase. This allows for more time to engage with the market, trial potential new concrete mixes, challenge the status quo concrete mix, and specify the required concrete mix design early on as a requirement for the project.
- A multidisciplinary team such as durability engineers/concrete technologists, designers, structural engineers, the client, sustainability professionals, representatives from the market (i.e. potential suppliers) and the construction team must provide input into the use of increased SCMs—specifically regarding local SCM content availability, technical specification and testing requirements, workability, curing time and the timeframes to be achieved from a program perspective. A multidisciplinary team assist to achieve a practical outcome, where most potential issues can be identified and addressed early.
- As demand increases for concrete with high SCM content, it is important to conduct trials to assess the performance characteristics of new concrete mixes, including workability and curing times, to identify suitable applications. Increased curing times can be planned for in the project schedule.
- Ensure that contracts require that all concrete elements are assessed for increased SCM content, not highlighting any one opportunity. This allows for the most material elements (highest opportunity) to be targeted. It is important to note that BAUs for pre-cast concrete are currently lower than in-situ elements due to additional constraints. This may include significant curing times, the need for additional space at casting yards and potential more energy required for curing.
- Work with regulatory bodies, the client, subcontractors and suppliers to identify potential increased SCM content in concrete, or low carbon concrete, opportunities. New products

may be available; or additional trials may have been undertaken providing additional confidence in the use of a specific concrete mix with increased SCM content. Collaboration is key as working in silos often result in blockers such as:

- Potential insufficient SCM content availability locally,
- An increased SCM percentage in the concrete mix is not approved and still need to be tested—the project program does not allow extra time for additional testing,
- It is easy and poses less risk to continue to use historical well tested concrete mixes with lower SCM content, than to increase the SCM content in concrete mixes,
- Potential increased costs for increased SCM content in concrete mixes,
- Limited awareness on potential increased SCM content in concrete or low carbon concrete alternatives and the implication on embodied carbon reduction.

The SCM percentage outcomes for the R2P Project are presented in Table 18.7 below. Note that the SCMs achieved at the R2P Project should not be considered as the maximum SCM content that can be achieved. It is possible to further increase the SCM content, given that the above lessons learnt are addressed. The Department's Sustainability Manual (Department for Infrastructure and Transport, 2021c) now recognises a suggested 10% increase in SCM on the project baseline, which requires that project teams maximise Portland cement reduction as a means of reducing embodied carbon.

#### 18.3.4.11 Beneficial Outcomes

Beneficial outcomes were assessed as part of cost benefit analysis, which includes additional costs incurred, savings and sustainable outcomes achieved. A summary of the cost benefit analysis is provided in Table 18.8.

Sustainable outcomes do not always come at an additional price. However, in selected instances when sustainable outcomes are more expensive than the baseline, one should evaluate the sustainable outcomes that will be achieved in comparison with the additional cost to determine if the increase in cost incurred is reasonable.

## 18.4 Conclusion and Way Forward

There is a worldwide drive to achieve more sustainable practices—including improved circular economy outcomes. As per this case study, the R2P Alliance demonstrated that it is indeed possible to improve the circular economy outcomes in civil infrastructure projects through the identification and implementation of circular economy initiatives.

The key measures for achieving key circular economy measures at the R2P Alliance are:

- Contractual requirements from the client.
- Hands-on approach and awareness created by the R2P Alliance's Sustainability Lead for design and the Environmental and Sustainability Manager during design and construction, with the support of the management team.
- Following the IS rating scheme (the resource efficiency credits) that resulted in the development of the resource efficiency strategy, which was used to develop a resource efficiency action plan that considered potential circular economy opportunities for the key project inputs and outputs. This process was driven through workshops with the relevant responsible teams such as design, engineering and sustainability representatives.

**Table 18.7** BAU and R2P project achieved SCM content in concrete

| Component | Kerbs | Slipform barriers | Footpaths | Selected structural concrete sections |
|---|---|---|---|---|
| BAU SCM% | 25% | 25% | 25% | 25% |
| R2P SCM% | 30–40% | 25% | 30% | 30% |

**Table 18.8** Cost Benefit Analysis Summary for Sustainable Initiatives Implemented at the R2P Project

| Initiative description | Approximate additional cost (AUD $) | Approximate savings (AUD $) | Sustainable outcomes |
|---|---|---|---|
| Recycled PM2/20 in lower base layers of pavement | 0 | 0 | Diversion of waste from landfill; reduction in use of natural material (circular economy outcomes) |
| Reuse of construction material on site (waste fill; LLCW; intermediate waste fill) | 0 | 275 400 | Reduction in GHG (reduced transport distance); reduction in natural material use |
| 100% recycled asphalt in main compound yard; use of soft plastics in asphalt (trail on arterial road) and in shared user paths | 0 | 0 | Reduction in embodied carbon emissions; materials diverted from landfill through circular economy outcomes; water reduction for dust suppression in a paved versus soil parking area |
| Capture rainwater in tanks for toilet flushing | 14 200 | 900 | Reduction in mains water use |
| Use of recycled aggregate in kerbs | 10 870 | 0 | Reduction in natural material use; and increased diversion of material from landfill (circular economy outcomes) |
| Reuse of ITS gantries, light poles and VMS | 0 | 111 370 | Embodied GHG reduction; circular economy outcomes |
| Reuse of noise walls | 0 | 7 000 | Embodied GHG reduction; circular economy outcomes |
| Increased use of SCMs | 0 | 0 | Embodied GHG reduction; circular economy outcomes |
| Reuse of mulch from site removed trees | 0 | 0 | Circular economy outcomes |
| Use of 5% Recycled Glass in PM2/20 | 510 | 0 | Diversion of waste from landfill; reduction in use of natural material (circular economy outcomes) |

- Consultation with suppliers on availability of circular economy opportunities in the market.
- Continuous evaluation for additional opportunities e.g. is it possible to expand on the already implemented trial areas.

The initiatives considered as part of the R2P Project were used to create a checklist (Table 18.9), that can be utilised by future civil infrastructure projects as a prompt for potential improved circular economy outcomes.

There are various limitations and challenges to implement and drive circular economy and sustainable outcomes in the civil infrastructure industry—it is in essence a change management process. As per the R2P Alliance's example, it is suggested that circular economy and sustainable outcomes must be supported by all parties involved across the industry to achieve successful outcomes—not just one stakeholder. Suggested collaboration that can be provided by stakeholders involved in civil infrastructure to assist in overcoming some of these limitations and challenges, is provided in Fig. 18.10.

**Acknowledgements** The authors would like to acknowledge the following entities:

- The R2P Alliance Partners
- The South Australian Department of Infrastructure and Transport
- Various industry partners who were involved in the initiatives outlined in this chapter.

**Table 18.9** Checklist for potential circular economy initiatives for civil construction projects

| No. | Suggested initiative | Considerations | Viability for new project (circle the correct word) |
|---|---|---|---|
| *Site won spoil* | | | |
| 1 | Reuse as much as possible of site won spoil on site | Was the spoil tested for contaminants—what is the legal requirements for reuse of spoil or movement of spoil form one land parcel to the next?<br>If spoil is provided to third party landowners, are clear, signed agreements in place with a copy of soil contamination classification to indicate that the spoil provided is not contaminated? | Yes/No/Potential |
| 2 | Determine if close by projects or third-party landowners are interested in spoil for reuse | | Yes/No/Potential |
| 3 | If the spoil is not reused on site or by third party landowners, consider reuse or recycling facilities instead of disposal | | Yes/No/Potential |
| *Asphalt and pavement* | | | |
| 4 | Include recycled materials (e.g. recycled asphalt plannings, -plastic, -canola oil, -glass, -etc.) into the asphalt mix | Are the additional recycled materials compliant to the road specifications? Dispensation may be required from the client to use recycled content in asphalt | Yes/No/Potential |
| 5 | Is recycled PM2/20 available locally instead of natural PM2/20? | | Yes/No/Potential |
| 6 | Is there a PM2/20 product available with recycled glass content that can be used? | | Yes/No/Potential |
| *Concrete* | | | |
| 7 | What is the specified SCM percentage content, and can it possibly be increased? Alternatively, are there any low or green carbon concrete alternatives available through the local market? | The SCM content may impact the curing time and workability of the concrete. It is advisable to conduct small scale trials to determine the impact of increased SCM percentages. The project program can potentially be adjusted to allow for increased curing timeframes<br>Consult with recently completed or far progressed infrastructure projects to determine what SCM percentages were used for which concrete strength applications; and enquire about the use of recycled materials in the concrete mix<br>Ensure that any additive (e.g. recycled plastic or other recycled materials) is approved by the client and compliant with technical specifications—dispensation may be required | Yes/No/Potential |
| 8 | Can recycled aggregate be used as a substitute to natural aggregate? | | Yes/No/Potential |
| 9 | Can recycled content (e.g. recycled plastics) be added to the concrete mixture? | | Yes/No/Potential |
| *Steel* | | | |
| 10 | Can the steel required be manufactured from recycled steel? | Ensure that any alternative to steel is approved by the client and is compliant with technical specifications—dispensation may be required<br>Ensure that the Environmental Declarations are approved by independent bodies to ensure that the content is verified | Yes/No/Potential |
| 11 | Compare Environmental Product Declarations of various steel products to determine which product has the lowest environmental impact, which can then be used to influence decision making on which steel to be used | | Yes/No/Potential |
| 12 | Use alternatives for steel where possible (e.g. e-mesh instead of reinforcing steel mesh) | | Yes/No/Potential |

(continued)

**Table 18.9** (continued)

| No. | Suggested initiative | Considerations | Viability for new project (circle the correct word) |
|---|---|---|---|
| *General* | | | |
| 13 | Second hand concrete barriers for road demarcation | Determine if secondhand concrete road barriers are available for use? Particularly evaluate the availability of concrete barriers from projects in close vicinity to the new project site, that are ramping down. | Yes/No/Potential |
| 14 | Is there an alternative for hazardous waste other than disposal at landfill (e.g. treatment; or encapsulation on the same land parcel that it was found)? | The outcome must be legally compliant—consult with the EPA for clarity<br>Obtain guidance from a soil contamination specialist if you have contaminated spoil on site<br>Consult with local waste collection companies—they may have an alternative solution for treatment available | Yes/No/Potential |
| 15 | Can inert- and office waste (timber, plastic, solid dry non-hazardous waste, paper, cardboard) be separated on site to make recycling opportunities more viable? | Consult with local waste collection companies—they may have an alternative solution available such as treating the waste and using it as a fuel source, instead of disposal at landfill<br>Availability of recycling facilities in proximity to the site in regional areas | Yes/No/Potential |
| 16 | Can trees cut down on site be reused e.g. placement of stumps and trucks as habitat for fauna; cutting up trunks to suitable sizes and provide to parks and local schools for nature play; mulch vegetative material and provide to local Council or third-party landowners for reuse—it can even be reused on site as part of the final project landscaping? | Proximity of the construction site to potential third-party landowners, and cost involved in transportation of wood/mulch<br>Time required for tree loppers to create specific trunk sizes for nature play comes at an extra cost | Yes/No/Potential |
| 17 | Can existing Intelligent Transportation System (ITS) and road lighting equipment be reused, or potentially be refurbished and reused on site? | The ITS must comply with technical specifications | Yes/No/Potential |
| 18 | Can site compound offices and tanks for harvesting rainwater be sourced from another project and reused instead of purchasing new? | Consider the distance from the other project to the new project, and plan for transport costs | Yes/No/Potential |

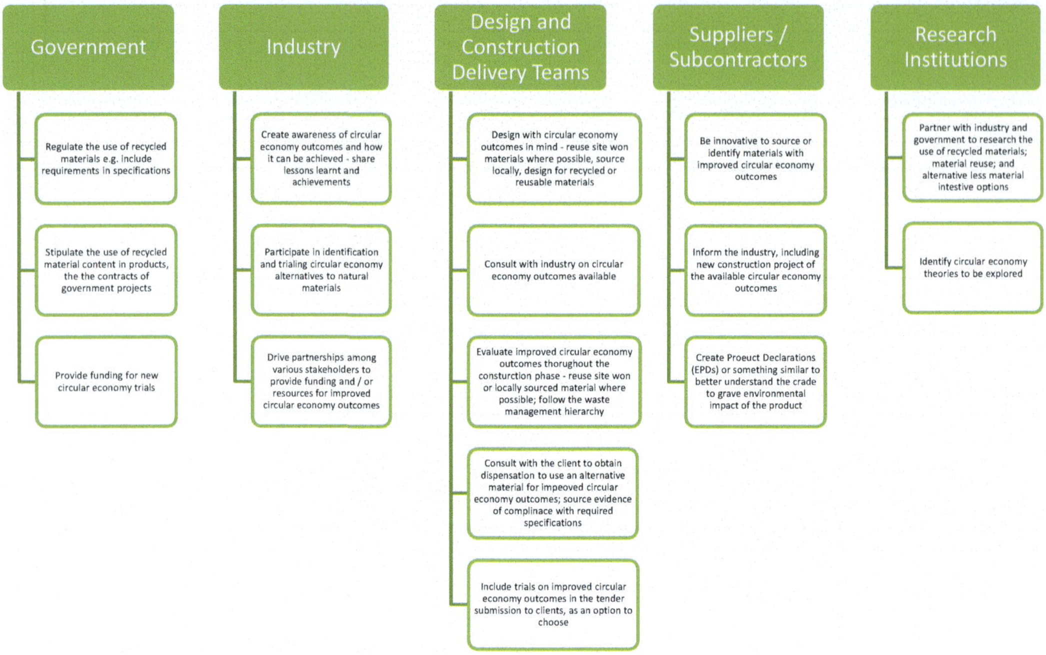


**Fig. 18.10** Suggested responsibilities for driving improved circular economy and sustainability outcomes in civil infrastructure

## Appendix 18.1: REAP Template

| No | Inputs / Outputs | Resource Management Initiative | Objective / Goal as per waste hierarchy | Justification for selecting waste hierarchy level | Constraints / Risks / Opportunities / Costs | Project Lifecycle Applicable (e.g. design, construction, operational phases) | Business as Usual (BAU) | Project Specific Performance Target | Action Plan | Responsible Person / Role | Due Date |
|---|---|---|---|---|---|---|---|---|---|---|---|
| | | | | | | | | | | | |
| | | | | | | | | | | | |
| | | | | | | | | | | | |
| | | | | | | | | | | | |

## References


Australian Government. (2021). *Recycling and clean energy National Manufacturing Priority road map*. Australian Government.

Betancourt Morales, C. M., & Zartha Sossa, J. W. (2020). Circular economy in Latin America: A systematic literature review. *Business Strategy and the Environment, 29*(6), 2479–2497.

Blewitt, J. (2015). *Understanding sustainable development* (2nd ed.). Routledge.

Boulding, KE 1966, The economics of the coming spaceship earth, .

CIRCULAR ECONOMY, UNECE, viewed 26/2 2022. https://unece.org/trade/CircularEconomy

Commonwealth Environment Protection Agency. (1992). *National Waste Minimisation and recycling strategy 1992, Canberra*. Commonwealth Environment Protection Agency.

Commonwealth of Australia. (2018a). *Analysis of Australia's municipal recycling infrastructure capacity*. Commonwealth of Australia.

Commonwealth of Australia. (2018b). *National waste policy:Less waste, more resources*. Commonwealth of Australia.

Department for Infrastructure and Transport. (2020, August). *Roads. Master specification – Supply of pavement materials (RD-PV-S1)*. MASTER SPECIFICATION – PART RD-PV-S1 – SUPPLY OF PAVEMENT MATERIALS.DOCX (dit.sa.gov.au)

Department for Infrastructure and Transport. (2021a). *Project Controls. Master Specification – Sustainability in Design (PC-ST1)*. MASTER SPECIFICATION – PART PC-ST1 – SUSTAINABILITY IN DESIGN. DOCX (dit.sa.gov.au)

Department for Infrastructure and Transport. (2021b). *Project controls. Master specification – Sustainability in construction (PC-ST2)*. MASTER SPECIFICATION – PART PC-ST2 – SUSTAINABILITY IN CONSTRUCTION.DOCX (dit.sa.gov.au)

Department for Infrastructure and Transport. (2021c). *Sustainability manual*. SUSTAINABILITY MANUAL.DOCX (dit.sa.gov.au)

Environment Protection and Heritage Council. (2009). *National Waste Policy: Less waste, more resources*. Environment Protection and Heritage Council.

Green Industries, S. A. (2021). *South Australia's recycling activity survey 2019–20 report*. Rawtec Pty Ltd.

Kirchherr, J., Reike, D., & Hekkert, M. (2017). Conceptualizing the circular economy: An analysis of 114 definitions. *Resources Conservation and Recycling, 127*, 221–232.

Levitzke, P. S. M. V. (2020). In S. K. Ghosh (Ed.), *The development of a circular economy in Australia*. Springer.

Morseletto, P. (2020). Targets for a circular economy. *Resources, Conservation and Recycling, 153*, 104553.

Nikolaou, I. E., & Tsagarakis, K. P. (2021). An introduction to circular economy and sustainability: Some existing lessons and future directions. *Sustainable Production and Consumption, 28*, 600–609.

OECD. (2020). *The circular economy in cities and regions: Synthesis report* (OECD Urban studies). OECD Publising.

Schandl, H., King, S., Walton, A., Kaksonen, A., Tapsuwan, S., & Baynes, T. (2020). *NATIONAL circular economy roadmap for plastics, glass, paper and tyres*. CRISO.

Singhal, D., Tripathy, S., & Jena, S. K. (2020). Remanufacturing for the circular economy: Study and evaluation of critical factors. *Resources, Conservation and Recycling, 156*, 104681.

Yuan, X., Liu, M., Yuan, Q., Fan, X., Teng, Y., Fu, J., Ma, Q., Wang, Q., & Zuo, J. (2020). Transitioning China to a circular economy through remanufacturing: A comprehensive review of the management institutions and policy system. *Resources, Conservation and Recycling, 161*, 104920.

Zhu, J., Fan, C., Shi, H., & Shi, L. (2018). Efforts for a circular economy in China: A comprehensive review of policies. *Journal of Industrial Ecology, 23*(1), 110–118.

# Life Cycle Assessment of Construction and Demolition Waste: A Case Study of Recycled Aggregate Products in Australia

19

Daniel Oteng, Mitchell Bacon, Andrew Silvestri, and Jian Zuo

## 19.1 Introduction

The amount of concrete used in construction is twice as much as other building materials such as aluminium, steel or wood. The production of concrete annually is about 830Mt worldwide (Mah et al., 2018b). Building and civil engineering works produces a massive amount of construction and demolition waste (C&DW) through the construction process and at the time of demolition (Bovea & Powell, 2016). According to Wu et al. (2019), C&DW amounting to over ten billion tons are generated yearly in the construction industry. This causes environmental degradation (through illegal dumping) if not appropriately managed (Estanqueiro et al., 2018). However, they can represent new resources and high value materials when recovered or recycled for new construction (Simion et al., 2013).

Many developed countries have far surpassed 90% recycling rate due to laws and regulations which regulates the recycling of C&DW. A recycling rate of 99.9% is attributed to Singapore with the highest rate of C&DW recycling. In 2012, Japan recorded a recycling rate of 99.55% (Mah et al., 2018b). The C&DW generated in Australia is 90% recycled (Mah et al., 2018b), however, it changes from state to state (Wu et al., 2020). Australia generates about 20 million C&DW annually, which comparing to other countries may not be as huge, however, it remains the most significant solid waste stream that is generated in Australia (Wu et al., 2021).

Recently, the sustainable agenda has created the requirement towards designing sustainable buildings, which has also created the need for sustainable materials (Ernest et al., 2020). This has prompted researchers within the field to develop alternative recycling technologies for C&DW (Mah et al., 2018a; Zhang et al., 2020). These studies have established recycled aggregates (RA) as a feasible material for structural use (Colangelo et al., 2021). The recycled products can be used for different construction applications with the performance often comparable to natural aggregates (NA). The study, therefore, investigates a comparative assessment of recycled aggregate products and the production of natural aggregates using life cycle analysis. Scenarios are also created to examine the efficiency of the processes through transportation. The chapter is in four sections. Section one introduces the study background and highlights the research gap. Section two explains the methodology of the study and processes. The section

D. Oteng (✉)
School of Project Management, The University of Sydney, Sydney, NSW, Australia
e-mail: daniel.oteng@sydney.edu.au

M. Bacon · A. Silvestri
ResourceCo Pty Limited, Rose Park, SA, Australia

J. Zuo
School of Architecture and Civil Engineering, The University of Adelaide, Adelaide, SA, Australia



J. Zuo et al. (eds.), *Circular Economy for Buildings and Infrastructure*, Sustainable Development Goals Series, https://doi.org/10.1007/978-3-031-56241-9_19

three highlights the results of the study with a comparative discussion of past and current literature. The last section makes final conclusions and recommendations on the study.

## 19.2 Materials and Methods

Life cycle assessment (LCA) is an environmental assessment methodology recognised internationally for the evaluation of a product or systems environmental impacts. The consumption of energy and reagents, as well as emissions are accounted for in the products or systems entire life cycle (Butera et al., 2015). The ISO 14040 and 14044 defines the four phases of an LCA study as applied in this study (ISO 14040, 2006; ISO 14044, 2006). The management practice of C&DW can be improved through life cycle impact assessment (LCIA), making it a good decision tool for policy makers and industry (Mah et al., 2018b). This study, therefore, uses the LCA to assess the impact of the production of recycled and natural aggregate.

### 19.2.1 Case Study, Goal, and Scope Definition

The study is conducted in one of the biggest C&DW recycling facilities in Australia. The Wingfield facility is the most efficient facility among the others used by ResourceCo in Australia. The plant receives different forms of construction waste. Plastics and wood are removed and carted to their joint venture companies for alternative fuels. Materials such as steel, copper, aluminium, and other metals are also removed and sent out to other recyclers. The main business at the plant is processing C&DW mainly consisting of asphalt, concrete, bricks, and rubble. The scope of this LCA covers the product stage (cradle-to-gate) which modules A1 to A3. The contribution of capital goods (production equipment and infrastructure) and personnel is outside the scope of the LCA, which is in line with the Product Category Rules (PCR) for construction products (Ingwersen & Subramanian, 2014).

### 19.2.2 Functional Unit and System Boundary

The declared unit is 1 tonne (1000 kg) of final aggregate products from construction and demolition waste processes. The declared unit covers the whole range of secondary aggregate products that are produced at the plant. The source of the waste for recycling comes from all parts of the state (South Australia). The dropped of materials are considered to be at their end-of-life when they are transported to ResourceCo recycling facility. Consequently, all processes that accounts for the manufacturing and production of the materials before it reaches the facility are excluded from these processes which is in line with the polluter pays principle (Fig. 19.1).

In line with the polluter pays principle, the transport part (from construction site to recycling centre) of this section is excluded from the assessment because the polluter pays for the transport in disposing of the waste. The core process included in the life cycle assessment is the operation of the recycling plant and any movement of the waste materials on the site. The waste when it reaches the site, is weighed and checked if it is the right waste stream coming to the site. The construction and demolition waste are piled in front of the plant and are pulverised to separate the steel from the concrete. The materials are then fed into the plant (feed hopper or primary jaw) by a WA 500 Loader. As the materials go through the picking station one, screen one, picking station two, Symons's crusher and another screen; the materials are crushed, scrubbed, screened, washed, and stockpiled into different recovered aggregates and sands.

**Inventory Analysis**

The processes within the system boundary needs to be allocated at the Life Cycle Inventory (LCI) stage. The inputs and outputs for the processes are compiled and quantified in relation to the defined system boundary (ISO 14044, 2006). Background data are sourced from AusLCI plus the AusLCI shadow database (AusLCI, 2019) and ecoinvent v3. However, these inventories have been updated to the current AusLCI backup file

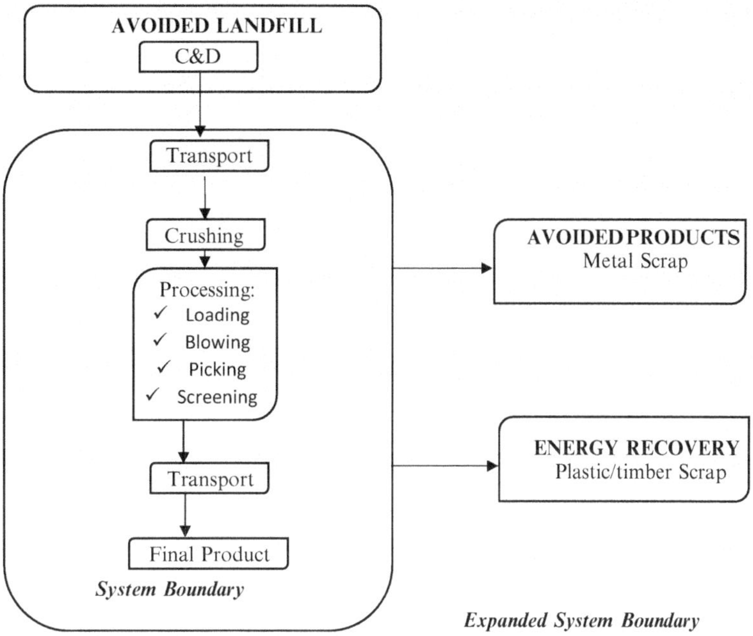


**Fig. 19.1** Product system boundary

Version 1.36 released on 2 March 2021. Thus, background data used are less than 10 years old. Foreground or primary data of the recycling plant and its processes were collected within a one-year period, from July 2020 to June 2021. Site inspection and interviews were conducted to collect details of processes, machinery, energy, waste input, fuel consumption and trucks. Processes are updated as and when needed. The life cycle data used for the recycling plant processes are less than 5 years old. Methodological choices have been applied in line with EN 15804:2012+A1: 2013 (EN 15804, 2013); and any deviations have been recorded. The SimaPro software version 9.1.0.11 is used for the impact assessment.

ResourceCo produces several mineral aggregate products (aggregates and sands with classifications different from each other). Since all these products undergo the same recycling plant process, mass (causal relationship) was chosen for the allocation and thus the products receive the same environmental impacts per tonne of product. If economic allocation had been used, the aggregates would receive a higher portion of the impacts and the sands would receive a smaller portion. This result reflects the difference in respective values, which varies by +/− 10% from the median value. The processes are not separated because the energy fed to the process is not divided and data was not present. Therefore, there is no separate energy (power) for each section of the plant. The process based LCA is used because of its detailed process-level information which reflects how the unit process of the products are interconnected (Yang et al., 2017).

A small amount of ferrous metal scrap can be present in the materials that enter the site (approximately 1.5%). Metals are separated before the C&D waste materials enter the recycling plant and the scrap is sold-off to metal recyclers. Economic allocation will be applied in the LCA to attribute the plant's environmental impacts to scrap and recovered aggregate co-products. If the amount of ferrous scrap in the received materials changes from 1.5%, the results of the LCA would be impacted. Electricity used by the recycling plant has been modelled on the South Australia state electricity grid. The site as of the time data was collected did not use any solar panels or any other sources of electricity. The transport distance for the operation of the loaders and other trucks are assumed to be 20 tkm in the production of 1 tonne recycled aggregate. The change in distance will affect the results of the LCA.

## 19.3 Results and Discussion

### 19.3.1 Impact Assessment and Emission Modelling

The results of the LCA (see Table 19.1) are presented as the aggregated value for modules A1-A3, based on the declared unit of 1 tonne of final products from C&D waste recycling processes. Australian Life Cycle Assessment Society (ALCAS) Best Practice Life cycle impact assessment (LCIA) carbon neutral Method is used for the assessment.

The most relevant indicator which affects climate change is the global warming potential or greenhouse gases (GhG). The greenhouse gas emission was reported as *4.88E+00* $kgCO_{2\text{-eq}}$ which is lower than LCA reports on quarry products for aggregates. This brings a lot of gains as it offsets some GhG emissions and reduce the embodied energy if the products are used for construction without solely relying on primary or virgin materials. The Fig. 19.2 below shows the characterisation of the impact assessment of the processes of the recycled C&D waste materials.

Even though the global warming potential (GWP) value is not as high as that of quarry production, other alternatives to the application of the unit processes of recycling the aggregates can still help in reducing the value drastically. It is recommended that, alternatives such as solar power and electric vehicles (as seen in the scenarios in Table 19.3) or hydrogen fuelled vehicles can provide less impact on the recycling processes. The GhG emissions for renewables are negligible which reduces the amount of energy needed when processing waste materials. This will further offset the carbon footprint of the recycling process.

### 19.3.2 GWP Comparison with Quarry Products

Recycled aggregates provide several benefits when it comes to construction, with the most relevant one, which is, offsetting carbon footprint of the use of virgin aggregate materials. The global warming potential for the two products that is the ResourceCo recycled aggregates and aggregates from a typical Australian quarry were compared as shown in the Table 19.2 below. The declared unit for both products is 1 tonne of aggregate for both production processes. The ALCAS LCIA carbon neutral method was used for the assessment because of the geographical location of the studies to aid in achieving accurate results (Oteng et al., 2023).

**Table 19.1** Environmental indicators of 1 tonne recycled C&D waste

| Environmental indicator | Unit | Results (A1-A3) |
|---|---|---|
| Global Warming Potential (GWP) | kg $CO_{2\text{-eq}}$ | 4.88E+00 |
| Abiotic Depletion Potential—Elements (ADPE) | kg $Sb_{\text{-eq}}$ | 5.37E-04 |
| Abiotic Depletion Potential—Fossil Fuels (ADPF) | MJ NCV | 9.34E+01 |
| Ozone Depletion Potential (ODP) | kg CFC-$11_{\text{-eq}}$ | 1.12E-06 |
| Photochemical Ozone Creation Potential (POCP) | kg $C_2H_{4\text{-eq}}$ | 1.60E-03 |
| Acidification Potential (AP) | kg $SO_{2\text{-eq}}$ | 2.04E-02 |
| Eutrophication Potential (EP) | kg $PO_{4\text{-eq}}$ | 2.74E-03 |
| Particulate matter | kg $PM_{2.5}$ | 2.15E-03 |
| Human toxicity, cancer | CTUh | 1.41E-07 |
| Human toxicity, non-cancer | CTUh | 3.73E-07 |
| Freshwater ecotoxicity | CTUe | −1.89E+05 |
| Ionizing radiation HH | kBq $U235_{\text{-eq}}$ | 1.25E-02 |
| Water Scarcity | $m^3H_2O_{\text{-eq}}$ | −3.13E-01 |

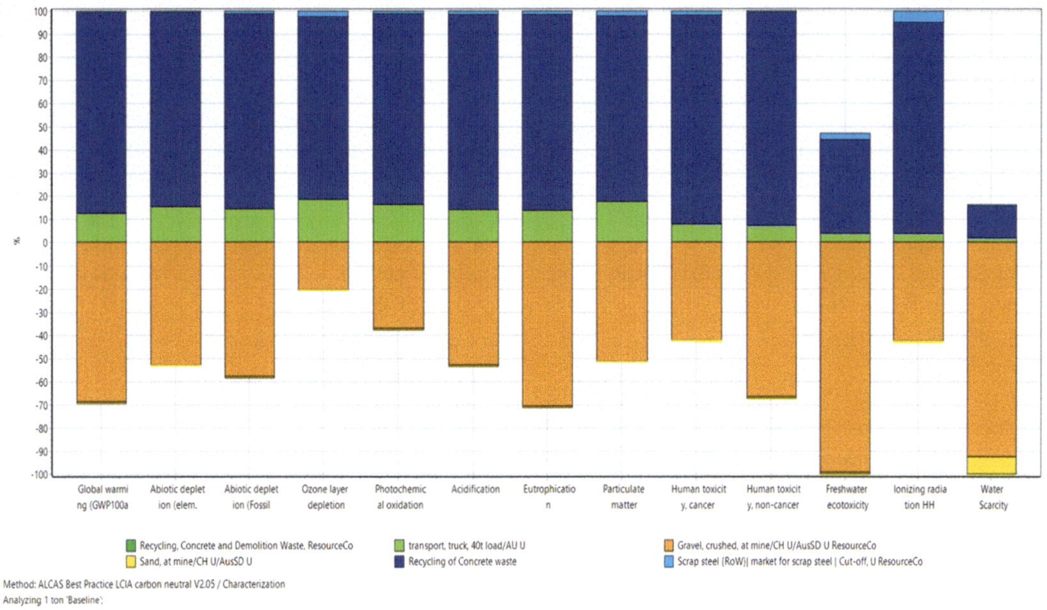


**Fig. 19.2** Impact assessment of 1 tonne C&D waste recycling

The greenhouse gas emission was reported as *4.88E+00* $kgCO_{2\text{-eq}}$ which is lower than LCA reports on quarry products for aggregates in Australia which is *11.3E+00* $kgCO_{2\text{-eq}}$ according to AusLCI database.[1] Thus, recycled aggregate from ResourceCo reduces the carbon footprint by approximately *57%* as compared to that of the quarry production. According to research on quarry production in Malaysia and Portugal (Dias et al., 2021; Mah et al., 2018b), it was revealed that, the GWP ($kgCO_{2\text{-eq}}$) was around *8.10E+00 and 28.6E+00* respectively, which shows a huge increase in the carbon footprint of quarry products in other countries compared to the recycled process of ResourceCo. This brings many gains as it offsets some GhG emissions and reduce the embodied energy if the products are used for construction without solely relyingy on primary or virgin materials.

The difference as observed from Fig. 19.3, shows a small value for GWP of the Recycled aggregate as compared to that of the Quarry. This information gives buyers of recycled aggregates confidence as their projects are rewarded for using low embodied energy materials as well as get good marks when applying for sustainable ratings. Apart from the carbon footprint, environmental indicators like, *Abiotic Depletion Potential—Elements, Abiotic Depletion Potential—Fossil Fuels, Acidification Potential, Eutrophication Potential, Particulate matter, Human toxicity—non-cancer, Freshwater ecotoxicity, Water Scarcity* are less compared to that of the quarry production. This shows the importance of recycling construction and demolition waste and its benefits to the environment.

### 19.3.3 Scenario Analysis

Two other scenarios are generated to understand the impact of transport distance on the recycling processes at the site. The first scenario was the use of the electric or hybrid trucks in the movement of waste materials and the second was the use of

[1] Note: This data is an un-modified version of ecoinvent 2.2-unit process, however its upstream processes are linked to Australian data from the AusLCI database so the calculated result will be different to ecoinvent 2.2. However, data such as water has been modified to suit the South Australian domestic water supply.

**Table 19.2** Environmental indicators of 1 tonne recycled C&D waste

| Environmental Indicator | Unit | Results (A1-A3) | |
|---|---|---|---|
| | | Recycled aggregate production (ResourceCo) | Quarry production (Australia) |
| Global Warming Potential (GWP) | kg $CO_{2\text{-eq}}$ | 4.88E+00 | 1.13E+01 |
| Abiotic Depletion Potential—Elements (ADPE) | kg $Sb_{\text{-eq}}$ | 5.37E-04 | 6.29E-04 |
| Abiotic Depletion Potential—Fossil Fuels (ADPF) | MJ NCV | 9.34E+01 | 1.35E+02 |
| Ozone Depletion Potential (ODP) | kg CFC-$11_{\text{-eq}}$ | 1.12E-06 | 2.98E-07 |
| Photochemical Ozone Creation Potential (POCP) | kg $C_2H_{4\text{-eq}}$ | 1.60E-03 | 9.94E-04 |
| Acidification Potential (AP) | kg $SO_{2\text{-eq}}$ | 2.04E-02 | 2.39E-02 |
| Eutrophication Potential (EP) | kg $PO_{4\text{-eq}}$ | 2.74E-03 | 6.95E-03 |
| Particulate matter | kg $PM_{2.5}$ | 2.15E-03 | 2.36E-03 |
| Human toxicity, cancer | CTUh | 1.41E-07 | 1.07E-07 |
| Human toxicity, non-cancer | CTUh | 3.73E-07 | 7.83E-07 |
| Freshwater ecotoxicity | CTUe | −1.89E+05 | 3.68E+05 |
| Ionizing radiation HH | kBq $U235_{\text{-eq}}$ | 1.25E-02 | 9.60E-03 |
| Water Scarcity | $m^3H_2O_{\text{-eq}}$ | −3.13E-01 | 1.38E+00 |

**Table 19.3** Environmental indicators of different scenarios

| Environmental Indicator | Unit | Scenario 1 | Scenario 2 | Scenario 3 |
|---|---|---|---|---|
| | | Baseline (Current production) | Electric (Transport) | High vehicle movement (Transport) |
| Global Warming Potential (GWP) | kg $CO_{2\text{-eq}}$ | 4.88E+00 | 2.87E+00 | 1.29E+01 |
| Abiotic Depletion Potential—Elements (ADPE) | kg $Sb_{\text{-eq}}$ | 5.37E-04 | 3.60E-04 | 1.25E-03 |
| Abiotic Depletion Potential—Fossil Fuels (ADPF) | MJ NCV | 9.34E+01 | 6.09E+01 | 2.23E+02 |
| Ozone Depletion Potential (ODP) | kg CFC-$11_{\text{-eq}}$ | 1.12E-06 | 8.58E-07 | 2.17E-06 |
| Photochemical Ozone Creation Potential (POCP) | kg $C_2H_{4\text{-eq}}$ | 1.60E-03 | 1.18E-03 | 3.28E-03 |
| Acidification Potential (AP) | kg $SO_{2\text{-eq}}$ | 2.04E-02 | 1.42E-02 | 4.51E-02 |
| Eutrophication Potential (EP) | kg $PO_{4\text{-eq}}$ | 2.74E-03 | 1.41E-03 | 8.05E-03 |
| Particulate matter | kg $PM_{2.5}$ | 2.15E-03 | 1.36E-03 | 5.29E-03 |
| Human toxicity, cancer | CTUh | 1.41E-07 | 1.21E-07 | 2.20E-07 |
| Human toxicity, non-cancer | CTUh | 3.73E-07 | 2.90E-07 | 7.05E-07 |
| Freshwater ecotoxicity | CTUe | −1.89E+05 | −2.03E+05 | −1.33E+05 |
| Ionizing radiation HH | kBq $U235_{\text{-eq}}$ | 1.25E-02 | 1.16E-02 | 1.58E-02 |
| Water Scarcity | $m^3H_2O_{\text{-eq}}$ | −3.13E-01 | −3.21E-01 | −2.82E-01 |

diesel driven trucks but with a total distance of 100 tkm. The scenarios clearly identify the dynamics of transport in the processing of waste C&D materials.

The results of the additional scenarios to the baseline are shown in Fig. 19.4. The dynamics of the impact of transportation distance is well highlighted in Table 19.3 and Fig. 19.4. The movement of loaders and trucks when processing the waste materials plays an important role in the release of carbon emissions to the environment. An efficient solution to this is to use hybrid/

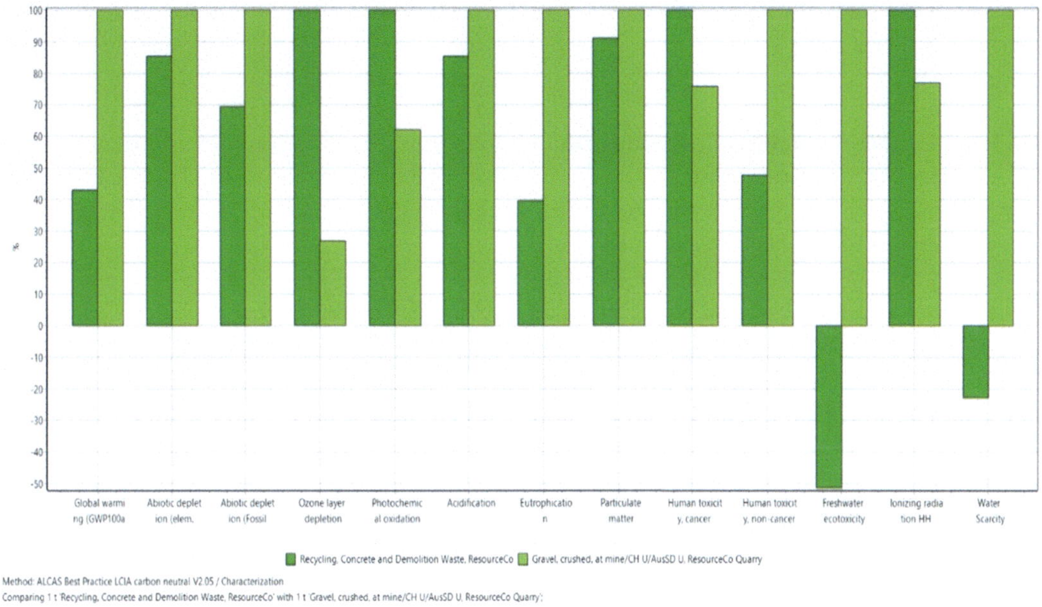


**Fig. 19.3** Impact assessment of 1 tonne C&D waste recycling and quarry production

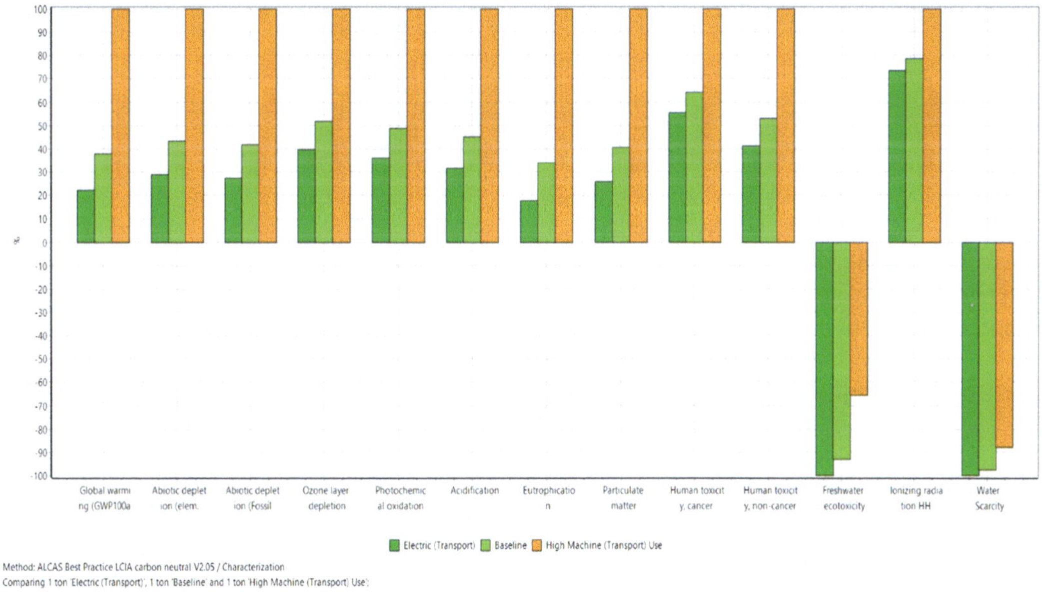


**Fig. 19.4** Impact assessment of different scenarios

electric or hydrogen fuelled trucks or loaders. This will significantly reduce the carbon footprint and its associated emissions. Reducing emissions from trucks is very important as it contributes to the net zero emission targets for Australia by 2050 (Oteng et al., 2022).

The results reveal that, alternative sources to transportation can severely affect the carbon footprint of ResourceCo's recycled products. Another suggestion that can aid in the reduction of the emissions, is the use of solar panels. This is not captured in the study because of the absence of solar panels on the site.

### 19.3.4 C&DW Recycling Contribution to the Environment

There are some associated benefits that come with recycling construction and demolition waste at the ResourceCo facility. The site receives materials from in and around the Adelaide area making it good value for money, when it comes to the transportation of these waste to the landfill and travelling longer distances to procure virgin materials from quarries. The process therefore saves transportation cost and reduce emissions to the environment (Wu et al., 2021). The benefits, however, applies to each project differently and the process is not directly included in the scope of this LCA.

The largest contribution of this process is the diversion of waste materials to landfill. Landfilling is associated with a lot of negative impacts such as that on the social life of the people and the emissions to air and land of the located area. This comes with a lot of responsibility to manage the safety, environment and health of the people as well as occupying viable lands. Recycling of C&D waste is very effective as it diverts about 97% of the waste from landfill. It also reduces the impacts of waste on the environment and its ability to occupy a lot of landfill space. This process also saves a lot of virgin material (Mah et al., 2018b) production which has huge number of impacts when it comes to its carbon footprint (transport, loss of biodiversity, loss of land, deforestation, emissions etc). The recycled materials can replace and offset some of the virgin materials such as aggregates, sand, scrap metal for use as road base, in pavements, and other construction purposes which will help reduce their carbon footprint.

There is little information on water use as the site uses bore water for the processes. It was difficult to collect information on the use as it is not metered. Again, information on cost and prices were limited, therefore, the life cycle costing (LCC) assessment was not performed for this project. However, the LCC of recycling the C&D waste is the net profit of summing the recycled aggregate price, and the avoided cost of sending it to landfill, and the recycling cost. This should be investigated further to provide a complete assessment of the current economic and social situation related to the life cycle of C&D waste at the plant.

## 19.4 Conclusions

The study presents a comparative assessment of recycled aggregate products from ResourceCo recycling processes and the production of Virgin materials. Several scenarios are also conducted to examine the impacts of the processes through transportation. The collected process data is from the operation of the Wingfield recycling plant combined with the ecoinvent database. The study demonstrated that, the greenhouse gas emission reported as 4.88E+00 $kgCO_{2\text{-}eq}$ is lower than the LCA results on quarry (natural aggregate) products in Australia which is 11.3E+00 $kgCO_{2\text{-}eq}$ according to the AusLCI database. Hence, recycled aggregate from ResourceCo reduces the carbon footprint by approximately 57% as compared to that of the quarry production.

The results also revealed that, alternative sources to transportation can severely affect the carbon footprint of ResourceCo's recycled products. The use of solar panels can also affect the results of the study as it produces clean energy during operation. It is recommended that; future collection of data focuses on the information gathered from water use, as it was bore water used on the current site which was not metered making it difficult to collect primary data. Moreover, information on cost and prices were limited, therefore, the life cycle costing (LCC) assessment was not performed for this project and should be explored later.

**Acknowledgement** The research is conducted in co-operation and with funding from ResourceCo Pty Ltd. The authors wish to acknowledge resource support also from the University of Adelaide, in the School of Architecture and Civil Engineering.

## References


AusLCI. (2019). *Australasian life cycle inventory database version 2019, V1.36*. AusLCI, Melbourne, Strategies, Life Cycle.

Bovea, M. D., & Powell, J. C. (2016). Developments in life cycle assessment applied to evaluate the environmental performance of construction and demolition wastes. *Waste Management, 50*, 151–172. https://doi.org/10.1016/j.wasman.2016.01.036

Butera, S., Christensen, T. H., & Astrup, T. F. (2015). Life cycle assessment of construction and demolition waste management. *Waste Management, 44*, 196–205. https://doi.org/10.1016/j.wasman.2015.07.011

Colangelo, F., Petrillo, A., & Farina, I. (2021). Comparative environmental evaluation of recycled aggregates from construction and demolition wastes in Italy. *Science of the Total Environment, 798*, 149250. https://doi.org/10.1016/j.scitotenv.2021.149250

Dias, A. B., Pacheco, J. N., Silvestre, J. D., Martins, I. M., & de Brito, J. (2021). Environmental and economic life cycle assessment of recycled coarse aggregates: A Portuguese case study. *Materials, 14*(18). https://doi.org/10.3390/ma14185452

EN 15804. (2013). *EN 15804: 2012+ A1: 2013. Sustainability of construction works-environmental product declarations-core rules for the product category of construction products*. European Committee for Standardization. Publications Office of the European Union.

Ernest, K., Samuel, A. S., Agyemang, D. Y., Daniel, O., & Caleb, D. (2020). Identification of factors influencing the pricing of sustainable construction materials in developing countries: Views of Ghanaian quantity surveyors. *International Journal of Construction Management*, 1–10. https://doi.org/10.1080/15623599.2020.1768462

Estanqueiro, B., Dinis Silvestre, J., de Brito, J., & Duarte Pinheiro, M. (2018). Environmental life cycle assessment of coarse natural and recycled aggregates for concrete. *European Journal of Environmental and Civil Engineering, 22*(4), 429–449. https://doi.org/10.1080/19648189.2016.1197161

Ingwersen, W. W., & Subramanian, V. (2014). Guidance for product category rule development: Process, outcome, and next steps. *The International Journal of Life Cycle Assessment, 19*(3), 532–537. https://doi.org/10.1007/s11367-013-0659-0

ISO 14040. (2006). *Environmental management: Life cycle assessment; Principles and framework* (Vol. 14040). International Organization for Standardization (ISO).

ISO 14044. (2006). *Environmental management: Life cycle assessment; requirements and guidelines* (Vol. 14044). International Organization for Standardization (ISO).

Mah, C. M., Fujiwara, T., & Ho, C. S. (2018a). Environmental impacts of construction and demolition waste management alternatives. *Chemical Engineering Transactions, 63*, 343–348.

Mah, C. M., Fujiwara, T., & Ho, C. S. (2018b). Life cycle assessment and life cycle costing toward eco-efficiency concrete waste management in Malaysia. *Journal of Cleaner Production, 172*, 3415–3427. https://doi.org/10.1016/j.jclepro.2017.11.200

Oteng, D., Zuo, J., & Sharifi, E. (2022). Environmental emissions influencing solar photovoltaic waste management in Australia: An optimised system network of waste collection facilities. *Journal of Environmental Management, 314*, 115007. https://doi.org/10.1016/j.jenvman.2022.115007

Oteng, D., Zuo, J., & Sharifi, E. (2023). An evaluation of the impact framework for product stewardship on end-of-life solar photovoltaic modules: An environmental lifecycle assessment. *Journal of Cleaner Production, 411*, 137357.

Simion, I. M., Fortuna, M. E., Bonoli, A., & Gavrilescu, M. (2013). Comparing environmental impacts of natural inert and recycled construction and demolition waste processing using LCA. *Journal of Environmental Engineering and Landscape Management, 21*(4), 273–287. https://doi.org/10.3846/16486897.2013.852558

Wu, H., Zuo, J., Zillante, G., Wang, J., & Yuan, H. (2019). Construction and demolition waste research: A bibliometric analysis. *Architectural Science Review, 62*(4), 354–365. https://doi.org/10.1080/00038628.2018.1564646

Wu, H., Zuo, J., Yuan, H., Zillante, G., & Wang, J. (2020). Cross-regional mobility of construction and demolition waste in Australia: An exploratory study. *Resources, Conservation and Recycling, 156*, 104710. https://doi.org/10.1016/j.resconrec.2020.104710

Wu, H., Zuo, J., Zillante, G., Wang, J., & Duan, H. (2021). Environmental impacts of cross-regional mobility of construction and demolition waste: An Australia study. *Resources, Conservation and Recycling, 174*, 105805. https://doi.org/10.1016/j.resconrec.2021.105805

Yang, Y., Heijungs, R., & Brandão, M. (2017). Hybrid life cycle assessment (LCA) does not necessarily yield more accurate results than process-based LCA. *Journal of Cleaner Production, 150*, 237–242. https://doi.org/10.1016/j.jclepro.2017.03.006

Zhang, J., Ding, L., Li, F., & Peng, J. (2020). Recycled aggregates from construction and demolition wastes as alternative filling materials for highway subgrades in China. *Journal of Cleaner Production, 255*, 120223. https://doi.org/10.1016/j.jclepro.2020.120223

# Towards Circularity in Roads Infrastructure: A Critical Review

# 20

Mingyuan Guo, Sepani Senaratne, Laura Almeida, and Srinath Perera

## 20.1 Introduction

Studies on circular economy (CE) in road infrastructure development represent a growing field. Although its major goal is to minimize the negative environmental impact, economic and strategic reasons have also been embedded. It is also known that the consumption of natural resources has been deemed the standard approach that is part of a linear economy. However, as resources are finite, and the consequences of climate change and disaster resilience cannot be unseen, it is becoming inevitable and critical to seek alternative solutions for sourcing and using materials.

This chapter uses a critical literature review to explore the CE strategies, enablers, and barriers to implementing a CE in road infrastructure projects. There is a research gap in understanding circular economy implementations for road projects, particularly on CE strategies, enablers and barriers for CE implementation in road infrastructure.

The chapter is structured in six sections. Following this introduction, the second section presents the research method on the systematic literature review undertaken. The next three sections discuss strategies, barriers and enablers found through the systematic review. Finally, the chapter summarizes the key findings and offers the way forward. Overall, the chapter highlights the importance of implementing circular principles aligned with industrial ecology to conserve energy and materials and reduce waste production.

M. Guo · S. Senaratne · L. Almeida · S. Perera (✉)
Centre for Smart Modern Construction (c4SMC), School of Engineering, Design and Built Environment, Western Sydney University Penrith, Penrith, NSW, Australia
e-mail: y.guo4@westernsydney.edu.au; s.senaratne@westernsydney.edu.au; l.almeida@westernsydney.edu.au; srinath.perera@westernsydney.edu.au

## 20.2 Research Method

This research used a systematic literature review approach due to its comprehensive, structured, reproducible, transparent, and quantitative nature (Pickering & Byrne, 2014). As topics and domains related to the scope of this research are numerous, the systematic literature review approach helped to find the most relevant publications, extract knowledge areas, and categorize CE barriers, after some filtering stages. The publication search was conducted in Scopus and Web of Science libraries, as by the result of a preliminary search, they had more relevant publications to the research theme.

The search rule was ( ALL ( energy ) AND TITLE ( road* ) AND ALL ( efficiency ) OR ALL ( saving ) OR ALL ( harvesting ) OR ALL ( bridge* ) OR ALL ( tunnel* ) ) AND ( LIMIT-TO (PUBSTAGE, "final" ) ), which was limited



J. Zuo et al. (eds.), *Circular Economy for Buildings and Infrastructure*, Sustainable Development Goals Series, https://doi.org/10.1007/978-3-031-56241-9_20

to Engineering domain, English language and Review paper type, as a result of which 54 documents remained. 37 out of 54 have been filtered out as potential Review Papers for this research. Figure 20.1 presents the literature search scheme, the findings of which serve as the source papers to be used for finding the strategies, enablers and barriers for implementing a CE in road infrastructure projects.

## 20.3 Strategies for a Circular Economy in Road Infrastructure

The social and physical environment of applying CE in road construction cannot be overlooked. As governments are the major stakeholders in most road projects, industrial standards, policies, and rules will be developed accordingly and parallelly. However, the linear economy habits are still embedded within the mindset of these stakeholders by observing the pre-set linear lifespan design requirement of levees (Guo et al., 2021). Therefore, a deeper understanding of CE and circular design for new road projects necessitate development. The circular design may not directly target the consequences of climate change; however, the entrenched circular strategies from the perception of industrial ecology will help establish a systematic framework to conserve and circulate limited energy resources and materials within the road infrastructure sector. During the past 50 years, much more information has become available on industrial ecology than this CE perception. Its principles have become applicable to varieties of different industries, including road construction (Ellen MacArthur Foundation, 2017).

One of the most significant current discussions in the road construction sector is the high energy resources and materials consumption. The main challenges of this phenomenon are that around 25–40% of energy consumption is associated with 40% $CO_2$ emission in the construction industry (Papachristos et al., 2020). Common materials such as asphalt, concrete, and steel significantly impact the environment during their life cycle phases. The construction industry alone accounts for 50% of global steel production and consumes more than three billion tons of raw materials annually (Pacheco-Torgal & Labrincha, 2013). It is worth emphasizing that most raw material consumption is not recoverable and goes to landfills after production. In Europe, the construction industry is responsible for more than 40% of the produced (weight) waste compared to the municipal waste of only about 12% (Pepe, 2015). Cement production alone is estimated to be responsible for 5% of the world's total $CO_2$ emissions (Damtoft et al., 2008).

Furthermore, material waste occurs during their production and all the lifecycle phases of road infrastructure. According to the US Green Building Council and US Environmental Protection Agency (United States Environmental Protection Agency, 2018), about 10% of these materials are only used for construction and deposited as waste afterwards. According to the Department of State Growth (2020), road infrastructure assets can be categorized into road, bridge, property, traffic services and data assets. The critical problem with this explanation is that the Department of State Growth (2020) does not classify the tunnel as one of its assets. However, it is almost certain that tunnels are vital components of road infrastructure in some cases (Department of Main Roads, 2006). Therefore, road infrastructure is a critical focus in considering circular economy principles.

To promote Australia's road infrastructure towards achieving a CE, various strategies for using road-related energy-saving solutions should be explored further and closely investigated in line with international standards. As a result, evaluating the specific products in each industry can provide important technical insights as it enables the analysis of energy-saving solutions with both considerations of economy and environment. Besides, developing an energy-saving framework for road assets under CE can also simulate the impact of different materials, such as asphalt and steel, on environmental sustainability and energy efficiency. Such studies would help stakeholders assess and select the most appropriate energy-saving solution for road infrastructure assets. The categorizations with associated economic practices within the

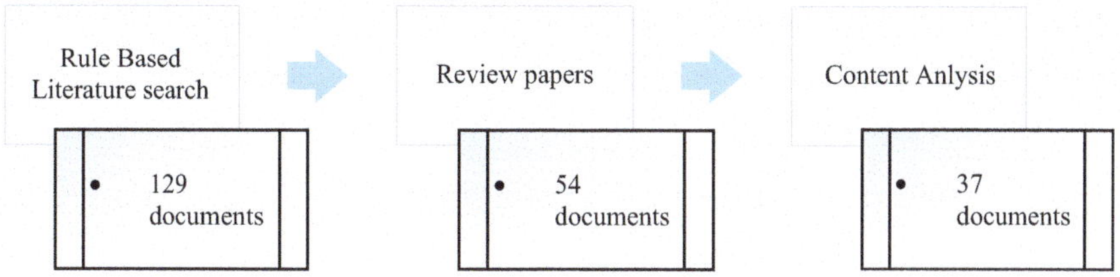


**Fig. 20.1** Literature search flowchart

framework would also be able to assess the performance of the selected energy-saving solutions under the concept of CE. Considering the above perspectives, CE performance indicators may motivate Australia and the world to explore and transform a linear economy into a CE for road infrastructure.

According to Potting et al. (2016), various circularity strategies have been proposed in the literature. Table 20.1 is created using the original work of Potting et al. (2016) on their work in the production chain and relating it to road infrastructure. For road infrastructure, however, applying renewable resources within the system boundary of CE has been added as an extra 'R' to the different strategies (often referred to as the '9 Rs') to accommodate contemporary technological developments. Therefore, a total of '10 Rs' have been listed in Table 20.1.

Several circularity strategies exist to reduce the consumption of natural resources and materials and minimize waste production. They can be ordered for priority according to their levels of circularity (Table 20.1) (Potting et al., 2017a, b). For example, by reusing the product from road components, smarter product manufacturing and use are generally preferred to overextend the lifetime of road infrastructure because these road components are being used for the same product function or more users are being served by one product (a strategy with high circularity). Lifetime extension is the next option, followed by recycling road materials through recovery. Incineration from which energy is recovered has the lowest priority in a circular economy because the materials can no longer be applied to other products (low-circularity strategy). As a rule of thumb, more circularity equals more environmental benefits (Potting et al., 2017a, b).

A higher level of circularity of materials in a product chain means that those materials remain in the chain for a longer period and can be applied again after a product is discarded, preferably retaining its original quality. As a result, fewer natural resources are needed to produce new materials for manufacturing products and their subsequent use (Potting et al., 2017a, b). Avoided resource extraction and production of materials benefit the environment. There are exceptions to this rule of thumb. For example, making a product chain more circular may lead to increased natural resource consumption, usually in the form of (fossil) fuels. If the fossil fuel requirements of the non-road construction equipment fleet are reduced by 15–25% using biofuel or biofuel blends as a replacement for fossil fuel, a limited scope life cycle assessment (LCA) of the construction of one lane mile of Portland cement concrete roadway using a generic set of non-road construction equipment required to place the concrete (i.e. a paving machine and texture/curing machine) indicated that the production and utilization of biodiesel consume more energy than that required to produce and utilize conventional or ultra-low Sulphur diesel fuel (Muench et al., 2011).

Due to recent technology updates and developments in renewable energy, solar energy offers possibilities of additionality. Solar energy is a natural energy source that is an important part of the biological cycle but can also be converted into operational energy through solar panels (McDonough & Braungart, 2013). It can also be said that the energy in the Cradle-to-Cradle system accommodates circularities across from the biological cycle to the technical cycle, with potentially many additionalities to be explored.

**Table 20.1** Design and Construction Strategies of circularity for road infrastructure

| Design and construction focus | Strategies | CE principles |
|---|---|---|
| Smart production and use | Preventing the use of resources by changing functions or delivering functions through different radical resources | R0 Refuse |
| | Intensify the use of resources for products by sharing or multi-functional use | R1 Rethink |
| | Efficient production and use by using fewer raw materials and resources | R2 Reduce |
| Extending the life of road infrastructure and components | Reuse of products for the same function by another user | R3 Reuse |
| | Repair and maintenance of broken products for use in the current function | R4 Repair |
| | Refurbish and modernize old product | R5 Refurbish |
| | Use of parts of the discarded product in a new product with the same function | R6 Remanufacture |
| | Use of discarded products or components in new products with different functions | R7 Repurpose |
| Useful use of materials | Processing material to the same (high-grade) or lesser (low-grade) quality | R8 Recycle |
| | Incineration of residual materials with energy recovery | R9 Recover |
| Technological developments | Use of 'additionalities' to accommodate circularities | R10 Renewable |

Source: Potting et al. (2017a, b)

**Table 20.2** Strategies of circularity for road infrastructure based on the road's life cycle phases

| Life cycle | Strategies | CE principles |
|---|---|---|
| Initiation & Planning | Design for economic lifespan rather than pre-set lifespan requirement<br>Make the road system adaptive (modular design) | Rethink (R1)<br>Reuse (R3), Repurpose (R7) |
| Design & Conditioning | Design for multiple life cycles and minimize potential locked-in<br>Design for recycling<br>Design for lifetime extension within the first life cycle<br>Design for easy maintenance and management<br>Design standardized reusable modules (modular design)<br>Use the physical and social environment of the road system | Renewable (R10)<br>Recycle (R8)<br>Repair (R4), Refurbish (R5)<br>Repair (R4)<br>Reuse (R3), Repurpose (R7)<br>Renewable (R10) |
| Sourcing | Design for material-poor constructions (resources)<br>Design for maximum use of recycled raw materials (resources)<br>Maximum utilization of raw materials (resources) or components from a previous life cycle<br>Design for employability of renewable raw materials (bio-based) (resources)<br>Develop a 'materials passport' linked-to materials (resources) and material flows | Reduce (R2)<br>Recycle (R8)<br>Reuse (R3)<br>Renewable (R10)<br>Recycle (R8) |
| Construction | Design for recycling, i.e., prevent un-recyclable mixing and kitting.<br>Design for the deconstruction of components (modules) | Recycle (R8)<br>Reuse (R3), Repurpose (R7) |
| Operation | Reduce energy consumption and energy needed should be self-generated | Renewable (R10) |
| Renewal | Design aimed at simple and effective maintenance and management, focusing on value retention over replacement | Repair (R4), Refurbish (R5) |
| Disassembly | Match supply and demand within a life cycle and over different life cycles | Refuse (R0), Rethink (R1) |

Source: Leendertse et al. (2018)

The following further strategies represented in Table 20.2 for road infrastructure have been established based on a road's life cycle stages to match the circularity strategies (Leendertse et al., 2018). These strategies with CE principles can be used to locate, analyze and develop potential circular economy practices for road infrastructure. For example, from Table 20.2 to design for economic lifespan rather than pre-set lifespan requirement, lighting, traffic management, and the road can be the perspectives to be considered to improve road infrastructure at initiation and planning stages. Regarding lighting, for example, traffic lights and vehicle driving speed planning can be considered as one of the CE practices to improve road traffic congestion, as integration between traffic lights and vehicles can save operational energy and reduce the impact of human behavior on the environment (Estaji et al., 2018; Nagamani et al., 2019). Therefore, from Table 20.2, a procedure of seeking the majority of the CE practices in road infrastructure can be used and developed further to achieve the goal of circularity in road infrastructure.

A real-life road project can be analyzed according to the previous seven practices. For example, a Circular Designed Highway Viaduct is the first circular bridge in the Netherlands. As the first circular bridge in the Netherlands, the initial project goal was simple: 'there is no waste, reuse requires no new raw materials and used raw materials are put to new use in the most high-value way possible' (RMoIWM, 2016). The original circular design principle of the project was to follow 'design for economic lifespan rather than pre-set lifespan requirement' to achieve a closed-loop design for deconstruction activities. As the project was built based on a modular design, it also satisfies the 'make the road system adaptive' principle. The modular design features of the project are also in line with the principle of designing for easy maintenance and management and designing standardized reusable modules at the design and conditioning stages.

At the sourcing stage, the project was designed to find solutions that minimize the consumption of raw materials and maximize the use of recycled raw materials while considering maximizing the utilization of raw materials or components across different life cycles. Besides, the project team has been introduced to the concept of a 'material passport'; however, it has not been elaborated further at later stages of the project. Since the project was tested between December 2018 and September 2019 and has not initiated the construction stage, it is believed only assumptions can be made for the rest of the project stages: construction, operation, maintenance and disposability. However, the project's highlight is the support from the Netherlands' government and key stakeholders who enable the idea of an open learning environment. This CE practice shares knowledge from the various market parties, government agencies and research institutes, potentially benefiting future circular-designed road projects. Due to the 'Renewable' feature of this CE practice, the project also enables formal corporate-level policies and targets for ensuring physical resources can be used most efficiently in the operation of the works.

Several lines of evidence suggest that the actual project design concentrated on the modular composition of the viaduct, especially a demountable bridge. Due to the project's pioneering features and the variety of stakeholders' involvement, the project took longer to operate the trial test and initiate the final construction. Therefore, it seems that the design principles of CE are fragmented and not appropriately integrated into the road infrastructure sector. Besides, although the Netherlands is a European pioneer in implementing new technology, a systematic modelling framework has not been appropriately developed in the road infrastructure sector.

## 20.4 Barriers for a Circular Economy in Road Infrastructure

Despite the importance of suggested CE practices in the construction industry, there remains a paucity of consideration on barriers to achieving a CE in road projects. Table 20.3 summarizes the barriers under four categories: cultural, regulatory, financial and sectorial, based on the

**Table 20.3** Barriers to CE adoption in road-related sectors within the construction industry

| Barrier category | Barriers | How to overcome |
|---|---|---|
| Cultural | Lack of interest, knowledge/skills and engagement throughout the value chain | Need framework—Service economic model |
| | Entrenched linear economy mindsets and habits | Need framework |
| | Lack of collaboration between businesses and business functions | Collaborative economic model |
| | Conservative, uncollaborative, adversarial industry | Need Collaborative economic model |
| Regulatory | Lack of consistent, supportive policies and regulatory framework | Need framework |
| | Lack of incentives to facilitate CE in public perception | Need framework |
| | Lack of regulatory disposal restrictions where EoL modules pass hazardous waste determination tests | Need framework |
| | Critical regulatory issues in glass recycling | Need framework |
| | Lack of policies for distinguishing hazardous waste exports from materials that can be used in recycling operations | Need framework |
| | Complexity of the increasing quality standards | Need framework |
| Financial | High upfront investment costs in recycling technologies and specialty infrastructure | Grey literature driver- short vision from business- Service Economy |
| | Poor business case and unconvincing case studies | Disconnection between academic and grey |
| | Low virgin material prices | Need framework to have a long vision |
| | Limited number of high-value materials for recovery | Seeking balance based on framework |
| | Fluctuating market prices | Material, labor, energy |
| | High cost in mining remediation | Material, labor, energy |
| Sectoral | Lack of coherent CE vision or road map for the industry | Need framework |
| | Lack of standards for CE practices in energy and materials | Need framework |
| | Insufficient use or development of CE-focused design and collaboration tools, information and metrics | Need framework to integrate, LCA is assistant but not enough |
| | Long product lifecycles | Road infrastructure and road transportation |
| | Technical challenges for materials and resources extraction, reusing, upcycling and recycling | Restorative economy |
| | Technical challenges for energy reducing, regenerating and recovering | Restorative economy |
| | Complexity of emerging composites | Need framework |
| | Complexity of recycling glass into high-value feedstocks | Restorative economy |
| | Low recovery rate from waste collection and extraction | Restorative economy |
| | Complexity and variety of products to disassemble or shred | Need framework |
| | Complexity of the wide-ranging product volumes reaching EoL | Restorative economy |
| | Insufficient use of freshwater from the extraction processes | Need framework |
| | Lack of water conservation | Need framework |
| | Low recycling rate of plastic waste and microplastics | Restorative economy |

Sources: Hart et al. (2019), Mulvaney et al. (2021), Ratnasabapathy et al. (2021), Shooshtarian et al. (2023)

ranking of the main barriers to CE adoption in the AEC industry by researchers (Hart et al., 2019; Mulvaney et al., 2021; Ratnasabapathy et al., 2021; Shooshtarian et al., 2023). These barriers were identified by considering the circular economy practices for road infrastructure. From Table 20.3, it was found that distinguishing all the barriers before the construction stage for CE, at different lifecycle stages is very important. Therefore, the different barriers were identified for each of the four categories, and the correspondent availability or not of an existing model can be used as an enabler for CE in road infrastructure projects.

As it is possible to see in Table 20.3, the most common economic models that can be used in CE for each one of the different barriers identified previously are the collaborative, services and restorative economy. However, most of the identified barriers still lack an appropriate methodology/model to enable the concept of CE in road infrastructure.

## 20.5 Enablers of CE Adoption in Road Infrastructure

The evidence reviewed here suggests a pertinent role for enablers to bypass or overcome the barriers to achieve a CE in road-related sectors. These enablers are discussed under the following sub-topics.

### 20.5.1 Economic Models under CE Framework

One well-known early study often cited in research on CE is that the construction industry predominantly adopts the linear economic model of "take, make, dispose" of materials (Benachio et al., 2020). Consequently, the heavy negative impact on energy or resources, in general, is also inevitable following this economic model. It can be seen from the information in Table 20.3 that the groups from cultural, financial and sectoral barrier categories present the intercorrelations among the three economic models, which are restorative, collaborative and service economic models under the CE framework.

To overcome the barriers to achieve a CE in road-related sectors, further enablers for CE adoption in road-related sectors within the construction industry have been identified in Table 20.4.

In an analysis of the restorative economy, Ellen MacArthur Foundation (2015) found that this economic model ultimately seeks to disconnect global economic development from finite resource consumption. Ellen MacArthur Foundation (2015) carried out a series of case studies that show a CE in practical applications to further investigate the role of this economic model. However, strategies to enhance the concept of CE might also involve Service Economy. A detailed examination of the service economy by the Ellen MacArthur Foundation (2014) showed that urbanization favors CE to close the reverse loop with lower costs. Therefore, it can be said that the service economy concentrates demand and allows tighter forward and reverse cycles.

Further, Stahel (1997) demonstrated that the industrial economy is not compatible with the concept of sustainability. One well-known early study often cited in research on the service economy is his philosophy: 'where most of the value is created by services, and most jobs are in service activities' (Stahel, 1997). The service economy measures the total wealth and welfare, considering the long-term optimization of system functioning. This motivates the service providers to reduce resource expenditure and energy consumption in the manufacturing process.

A collaborative economy has similar features to a restorative economy as it tends to decentralize and contain networks of individuals and communities instead of institutions. For example, there is a potential connection between vehicle consumers and vehicle users. The research to date has not been able to convincingly show whether examining end-of-first-life applications of electric vehicles (EVs) batteries only in a low-carbon circular economy can truly drive the global transition to EVs and a net-zero economy (Bonsu, 2020).

It is known that more than 85% of $CO_2$ emissions were accounted for the road transportation in China in 2007, which means the focus cannot only be on the road infrastructure sector regardless of the different phases of the road infrastructure assets (Cai et al., 2012; Leendertse et al., 2018). The shifted focus should not only be concerned with emitting carbon but also pay more attention to the vast number of resources in waste. According to Pepe (2015), more than 40% of the produced (weight) amount of waste has been generated within the construction industry in Europe. Thus, the balance between restorative and collaborative economy must be maintained

**Table 20.4** Enablers for CE adoption in road-related sectors within the construction industry

| Enabler category | Enablers | Related barriers |
|---|---|---|
| Economic models | Restorative, collaborative and service economic models | Cultural, financial and sectoral barrier categories (refer to Table 20.3) |
| Circular supply chain and business models | Industrial ecology and environmental economics; social assessment of businesses | Lack of collaboration between businesses and business functions<br>Cultural barriers<br>High upfront investment costs in recycling technologies and specialty recycling infrastructure<br>Energy consumption for processing<br>Energy consumption for recycling waste from material |
| Existing Government policies and regulations for promoting Road CE Practice | ISO standards; EU Standards | Material efficiency and emission efficiency<br>Industrial decarbonization |

Sources: Benachio et al. (2020), Wiesmeth (2020)

to comprehend the concept of CE. Vehicle sharing can be a good illustration for both economic models. From a technical perspective, vehicle consumers can also be vehicle users. For example, Renault, the first car maker to lease EV batteries, helps retain the value of EVs and makes batteries fully traceable for closed-loop re-engineering. These CE principles-based knowledge benefits are shared among car-sharing platforms such as Uber (Ellen MacArthur Foundation, 2014).

On the other hand, the above two economic models also connect with the service economy. Due to the Covid-19 global pandemic, the construction sector in Australia has been impacted by a series of materials shortage problems. For example, because of the impact of bushfires and high steel prices, the shortage of timber has triggered the risk to supply security (Coulter, 2021). However, insulation material for buildings in Australia is an exception. PINK BATTS and EARTHWOOL are two types of insulation materials that are both top-quality insulation products. Both PINK BATTS and EARTHWOOL offer the same thermal protection, while PINK BATTS are made in Australia. Under Covid-19 restrictions, while global supply chains are disrupted, using PINK BATTS as an alternative insulation solution minimizes the risk in insulation material supply (No Gap Insulation, 15 November 2018).

As a service provider in the construction industry, an insulation material supplier is responsible for redistributing material from the manufacturer and providing delivery, installation and maintenance services. Although the service costs more, the philosophy of the service economy proves that the consumers who stay with these suppliers are because of the good customer service (Stahel, 1997). Thus, the higher customer adhesion, the steadier cash flow and better yield. Compared with a cutting fluid supplier for the French car manufacturer Renault, the supplier is responsible for manufacturing material and providing maintenance services for the cutting equipment after applying the philosophy of service economy (Ellen MacArthur Foundation, 2014). Consequently, the latter action stimulates Renault's motivation in research and development, leading to cost-saving on ownership, wastewater treatment, and reduced wastewater discharge volume. It has been observed that this transition from a supplier to a service provider concentrates demand and tightens the loop between manufacturing and service providing, saving cost and reducing the environmental impact.

Moreover, both insulation material suppliers and cutting fluid suppliers are related to assets. The previous is categorized into the construction sector, and the latter is from the road transportation sector. Thus, they are either directly or indirectly related to the infrastructure sector. From a macro perspective, after the transition to cutting liquid suppliers, the company has turned a commodity product into a differentiated solution under the concept of the service economy, which provides the company with more strategies in the market. Although these two examples focus on different products in different industries, a trend based on the concept of service economy under CE principles has gradually shown globally (Ellen MacArthur Foundation, 2014). However, to enable the proof-of-concept, the development of a framework in road infrastructure must be established, which demonstrates the necessity to include materials, labor, energy, carbon emission and trade balance effects in the framework. The concept of CE through different industries worldwide has been consolidated to reiterate the importance of developing such a framework in road infrastructure. The connections among other case studies within the infrastructure sector have also been explicitly and elaborately analyzed.

### 20.5.2 Circular Supply Chain and Business Models

Due to different perceptions, there are challenges in the interaction between humans and technology, indicating discrepancies among the various perceptions of a circular economy (CE). There are two significant perceptions regarding the concept of CE: industrial ecology and environmental economics (Wiesmeth, 2020). During the past 50 years, much more information has become available on industrial ecology as this perception and its principles have become applicable to different industries (Ellen MacArthur Foundation, 2017).

One of the main barriers to CE is the lack of collaboration between businesses and business functions. As the major supply chain actors, businesses have decisive opportunities to apply sustainability assessment approaches from industrial ecology and supply chain management to analyze the network of actors. However, cultural barriers are often associated with human biases from entrenched linear economy mindsets and habits. While Walker et al. (2021) analyzed the social sustainability of the CE, the type of analyzed firms were 'frontrunners' instead of specific industries. It has also been observed that most frontrunner companies that implemented social assessments have either done it qualitatively or used industry-based social sustainability indicator frameworks. The article was analyzed based on Social Life Cycle Assessment (S-LCA). Besides, S-LCA guidelines are lacking in industrial ecology (IE) and supply chain management (SCM) literature. To strengthen the interconnection between supply chain and business models, it is believed that the social assessment of businesses is not the only enabler that applies the Life Cycle Assessment (LCA) as there are three types of sustainability: environmental, economic and social exist within the CE framework.

A growing body of literature recognizes the critical role played by industrial ecology, which implies the realization of practical differences to restore the interactions between the environment and economy cannot be ignored under the concept of CE. Therefore, modifying the perceptions of businesses by presenting evident case studies is likely to accelerate the CE uptake: adapting high upfront investment costs in recycling technologies and specialty recycling infrastructure. Regarding solid waste at the road-related natural resource extraction stage, energy consumption for processing can be reduced if solid waste is recovered and added to the system as material feedstock. Generally, as recycling waste from material extraction also requires a significant amount of energy consumption which is 4% world's electrical energy across all industries, the technological innovation for this sector is unavoidable to be used to efficiently offset this energy loophole (Jeswiet & Szekeres, 2016). Potential comminution technology advancements can be an efficient measurement to balance energy demand and consumption for processing waste, such as waste rock mass. Theoretically, through thermal conversion or heat transfer technologies, a large amount of heat can be restored and recovered back into the CE framework as energy recovery. However, some solid wastes contain critical materials. Firms actively seek technological advancements because the greater energy investment conserves more non-renewable materials and creates more economic value as an incentive for businesses to participate in and practice economic activities under CE (Calvo et al., 2016). For example, copper mines at the Buenavista del Cobre helped achieve more than 3% of high-grade copper concentrate recovery by applying sensor technologies from TOMRA to efficiently sift through barren ore deposits and detect critical metals and minerals using X-ray transmission (XRT) technology (Tomra, 2018).

### 20.5.3 Existing Government Policies and Regulations for Promoting Road CE Practice

Regarding industrial standards, it is observed that standards published by European Union (EU) and the U.S. are used by most industries regardless of country. However, ISO standards are frequently appearing as guidelines for E-LCA. Many of these studies have demonstrated their interest in resource efficiency under CE. Much of the literature on CE deals with the question of technological innovation with IE perception. However, Axelson et al. (2021) are more concerned with the relationship between material efficiency and emission efficiency. Axelson et al. (2021) have claimed that applying material efficiency strategies outweighs the advantages of using emission efficiency strategies to decarbonize EU steel, which implies more radical changes to business models (BM). Among the role of business model innovation (BMI) in industrial decarbonization literature, BM theory and literature in the empirical field of industrial decarbonization, Axelson et al. (2021) have attempted to find the relevance between material efficiency and steel producers and concluded that it possibly provides new business opportunities and incentivize firm-level BMI.

On the other hand, in the ceramic tile production industry, Dondi et al. (2021) believe that the frame of circular economy can be further developed by transitioning from natural raw materials to partially using secondary raw materials. They also believe that more efficient exploitation of the European deposits of ceramic raw materials can reduce the ceramic tile production's environmental impact and carbon footprint, especially for companies that have not utilized technologies to align this new concept of ceramic batches under CE. Further, Dondi et al. (2021) have pointed out that resource efficiency within the supply chain of ceramic production is not the only driver of environmental sustainability, and the interferences from technological innovation and market trends in the ceramic production industry always exist.

The complexities and uncertainties associated with the high cost of tracing recycled materials compared to low virgin material prices demonstrate the need for seeking balance within the CE framework. In between EU ceramic and steel industries, either the use of natural raw materials for ceramic or the primary production processes for steel embeds with the thinking of material flow. Both studies indicate the importance of adopting the materials efficiency concept at the beginning of the production process, which must be aligned with the EU Standards in the steel and ceramic production industry. Furthermore, it can also be seen that the secondary CE practices for steelmaking have been rooted in helping the primary steel production processes, which are believed to be easier for the EU steel industry to achieve full decarbonization. Considering all the evidence so far, tracing all the raw materials flow globally requires systematic analysis and approaches.

Due to poor and unconvincing business case studies, disconnections between academic and grey literature often happen within the CE framework, making material, labor and energy costs easily influenced by fluctuating market prices. However, to simplify the complex process, representative materials can be extracted and used as key materials as connections to demonstrate the future trend and reflect the situations in the real world. Therefore, synthesizing decarbonization, clean energy, and technological innovation can be vital and critical to facilitating technological development and policy innovations under CE with industrial ecology ideas in practice. Although Axelson et al. (2021) first contributed to the systematic assessment of industrial decarbonization strategies from a BM perspective, overall, these studies mentioned above provide converging evidence for the implications of analyzing raw material flow. It also indicates that the reduction in consumption of virgin materials at the planning stage of the circularity practices is always preferred, and technological innovation is unavoidable while developing a CE to achieve full decarbonization.

## 20.6 Chapter Summary

There is still a need for more understanding of circular design and economy for new road projects. Due to the construction sector‘s impact on resource consumption, implementing circular principles aligned with industrial ecology will help conserve energy and materials and reduce waste production. In terms of energy, strategies such as the development of a new framework for energy-saving solutions for road infrastructure assets (roads, bridges, property, traffic services or data assets) or analyzing the impact of different materials may help stakeholders in the assessment and selection of the most effective solutions for road infrastructure projects. Including economic practices will allow an optional categorization of all the available solutions in light of CE. Other strategies, such as the ‘10 Rs’ and the implementation of renewable resources, may be seen from a life cycle perspective. These should be included and analyzed in all the main seven stages of a road infrastructure project, from the initiation and planning, moving through design, construction, and operation, until the disassembly stage. These strategies have already started to be implemented in pilot projects where the minimization of raw materials and the maximization of recycling across different life cycle phases were targeted. One of the outcomes of the project was creating an open learning environment that can be used in future CE projects and as leverage for new circulatory policies for the efficient use of resources.

Moreover, implementing CE principles in road infrastructure projects may face cultural, sectoral, regulatory and financial challenges. These challenges may not only be due to misinformation or the complexity of the new practices but also in relation to the lack of CE policies, high initial investments or technical challenges, among others. However, there are several enablers for adopting CE in road infrastructures, such as the service economy, which promotes a closed-loop approach and lowers costs, resource consumption and energy use. Another enabling factor may be the collaborative economy approach, which focuses on individuals and communities instead of institutions. The new electric vehicle paradigm or car-sharing is pointed to as some solutions aligned with a collaborative economy approach. There is a need to change the analysis of not only the carbon emissions but also the waste produced in all the different phases of road infrastructure projects. The occurrence of unpredictable events, such as the COVID-19 pandemic, is another enabling factor. During the pandemic, it experienced a materials shortage and enabled the development and research of new cost-efficient strategies. Therefore, developing a CE framework in road infrastructure that includes economic and environmental components and social ones will help its consolidation. The social assessment is crucial to enable CE in road infrastructure and strengthen the interconnection between the supply chain and business models, along with the Industrial Ecology concept. Finally, the standardization and harmonization of regulations and standards among different sectors and regulatory bodies are key to the successful implementation of CE in road infrastructure projects.

Overall, the implementation of CE practices is influenced by a variety of enablers and barriers. In terms of enablers, the use of restorative, collaborative, and service-based economic models, circular supply chain and business models, and industrial ecology and environmental economics have the potential to promote CE implementation. However, several barriers to CE implementation need to be addressed. These include the lack of collaboration between businesses and business functions, cultural barriers, high upfront investment costs in recycling technologies and specialty recycling infrastructure, energy consumption for processing, and energy consumption for recycling waste from the material. To overcome these barriers, existing government policies and regulations can play an important role. For instance, ISO and EU standards can promote material efficiency and emission efficiency, as well as industrial decarbonization. By addressing these enablers and barriers, it is possible to improve the level of CE implementation in practice, leading to more sustainable and efficient use of resources. To advance research and conduct future studies, experts in the field or future case studies can examine and validate the provided comparative table to assess its applicability and compare its outcomes.

## References


Axelson, M., Oberthür, S., & Nilsson, L. J. (2021). Emission reduction strategies in the EU steel industry: Implications for business model innovation. *Journal of Industrial Ecology, 25*(2), 390–402. https://doi.org/10.1111/jiec.13124

Benachio, G. L. F., Freitas, M. C. D., & Tavares, S. F. (2020). Circular economy in the construction industry: A systematic literature review. *Journal of Cleaner Production, 260*, 121046. https://doi.org/10.1016/j.jclepro.2020.121046

Bonsu, N. O. (2020). Towards a circular and low-carbon economy: Insights from the transitioning to electric vehicles and net zero economy. *Journal of Cleaner Production, 256*, 120659. https://doi.org/10.1016/j.jclepro.2020.120659

Cai, B., Yang, W., Cao, D., Liu, L., Zhou, Y., & Zhang, Z. (2012). Estimates of China's national and regional transport sector $CO_2$ emissions in 2007. *Energy Policy Journal, 41*, 474–483. https://doi.org/10.1016/j.enpol.2011.11.008

Calvo, G., Mudd, G., Valero, A., & Valero, A. (2016). Decreasing ore grades in global metallic mining: A theoretical issue or a global reality? *Research Journal, 5*(4), 36. https://doi.org/10.3390/resources5040036

Coulter, E. (2021). Australian construction materials shortage causing delays for Tasmanian home builders. *ABC News*. https://www.abc.net.au/news/2021-05-22/building-material-shortage-causes-delays/100157774. Accessed 3 June 2021.

Damtoft, J. S., Lukasik, J., Herfort, D., Sorrentino, D., & Gartner, E. M. (2008). Sustainable development and climate change initiatives. *Cement and Concrete Research Journal, 38*(2), 115–127. https://doi.org/10.1016/j.cemconres.2007.09.008

Department of Main Roads. (2006). *Road planning and design manual: Chapter 23 Tunnels*. Queensland Department of Main Roads. https://www.tmr.qld.gov.au/-/media/busind/techstdpubs/Road-planning-and-design/Road-planning-and-design-manual/Current-document/RPDM_Chapter23.pdf?la=en. Assessed 16 Apr 2021.

Department of State Growth. (2020) *State roads infrastructure asset management policy*. Tasmania Department of State Growth. https://www.transport.tas.gov.au/__data/assets/pdf_file/0004/114439/Road_Management_Infrastructure_Asset_Management_Policy.pdf. Assessed 16 Apr 2021.

Dondi, M., García-Ten, J., Rambaldi, E., Zanelli, C., & Vicent-Cabedo, M. (2021). Resource efficiency versus market trends in the ceramic tile industry: Effect on the supply chain in Italy and Spain. *Resources, Conservation and Recycling Journal, 168*, 105271. https://doi.org/10.1016/j.resconrec.2020.105271

Ellen MacArthur Foundation. (2014). *Towards the circular economy volume 3: Accelerating the scale-up across global supply chains* (Vol. 3, pp. 1–64). https://www.ellenmacarthurfoundation.org/assets/downloads/publications/Towards-the-circulareconomy-volume-3.pdf. Assessed 3 June 2021.

Ellen MacArthur Foundation. (2015). *Towards a circular economy: Business rationale for an accelerated transition* (pp. 4–20). https://www.ellenmacarthurfoundation.org/assets/downloads/TCE_Ellen-MacArthur-Foundation-9-Dec-2015.pdf. Assessed 3 June 2021.

Ellen MacArthur Foundation. (2017) *Circular economy and curriculum development in higher education* (pp. 2–21). https://www.ellenmacarthurfoundation.org/assets/downloads/EMF_HE-Curriculum-Brochure-290917.pdf. Assessed 7 June 2021.

Estaji, A., Meisel, M., Novak, T., & Sauter, T. (2018). Integrated networked streetlighting infrastructure simulation with crossing as use case. In *IECON 2018-44th annual conference of the IEEE industrial electronics society* (pp. 3562–3567).

Guo, M., Gurmu, A., & Tivendale, L. (2021). Development of a framework for flood mitigation in Fiji. In *The 9th world construction symposium* (pp. 244–255).

Hart, J., Adams, K., Giesekam, J., Tingley, D. D., & Pomponi, F. (2019). Barriers and drivers in a circular economy: The case of the built environment. *Procedia CIRP Journal, 80*, 619–624. https://doi.org/10.1016/j.procir.2018.12.015

Jeswiet, J., & Szekeres, A. (2016). Energy consumption in mining comminution. *Procedia CIRP Journal, 48*, 140–145. https://doi.org/10.1016/j.procir.2016.03.250

Leendertse, W., Hendriksen, J., & Kerkhofs, S. (2018). Introducing the circular economy in road infrastructure development. Challenges and dilemmas in designing circular roads. In *The 7th Transport Research Arena TRA 2018 (TRA 2018)* (pp. 1–10).

McDonough, W., & Braungart, M. (2013). *The upcycle: Beyond sustainability – Designing for abundance*. Macmillan.

Muench, S., Anderson, J., Hatfield, J., Koester, J., Söderlund, M., & Weiland, C. (2011). *Greenroads Manual (1. 5)*. University of Washington.

Mulvaney, D., Richards, R. M., Bazilian, M. D., Hensley, E., Clough, G., & Sridhar, S. (2021). Progress towards a circular economy in materials to decarbonise electricity and mobility. *Renewable and Sustainable Energy Reviews Journal, 137*, 110604. https://doi.org/10.1016/j.rser.2020.110604

Nagamani, S., Reddy, J. M., Lavanya, G., & Chandana, A. L. H. (2019). Smart street light management system using Internet of Things. In *2019 international conference on intelligent computing and control systems (ICCS)* (pp. 103–107).

Pacheco-Torgal, F., & Labrincha, J. (2013). The future of construction materials research and the seventh UN Millennium Development Goal: A few insights. *Construction Building Materials, 40*, 729–737.

Papachristos, G., Jain, N., Burman, E., Zimmermann, N., Mumovic, D., Davies, M., & Edkins, A. (2020). Low carbon building performance in the construction industry: A multi-method approach of project management operations and building energy use applied in a UK public office building. *Energy Buildings, 206*, 109609.

Pepe, M. (2015). *A conceptual model for designing recycled aggregate concrete for structural applications*. Springer.

Pickering, C., & Byrne, J. (2014). The benefits of publishing systematic quantitative literature reviews for PhD candidates and other early-career researchers. *Higher Education Research Development, 33*(3), 534–548.

Potting, J., Hekkert, M., Worrell, E., & Hanemaaijer, A. (2016). *Circulaire economie: Innovatie meten in de keten*. Planbureau voor de Leefomgeving (PBL).

Potting, J., Hekkert, M., Worrell, E., & Hanemaaijer, A. (2017a). *Circular economy: Measuring innovation in the product chain*. Planbureau voor de Leefomgeving 2544.

Potting, J., Hekkert, M., Worrell, E., & Hanemaaijer, A. (2017b). *Circular economy: Measuring innovation in the product chain*. Planbureau voor de Leefomgeving, no. 2544.

Ratnasabapathy, S., Alashwal, A., & Perera, S. (2021). Exploring the barriers for implementing waste trading practices in the construction industry in Australia. *Built Environment Project Asset Management, 11*(4), 559–576.

RMoIWM. (2016). *Construction of a circular bridge*. Rijkswaterstaat Ministry of Infrastructure and Water Management (RMoIWM). https://www.rijkswaterstaat.nl/en/environment/circular-economy/construction-of-circular-bridge. Assessed 21 July 2021.

Shooshtarian, S., Hosseini, M. R., Kocaturk, T., Arnel, T., & Garofano, N. (2023). Circular economy in the Australian AEC industry: Investigation of barriers and enablers. *Building Research Information, 51*(1), 56–68.

Stahel, W. R. (1997). The service economy: 'Wealth without resource consumption'? *Philosophical Transactions of the Royal Society London Series A: Mathematical, Physical Engineering Sciences, 355*(1728), 1309–1319. https://doi.org/10.1098/rsta.1997.0058

Tomra. (2018). *New TOMRA x-ray sorter a game changer for mineral processing*. https://newsroom.tomra.com/new-tomra-x-ray-sorter-a-game-changer-for-mineral-processing/

United States Environmental Protection Agency. (2018). *Sustainable materials management options for construction and demolition debris*. National Service Center for Environmental Publications (NSCEP).

Walker, A. M., Opferkuch, K., Lindgreen, E. R., Simboli, A., Vermeulen, W. J., & Raggi, A. (2021). Assessing the social sustainability of circular economy practices: Industry perspectives from Italy and The Netherlands. *Sustainable Production and Consumption, 27*, 831–844. https://doi.org/10.1016/j.spc.2021.01.030

Wiesmeth, H. (2020). Chapter 2: The circular economy – Understanding the concept. In *Implementing the circular economy for sustainable development*. Elsevier.

# Develop a New Revit Plug-in Enabling Building Components-Based Construction and Renovation Waste Quantification: An Australia Study

# 21

Ruidong Chang, Wei Emma Zhang, Xiangnan Song, Jian Zuo, and Huanyu Wu

## 21.1 Introduction

Construction and demolition (C&D) waste is one of the most significant waste streams accounting for 30–40% of the total urban waste generation (Wu et al., 2019). Globally over ten billion tons of C&D waste is generated every year (Wu et al., 2019). Proper ways to manage waste should follow a waste hierarchy, namely reduce, reuse, recycle, recover and landfill. Management of CD waste should also follow this waste hierarchy, thereby ultimately achieving circular construction. In a circular economy, we keep resources in use for as long as possible, then recover and regenerate products and materials at the end. Ideally, the construction industry should also operate based on the circular economy principles to close material loops by reusing, sharing, recycling materials rather than continuing the traditional take-make-consume-dispose linear process. To achieve this ideal status of circular construction, information about material quantities and waste generation should be available across the whole project lifecycle including the early design stage, thereby facilitating decision-making such as material selection at the design stage to reduce waste generation at later stages.

However, waste data of construction projects currently is of low quality and the information is poorly managed across the project lifecycle. Despite available drawings and procurement information, construction waste estimation is highly inaccurate due to the deficiencies of existing methods. Some studies have proposed methods to estimate waste generation at the construction stage during the early design stage (e.g. Ding & Xiao, 2014). These studies suggest that construction waste can be estimated based on waste generation rates per floor area or total material stock, which is currently the mainstream waste estimation approach (Wu et al., 2019). However, as this method estimates waste based on waste generation rates per floor area or total material stock, it can only estimate the total amount of waste without knowing which kind of waste is coming from which building components (Wu et al., 2019). Without the capability of

R. Chang (✉) · J. Zuo
School of Architecture and Civil Engineering, The University of Adelaide, Adelaide, Australia
e-mail: ruidong.chang@adelaide.edu.au; jian.zuo@adelaide.edu.au

W. E. Zhang
School of Computer and Mathematical Sciences, The University of Adelaide, Adelaide, Australia

X. Song
School of Management, Guangzhou University, Guangzhou, China

H. Wu
School of Civil and Transportation Engineering, Shenzhen University, Shenzhen, China
e-mail: wuhuanyu@szu.edu.cn



J. Zuo et al. (eds.), *Circular Economy for Buildings and Infrastructure*, Sustainable Development Goals Series, https://doi.org/10.1007/978-3-031-56241-9_21

identifying the waste sources, even though this method can estimate the total amount of waste, it provides little insight into reducing waste and improving recycling.

Building information modelling (BIM) emerges as a potential solution to incorporate waste management considerations into the design from the beginning of projects. Some pioneer research has proposed the agenda of integrating BIM models for extracting the volumes of salvage, sorting, testing, reprocessing and integrating that information for optimised resources recovery (Chileshe et al., 2019). However, few studies developed the actual BIM tool which can be directly used by practitioners to estimate construction waste, and associate the estimated waste with corresponding building components in the early design stage. Responding to this gap of knowledge, this study develops a prototype Revit plugin to achieve the function of building components-based waste estimation, by using relevant construction data in Australia as an example. Different from the traditional floor-area based waste estimation, this newly developed building components-based estimation identifies the sources of waste generation. It shows how much and what type of waste is generated from which building components, enabling designers to make targeted decisions optimising material selection and building component design to reduce the waste generated. The remaining contents of this chapter are organised as follows. First, reviews of waste management in Australia and Revit plug-in development will be presented. Next, the detailed steps of developing the plug-in for waste quantification are presented. Subsequently, the various functions are illustrated and discussed, from data retrieval to result display, followed by a summary and conclusion of this chapter.

## 21.2 Construction and Demolition Waste Management in Australia

In Australia, 74 million tons of waste are generated during 2018–19 and only 60% of them are recycled. Every year, around 28 million tonnes of materials are disposed into landfill. Australia bans the export of unsorted waste plastic, paper, glass and tires from July 2021, which means Australia has to increase its capabilities to recycle and manage wastes locally. "Australia has a once-in-a-generation opportunity to improve waste management and recycling", stated by the Hon Sussan Ley MP, Minister for the Environment, Australia.

C&D waste stream is the largest, contributing to 36% of the total waste in Australia. C&D waste is the only waste stream that increases per capita in Australia. Over the past 13-year period, total waste generation increased by 11.3 Mt. (18%). Municipal Solid Waste (MSW) generation fell by 20% per capita and Commercial and Industrial (C&I) waste by 15% per capita. In contrast, Construction and Demolition (C&D) waste grew by 32% per capita. In 2017, Australia's population is 24.6 million, and it is projected to reach between 37.4 and 49.2 million people by 2066. The growing population combined with the increased per capita C&D waste generation presents a huge challenge to reduce the C&D waste and increase its recycling rate, which requires a rather rapid transition towards circular construction. Waste reduction, reuse and recycling should be considered at the early planning and design stage. However, due to the lack of approaches to evaluate the impacts of design on reducing C&D waste, architects generally do not view C&D waste as a priority in their design process (Wu et al., 2019). Due to a lack of information on waste generation types and quantities at the early project lifecycle stages (Wijewickrama et al., 2021), detailed plans on how to treat C&D waste are usually proposed after those wastes are generated (Wu et al., 2019).

## 21.3 Review of Revit Plug-in Development

In order to solve practical problems in the actual construction projects, the secondary development of Revit is favoured. Secondary development refers to loading a new program into Revit through the application programming interface without changing the kernel of Revit itself to realize the required functions (Chen et al., 2016;

Peng et al., 2021). Furthermore, the plug-ins based secondary development of Revit specifically refers to the Revit API, which adds additional plates and creates new functional modules for Revit according to the actual project requirements (Xiao & Cao, 2021).

Revit secondary development has become a research hotspot in the construction area. Many scholars have attempted to use functions based on Revit-the most commonly used BIM platform-to carry out secondary development, such as building energy analysis, construction risk management, cost estimation, project schedule optimization, etc. For example, Lu et al. (2021) developed a plug-in to link BIM with construction safety risk data in Revit to help architects and structural designers quickly select design alternatives. Around the prefabricated buildings that are popular at present, a Design for Manufacture and Assembly (DFMA-oriented BIM secondary development process is explored by Yuan et al. (2018) to make DFMA-oriented parametric design implemented better. Besides, through secondary development of the Revit platform, the construction hazard source safety management module (Deng et al., 2019), the embedding of the security risk early warning model (Peng et al., 2021), and the construction information visualization monitoring system (Jiang & Jiang, 2022) have been developed as well. From the above research, it can be seen that there are very few plug-ins based on Revit to quantify construction waste (Stange & Jose, 2014).

## 21.4 Development of the Revit Plug-in for Waste Quantification

### 21.4.1 Waste Calculations (Fig. 21.1)

In practice, there are two scenarios where C&D waste can be generated: the first one is from existing projects during the demolition or renovation processes, and the second one is from new projects during the construction process. In the first scenario of demolishing existing buildings, all building material becomes waste to be treated, and thus the waste quantity equals the total material stock of the building. In contrast, as the components of buildings all have a life expectancy, during the operation of the buildings, building components need to be replaced when they reach their life expectancies. This renovation process generates waste in different periods before the final demolition of the buildings. Based on this situation, The amount of building material lost at a given time can be calculated according to the formula in Eq. 21.1 below. In short, this is achieved by multiplying the amount of material in each component of the building by their loss factor, and then multiplying by 1 or 0 depending on the replacement life of the materials in each component (Rutkowski et al., 2014). The role of the right-hand side of the equation is to determine whether the building construction time and material replacement life have an integer multiple. For instance, for the period from now to 2025, wooden doors with a

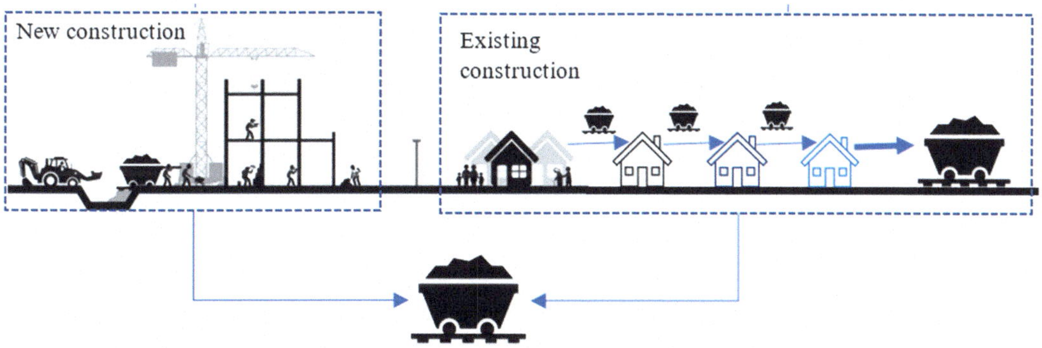


**Fig. 21.1** Waste generations of building projects

service life of 10 years in a house built in 2012 will be replaced in 2022. This means that in 2022, the value is 1. In all other years of this period, the value is 0.

$$CRF_{m,y} = \sum_{b=1}^{B} \sum_{a=1}^{A} Q_{m,a,b} \times w_{m,a,b} \times \begin{cases} \delta = 1 \Leftrightarrow \frac{(y - CY_b)}{SL_{m,a,b}} \in \mathbb{Z}^+ \\ \delta = 0 \Leftrightarrow \frac{(y - CY_b)}{SL_{m,a,b}} \notin \mathbb{Z}^+ \end{cases} \forall CY_b < y \leq TH \quad (21.1)$$

where $y$ is the year we specify and $CRF_{m,y}$ is the waste material m generated in year $y$. $CY_b$ is the year in which the building was constructed. $SL_{m,a,b}$ is the life of material m in part a of the building $b$. $Q_{m,a,b}$ indicates the quantity of material $m$ in part a of the building $b$. $W_{m,a,b}$ is the waste factor of the building material in part a of the building $b$.

### 21.4.2 Plug-in Development

At present, there are two mainstream development modes based on the Revit platform. The first one is to use a programming language (C# or VB.NET) on the Visual Studio platform for secondary development to form .dll file on Visual Studio platform and load it into Revit's add-on module in the form of a plug-in. The other is to use the open source plug-in Dynamo for secondary development directly on the Revit platform, by connecting Dynamo nodes with different roles to achieve the development functions. This project uses the first method for the development of plug-ins. Revit is a Windows web framework environment based on .NET, which is written in C#. Thus, the secondary development for Revit plug-ins is also written in C#. C# is an object-oriented programming language, and the development process requires the classification and presentation of program functions. C# is simple in syntax, flexible and compatible, and can interact with Revit API functions. The computer running environment is Win10. Visual Studio is an intelligent programming language compiler that can intelligently indicate errors based on the user's code, making it easy for the programmer to make changes, and at the same time, the programmer can create "breakpoints" at any time to see how the code is running.

The Revit API is a function that can be called by Autodesk to meet the secondary development of the software. Users can edit and call the function according to their own functional development needs to achieve functions that cannot be achieved by Revit software itself, and this project combines C# language and Revit API functions to call API functions and write methods for the actual needs of the project. This project combines C# language and Revit API functions, and writes API functions and methods to quantify and visualise the amount of construction material waste. The development steps of the plug-in include 4 main parts. First, we need to install the Revit plug-in manager or SDK, and additionally configure the plug-in DLL and file placement location. Second, we need to read the data related to construction materials in the drawing list of Revit. Thirdly, the design of the Firebase database structure, the implementation of a series of functions such as adding, deleting, saving, uploading and retrieving updates of construction material data and data optimisation reading need to be achieved. Fourth, the interface design of the plug-in, such as the adjustment of the coefficient, the display of construction materials, etc. needs to be implemented.

Specifically, we created a new project in Visual Studio 2019 and create a class library. Then, we added the Revit API assembly files Revit API.DLL and Revit API UI.dll to the project and add namespace references in the program header. After that, we referred to the external command interface IExternal Command, wrote the Execute code and debugged it to implement the functionality designed for the project. To solve the problem of incomplete and uncertain construction material data, when the users agree that the data can be collected, the plug-in will automatically save the construction material information entered by the users in the plug-in, including material name, service life, material loss factor, etc. The plug-in will then transfer the data to Firebase for preservation. The more users, the more data stored in the Firebase which will be used to update the material loss factor constantly. Firebase is a cloud platform (BaaS) that uses the backend as a service (Lin et al., 2017). It is designed for developers to separate the back end and deploy it in the cloud, helping developers solve various problems faced by the back-end architecture, such as stability, security, scalability, etc. In addition, it provides a unified API and SDK, allowing users to deploy applications using multiple frameworks and languages, covering Web, iOS, Android and other platforms (Lin et al., 2017). Firebase provides the following features for backend development: real-time database, authentication, cloud function, cloud messaging, hosting, and cloud storage database function.

## 21.5 Functions of the Developed Plug-in

### 21.5.1 Data Retrieval, Input and Storage

Whenever users use the plugin, a pop-up window will ask users if they would like to accept the terms of the plug-in to save their data. Users can use the plug-in normally regardless of whether users accept the terms and conditions. If they agree to have their data saved by the plugin, the building material information filled in by users will be saved to the real-time database of Firebase, especially the value of the waste coefficient. Then, according to the saved material data, Firebase will return the collected data to the plug-in and form a probability distribution chart of the material loss coefficient within the plug-in. Figure 21.2 shows that on Firebase, we save the data in the form of material root-username-material. The material data in turn shows the time when the user added the data, the material name, the user name, and the value of the waste coefficient. For instance, when we saved information about the building material softwood in the plug-in, the data is uploaded to Firebase. In the real-time database of Firebase, the data is saved sequentially according to material name: softwood, user name: my computer ID, and construction material wastage coefficient: softwood wastage coefficient. It has the advantage that not only can the material information entered by different users be clearly seen, but also the plug-in can find all the waste factors under the same material on Firebase. Figure 21.3 shows the starting page of the plug-in after users click the pop-up window.

When users import architectural drawings into Revit, the plug-in will read the relevant data from Revit, such as the name of the building material, the volume of the material, and the density of the material. The plug-in will convert the volume and density of the material to get the weight of the material and display the information of material

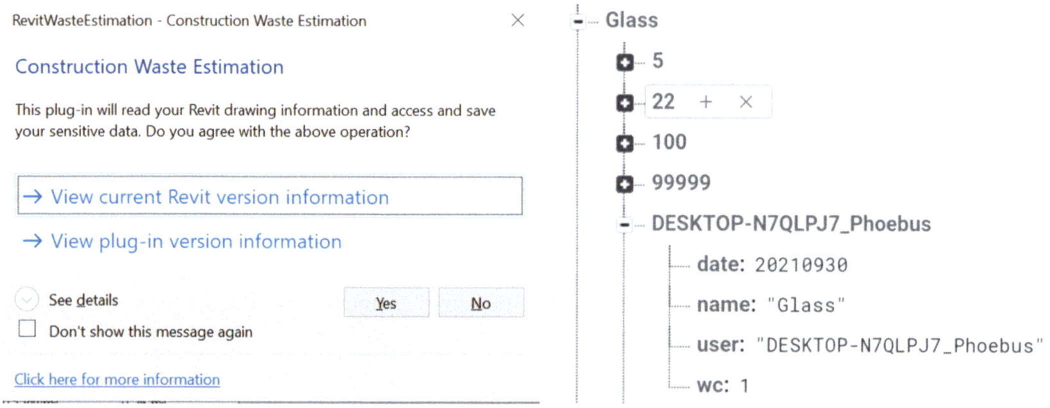


**Fig. 21.2** Data collection through the plug-in to Firebase

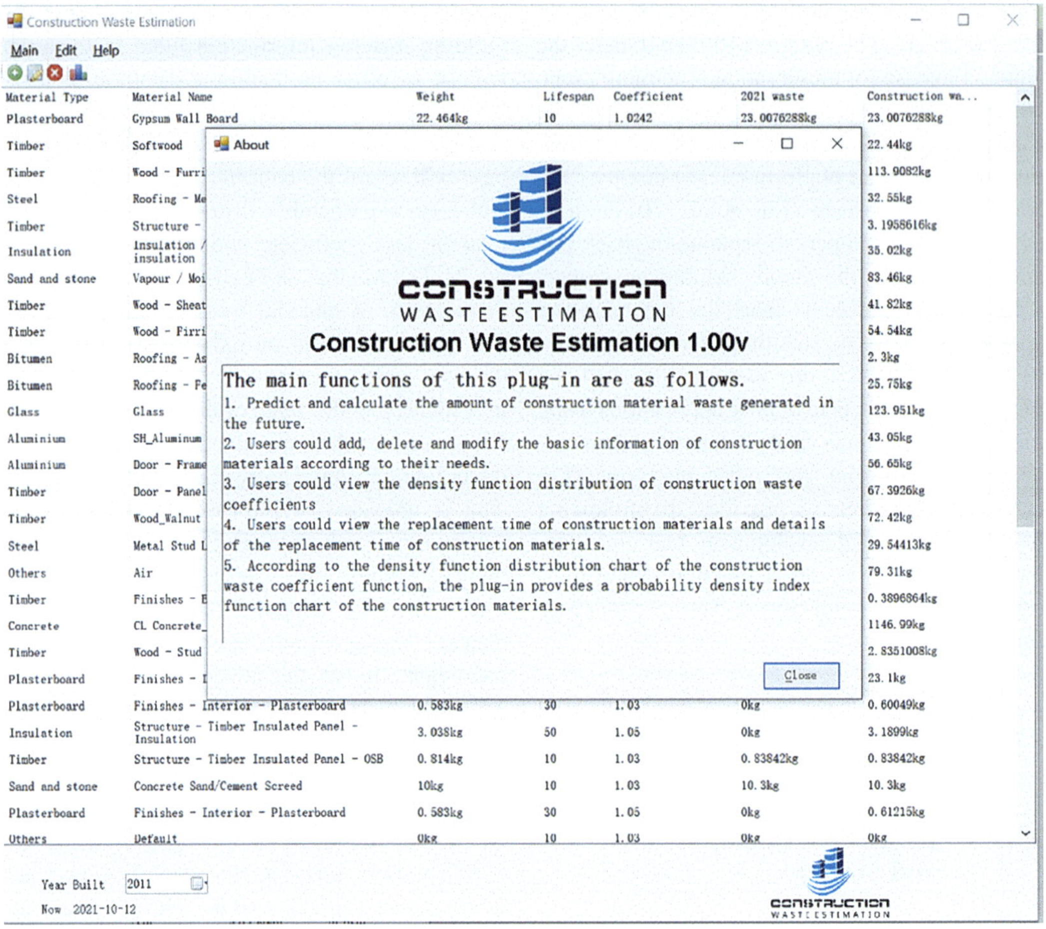


**Fig. 21.3** Starting page of the plug-in

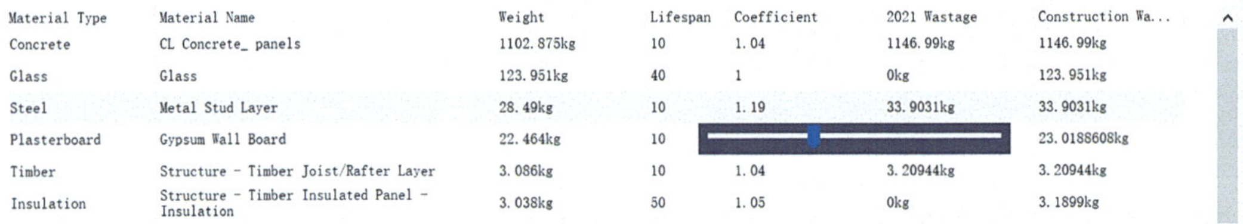


**Fig. 21.4** Slider to select the material loss coefficient

type, name, weight, life cycle, loss coefficient, estimated wastage, and wastage during construction on the main page of the plug-in. The Construction Wastage column calculates the material loss of a building during the construction process. In addition, when users click on the table names Material Type and Material Name, the list is sorted alphabetically. If users click on the table names Weight, Lifespan, Coefficient, and Wastage, the list is automatically sorted according to the size of the data. As the loss coefficient of building material is not normally a definite value, we have designed a slider for users to select the value of the building material coefficient, as shown in Fig. 21.4.

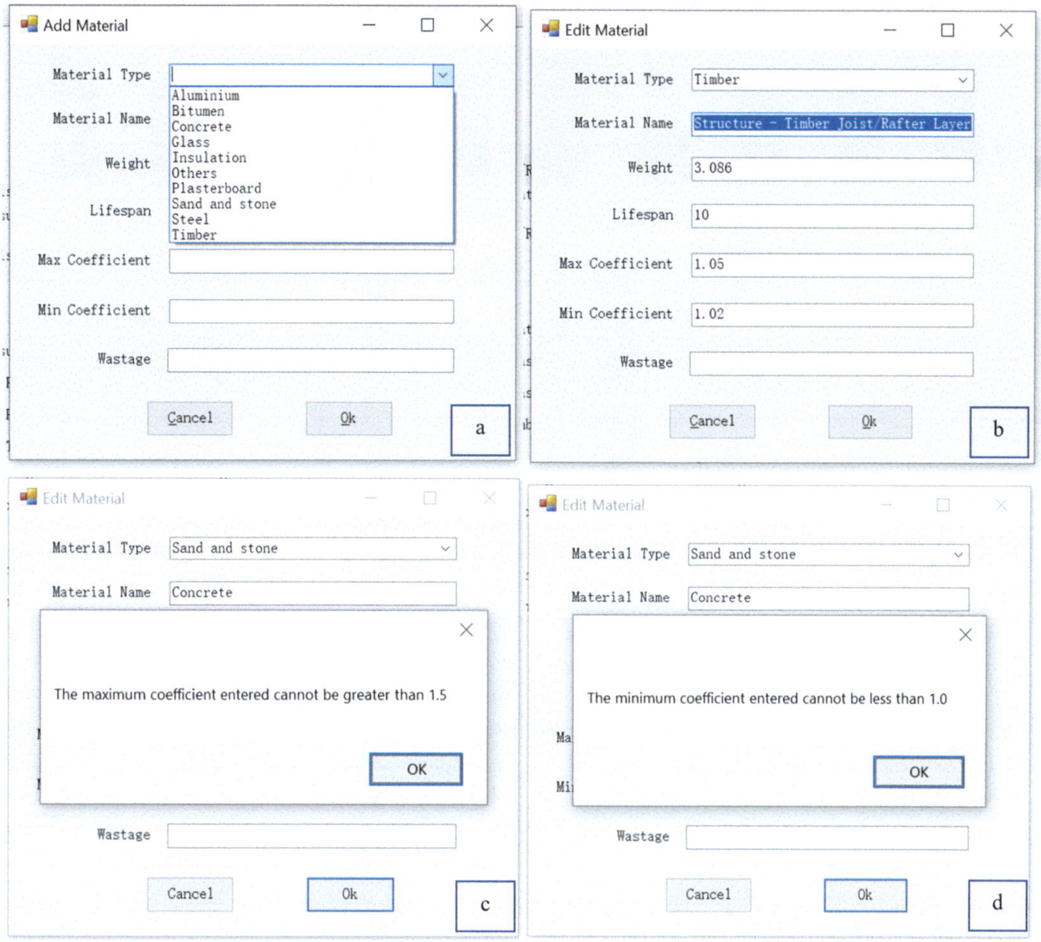


**Fig. 21.5** Material add and editing functions

The Add function allows users to add data for the materials as shown in Fig. 21.5. We have summarised the main material types for users to choose from and arranged them in alphabetical order, which is also easy for users to use. In the editing screen, users can select existing building materials to edit, and users can edit the coefficient as well. However, based on an extensive literature review, we have specified in the plug-in that the value of the coefficient should be greater than 1 and the maximum value should be less than 1.5 (Fig. 21.5c, d). If the data entered by the users do not meet the requirements, the plug-in will pop up a reminder window to alert the users. If users select OK, the wastage value will be changed according to the users' filling. Conversely, it will be calculated according to the rules of the formula.

## 21.5.2 Result Display and Testing

In Fig. 21.6 below, the amount of material wastage in a building over 100 years is shown. When the users click the Query button, they can query the building material losses for a particular year in the plug-in. For example, if we enter 30, the plug-in will show the material waste in the 30th year of the building as shown in Fig. 21.6. To test the reliability of the plug-in, we selected several

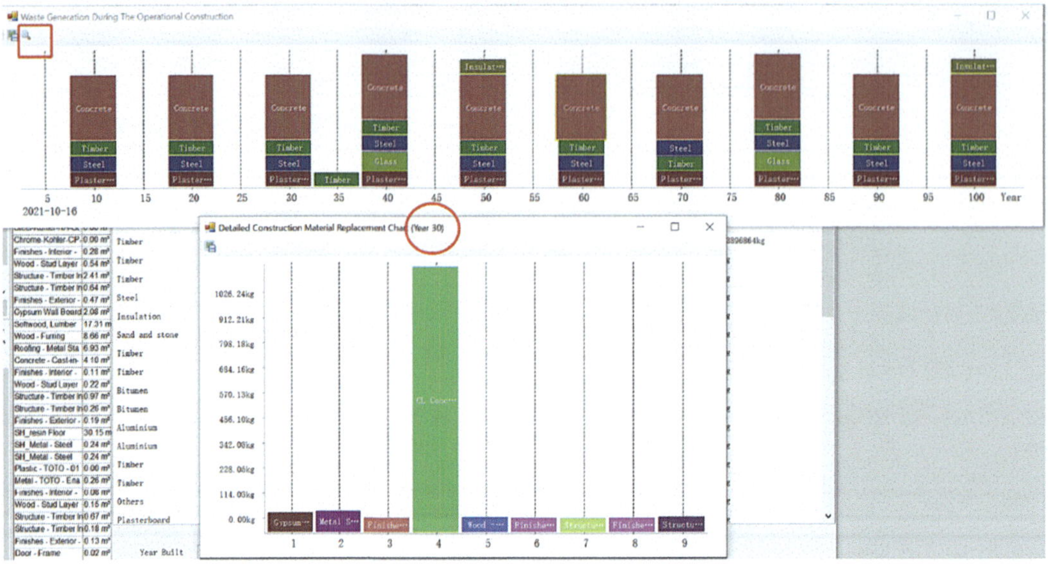


Fig. 21.6 Result display and testing

different models and imported them into the plug-in. The test was successful as it was able to identify the building materials and successfully read the material information in different models. Lastly, the plug-in was tested in several versions of Revit to ensure its suitability, for example, in Revit 2019 and Revit 2020. The plug-in can be used in Revit versions of 2019 and after. It is worth noting that the plug-in will become more accurate with more professionals using it, as it will store the waste generation rates input by users. At this stage, the default waste generation rates are confirmed based on the literature review, and the rates database will keep updated with more users' inputs. In addition, regarding the requirement for hardware, the developed plug-in requires minimum storage as it only shows digits and plots. So if the computer meets the requirement of running Revit, it will be ok for the use of this plug-in.

## 21.6 Conclusions

This paper proposes a Revit plug-in for building components-based construction and renovation waste quantification. The component-based method is applied to quantify construction waste which yields more accurate results than the traditional area-based waste rate method. The model is then combined with the BIM tool Revit, and we used C# and the Revit API for secondary development of the plug-in. The plug-in can quantify and visualize the amount of material waste. More importantly, it is based on a distribution of the selection probability of the construction material loss factor with the inputs from the Revit user. The accuracy will be improved with more user inputs added to moderate the distribution curve of the waste generation factor of building components. Similarly, plug-in users can also enter more realistic and valid material data, such as material quantities, and service life, into the plug-in for storage. Only by reducing the uncertainty in the data relating to construction material waste can the accuracy of the model and the use of the plug-in be significantly improved. This will then make a significant contribution to reducing the uncertainty in the parameters. To conclude, by storing the user inputs on the component-based waste generation factor, this Revit plug-in will not only automate the waste calculation of construction projects, but also improve the accuracy significantly as the traditional fixed area-based waste factors only rely on expert knowledge without the chance of incorporating user inputs from the industry professionals.

## References


Chen, N., Kan, F., Wang, X., Wang, C., Qi, N., Liu, X., et al. (2016). Research on the secondary development of Revit software. In *International conference on education, management, computer and society* (pp. 1130–1133). Atlantis Press.

Chileshe, N., Jayasinghe, R. S., & Rameezdeen, R. (2019). Information flow-centric approach for reverse logistics supply chains. *Automation in Construction, 106*, 102858.

Deng, L., Zhong, M., Liao, L., Peng, L., & Lai, S. (2019). Research on safety management application of dangerous sources in engineering construction based on BIM technology. *Advances in Civil Engineering, 2019*, 1–10.

Ding, T., & Xiao, J. (2014). Estimation of building-related construction and demolition waste in Shanghai. *Waste Management, 34*, 2327–2334.

Jiang, H., & Jiang, A. (2022). Monitoring system for the construction of arch cover method Subway Station based on DT and IoT. *Journal of Sensors, 2022*, 1–16.

Lin, J. C.-W., Yang, L., Fournier-Viger, P., Hong, T.-P., & Voznak, M. (2017). A binary PSO approach to mine high-utility itemsets. *Soft Computing, 21*, 5103–5121.

Lu, Y., Gong, P., Tang, Y., Sun, S., & Li, Q. (2021). BIM-integrated construction safety risk assessment at the design stage of building projects. *Automation in Construction, 124*, 103553.

Peng, Y., Li, M., Li, Z., Hu, T., Xiao, Y., Wen, J., & Liu, Y. (2021). Bridge integrated early warning system based on BIM secondary development. In *2021 3rd international conference on artificial intelligence and advanced manufacture* (pp. 935–939).

Rutkowski, L., Jaworski, M., Pietruczuk, L., & Duda, P. (2014). The CART decision tree for mining data streams. *Information Sciences, 266*, 1–15.

Stange, R. L., & Jose, J. (2014). Applying adaptive technology in machine learning. *IEEE Latin America Transactions, 12*, 1298–1306.

Wijewickrama, M., Chileshe, N., Rameezdeen, R., & Ochoa, J. J. (2021). Information sharing in reverse logistics supply chain of demolition waste: A systematic literature review. *Journal of Cleaner Production, 280*, 124359.

Wu, H., Zuo, J., Zillante, G., Wang, J., & Yuan, H. (2019). Status quo and future directions of construction and demolition waste research: A critical review. *Journal of Cleaner Production, 240*, 118163.

Xiao, J., & Cao, J. (2021). Secondary development and application of Revit based on information management and interaction. In *2021 7th international conference on hydraulic and civil engineering & smart water conservancy and intelligent disaster reduction forum (ICHCE & SWIDR)* (pp. 203–206). IEEE.

Yuan, Z., Sun, C., & Wang, Y. (2018). Design for Manufacture and Assembly-oriented parametric design of prefabricated buildings. *Automation in Construction, 88*, 13–22.

# Index





J. Zuo et al. (eds.), *Circular Economy for Buildings and Infrastructure*, Sustainable Development Goals Series, https://doi.org/10.1007/978-3-031-56241-9

SPRINGER NATURE

# GPSR Compliance

*The European Union's (EU) General Product Safety Regulation (GPSR) is a set of rules that requires consumer products to be safe and our obligations to ensure this.*

*If you have any concerns about our products, you can contact us on ProductSafety@springernature.com*

In case Publisher is established outside the EU, the EU authorized representative is:

Springer Nature Customer Service Center GmbH
Europaplatz 3
69115 Heidelberg, Germany

The manufacturer's authorised representative in the EU is Springer Nature Customer Service Centre GmbH, Europaplatz 3, 69115 Heidelberg, Germany. If you have any concerns regarding our products, please contact ProductSafety@springernature.com

Printed and bound by CPI Group (UK) Ltd, Croydon, CR0 4YY
01/10/2026
02225935-0020